ANNUAL REVIEW OF NUCLEAR SCIENCE

ANNUAL REVIEW OF NUCLEAR SCIENCE

EMILIO SEGRÈ, *Editor*
University of California, Berkeley

J. ROBB GROVER, *Associate Editor*
Brookhaven National Laboratory

H. PIERRE NOYES, *Associate Editor*
Stanford University

VOLUME 27

1977

ANNUAL REVIEWS INC.   4139 EL CAMINO WAY   PALO ALTO, CALIFORNIA 94306

ANNUAL REVIEWS INC.
Palo Alto, California, USA

International Standard Book Number: 0-8243-1527-8
Library of Congress Catalog Card Number: 53-995

REPRINTS

The conspicuous number aligned in the margin with the title of each article in this
volume is a key for use in ordering reprints. Available reprints are priced at the
uniform rate of $1 each postpaid. The minimum acceptable reprint order is 10
reprints and/or $10.00, prepaid. A quantity discount is available.

FILMSET BY TYPESETTING SERVICES LTD, GLASGOW, SCOTLAND
PRINTED AND BOUND IN THE UNITED STATES OF AMERICA

CONTENTS

ERRATA

Volume 25 (1975)

In Electron Scattering and Nuclear Structure, by T. W. Donnelly and J. D. Walecka:

page 340, Equation 1.29 should read

$$F^2 \equiv (q\mu^2/q^2)^2 F_L + \left[\tfrac{1}{2}(q\mu^2/q^2) + \tan^2 \frac{\theta}{2} \right] F_T^2;$$

page 388, the left side of Equation 4.7 should be multiplied by $-i$;

page 390, in Figure 4.6, the phase convention employed in plotting $p(y)$ for the inelastic form factor is discussed in (174).

page 343, the last factor in Equation 1.36 should read $F_1 + 4MF_2$.

ANNUAL REVIEWS INC. is a nonprofit corporation established to promote the advancement of the sciences. Beginning in 1932 with the *Annual Review of Biochemistry*, the Company has pursued as its principal function the publication of high quality, reasonably priced Annual Review volumes. The volumes are organized by Editors and Editorial Committees who invite qualified authors to contribute critical articles reviewing significant developments within each major discipline. Annual Reviews Inc. is administered by a Board of Directors whose members serve without compensation.

Annual Reviews are published in the following sciences: Anthropology, Astronomy and Astrophysics, Biochemistry, Biophysics and Bioengineering, Earth and Planetary Sciences, Ecology and Systematics, Energy, Entomology, Fluid Mechanics, Genetics, Materials Science, Medicine, Microbiology, Nuclear Science, Pharmacology and Toxicology, Physical Chemistry, Physiology, Phytopathology, Plant Physiology, Psychology, and Sociology. The *Annual Review of Neuroscience* will begin publication in 1978. In addition, two special volumes have been published by Annual Reviews Inc.: *History of Entomology* (1973) and *The Excitement and Fascination of Science* (1965).

Ann. Rev. Nucl. Sci. 1977. 27:1–35

SPONTANEOUSLY FISSIONING ISOMERS

×5580

R. Vandenbosch[1]

Department of Chemistry, University of Washington, Seattle, Washington 98195

INTRODUCTION

In 1962 the first report (1) on the observation of a spontaneously fissioning species with an anomalously short lifetime appeared. Further work revealed that the excitation energy of one of these isomers was about 3 MeV and that its spin was considerably lower than the value required to account for the necessary γ-ray retardation in terms of a spin-forbiddenness effect. A number of different isomers were characterized and a systematics of the half-lives began to emerge. These developments led to increased interest in the idea that spontaneous-fission isomers were actually

[1] Work supported in part by US Energy Research and Development Administration.

1

shape isomers, corresponding to a second minimum in the potential energy along the elongation degree of shape deformation. A schematic representation of such a potential energy function is given in Figure 1, where we define a few of the quantities used throughout this review.

A very significant theoretical development was taking place concurrently with the experimental characterization of shape isomers. This development had its origins partly in an attempt to understand the nature of single-particle effects on nuclear binding energies (2) and partly in the failure of early purely microscopic calculations to give a reasonable description of the potential energy at large deformations. In a macroscopic-microscopic approach as first developed by Strutinsky (3,4), the smooth trends of the potential energy with respect to particle numbers and deformation are taken from a macroscopic (liquid-drop) model and the local fluctuations due to quantization of the single-particle motion are taken from a microscopic (single-particle) model. The latter fluctuations are sometimes designated the shell corrections and the method of calculating the potential energy is known as the shell-correction method. The second minimum in the potential energy curve for certain actinide nuclei is found to result from a superposition of a second minimum in the shell correction with the relatively flat maximum in the macroscopic part of the deformation energy.

In addition to providing an understanding of the origin of the anomalous spontaneous-fission lifetimes, the double-humped barrier also provided an explanation for some previously puzzling features of induced fission. Resonant-like structure in sub-barrier fission-excitation functions could now be attributed to transmission resonances associated with the dependence of the penetration of a double-humped fission barrier on energy. Intermediate structure observed in low-energy neutron-induced resonance studies could be attributed to coupling of the more numerous states in the first well with the smaller number of states correspond-

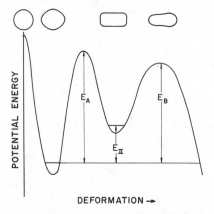

Figure 1 Schematic view of double-humped barrier and of nuclear shapes at selected deformations.

ing to excitation of the shape isomer. Anomalies in the level densities for deformations corresponding to the inner and outer barriers were identified with the breaking of certain symmetries of the nuclear shape (5).

In this review we concentrate on the experimental information relating to the properties of shape isomers. We do not attempt to provide a comprehensive review of theoretical developments, but primarily draw only on the results of these investigations as they bear on the experimental observations. For fairly recent summaries of theoretical aspects of this subject, the reader is referred to the papers of Brack et al (6), Pauli (7), and Möller & Nix (8). We also do not discuss in much detail resonance effects in near-barrier induced fission. The Proceedings of the 1973 IAEA Symposium on the Physics and Chemistry of Fission (9) provide valuable references concerning all of these aspects. A fairly recent review (10) and a book (11) discuss shape isomerism, as well as other aspects of nuclear fission.

STATIC PROPERTIES

Isomer Excitation Energies

The excitation energy E_{II} of the isomeric state relative to the ground state is usually obtained from the threshold behavior of the excitation function for isomer production. Most of the measurements are performed using reactions in which two or more neutrons are emitted prior to reaching the nucleus of interest, leading to appreciable slopes of the excitation functions near threshold. Thus an extrapolation from the lowest energy at which measurements are made to the threshold energy is required. A statistical model is employed to guide this extrapolation, and an absolute accuracy of perhaps 0.3 MeV can be achieved. A considerable body of data has been analyzed by Britt and co-workers (12). Values between 1.6 and 2.6 MeV have been obtained for a number of plutonium, americium, and curium isomers. (We exclude from discussion here excitation energies of isomers believed to be excited states of shape-isomeric nuclei; see discussion in section on Excitations of Shape Isomers.) One very precise isomer excitation energy, 2.56 MeV, is known from the observation of the γ decay branch from the ^{238}U isomeric state back to the ground state (13). Unfortunately, this isomer cannot be populated in reactions suitable for threshold measurements and hence a quantitative test of the statistical model analyses is not possible.

Barrier Heights

It is of considerable interest to characterize the inner and outer barriers that support the isomeric states. The height of the highest of the two barriers can usually be determined fairly easily from prompt fission–probability excitation functions. If transmission resonances (see section on Vibrational Excitations) are observed, then information about both barriers can be obtained. Additional information comes from analysis of isomer excitation functions. Back and co-workers (14, 15) have performed a very comprehensive statistical model analysis of fission-probability distributions for nuclei of actinide elements from thorium through californium. The model takes into account competition between fission, neutron, and γ emission in the decay of the compound nucleus. It uses resonant fission penetrabilities for a double-humped fission barrier, with damping of the resonances accomplished by introducing an imaginary component in the potential at the position of the second well. The results of analyses with this model are shown in Figure 2, where it is seen that the value of the height of the inner barrier E_A remains remarkably constant throughout this large range of Z and A. The height of the outer barrier E_B, however, decreases considerably as Z is increased.

Comparison of Excitation Energies and Barrier Heights with Theoretical Potential Energy Surfaces

The development by Strutinsky (3,4) of the macroscopic-microscopic method for calculating potential energy surfaces has made it possible to calculate fission barriers and isomer excitation energies that are in fairly good agreement with experimental trends. The basis of this method is to use a macroscopic model to determine the smooth trends of the potential energy with deformation and particle number, and a microscopic model to determine the fluctuations about these trends due to single-particle (shell) effects. The macroscopic model employed is the liquid-drop or liquid-droplet model.

A full exposition of the method is given in a review paper (6). There have been several tests of this method by comparison with Hartree-Fock calculations (16–18). The results seem to confirm the validity of the macroscopic-microscopic approach. Only one (18) of these calculations, however, attempted to reproduce the potential energy curve for a heavy nucleus exhibiting a double-humped barrier, and in this case only reflection symmetric shapes were considered. Thus, for the present one must rely on the results of the macroscopic-microscopic method when comparisons with experiment are attempted.

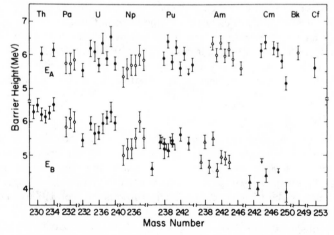

Figure 2   Fission-barrier heights obtained from fits to the fission probability distributions. Note the separate scales for the E_A and E_B values. From Back et al (15).

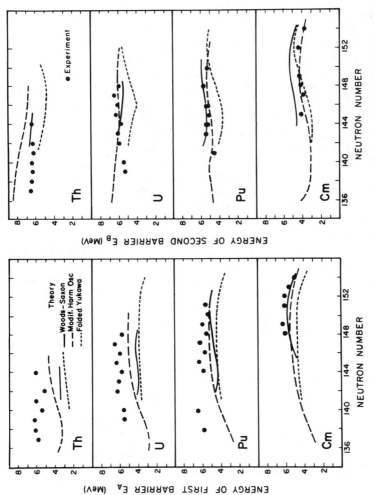

Figure 3 Measured heights of the first and second barrier as a function of neutron number and theoretical predictions from shell-correction calculations based on different single-particle potentials. After Specht (29), as updated by Metag. See Specht (29) for references to original work, except for the two new points for ^{232}Pu and ^{234}Pu, representing unpublished work of Metag et al (private communication).

The potential used in the macroscopic-microscopic method to generate the single-particle energies from which the shell and pairing corrections are derived may be of oscillator, Woods-Saxon, or folded Yukawa form. The shape parameterization must be general enough to describe the necking-in of the nuclear shape at the later stages, particularly at the second barrier. It has also been found that it is very important to include reflection-asymmetric shapes, as the outer barrier height E_B is lowered several MeV with inclusion of this degree of freedom (19–22). For the higher-Z actinides (Pu and beyond) a significant lowering of the inner barrier occurs when axially asymmetric degrees of freedom are included (23–26).

The most recent theoretical calculations (8, 27, 28) reproduce most of the important features of the observed dependencies of the inner and outer barriers on Z and A. Specht (29) has compared the results of several calculations with experimental results as shown in Figure 3. The decrease of the outer barrier height with increasing Z is in large measure due to the shift of the liquid-drop part of the barrier to smaller deformations, as illustrated by Metag (30) in Figure 4. The major anomaly revealed by the comparison between the calculated and experimental results in Figure 3 is that the height of the inner barrier is greatly underestimated in the theoretical calculations. One possible resolution of this anomaly has been suggested by Möller & Nix (8), who have found that for nuclei with small neutron numbers the outer barrier is split into two barriers, separated by a third minimum. They suggest that it is this third minimum that supports the transmission resonances observed (31), and that the analyses of the fission probabilities have located these two outer barriers, rather than the inner and outer barriers as assumed. It is not clear, however, that this suggestion can explain the discrepancy for heavier nuclei such as plutonium.

A previous (32) comparison of the experimental and theoretical isomer excitation energies is given in Figure 5. The theoretical calculations (8, 33, 34) give approximately the correct energies, except for the lighter elements, where the calculations of Möller & Nix (8) appear to seriously underestimate the energies.

Spins and Moments

SPIN DEPENDENCE OF PROJECTILE-FRAGMENT ANGULAR CORRELATIONS   There have been several attempts to learn about the spins of specific isomeric states by studying

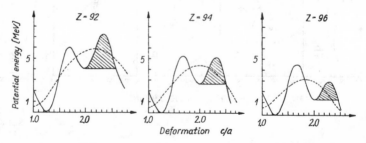

Figure 4   The influence of the liquid-drop part (*dashed curve*) of the double-humped fission barrier (*solid curve*) on the relative heights of the inner and outer barriers. From Metag (30).

the angular distribution of fission fragments with respect to the beam (35–40). The principle of this method is that a reaction such as $\alpha, 2n$ is used to form compound nuclei with their angular momenta aligned in the plane perpendicular to the beam direction. The fragment anisotropy is then determined by the angular momentum I of the isomeric state and by the value of the projection of this angular momentum on the nuclear symmetry axis, K_B, at the second barrier. Here the usual assumption, confirmed in prompt fission studies, has been made that it is the available K bands at the barrier that determine the anisotropy. In the present circumstance of barrier penetration rather than over-the-barrier fission, the rationalization for this assumption is that the penetrability integral for each particular K-channel is most affected by the maximum in the potential energy curve for that channel, i.e. at the top of the barrier. Contributions from several K_B states cannot be excluded, although the

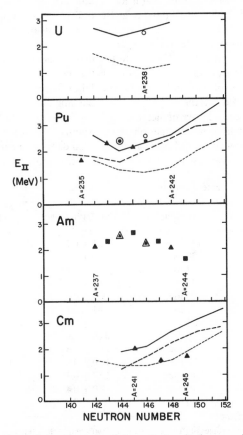

Figure 5   Comparison of experimental and theoretical excitation energies. The solid, long-dashed, and short-dashed lines are the theoretical results of Pauli & Ledergerber (33), Möller (34), and Möller & Nix (8). From (32).

exponential dependence of the penetrability on potential energy will usually result in the lowest state dominating. The fragment angular distribution for fission from a state with angular momentum I and projection K_b can be written (35) as

$$W_{K_b}^I(\theta) = \sum_\lambda A_\lambda G_\lambda(t) P_\lambda(\cos\theta), \qquad \lambda = 0, 2, \ldots, 2I \qquad\qquad 1.$$

with

$$A_\lambda = \frac{2I+1}{2} \sum_M (-1)^{K_b - M} f(M) C_{-MM0}^{II\lambda} C_{-K_b K_b 0}^{II\lambda}. \qquad\qquad 2.$$

In these expressions, P_λ are the Legendre polynomials, the C are Clebsch-Gordan coefficients, the $f(M)$ the occupation probabilities of the magnetic substates M of the spin relative to the beam axis, and the $G_\lambda(t)$ the attenuation factors for any perturbation of the correlation by extranuclear fields. A summary of the available data on the experimental ratios of $W(\theta)/W(90°)$ is given in Table 1. We note that the mere existence of an anisotropy requires that $I \neq 0$ or $\frac{1}{2}$. Thus we have the significant immediate conclusion that the even-even 30-nsec ^{236m}Pu and 6-nsec ^{238m}Pu isomeric states are not the superfluid $I = 0$ lowest states at the second minimum. The uncertainties in the use of the above equations in the analysis of fragment anisotropies lies in the estimation of $G_\lambda(t)$ and $f(M)$. In two of the experiments (35, 37), the recoils were stopped in Pb, in the hope that the spin alignment could be preserved $(G_\lambda(t) \approx 1)$ in the cubic lattice of metallic Pb. Recoiling into crystalline Pb does not assure that there will be no attenuation. Both instantaneous dislocations due to the stopping of the recoil atom and cumulative radiation damage due to the stopping of the fission fragments can give rise to attenuations. These problems are exacerbated by

Table 1   Fragment anisotropies for isomeric fission

Isomer	Half-life	$W(0°)/W(90°)$[a]	Reference
^{235m}Pu	30 nsec	1.1 ± 0.1	38
^{236m}Pu	30 nsec	0.70 ± 0.15	36
^{237m_1}Pu	114 nsec	0.90 ± 0.15	36
	45 nsec	0.58 ± 0.16	35
	100 nsec	0.4 ± 0.1	39
^{237m_2}Pu	1.1 μsec	1.41 ± 0.14	35
	1.1 μsec	1.9 ± 0.4	37[b]
^{238m}Pu	6 nsec	0.69 ± 0.09	35
^{239m}Pu	8 μsec	1.13 ± 0.04	40
^{240m}Pu	3.8 nsec	1.50 ± 0.8	36
^{241m}Cm	15 nsec	1.87 ± 0.40	36
	10 nsec	2.0 ± 0.4	35
^{243m}Cm	40 nsec	1.2 ± 0.4	36

[a] The anisotropies of Ref. 35 are $W(13°)/W(90°)$, rather than $W(0°)/W(90°)$.

[b] The anisotropy value of 1.9 given in this table differs from that reported in this reference, due to an additional correction for an A_4 term (J. Pedersen, private communication 1974).

the large quadrupole moments of shape isomers. The difficulty several workers (V. Metag, J. Pedersen, G. Sletten, private communication) have had in reproducing results suggests that the solid-state effects are not sufficiently well-understood at the present time to make reliable statements about the value of $G_\lambda(t)$. In the remaining experiment, the recoils were allowed to decay in flight in vacuum. Metag et al (41) conclude that 80% of the original alignment is preserved within the first 60 psec for recoil into vacuum. Most anisotropy measurements, however, have been performed for isomers with half-lives several orders of magnitude longer-lived than this. Comparison of the results of the different experiments in Table 1 indicates a lower value of the anisotropy for ^{237m}Pu when decaying in a vacuum compared to decaying in Pb, whereas similar anisotropies were observed for ^{241m}Cm. For spinless projectiles and targets, one would have $f(M = 0) = 1$ and $f(M \neq 0) = 0$ if the neutrons and γ rays did not carry off any angular momentum. The emission of neutrons and γ rays introduces a spread in $f(M)$. Since each emission carries off only a few units of angular momentum, and since the directions of emission are nearly isotropic and largely uncorrelated with one another, a statistical description should suffice for an approximate estimate of this distribution. The most serious shortcoming of this approximation may be contributions from stretched $E2$ transitions at the end of the γ cascade. This could be more important for α, $3n$ rather than α, $2n$ reactions, since the angular momentum input in the former is larger. Specht et al have assumed a Gaussian distribution of final M values, and estimated the dispersion of the distribution to be between 2 and 3. Hamamota & Ogle (42) have treated the γ cascade somewhat differently, and obtain considerably larger anisotropies for a given K and I combination. The calculations of Hien (43) give values qualitatively similar to those of Specht et al, but are sufficiently different to lead to different spin assignments for a given anisotropy. The spreading of the M distribution and the uncertainty in its width has a large effect on the expected anisotropies in all models. This is illustrated for the calculation of Specht et al in Figure 6. This uncertainty, together with the rather similar anisotropies expected for a number of different I and K_b combinations, means that unique spin assignments are not possible. In addition, one has uncertainties about the attenuation factors $G_\lambda(t)$, which have been assumed to be unity in all calculations. However, some valuable information not dependent on the details of the calculations or on the complete preservation of the alignment can be gleaned from anisotropy results.

We first discuss the interesting case of ^{237}Pu, where two different spontaneous-fission isomers have been discovered. The first attempt to gain information about the spins of the isomeric states was based on the relative populations of the two isomeric states in ^{237}Pu as a function of the angular momentum deposition·in the reactions that produced them (44). The yield of the long-lived (1100-nsec) state relative to the short-lived (82-nsec) state increased with increasing angular momentum, showing that the long-lived isomer has a higher spin than the short-lived isomer. From Table 1 we see that the anisotropy for the short-lived state is less than unity, whereas that for the long-lived isomer is greater than unity. This difference implies that the upper (1100-ns) state decays primarily directly by fission rather than by a γ transition to the lower state followed by fission. It also demonstrates that the K value of the second minimum state is not preserved during penetration through the barrier. From the absolute value of the anisotropies, the I value of the long-lived state would appear to be $\geq \frac{5}{2}$, with K_b small compared to I. The I value of the short-lived state is poorly defined, with the K_b values having to be close to I for this state. If one requires the K_b value to be the same for the two states and uses the calculated anisotropies of Specht et al or of Hien [which include the dubious assumption that $G_\lambda(t) = 1$], this would limit the spin assignments to $I = \frac{11}{2}$ for the 1.1-μsec state and $I = \frac{5}{2}$ for the short-lived state, with $K_B = \frac{5}{2}$. This result is consistent with a hypothesis (32) that the upper and lower states correspond to the Nilsson orbitals $\frac{11}{2} + [615]$ and $\frac{5}{2} + [862]$. These orbitals are close to the Fermi surface in most theoretical calculations (7, 8, 42, 45), as shown in Figure 7. They are not, however, consistent with the conclusions reached in an analysis of a g-factor measurement to be described later. If one makes the same requirement (that K_B be the same for both states) and uses the calculated anisotropies of Hamamoto & Ogle, assignments of $I = \frac{7}{2}$ and $\frac{3}{2}$ with $K_B = \frac{3}{2}$ would be indicated.

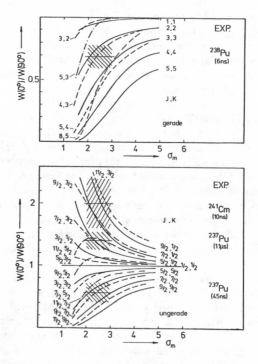

Figure 6   Calculated anisotropies $W(0°)/W(90°)$ as a function of the Gaussian width σ_M of the substate occupation number distribution function $f(M)$ for all combinations of I and K (up to $\frac{11}{2}$), for even (a) and odd (b) nuclei. The shaded areas represent the experimental anisotropies of the four isomers with their experimental errors and the uncertainty in the width σ_M. From Specht et al (35).

The anisotropy for ^{241m}Cm requires that $I \geq \frac{7}{2}$ with K_B small. The small anisotropy for the ^{243m}Cm isomer does not allow anything quantitative to be concluded.

NUCLEAR G-FACTORS   Further information about the properties of single-particle states of shape-isomeric nuclei can be obtained if the nuclear g-factor can be measured. Although such measurements are very difficult, preliminary results have been obtained for both the 1.1- and 0.1-μsec isomers of ^{237m}Pu. The technique used is to measure the time-dependent angular distribution of delayed-fission fragments when an external magnetic field is applied perpendicular to the plane of the detectors and target. The time-dependent angular distribution is obtained from an extension of Equation 1,

$$W_{K_b}^I(\theta, t) = \sum_\lambda A_\lambda G_\lambda(t) P_\lambda \cos(\theta - \omega_L t),$$ 3.

with the Larmor precession frequency

$$\omega_L = -g B_{\text{eff}} \mu_N / h.$$ 4.

If only the $\lambda = 0$ and $\lambda = 2$ terms are significant, the difference in counts between one detector at $+\theta$ and another at $-\theta$ with respect to the beam has an exponentially decaying time dependence modulated by twice the Larmor precession frequency. If, in addition to being placed at $\pm\theta$ with respect to the beam, the detectors are also placed at 90° with respect to each other, the difference in counts between the detectors divided by the sum of the counts gives a simple sinusoidal time dependence. A

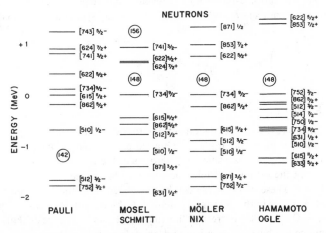

Figure 7   Neutron single-particle levels at the deformation corresponding to the second minimum. This diagram has been constructed from calculations reported by Pauli (7), Mosel & Schmitt (45), Möller & Nix (8), and Hamamoto & Ogle (42). Occasionally, significant mixing between orbitals occurs and the orbital labels are no longer very meaningful, for example the $[633]\frac{5}{2}^+$ and $[862]\frac{5}{2}^+$ orbitals, which are strongly mixed at this deformation in the calculation of Hamamoto & Ogle.

knowledge of the sign of the effective magnetic field B_{eff} (including both external and internal components) together with the sign of A_2, which can be obtained from an independent measurement of the time-integrated anisotropy $W(0°)/W(90°)$, enables one to obtain the sign of g from the phase of the oscillations. The g-factors for both the 1.1-μsec (37) and 0.1-μsec (39) isomers have been determined to be positive.

The absolute magnitude of g depends on a knowledge of $B_{\text{eff}} = \beta B_{\text{ext}}$, where β is the paramagnetic correction factor. This correction factor depends on the ionic state of the recoil Pu ion in the stopping Pb environment and, with certain assumptions, can be calculated for an assumed ionic charge. As defined, β has contributions from both a temperature-dependent part from the partial polarization of the 5f electrons and a temperature-independent part from the direct effect of the external field. The relative magnitudes of the two terms are different for each ionic state, so that a measurement of the temperature dependence of the precession frequency in principle enables one to determine β experimentally. In the experiment of Kalish et al, a measurement of the temperature dependence was used to decide between calculated values of β for 3+ and 4+ ionization states, with the results indicating the 4+ ionization state was involved. With this determination, g values of 0.14 ± 0.02 and 0.44 ± 0.05 were obtained for the 1.1- and 0.1-μsec isomeric states, respectively. The errors indicated do not include uncertainties in the β value, which might be as large as 30%.

In comparing the experimental values with theory, it is important to note that the isomeric state is presumed to have $K = I$ while precessing in the magnetic field, even if, during the leakage through the barrier, the effective value K_B is different than that of K. The theoretical estimate for the g-factor for a single-particle state described by the quantum numbers K and I moving in an axially symmetric potential can be written as

$$g = \left(1 - \frac{K^2}{I(I+1)}\right)g_R + \frac{K^2}{I(I+1)}g_\Omega, \quad K \neq \tfrac{1}{2}, \qquad 5.$$

with

$$\Omega g_\Omega = g_l\langle l_3\rangle + g_s\langle S_3\rangle. \qquad 6.$$

Here g_R represents the magnetic moment due to the collective flow of protons and neutrons, whereas g_Ω results from the unpaired particle. The terms g_l and g_s arise from the orbital and intrinsic spin contributions to the g-factor, with l_3 and S_3 the components of the orbital and intrinsic spin angular momenta along the nuclear symmetry axis. These components can be calculated from the wave functions for single-particle states in the deformed potential. Orbitals with the Nilsson quantum numbers $\Sigma = \Omega - \Lambda = +\tfrac{1}{2}$, corresponding to parallel coupling of the orbital and spin angular momenta, have S_3 positive, and orbitals with $\Sigma = \Omega - \Lambda = -\tfrac{1}{2}$ have S_3 negative. The absolute values of S_3 can be reduced significantly from $\tfrac{1}{2}$ (reductions of up to 50% are calculated and observed for rare-earth nuclei (46) since the spin-orbit part of the potential can result in significant admixtures of components with the opposite sign. This has the consequence that the expected g_Ω values for different Ω values can overlap. Additional uncertainties arise from inadequate knowledge about

g_R, g_l, and g_S. For deformed rare-earth nuclei, empirical effective values of $g_R =$ 0.35 $\pm$ 0.4, $g_l^{eff}(n) = -0.03 \pm 0.04$, $g_l^{eff}(p) = 1.07 \pm 0.04$, $g_S^{eff}(n) = -2.4 \pm 0.6$, $g_S^{eff}(p) = 3.7 \pm 0.6$ have been deduced (47, 48). These g_S values are significantly reduced (by a factor of about ~0.65) from the free-particle values.

The reported positive sign of g for both the 1.1- and 0.1-μsec ^{237m}Pu isomeric state implies that the relevant Nilsson orbitals have quantum numbers $\Omega - \Lambda = -\frac{1}{2}$. This is in disagreement with the earlier suggestion that these states correspond to the $\frac{11}{2} + [615]$ and $\frac{5}{2} + [862]$ orbitals. Hamamoto & Ogle (42) have calculated g values for several orbitals close to the Fermi surface for the second minimum. These results are given in Table 2.

Some readjustments of their standard parameters are required to reproduce the experimental g = 0.14 value for the 1.1-μsec state, with $\frac{5}{2} + [633+862]$, $\frac{7}{2} - [514]$ and $\frac{3}{2} - [752]$ orbitals being possible candidates. The $\frac{3}{2} - [512]$ and $\frac{7}{2} - [514]$ orbitals could be candidates for the 0.1-μsec state with g = 0.44. These suggestions must be considered very tentative at this time, in view of the preliminary nature of the experimental results and the fact that the theoretical calculation used a spin-orbit strength stronger than that indicated by the experimental level ordering at smaller deformations.

MOMENTS OF INERTIA   As discussed later, the lowest-lying levels of even-even nuclei are rotational in character. If the energy spacings between these levels can be determined for a shape-isomeric nucleus, the moment of inertia can be deduced. The transitions between these low-lying levels are highly converted, yielding low-energy

Table 2   Estimated g-factors for four states in the region of the second minimum for ^{237}Pu[a]

I	Orbital	Calculated		
		Parameters	g_{min}	g^{max}
$\frac{7}{2}$	$\frac{7}{2}-$	514	standard	0.22	0.41
$\frac{5}{2}$	$\frac{5}{2}+$	(633)	standard	−0.04	0.05
			$\eta_2 = 0.70$	0.12	0.24
			$\eta_4 = 0.08$	0.05	0.13
$\frac{3}{2}$	$\frac{3}{2}-$	512	standard	0.42	0.70
$\frac{3}{2}$	$\frac{3}{2}-$	752	standard	0.00	0.10

Experimental

$g_{exp}(1.1\ \mu sec) = 0.14 \pm 0.05$[b]

$g_{exp}(0.1\ \mu sec) = 0.44 \pm 0.14$

[a] The values were calculated (42) for a variation of the g_R, g_l, and g_S factors given by $0.25 \leq g_R \leq 0.40$, $-0.08 \leq g_l \leq -0.04$, and $0.65 \leq (g_S)_{eff}/(g_S)_{free} \leq 1.0$.
[b] Experimental values from Kalish et al (37) and Shackleton et al (39). The errors are our estimates, which include the uncertainties in β, as well as the statistical errors in determining the Larmor frequencies.

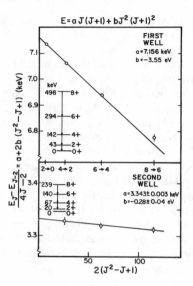

$$E = a\,J(J+1) + b\,J^2\,(J+1)^2$$

Figure 8 Fit of the transition energies to the energy expression $E = aJ(J+1) + bJ^2(J+1)^2$ (a) for the ground-state band of the first well and (b) for the isomeric-state band of the second well. The intercept gives the rotational constant a and the slope the nonadiabaticity parameter b. (After Specht and co-workers (49) and including corrections for the high–ionic charge states (see footnote 2).

conversion electrons. The observation of these electrons is very difficult, as the feeding of the isomer rotational band is usually at least 10^4 times smaller than that of the ground-state band, and there are also many other sources of low-energy electrons in the experiments where the isomers are produced. In spite of these difficulties, conversion electron groups from several transitions in each of two shape-isomeric nuclei have been observed. Specht et al (49) have studied ^{240m}Pu produced in an $(\alpha, 2n)$ reaction, while Borggreen et al (50) have studied ^{236m}U produced in a (d, p) reaction. The identification of electrons associated with the de-excitation of the rotational band built on the shape-isomeric state was performed using a delayed coincidence between the electrons and the fission fragments from the subsequent decay of the isomer. The level scheme deduced for ^{240}Pu is shown in the inset of Figure 8. The value of $\hbar^2/2\mathscr{I}$ obtained for ^{240}Pu is 3.343 ± 0.003 keV,[2] and for ^{236}U is 3.36 ± 0.01 keV. These values are less than half the values exhibited by the ground-state rotational bands, and provide fairly direct evidence that the shape isomer has a distortion qualitatively different from that of the ground state.

The moments of inertia depend not only on the shape (deformation) of the

[2] This value is slightly different than the value given in the original paper, due to a correction for the high ionic charges of the atoms. The new values are 46.72 ± 0.09 keV, 73.12 ± 0.12 keV, and 99.35 ± 0.13 keV for the $4+ \rightarrow 2+$, $6+ \rightarrow 4+$, and $8+ \rightarrow 6+$ transitions (H. J. Specht, private communication).

nucleus, but also on the nature of the mass motion. The moments of inertia of ground-state deformed nuclei are considerably reduced from rigid-body moments. This reduction can largely be reproduced in microscopic calculations using the cranking model and including pairing interactions (51). Several such calculations have been extended to include deformations corresponding to the shape-isomeric state (52–56). Both the calculated and experimental values of the ratio of the moment of inertia of the shape-isomeric state to that of the ground state is larger than that for a rigid rotator, reflecting the fact that the true moment of inertia more nearly approaches the rigid-body limit at large deformations. This result is illustrated in Figure 9. Most calculations tend to underestimate slightly ($\sim 15\%$) the moments of inertia at the ground state. The discrepancy is reduced in the calculations of Hamamoto (56), which include an additional term arising from the rotational motion on the pairing correlation. This calculation, however, overestimates the moment or inertia of the shape isomer. All of the calculations are somewhat sensitive to the assumed dependence of the pairing-strength parameter G on deformation. Calculations with both a constant pairing strength ($G = $ const.) and with a pairing strength proportional to the surface area S ($G \propto S$), have been performed. It is unlikely (57) that the dependence of the pairing strength on surface area S is as strong as $G \propto S$.

QUADRUPOLE MOMENTS   The quadrupole moment of a nuclear state also provides information on the nuclear deformation. In fact, the relation between the static quadrupole moment and the shape is less model-dependent than is the relation between the moment of inertia and the shape. Two rather different experimental approaches (58, 59) have been used recently to measure or deduce the lifetimes for rotational transitions feeding a fission isomeric state. If the transitions are primarily $E2$ in character, it is possible to deduce the quadrupole moment from the observed lifetimes. Since the expected lifetimes are sub-nanosecond, direct electronic timing experiments are not feasible.

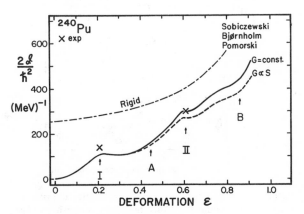

Figure 9   Comparison of theoretical and experimental values of the moment of inertia as a function of deformation. After Sobiczewski et al (54).

An ingenious "charge plunger" technique, capable of working in the same approximate time range as the Doppler-shift plunger technique, has been developed at Heidelberg (58). It is based on the fact that the rotational transitions in highly deformed nuclei are largely converted, that is, decay by internal conversion. The fast Auger cascades following such transitions lead to ionic charge states typically an order of magnitude larger than the equilibrium charge states (1^+-2^+) at the recoil velocities present. In the experiment, the recoiling ions are passed through a charge-resetting foil placed at various distances from the target. If the rotational de-excitation cascade is completed prior to arriving at the charge-resetting foil, the emerging ions will retain the small equilibrium charged attained for the rest of their flight, whereas if a converted transition occurs after reaching the foil, there will be a large increase in the ionic charge. The relative numbers of ions with low and high ionic charges can be determined by deflecting the ions in a magnetic field and recording the subsequent spontaneous fission in track detectors placed as recoil collectors. The time scale of the rotational de-excitation cascade can be determined by measuring the intensity ratio of the high-charge and low-charge components as a function of the distance between the target and the charge-resetting foil.

This technique has been applied to a study of the rotational band feeding the 8-μsec isomer of ^{239}Pu produced by a $\alpha, 3n$ reaction. In this first experiment an odd-A nucleus was chosen because of yield and half-life considerations. For an odd-A nucleus, however, the lifetimes are sensitive to the K value of the band as well as to the quadrupole moment. For small K- values, the transitions become very low in energy and are slowed down. Without a knowledge of the K value it is not possible to determine the quadrupole moment uniquely. Comparison of the observed decay curve (corrected for a long-lived component) with rotational-model calculations for various K values and an assumed quadrupole moment $Q_0 = 36.0$ b is shown in Figure 10. It is claimed that for K values of $\leq \frac{3}{2}$, the calculated decay curves are

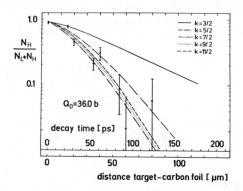

Figure 10   Fraction of highly charged recoil ions as a function of distance (or flight time) between target and charge-resetting foil. Corrections for a longer-lived 12-nsec component have been made. The theoretical curves represent cascade calculations for different K values of the rotational band and a quadrupole moment Q_0 of 36 b. From Habs et al (58).

inconsistent with the experimental data. For K values $K \geq \frac{5}{2}$, the quadrupole moment is determined to be $Q_0 = (36.0 \pm 4.4)$ b.

A rather different approach has been used by Metag & Sletten (59) at Copenhagen to deduce the rotational-state lifetimes. The branching ratios for spontaneous fission and electromagnetic decay of the rotational states are inferred from the angular distribution of delayed fragments. The lowest $0+$ state decays isotropically, but the higher states contribute increasingly to an anisotropic decay pattern. The expected anisotropy associated with decay from each rotational state is calculated in a manner similar to that described previously. The relative population of the different rotational levels and their partial spontaneous-fission lifetimes must be estimated in order to deduce the quadrupole moment from the anisotropy. The result, however, is not sensitive to side-feeding times into the rotational band, as is the previous method discussed. The method requires an isomer with a very short half-life so that the rotational de-excitation and fission lifetimes will be comparable. The 37-psec isomer of ^{236}Pu has been investigated. The value of the quadrupole moment deduced from the anisotropy of 1.48 ± 0.15 is 37^{+14}_{-8} b.

These values of the quadrupole moment are in very good agreement with theoretical estimates (18, 55, 60, 61). If the shape of the nucleus in the isomeric state is approximated by a prolate spheroid, a quadrupole moment of 36 ± 4.4 b corresponds to an axis ratio of (1.95 ± 0.1) for a radius parameter of $r_0 = 1.20$ fm (58).

DECAY PROPERTIES

Half-lives and Barrier Penetrabilities

The half-lives of known spontaneously fissioning isomers span the range from 10^{-11} to 10^{-2} sec. The lower end of the range is determined by the limitations of current experimental techniques. The development of a projection method (62) for determining half-lives as short as 10^{-11} sec is in fact a remarkable achievement,

Table 3 New isomer half-life data since the compilation of Britt (67)

Nuclide	Half-life	Reference
^{236m}U	116 ± 7 nsec	64
^{236m}Pu	37 ± 4 psec	63, V. Metag, private communication
^{238m}Pu	0.6 ± 0.2 nsec	63, 65
^{239m_2}Pu	12 ± 7 nsec	58
^{242m}Pu	3.5 ± 0.6 nsec	63
^{244m}Pu	380 ± 80 psec	63
^{240m_1}Cm	55 ± 12 nsec	66
^{240m_2}Cm	10 ± 3 psec	66
^{242m}Cm	40 ± 15 psec	66
^{244m}Cm	≤ 5 psec	66

particularly in consideration of the low yields of delayed fission relative to prompt fission encountered (10^{-5}–10^{-6}). The new results (63) obtained with this technique, as well as other new half-life information (64–66) obtained since the compilation of Britt (67), are given in Table 3. A plot of isomeric half-lives as a function of neutron number for different elements is given in Figure 11. A few even-even isomers in Pu and Cm, believed to be two-quasiparticle excitations, have been omitted from this plot. The paucity of information on spontaneous-fission isomers of Np and U is attributed to predominant decay by γ decay rather than by fission decay. The sensitivity of detection for γ decay is much poorer than for fission decay, and γ radiation is less characteristic of shape isomerism. The observed dependence on neutron number makes it unlikely that isomers longer-lived than 14-ms ^{242}Am will be found, with the possible exception of odd-odd Np isomers, which might have very weak fission decay modes. A number of systematic features are revealed in

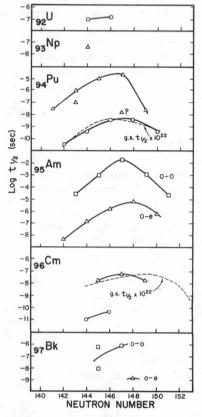

Figure 11 Spontaneously fissioning isomer half-lives as a function of neutron number. Circles, triangles, and squares represent values for even-even, odd-A, and odd-odd nuclei, respectively. Updated version of figure from (32).

Figure 11. The first of these is the maximum in the half-life at $N = 146$. This is qualitatively consistent with the variation of the calculated neutron-shell correction at the second minimum with neutron number, and corresponds approximately with gaps in the single-particle spectra calculated by various authors (see Figure 7).

An attempt to understand the effect of the interplay between the shell corrections at the second minimum and at the outer barrier on the neutron-number dependence of the half-lives has been given elsewhere (32). The variation in isomer half-lives with neutron number is surprisingly similar to that exhibited by the ground-state half-lives, although there is an upward shift in the neutron number corresponding to the longest half-lives as one goes from the isomeric states to the ground states. To the extent that the similarity arises from shell effects, one would expect it to be primarily due to effects at the outer barrier that are common to both ground and isomer decays.

A plot of the isomer half-lives versus proton number, rather than neutron number, yields points that fall on straight lines for a given neutron number (30). Three lines are obtained for a given neutron number, corresponding to even-even, odd-even, and odd-odd nuclei. The linear falloff of the logarithm of the half-life with increasing proton number is qualitatively consistent with the decrease of the liquid-drop part of the height of the outer barrier with increasing Z^2/A. The dependence of the shell correction on proton number is apparently considerably weaker than that for neutrons and does not appear in the form of a further modulation of the dependence of the half-lives on proton number. This is consistent with theoretical calculations that indicate that spontaneous-fission isomerism results primarily from a neutron-shell effect and not a proton-shell effect. This weakens the argument that fission isomers simply reflect the particular degeneracies of the single-particle states for a deformed potential corresponding to a shape whose semi-axes are in the ratio of 2:1.

Another feature revealed in Figure 11 is the large retardation of the half-lives for odd-A and odd-odd nuclei compared to even-even nuclei. This retardation also appears to be somewhat more regular than that exhibited by ground-state half-lives. Some tendency for smaller fluctuations in the isomer half-lives may arise from the fact that an alternate decay path is available, i.e. γ decay back to the first minimum. Thus very large retardations may not appear if the partial half-life for decay through the outer barrier becomes long compared to that for decay through the inner barrier. Returning to the average retardation of the decay for odd-A and odd-odd nuclei, two possible origins of this retardation can be identified by consideration of the expression for penetration through an idealized parabolic barrier. The penetration p is given by $p = \exp\left[-2\pi(E_B - E_{II})/\hbar\omega_B\right]$. The presence of unpaired particles may increase the height of the barrier that must be penetrated, $E_B - E_{II}$, by an amount sometimes designated as the "specialization" energy (68). Unpaired particles are also expected to increase the inertial mass and hence decrease the parameter $\hbar\omega_B$. In a previous examination (32) of this problem, we have determined the approximate magnitudes of each of these two effects that would be required to account for the observations if the other effect were absent. Specialization energies of 0.8 and 1.7 MeV for odd-A and odd-odd nuclei, respectively, would be required in the absence of any effect on $\hbar\omega_B$. Conversely, 25% and 45% decreases in $\hbar\omega$ would be required in the absence of specialization-energy effects. Metag (30) has performed a more quantitative examination of this problem by performing a 5-parameter least-squares fit to the half-lives of 33 isomers. In this fit, an odd-even effect on both the barrier height and on $\hbar\omega_B$ was allowed. For odd-mass nuclei, a specialization energy of 0.3 MeV and a decrease in $\hbar\omega_B$ of 16% was found. This decrease in $\hbar\omega_B$ of 16% corresponds to an increase in the inertial mass parameter of 30% if the curvature of the barrier is not changed by the addition of the unpaired particle. This result is in good agreement with theoretical estimates of the increase of the inertial mass parameter due to blocking effects of the odd particle (69–71).

The semi-empirical formula used by Metag to fit the half-lives is given by

$$t_{1/2}^{\text{S.F.}} = (\ln 2)(4 \times 10^{-21})\left\{1 + \exp\frac{2\pi}{\hbar\omega}[ax + b(N - N_0)^2 + d + S)]\right\},$$

with

$$\hbar\omega = \hbar\omega_0 \begin{cases} 1/\delta & \text{odd-odd} \\ 1 & \text{odd-even} \\ \delta & \text{even-even} \end{cases} \quad \text{and} \quad S = S_0 \begin{cases} 2S_0 & \text{odd-odd} \\ S_0 & \text{odd-even,} \\ 0 & \text{even-even} \end{cases}$$

and where the fissility parameter x was assumed to be

$$x = \frac{Z^2/A}{2(a_S/a_c)\{1 - [\kappa(N - Z)^2/A^2]\}},$$

with $\kappa = 1.87$, $a_S = 17.64$ MeV, and $a_c = 0.72$ MeV. Shell effects are introduced through the dependence on $(N - N_0)$, where the magic number N_0 was found to be 146. (Theoretical calculations give magic numbers of 142 or 148 as shown in Figure 7.) An attempt to include proton-shell effects showed that they were an order of magnitude smaller. A least-squares fit gave the following values of the parameters:

$$\frac{a}{\hbar\omega_0} = -49.4 \pm 5.2,$$

$$\frac{b}{\hbar\omega_0} = -(3.9 \pm 0.4) \times 10^{-2},$$

$$\frac{d}{\hbar\omega_0} = 46.0 \pm 4.4,$$

$$\frac{S_0}{\hbar\omega_0} = 0.43 \pm 0.26,$$

$$\delta = 1.16 \pm 0.08.$$

A comparison of the experimental half-lives with the calculated values from the five-parameter fit is shown in Figure 12. The success of this parameterization suggests that the curves in this figure will have fairly good predictive power for isomers that have yet to be identified.

There have been two recent attempts to calculate both ground- and isomer-state lifetimes from potential energy surfaces obtained by the shell-correction method and microscopic inertial parameters calculated with the cranking model (72–74). The adiabatic approximation is made and the penetrability is calculated using the WKB approximation. The mean lifetime for spontaneous fission may be written as

$$\tau = \frac{2\pi}{\omega_f}\exp(2S/h) \qquad\qquad 7.$$

where $\omega_f/2\pi$ is the frequency of barrier assaults and the action integral S is given by

$$S = \int_{r_1}^{r_2} [2B(r)(V_{(r)} - E_{\text{vib}})]^{1/2}\, dr. \qquad\qquad 8.$$

The action integral has to be calculated along the trajectory of least action, determined by $\delta S = 0$. This leads to a multidimensional problem for which both the potential energies and the inertial parameters have to be calculated as a function of all the relevant shape degrees of freedom. Ledergerber & Pauli (72) have pointed out some interesting features of this problem. As a practical matter, they have found the least-action path by trial. Although there is some correlation of the inertial parameter

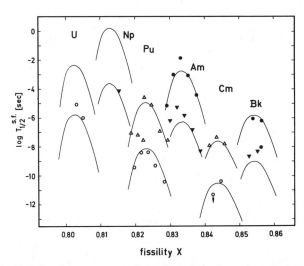

Figure 12   Half-lives for spontaneous fission from shape-isomeric states in U, Np, Pu, Am, Cm, and Bk isotopes as a function of the fissility parameter x. Open circles refer to isomers in even-even nuclei, triangles to isomers in odd-mass nuclei, and closed circles to isomers in odd-odd nuclei. Solid curves represent half-lives calculated with a 5-parameter semi-empirical formula whose parameters were determined by a least-squares fit to the experimental data and are given in the text. After Metag (30).

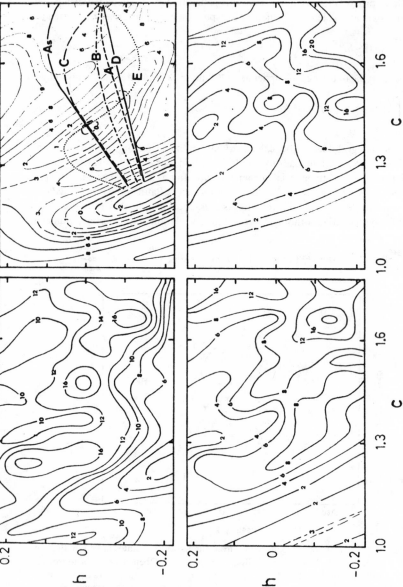

Figure 13 The deformation energy W (*upper right*) and the mass parameters B_{cc} (*upper left*), B_{ch} (*lower left*), and B_{hh} (*lower right*) of ^{240}Pu versus the two symmetric deformations c and $h(x = 0)$ are shown as contour plots. The mass parameters are given in units of $100 \, \hbar \cdot \text{MeV}^{-1}$; the energy is in MeV. In the plot of the deformation energy, some fission trajectories, leading from the ground-state well to the exit region, are shown by thicker lines. The values of the corresponding action integrals are: $S = 60.25 \, \hbar \, (A)$, $60.4 \, \hbar \, (B)$, $60.3 \, \hbar \, (C)$, $60.9 \, \hbar \, (D)$, $66.1 \, \hbar \, (E)$, $70 \, \hbar \, (As)$. The energy of collective motion is $E = 0.5 \, \text{MeV}$. From Ledergerber & Pauli (72).

with the shell-correction energy, the least-action path does not follow the apparent path of steepest descent. This is illustrated in Figure 13. One must remember, however, when examining the potential energy surface, that the shape of such a surface is dependent on the choice of parameters against which it is plotted. Certain features such as local maxima, local minima, and saddle points are invariant under coordinate transformations (75). Several paths are shown superimposed on the potential energy surface in the upper right-hand corner of this figure. The path E, which is closest to the path of steepest descent that would go through the saddle between the ground- and shape-isomer minima, has a considerably larger integral (less favorable penetrability) than the path A, which tunnels through a region where the potential energy is higher. A consequence of this result is the difficulty of defining a proper fission barrier. A "dynamic" barrier, which, following Ledergerber & Pauli, we define as the maximum in the potential energy along the path of least action, may be an MeV or so larger than the static barrier, defined as the saddle point in the deformation energy landscape. Thus the effective barrier height may depend on the collective kinetic energy with which the barrier is approached.

Another important result of these investigations is the decrease in the action integral obtained when reflection-asymmetric degrees of freedom are taken into account for deformations larger than the second minimum. For ^{240}Pu, the lowest action for a symmetric trajectory is 60.25 $\hbar$, whereas for an asymmetric trajectory it can be lowered to 52.8 $\hbar$, leading to a reduction in the calculated lifetime of a much needed factor of more than 10^7. Ledergerber & Pauli (72) have also noted a correlation between a parameter characterizing the asymmetry of the nuclear shape at the "dynamic" outer barrier defined above and the experimental mass asymmetry for nuclei from uranium to fermium.

Branching Ratios

The most easily detected and presumably the most important mode of decay for presently known shape isomers in the actinide region is spontaneous fission. An alternative mode of decay is by γ emission to lower-lying states in the first minimum. The relative probability for these processes depends both on the penetrabilities of the inner and outer barriers and on the hindrance of the γ decay process relative to fission (76). There are both theoretical and empirical indications that the inner barriers should become more penetrable relative to the outer barriers for the lower-Z elements. Theoretical calculations show that the inner barrier is the higher of the two barriers for plutonium and higher-Z actinides and that the outer barrier is the higher for the lower-Z actinides (27, 77). The systematic trend is in part the consequence of the inward shift in deformation of the liquid-drop part of the fission barrier as Z decreases. More difficult to estimate theoretically are the effects of the barrier width and inertial paramaters on the penetrability. The calculations of Ledergerber & Pauli (72) (see Table 1 of this reference) show that the inner barrier can be considerably more penetrable than the outer barrier even when the inner barrier is the higher of the two. In terms of a simple one-dimensional parabolic barrier, this would correspond to barrier curvature energies $\hbar\omega_A > \hbar\omega_B$. Experimental determinations (14) of the barrier heights E_A and E_B show $E_A < E_B$ for Pu and higher-Z even-even actinides, comparable to values of E_A and E_B for uranium, and $E_A < E_B$ for thorium isotopes. The analysis of the direct-reaction fission probabilities show that the $\hbar\omega$ values determining the penetrability and reflection near the top of the barrier are characterized by $\hbar\omega_A > \hbar\omega_B$ for almost all even-even actinide nuclei. These results are illustrated in Figure 14.

Motivated by these results, together with the indications from fission-yield systematics that there might be a sizable non-fission decay branch, Russo et al (13) searched for a γ decay branch for the ^{238}U shape isomer. The γ decay mode is much more difficult to observe experimentally than the fission decay mode because of high backgrounds of delayed γ rays from fission fragments and because of lower absolute detection efficiencies for γ rays. Such measurements are therefore only possible if there is a sizable branching ratio for γ decay relative to fission. Two γ-ray lines were found with half-lives in agreement with that observed for the fission branch. These γ rays have energies of 1.879 and 2.514 MeV, giving an energy difference consistent with the previously known difference between the lowest $1-$ and $2+$ levels of ^{238}U. These lines have therefore been attributed to decays leading to these states, as indicated by the decay scheme in Figure 15. The sum of the yields of these two states is more than 20 times larger than the yield for fission decay of the isomer. An estimate of the unobserved yield to higher states would increase this ratio by about a factor of 2. These results, together with a model-dependent estimate (78) of the hindrance of γ decay relative to fission, lead to a ratio of the penetrability of the inner barrier P_A to the outer barrier P_B of 4×10^7. Taking $E_A = 5.90$ MeV and

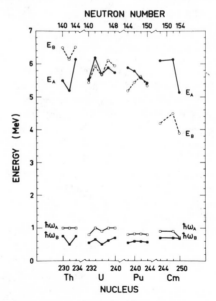

Figure 14 Fission-barrier parameters determined from the analysis of fission probabilities. From Back et al (14).

$E_B = 6.17$ (14), and assuming parabolic barriers leads to $\hbar\omega_A = 1.18$ MeV and $\hbar\omega_B = 0.63$. The predominance of γ decay over fission decay for ^{238m}U is more a consequence of the "thinner" inner barrier, reflected by $\hbar\omega_A > \hbar\omega_B$, than of the slightly lower value of E_A relative to E_B. The absolute values of $\hbar\omega_A$ and $\hbar\omega_B$ deduced are somewhat larger than the values deduced from direct-reaction fission probability data (14), perhaps reflecting the fact that the latter determination is only sensitive to the curvature near the top of the barrier, while decay of the shape isomer involves penetration 3 MeV below the top of the barrier.

There is a preliminary report from Heidelberg (79) on the observation of a γ-ray line at 2.215 MeV with a half-life of 104 ± 19 nsec produced in the ^{235}U(d, p) reaction. This half-life is in good agreement with the value 116 ± 7 nsec observed for delayed fission of ^{236m}U (64). If this transition populates the first rotational level of the ground state, the isomer excitation energy is 2.26 MeV. The population of this isomer

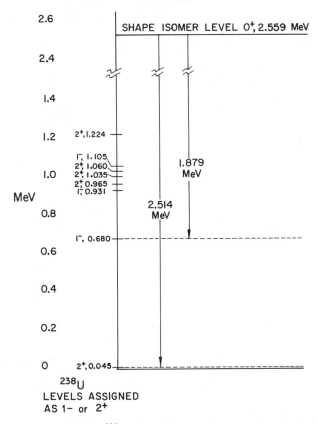

Figure 15   Decay scheme for the ^{238}U γ-branch based on the assumption that the 2.514-MeV transition is the decay of the shape-isomeric state to the first excited 2^+ state at 0.045 MeV in normally deformed ^{238}U. From Russo et al (13).

in neutron capture has been studied in another experiment, and from the measured ratio of delayed to prompt fission, an indirect estimate of 7 ± 2 for the ratio of γ decay to fission decay of the isomer has been made (80).

Although a γ branch has not been detected in any odd-A nuclei, it is believed that the failure to observe even-odd uranium isomers is a consequence of γ decay being the predominant mode of decay for these isomers. Similarly, the paucity of isomers for neptunium isomers and the very low yield of the one known isomer (81a) suggest that γ decay is predominant for the neptunium shape isomers as well.

Bowman (81b) has suggested that additional information about the barrier penetrabilities can be determined from measurements of photo-fission cross sections at very low energy, where a shelf in the photo-fission cross section is expected. Evidence for such a shelf has been seen in several nuclei (81c, 81d).

Fragment Kinetic Energy and Mass Distributions for Isomeric Fission

Differences between the fission characteristics of isomeric fission and either ground-state spontaneous fission or induced fission could arise from following different dynamical paths to scission depending on the initial excitation energy or depending

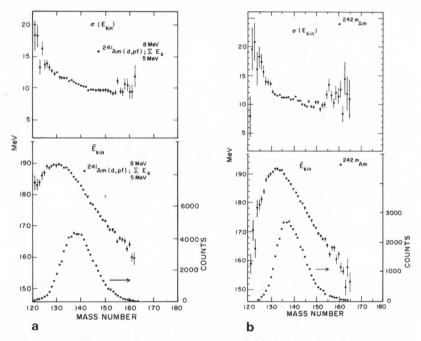

a   **b**

Figure 16 (a) The average total kinetic energy (*bottom*) and its standard deviation (*top*) as a function of the heavy fragment pre–neutron emission mass for the ^{241}Am(d,pf) reaction summed over the excitation energy region of 5–8 MeV in ^{242}Am. The bottom points in the figure show the relative mass-yield distribution for the ^{242}Am* fission. (b) The same quantities for ^{242m}Am isomeric fission. After Weber et al (85).

on whether the nucleus starts at its ground-state deformation or its isomeric-state deformation. If the excitation-energy effect were dominant, one would expect values for isomer fission intermediate between those of ground-state and induced fission.

Early measurements (82, 83) of single-fragment kinetic energies for 14-msec ^{242m}Am isomeric fission showed a double-peaked kinetic energy distribution. This implied that the mass distribution for isomeric fission is asymmetric, as it is for ground-state spontaneous fission and for induced fission at low excitation energies. Ferguson et al (84) extended these studies to other isomers and found that both the mass distributions and total kinetic energy distributions are similar to those found for normal spontaneous and low–excitation energy fission. In a more recent experiment (85) a careful comparison between the properties of coincident fission fragments for isomeric and excited-state fission of ^{242}Am has been performed. The mass distributions deduced from these measurements are shown in the bottom parts of Figure 16, together with the dependence of the total kinetic energy on the mass of the heavy fragments. Also shown in the upper part of the figure is the dependence of the standard deviations of the kinetic energy distributions on mass number. The qualitative similarities between the two kinds of fissions is clear. There are some differences, however. The average value of the total kinetic energy for isomer fission is 2.0 ± 0.4 MeV larger than for the induced fission. (Another, less accurate experiment (86) gives an opposite result of -1.85 ± 0.84 MeV.) A larger energy compared to induced fission is opposite to the trend (87–89) in which spontaneous fission from the ground state is found to have a lower average total kinetic energy than that for thermal neutron-induced fission from the same compound nucleus. It is, however, consistent with the observation that the average total kinetic energy in induced fission decreases with excitation energy at energies near ($E^* = 4.8$–9 MeV) (90–92) and above (93) the barrier. Weber et al (94) have found that the average kinetic energy for isomeric fission of ^{240}Pu is larger than that for either spontaneous fission or near-barrier fission of the same nucleus, in agreement with the observations for ^{242}Am.

The peak-to-valley ratio for isomeric fission from ^{242}Am is larger than that for the higher–excitation energy fission (85). This may simply reflect the known tendency for the peak-to-valley ratio to decrease with increasing excitation energy, or it may indicate a tendency to follow slightly different trajectories for through-the-barrier fission starting from the isomeric state as compared to over-the-barrier fission starting from the ground-state deformation.

EXCITATIONS OF SHAPE ISOMERS

Rotational Excitations

The lowest-lying states of normally deformed even-even nuclei are rotational states, with the lowest state occurring at an energy more than an order of magnitude smaller than the lowest-lying vibrational or two-quasiparticle states. For the more highly deformed shape-isomeric state, the rotational excitations will occur at even lower energies due to the larger moment of inertia. Rotational bands in two even-even shape-isomer nuclei have been identified, as has been discussed earlier in connection with moments of inertia. The rotational spacings for these isomers are indeed less than half the spacings for rotational bands based on the normally deformed ground state. If the rotational energies are represented as a power series in the angular momentum

$$E(J) = aJ(J + 1) + bJ^2(J + 1)^2,\qquad\qquad 9.$$

the coefficient b of the second correction term is much smaller for the isomer than for the ground state, as can be seen in Figure 8. This result can be qualitatively understood as a consequence of a better separation between the rotational degree of freedom and other degrees of freedom, such as vibrations and particle excitations, when the rotational energies are very low, as for the shape isomeric state. If the correction term is attributed to centrifugal stretching, it has been shown (32) that one can deduce the ratio of the stiffness in the two wells, C_I/C_{II} from the ratio of the values of the b coefficients in the above expression, if one makes the assumption that the dependence of the moment of inertia on deformation has the same functional dependence at the first and second minima. This assumption, however, is quite questionable in view of the significant modulations in the dependence of the microscopically calculated moments of inertia on deformation (see Figure 9).

An alternative expansion of the rotational energies in even powers of the rotational frequency (95), rather than angular momentum, gives coefficients for the second term that are within their errors the same for the ground-state and isomeric-state bands (10). This is consistent with our earlier analysis, as it can be shown that this model is equivalent to the VMI model and that the coefficient of this term depends only on the stiffness C and not on the deformation (96).

Theoretical calculations indicate that the stiffness at the first and second minima are comparable. The potential energy function for ^{240}Pu of Tsang & Nilsson (77) is approximately twice as stiff at the first minimum as at the second minimum. The potential energy surface of Ledergerber & Pauli (72) yields a slightly larger stiffness at the first minimum than at the second minimum.

Vibrational Excitations

There is no experimental information on the low-lying vibrational states at the second minimum. An approximate estimate of the expected vibrational-level spacings can be obtained by using the dependence of $\hbar\omega$ on the stiffness C and the inertial parameter B, $\hbar\omega = (C/B)^{\frac{1}{2}}$, to scale the known values of $\hbar\omega$ at the first minimum, $\hbar\omega \sim 1$ MeV, to the second minimum. Goldstone et al (97) have deduced a value of $\hbar\omega_{II}/\hbar\omega_I = 0.5$ for ^{240}Pu from the calculations of Ledergerber & Pauli (72). A somewhat larger value is indicated by the calculation of Randrup et al (74). Dudek (98) has calculated the transmission resonances for ^{241}Pu from potential and inertial functions resulting from a microscopic theoretical calculation. This calculation indicates that $\hbar\omega = 0.8$ MeV.

Although the lowest vibrational states are not known, there is information on higher vibrational states from the observation of resonances in the sub-barrier fission cross-sectional excitation functions for neutron, photon, and direct-reaction charged-particle–induced fission. Transmission resonances are expected to occur at energies closely corresponding to the collective β-vibrational levels of the shape isomer. These vibrations, sometimes designated fission vibrations, have $K^\pi = 0^+$ for even-even nuclei. Due to the strong dependence of the penetrability on energy relative to the barrier, only a few such resonances are likely to be experimentally observable. Additional resonances are expected, due to the presence of states that consist of a fission vibration coupled to other kinds of collective low-lying excitations such as rotations, $K = 2^+$ gamma vibrations, $K = 0^-$ octupole vibrations, etc. In simple models these states give rise to transmission resonances with strengths comparable to that for the $K = 0^+$ bandhead. For example, a strong resonance near 5-MeV excitation in ^{240}Pu (relative to the ground state) has been known for a long time (99–101). Angular correlation studies show that $K = 0$ fission channels are dominant for this resonance (102–104). The 5-MeV vibrational level is complex, with structures at 4.95 MeV and 5.1 MeV. In some analyses the weaker 4.95-MeV resonance is assumed to be $K = 0^+$, whereas in others the 5.1-MeV resonance is attributed to the $K = 0^+$ state. Less prominent structure has been seen at 5.25 and 5.45 MeV, and attributed (97, 105) to other collective states coupled to the 5-MeV vibrational state. The region below 5 MeV has been reinvestigated in more detail recently, revealing two peaks at 4.55 and 4.70 MeV (97). Angular correlation measurements (106) again show that this region is dominated by $K = 0$. (Such measurements cannot, however, distinguish between $K = 0^-$ and $K = 0^+$ contributions.) Goldstone et al have proposed that the 4.55-MeV level is the $K = 0^+$ fission vibration and the 4.70-MeV level is a $K = 0^-$ collective excitation based on this level. This yields an energy difference of 0.4 or 0.55 MeV between two vibrational levels, depending on which resonance near 5 MeV is attributed to $K = 0^+$. (One would expect it to be the lower, but model calculations of the fission probability significantly overestimate the strength if it is taken to be 0^+.) These results would indicate $\hbar\omega_{11} \sim 0.4$ or 0.55. However, the next higher $K = 0^+$ resonance has not been observed, and comparison of model calculations (107) with the rather structureless data at higher energies suggest that it has to be above 5.8 MeV in excitation energy. This is rather surprising, as one would expect anharmonic effects to make the spacing smaller rather than larger as one approaches the top of the barrier. The conclusion that there is no 0^+ resonance between 5.0 and 5.8 MeV has to be taken as somewhat tentative, as the same model calculations seriously overestimate the strength of the lower $K = 0^+$ resonance assumed to exist at 4.95 MeV. The very recent observation at Heidelberg (108) of a 4.21-MeV resonance with an anisotropy more indicative of a 0^+ or 0^- resonance than the 4.55- and 4.70-MeV resonances further complicates the interpretation of these results. If the 4.2-MeV, rather than the 4.55-MeV, resonance is the first 0^+ resonance below about 5 MeV, then $\hbar\omega_{11}$ would be about 0.8 MeV.

Single-Particle Excitations

ODD-A NUCLEI   For a deformed heavy nucleus, the average spacing between single-particle levels is 200–300 keV (see Figure 7). If two such levels happen to differ in K by more than a few units, then the rate for γ decay from the upper level to the lower level may be sufficiently slow that the decay of this state can be observed as a delayed fission component, either by virtue of a direct spontaneous-fission branch or as a result of γ decay to a lower state that then decays by spontaneous fission. In such circumstances one may expect to see two distinct fission half-lives. A number of such pairs of isomers have been observed in odd-A nuclei, as can be seen in Figure 11.

The most thoroughly studied pair of isomers is in ^{237}Pu. The longer-lived state has been found to lie 0.3 $\pm$ 0.15 MeV above the shorter-lived state (109). The spin of the upper state has been found to be higher than that of the short-lived state (44). Fragment angular distributions show that the upper state decays primarily directly by spontaneous fission rather than by spontaneous fission following γ decay to the lower-lying level. More quantitative information on the quantum numbers of the levels from angular distribution and g-factor measurements does not seem fully consistent at the present time. A resolution of this problem and the gathering of definitive information on other pairs would provide a valuable test of the theoretical single-particle level schemes shown in Figure 7.

TWO-QUASIPARTICLE EXCITATIONS   There are now three even-even Pu isotopes and two even-even Cm isotopes for which two isomers are known. The isomer half-lives

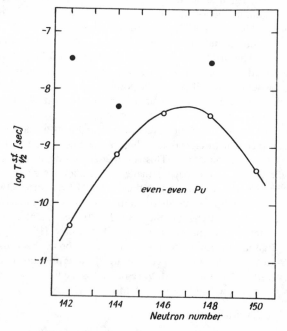

Figure 17   Logarithm of the half-lives for spontaneous fission from the shape-isomeric state in even-even Pu isotopes as a function of neutron number. Experimental points considered to represent the decay of the lowest state in the second minimum are connected by a solid curve [from (30)].

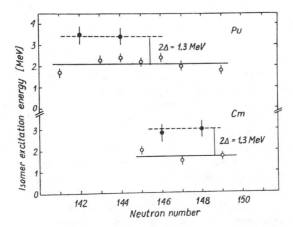

Figure 18   Excitation energies of isomeric states in Pu and Cm isotopes compared for ground states, open circles, and (presumably) two-quasiparticle states, closed points [from (29)].

for the even-even Pu isomers are plotted versus number in Figure 17. The isomers plotted with open circles have shorter half-lives, which vary smoothly with neutron number and have been attributed to the lowest state in the second minimum. The longer-lived states are attributed to two-quasiparticle excitations. In the case of ^{238}Pu there is direct experimental support for this view. The long-lived isomer has a threshold 1.3 MeV higher than the shorter-lived isomer and is also formed in lower yield (65). There are also short-lived isomers in ^{236}Pu and in ^{242}Cr and ^{244}Cm, which have excitation energies about 1.3 MeV higher than their neighboring odd-A isomers, as illustrated in Figure 18. An excitation energy of 1.3 MeV is very similar to that of two-quasiparticle excitations of ground-state plutonium and curium nuclei. This result does not allow us to determine whether the pairing strength G is independent of deformation. Some theoretical calculations of potential energy surfaces, moments, and inertial parameters have been performed, both with G constant and with G proportional to the surface area. Pairing theory predicts a dependence intermediate to these assumptions (110). In the latter case of G proportional to the surface area, one would expect two-quasiparticle excitations to be about 0.3 MeV greater at the second minimum than at the first minimum.

Another piece of experimental evidence that the longer-lived isomers are not the spin-zero ground states of the second minimum comes from anisotropy measurements discussed earlier. Both 30-nsec ^{236}Pu and 6-nsec ^{238}Pu have anisotropies deviating significantly from unity (see Table 1), with the sign of the deviation indicating that $K_b \approx I$. More quantitative statements about the spin of these states cannot be made at the present time. If these two-quasiparticle states were to have significant branching ratios to the ground state of the second minimum, then a determination of the population of the rotational levels could lead to unambiguous spin determinations.

CONCLUSION

During the last few years, the study of spontaneous fission isomers has come of age. The survey experiments defining the systematics of half-lives, excitation energies, and barrier parameters have been largely completed. The emphasis in recent years has been on more quantitative and spectroscopic aspects of shape isomerism. The evidence for spontaneous-fission isomers being shape isomers with a deformation more than twice that of the ground states has been firmly established with the determination of the moment of inertia and of the electric quadrupole moment of several isomers. A start has been made on the determination of spectroscopic properties of shape isomers, such as spins and moments. These experiments are very difficult and a number of ingenious methods are being developed and applied.

Our theoretical understanding of shape isomerism is in fairly good condition. There are, however, several areas where significant discrepancies between theory and experiment persist. Whether these are quantitative or qualitative remains to be seen. One of the more interesting theoretical developments has been the demonstration of the importance of inertial as well as potential energy effects on the dynamical path from the ground or isomeric state toward the scission point. Comparison of spins and moments with theoretical calculations should improve the definition of the single-particle potential and allow for a more reliable extrapolation to other deformations and mass regions.

The possible manifestation of shape isomerism in other regions of the periodic table is an interesting area for future exploration. It is highly unlikely that states of such "pure" deformation and high stability as the spontaneously fissioning isomers will be encountered. The shape isomers in the actinide region have two factors going for them that are fairly unique. In the first place, the potential energy surface on which the shell effects play their game is unusually flat because of the near cancellation of the shape dependence of the surface and Coulomb energies. Second, the number of particles involved, and hence the effective mass, is large, inhibiting the mixing of states of rather different deformation. In spite of these unique aspects, it is quite likely that there are shape isomers with sufficient purity to manifest themselves in a number of ways. There is already some evidence for such states in such diverse situations as the $4p$-$4h$ states of ^{16}O and in the crossings of rotational bands with quite different deformations in the rare-earth region. The full relationship between these phenomena and spontaneously fissioning shape isomers remains to be revealed.

ACKNOWLEDGMENTS

I am indebted to P. Paul, J. Pedersen, R. Sielemann, G. Sletten, H. J. Specht, and V. Metag for helpful comments and information about work in progress.

Literature Cited

1. Polikanov, S. M., Druin, V. A., Karnaukhov, V. A., Mikheev, V. L., Pleve, A. A., Skobelev, N. K., Subbottn, V. G., Ter-Akopyan, G. M., Fomichev, V. A. 1962. *Sov. Phys. JETP* 15: 1016
2. Swiatecki, W. J. 1964. *Nuclidic Masses,* p. 58. (Proc. Conf. Vienna 1963) Vienna: Springer-Verlag
3. Strutinsky, V. M. 1966. *Sov. J. Nucl. Phys.* 3: 449
4. Strutinsky, V. M. 1967. *Nucl. Phys. A* 95: 420
5. Bjørnholm, S., Bohr, A., Mottelson, B. R. 1974. See Ref. 9, I: 367
6. Brack, M., Damgaard, J., Pauli, H. C., Jensen, A. S., Strutinsky, V. M., Wong, C. Y. 1972. *Rev. Mod. Phys.* 44: 320
7. Pauli, H. C. 1973. *Phys. Rev. C* 7: 35
8. Möller, P., Nix, J. R. 1973. See Ref. 9, I: 103
9. Physics and Chemistry of Fission 1973. Proc. Symp., Rochester, NY, August 13–17. Vol. I, Vol. II. Vienna: IAEA. 578 pp., 525 00.
10. Specht, H. J. 1974. *Rev. Mod. Phys.* 46: 773
11. Vandenbosch, R., Huizenga, J. R. 1973. *Nuclear Fission,* New York: Academic. 422 pp.
12. Britt, H. C., Bolsterli, M., Nix, J. R. 1973. *Phys. Rev. C* 7: 801
13. Russo, P. A., Pedersen, J., Vandenbosch, R. 1975. *Nucl. Phys. A* 240: 13
14. Back, B. B., Hansen, O., Britt, H. C., Garrett, J. D. 1974. *Phys. Rev. C.* 9: 1924
15. Back, B. B., Britt, H. C., Hansen, O., Leroux, B., Garrett, J. D. 1974. *Phys. Rev. C* 10: 1948
16. Bassichis, W. H., Tuerpe, D. R., Tsang, C. F., Wilets, L. 1973. *Phys. Rev. Lett.* 30: 294
17. Brack, M., Quentin, P. 1973. See Ref. 9, I: 231
18. Flocard, H., Quentin, P., Vautherin, D., Kerman, A. K. 1974. See Ref. 9, I: 221
19. Möller, P., Nilsson, S. G. 1970. *Phys. Lett. B* 31: 283
20. Pauli, H. C., Ledergerber, T., Brack, M. 1971. *Phys. Lett. B* 34: 264
21. Pashkevich, V. V. 1971. *Nucl. Phys. A* 169: 275
22. Mustafa, M. G., Mosel, U., Schmitt, H. W. 1972. *Phys. Rev. Lett.* 28: 1050
23. Pashkevich, V. V. 1969. *Nucl. Phys. A* 133: 400
24. Larson, S. E., Ragnarsson, I., Nilsson, S. G. 1972. *Phys. Lett. B* 38: 269
25. Götz, U., Pauli, H. C., Junken, K. 1972. *Phys. Lett. B* 39: 436

26. Schultheiss, H., Schultheiss, R. 1971. *Phys. Lett. B* 34: 245
27. Möller, P., Nix, J. R. 1974. *Nucl. Phys. A* 229: 269
28. Larsson, S. E., Leander, G. 1973. See Ref. 9, I: 177
29. Specht, H. J. 1975. *Nukleonika* 20: 717
30. Metag, V. 1975. *Nukleonika* 20: 789
31. James, G. D., Lynn, J. E., Earwaker, L. G. 1972. *Nucl. Phys. A* 189: 225
32. Vandenbosch, R. 1974. See Ref. 9, I: 251
33. Pauli, H. C., Ledergerber, T. 1971. *Nucl. Phys. A* 175: 145
34. Möller, P. 1972. *Nucl. Phys. A* 192: 529
35. Specht, H. J., Konecny, E., Weber, J., Kozhuharov, C. 1973. See Ref. 9, I: 285
36. Galeriu, D., Marinescu, M., Poenaru, D., Vilcov, I., Vilcov, N. 1974. See Ref. 9, I: 297
37. Kalish, R., Herskind, B., Pedersen, J., Shackleton, D., Strabo, L. 1974. *Phys. Rev. Lett.* 32: 1009
38. Konecny, E. 1975. Private communication as quoted by P. Z. Hien, 1976. *Sov. J. Nucl. Phys.* 22: 489
39. Shackleton, D., Herskind, B., Kalish, R., Pedersen, J., Strabo, L. 1974. *Proc. Int. Conf. Hyperfine Interactions Studied Nucl. React. Decay,* ed. E. Karlson, R. Wappling, 3: 28–74. Amsterdam: North-Holland
40. Habs, D., Jaeschke, E., Just, M., Metag, V., Neumann, B., Repnow, R., Singer, P., Specht, H. J. 1974. Max Planck Inst. Kernphys., Jahresbericht, 1974. p. 1
41. Metag, V., Habs, D., Specht, H. J., Ulfert, G., Kozhuharov, C. 1976. *Hyperfine Interactions* 1: 405
42. Hamamoto, I., Ogle, W. 1975. *Nucl. Phys. A* 240: 54
43. Hien, P. Z. 1976. *Sov. J. Nucl. Phys.* 22: 489
44. Russo, P. A., Vandenbosch, R., Mehta, M., Tesmer, J. R., Wolf, K. L. 1971. *Phys. Rev. C* 3: 1595
45. Mosel, U., Schmitt, H. W. 1971. *Nucl. Phys. A* 165: 13
46. Bohr, A., Mottelson, B. R. 1975. *Nuclear Structure,* Reading, Mass: Benjamin 2: 228
47. See Ref. 46, p. 303
48. Nagamiya, S. 1972. *Sci. Pap. Inst. Phys. Chem. Res. (Jpn)* 66: 39–140
49. Specht, H. J., Weber, J., Konecny, E., Heunemann, D. 1972. *Phys. Lett. B* 41: 43
50. Borggreen, J., Pedersen, J., Sletten, G., Heffner, R., Swanson, E. 1977. *Nucl. Phys A* 279: 189

51. Griffin, J. J., Rich, M. 1960. *Phys. Rev.* 118:850
52. Damgaard, J., Pauli, H. C., Strutinsky, V. M., Wong, C. Y., Brack, M., Stenholm-Jensen, A. 1969. IAEA Symp. Phys. Chem. Fission, 2nd, IAEA, Vienna, p. 213
53. Krumlinde, J. 1971. *Nucl. Phys. A* 160: 471
54. Sobiczewski, A., Bjørnholm, S., Pomorski, K. 1973. *Nucl. Phys. A* 202: 274
55. Brack, M., Ledergerber, T., Pauli, H. C., Jensen, A. S. 1974. *Nucl. Phys. A* 234: 185
56. Hamamoto, I. 1975. *Phys. Lett. B* 56:431
57. See Ref. 11, p. 40
58. Habs, D., Metag, V., Specht, H. J., Ulfert, G. 1977. *Phys. Rev. Lett.* 38:387
59. Metag, V., Sletten, G. 1977. *Nucl. Phys. A* 282:77
60. Pomorska, B. N., Sobiczewski, A., Pomorski, K., *Proc. Int. Conf. Nucl. Phys., Munich, 1973.* I:598
61. Nerlo-Pomorska, B. 1976. *Nucl. Phys. A* 259:481
62. Metag, V., Liukkonen, E., Sletten, G., Glomset, O., Bjørnholm, S. 1974. *Nucl. Instrum. Methods* 114:445
63. Metag, V., Liukkonen, E., Glomset, O., Bergman, A. 1974. See Ref. 9, p. 317
64. Christiansen, J., Hempel, G., Ingwersen, H., Klinger, W., Schatz, G., Schatz, G., Withurn, W. 1975. *Nucl. Phys. A* 239: 253
65. Limkilde, P., Sletten, G. 1973. *Nucl. Phys. A* 199:504
66. Sletten, G., Metag, V., Liukkonen, E. 1976. *Phys. Lett. B* 60:153
67. Britt, H. C. 1973. *At. Nucl. Data Tables* 12:407
68. See Ref. 11, p. 54
69. Urin, M. G., Zaretsky, D. F. 1966. *Nucl. Phys.* 75:101
70. Sobiczewski, A., Szymanski, Z., Wycech, S. 1969. *Second IAEA Symposium Phys. Chem. Fission, Vienna, Austria,* p. 905
71. Sobiczewski, A., *Proc. Robert A. Welch Found. Conf. Chem. Res., Houston,* p. 13. Houston: Robert A. Welch Found.
72. Ledergerber, T., Pauli, H. C. 1973. *Nucl. Phys. A* 207:1
73. Pauli, H. C., Ledergerber, T. See Ref. 9, I:463
74. Randrup, J., Larsson, S. E., Möller, P., Nilsson, S. G., Pomorski, K., Sobiczewski, A. 1976. *Phys. Rev. C* 13: 229
75. Wilets, L. 1964. *Theories of Nuclear Fission,* Oxford: Clarendon
76. Nix, J. R., Walker, G. E. 1969. *Nucl. Phys. A* 132:60
77. Tsang, C. F., Nilsson, S. G. 1970. *Nucl. Phys. A* 140:275
78. Russo, P. A., Pedersen, J., Vandenbosch, R. 1974. See Ref. 9, I:271
79. Habs, D., Just, M., Metag, V., Mosler, E., Neumann, B., Paul, P., Singer, P., Specht, H. J., Ulfert, G. 1975. Max Planck Inst. Kernphys., Jahresbericht, p. 55
80. Andersen, V., Christensen, C. J., Borggreen, J. 1976. *Nucl. Phys. A* 269:338
81a. Wolf, K. L., Unik, J. P. 1973. *Phys. Lett. B* 43:25
81b. Bowman, C. D. 1975. *Phys. Rev. C* 12: 856
81c. Bowman, C. D., Schröder, I. G., Dick, C. E., Jackson, H. E. 1975. *Phys. Rev. C* 12:863
81d. Zhuchko, V. E., Ignatyuk, A. V., Ostapenko, Yu. B., Smirenkin, G. N., Soldatov, A. S., Tsipenyuk, Yu. M. 1975. *JETP Lett.* 22:118
82. Brenner, D. S., Westgaard, L., Bjørnholm, S. 1966. *Nucl. Phys.* 89:267
83. Erkkila, B. H., Leachman, R. B. 1968. *Nucl. Phys. A* 108:689
84. Ferguson, R. L., Plasil, F., Alam, G. D., Schmitt, H. W. 1971. *Nucl. Phys. A* 172: 33
85. Weber, J., Erdal, B. R., Gavron, A., Wilhelmy, J. B. 1976. *Phys. Rev. C* 13: 189
86. Fontenla, C. A., Fontenla, D. P. 1973. *Proc. Int. Conf. Nucl. Phys., Munich,* ed. J. DeBoer, H. J. Mang, I:588. Amsterdam: North-Holland & Am. Elsevier
87. Okolovich, V. N., Smirenkin, G. N. 1963. *Sov. Phys. JETP* 16:1313
88. Unik, J. P., Gindler, J. E., Glendenin, L. E., Flynn, K. F., Gorski, A., Sjoblom, R. K. 1974. See Ref. 9, II:19
89. Deruytter, A. J., Wegener-Penning, G. 1974. *Physics and Chemistry of Fission 1973,* 2:51. Vienna: IAEA
90. Unik, J. P., Loveland, W. D. 1971. Private communication quoted in Ref. 11, p. 300
91. Akimov, N. I., Vorobeva, V. G., Kabenin, V. N., Kolosov, N. P., Kuzminov, B. D., Sergachev, A. I., Smirenkina, L. D., Tarasko, M. Z. 1971. *Sov. J. Nucl. Phys.* 13:272
92. Milton, J. C. D., Fraser, J. S., Specht, H. J. 1977. *Proc. Eur. Conf. Nucl. Phys., Aix-en-Provence, 1972,* Pap. I–15
93. Ferguson, R. L., Plasil, F., Pleasanton, F., Burnett, S. C., Schmitt, H. W. 1973. *Phys. Rev. C* 7:2510
94. Weber, J., Specht, H. J., Konecny, E.,

Heunemann, D. 1974. *Nucl. Phys. A* 221:414
95. Harris, S. M. 1965. *Phys. Rev. B* 138: 509
96. Mariscotti, M. A. I., Scharff-Goldhaber, G., Buck, B. 1969. *Phys. Rev.* 178:1864
97. Goldstone, P. D., Hopkins, F., Malmin, R. E., von Brentano, P., Paul, P. 1976. *Phys. Lett. B* 62:280
98. Dudek, J. 1973. *Nucl. Phys. A* 203:121
99. Pedersen, J., Kuzminov, B. D. 1969. *Phys. Lett. B* 29:176
100. Britt, H. C., Burnett, S. C., Cramer, J. D. 1969. *Proc. IAEA Symp. Phys. Chem. Fission, 2nd, Vienna*, p. 375. Vienna:IAEA
101. Specht, H. J., Fraser, J. S., Milton, J. C. D., Davies, W. G. 1969. See Ref. 100, p. 363
102. Britt, H. C., Gibbs, W. R., Griffin, J. J.,

Stokes, R. H. 1965. *Phys. Rev. B* 139: 354
103. Specht, H. J., Fraser, J. S., Milton, J. C. D. 1966. *Phys. Rev. Lett.* 17:1187
104. Britt, H. C., Rickey, F. A. Jr., Hall, W. S. 1968. *Phys. Rev.* 175:1525
105. Back, B. B., Bondorf, J. P., Otroshenko, G. A., Pedersen, J., Rasmussen, B. 1969. See Ref. 100, p. 375
106. Glässel, P., Rösler, H., Specht, H. J. 1976. *Nucl. Phys. A* 256:220
107. Goldstone, P., Hopkins, F., Malmin, R. E., Paul, P. To be published
108. Just, M., Habs, D., Metag, V., Paul, P., Specht, H. J., Weber, J. 1976. Private communication
109. Vandenbosch, R., Russo, P. A., Sletten, G., Mehta, M. 1973. *Phys. Rev. C* 8: 1080
110. Kennedy, R. C., Wilets, L., Henley, E. M. 1964. *Phys. Rev. Lett.* 12:36

Ann. Rev. Nucl. Sci. 1977. 27:37–74

ELEMENT PRODUCTION IN THE EARLY UNIVERSE[1]

×5581

David N. Schramm
Enrico Fermi Institute and Department of Astronomy and Astrophysics,
University of Chicago, Chicago, Illinois 60637

Robert V. Wagoner
Institute of Theoretical Physics, Department of Physics, Stanford University,
Stanford, California 94305

CONTENTS

[1] Research supported in part by the National Science Foundation at the University of Chicago (AST 76-21707) and at Stanford University (PHY 76-21454).

"For violent fires soon burn out themselves."
(William Shakespeare, *King Richard II*)

1 INTRODUCTION

In the 1940s, Gamow and his associates (1, 2) began to recognize the significance that a hot big-bang model of the universe might have for the origin of the elements. They noted that if the universe was once at temperatures greater than 10^9 K, nuclear reactions could have built up heavier elements out of nucleons. The Gamow model began with only neutrons, but this assumption was later shown to be invalid (3, 4), since at sufficiently high temperatures ($T \gtrsim 10^{10}$ K), the weak reactions governing the interconversion of nucleons are in equilibrium, making neutrons and protons almost equally abundant. Nevertheless, the basic idea of big-bang nucleosynthesis began at that point. Initial attempts to make all the elements in the early universe were blocked by the stability gaps at masses 5 and 8. In order to synthesize heavier nuclei, sites of higher density and longer lifetime (stars) were required, where the mass gaps could be bridged by the triple-α process of Salpeter (5) and Hoyle (6). In fact, this realization prompted Gamow (7) to write a famous poem entitled "New Genesis," wherein the difficulty in producing heavy elements in the primeval fireball prompted the creator to create Hoyle in order to solve this problem within stars.

The current viewpoint concerning which elements can be produced in the big bang dates back to the 1960s, when calculations by Hoyle & Tayler (8), Peebles (9, 10), and Wagoner, Fowler & Hoyle (11) showed that 20–30% of the mass emerged in the form of ^{4}He, in good agreement with its observed abundance in a variety of objects. This agreement provided strong support for this process of nucleosynthesis. Of course, the strongest evidence for a hot big bang came with the discovery (12) of the microwave background radiation in 1965, which motivated the later investigations (9–11). The only natural explanation for its observed 3-K blackbody spectrum is for the early universe to have been sufficiently hot and dense that scattering was frequent in the uniform universal plasma (the primeval fireball). Other cosmological models (such as steady-state) that do not involve such a high-density "birth" cannot explain the existence of this radiation.

In addition to ^{4}He, it has been shown (11) that significant amounts of ^{2}H, ^{3}He, and ^{7}Li were also produced in the early universe. We show later that the abundance of deuterium may be particularly valuable in determining the geometry of the universe. It is significant that ^{2}H and ^{4}He are the two nuclei that appear most difficult to produce by other nuclear processes, such as those occurring in stars and in the interstellar medium. In fact, the simplest (standard) big-bang model produces these nuclei in amounts that agree best with their observed abundances.

The purpose of this review is to explore the nature and consequences of this process of nucleosynthesis. We begin (in Section 2) by discussing those observed properties of the universe that lead us to believe that it has expanded from a state of extreme temperature and density. We then introduce two sets of assumptions that define the models we consider. The first set of (basic) assumptions, which hold in any big-bang model, are taken to be valid throughout. The second set of (model) assumptions define the standard big-bang model, and are relaxed only in Section 5.

We then discuss the physical state of the early universe in Section 3, and the role the various relevant nuclear reactions play in the buildup of nuclear abundances. In particular, we analyze the factors that determine the helium abundance, which can be separated to some extent from the factors that determine the abundance of the other nuclei.

The observed abundances are discussed in Section 4, beginning with those of the most critical nuclei: ^{4}He and ^{2}H. The abundance determinations of ^{3}He and ^{7}Li are then analyzed, and shown to be difficult to interpret because of their additional production during the normal course of stellar evolution. Finally, we indicate why we believe that the abundances of other nuclei are not a result of the big bang.

In Section 5, we explore the sensitivity of the calculated abundances to violations of the five model assumptions. In general, it is found that the production of ^{4}He fails to agree with the observed amount when *any* of these assumptions is significantly violated. Examples of the nonstandard big-bang models that we investigate are those containing degenerate neutrinos, equal numbers of baryons and antibaryons, new types of neutrinos associated with heavy leptons, or free quarks. Models that do not assume the validity of the cosmological principle or general relativity are also discussed.

The implications of this remarkable agreement with the predictions of the simplest big-bang model are explored in Section 6. We show that the abundance of ^{4}He is consistent with all other observations, such as the ages of globular clusters. Once one accepts that the helium is of big-bang origin, then one can explore the consequences of the other results of the calculation. In particular, we show that the deuterium abundance indicates a present universal density that agrees with that associated with galaxies. Such a density is well below that required to close the universe if the cosmological constant is assumed to vanish. In view of this important implication of the deuterium abundance, we also investigate other sites for its production, and show why they run into considerable difficulties.

We conclude (Section 7) by summarizing our reasons for believing in the validity of the standard big-bang model, and the far-reaching consequences of such a belief.

For recent articles dealing with element production in the early universe, the reader is referred to papers by Wagoner (13, 14), Schramm & Wagoner (15), Fowler & Pasachoff (16), and Reeves (103). Recently, Weinberg (140) has discussed both physical and historical issues associated with big-bang cosmologies, at a less technical level.

2   STANDARD BIG-BANG MODEL

2.1   Evidence for a Big Bang

In the 1920s, Hubble and his colleagues made several discoveries that totally changed our perspective about the universe in which we live. They found that the faint spiral nebulae were actually galaxies much like, but very distant from, our own Milky Way galaxy (17, 18). Slipher (141) discovered that (with the exception of the closest ones) these galaxies were moving away from us. In addition, their velocities of recession v were proportional to their distances r (19). These observations were summarized by the famous Hubble law $v = H_0 r$, where H_0 is the Hubble constant. This law was consistent with a universe expanding in a uniform manner, the same law holding for an observer on any galaxy. In fact, Hubble noted that on large scales the galaxies appeared to be distributed uniformly throughout space.

There emerged two opposing viewpoints regarding the implications of the expansion:

1. The big-bang models described a universe that emerged from a state of much higher density in the past. These cosmological models were initially developed by Friedmann (20) and Lemaître (21) within the framework of general relativity, and later investigated by Gamow (22).

2. The steady-state model proposed by Bondi & Gold (23) and Hoyle (24) retained constant density in spite of the expansion, by postulating the creation of new matter. Thus the rate of creation of new matter was related to the expansion rate. Of course, the only places in such a universe where temperatures and densities are sufficient for nucleosynthesis is in the interiors of stars.

The steady-state model began to run into difficulties when it was discovered (25) that radio sources did not seem to remain uniformly distributed as one looked out to greater distances (i.e. further back into the past). Its demise came with the discovery of the microwave background radiation (12) and the subsequent proof that it has a blackbody spectrum. The density of a steady-state model of the universe was always what it is today, far too low to provide the photon interactions necessary for thermalization. When Penzias & Wilson (12) discovered that the background noise in their satellite-monitoring antenna was at a temperature of a few degrees in all directions, people began to believe that we did indeed live in a big-bang universe.

Many subsequent observations of this relic radiation at long wavelengths were in fact consistent with a 3-K blackbody (26). However, not for many years was it possible to make observations at wavelengths shorter than that of maximum intensity. As long as there existed measurements only in the low-frequency part of the spectrum, it was still possible to believe that the radiation was due to a superposition of other types of sources. It is difficult to make measurements at the required wavelengths $\lambda \lesssim 0.3$ cm because of atomspheric absorption, necessitating the use of rockets, balloons, or satellites. The background of the earth at 300 K is also a major problem. However, there now exist several measurements at these wavelengths (27), which clearly indicate a turnover of the spectrum. In addition, astronomers have noted that the molecular lines of CN that exist near the peak of the spectrum appear to be excited by an interstellar radiation field of the same temperature (28). All these measurements are consistent with a blackbody spectrum of temperature $T = 2.90 \pm 0.08$ K (27).

Additional evidence for the universal origin of this radiation is its isotropy. Its intensity is measured to be the same in all directions to within about 0.1% on all angular scales that have so far been investigated (26). No galactic-source models for this radiation have survived this constraint. In the big-bang models, this isotropy reflects the uniformity of the universal plasma at recombination, when this radiation decoupled from the matter (as is discussed in Section 3.1). Photons were able to freely propagate after that time, with their blackbody spectrum continuously shifted to lower temperatures by the universal expansion.

There is one type of anisotropy that should be observable. This is due to our motion relative to the frame in which this "photon fluid" is at rest. Since we are rotating about the center of our galaxy with a velocity $v \cong 250$ km/sec, and galaxies themselves have random velocities of the same order of magnitude, we should expect to see an increase in temperature in the direction of our motion (and a decrease in the opposite direction) of $\Delta T/T \cong v/c \sim 0.1\%$. Such an anisotropy may in fact have been detected (29). This predicted anisotropy is such a crucial feature that failure to detect it would force radical changes in our present conception of the universe.

It is a curious fact that, unbeknownst to its discoverers (12, 30), just such a universal sea of photons at a temperature of a few degrees had been predicted by Gamow (22) and Alpher & Herman (31) in the late 1940s, based on a belief that nucleosynthesis must have occurred in the early universe at a temperature of about 10^9 degrees. At temperatures higher than this, photodisintegration does not allow the formation of nuclei, while at lower temperatures, the Coulomb barrier greatly reduces the effectiveness of charged-particle reactions. As we show in Secion 3.1, the baryon density ρ_b remains approximately proportional to T^3 as the universe expands. Thus, knowing the present baryon density (10^{-32}–10^{-29} g cm^{-3}) allowed them to estimate the present temperature if they could estimate the baryon density during nucleosynthesis. Curiously, their estimate of this last factor was correct, but for the wrong reasons. In fact, using the correct physics produces the prediction $3 \times 10^{-2} \lesssim T_0 \lesssim 30$ K.

The Princeton group that was attempting to detect the microwave background radiation used a slightly different argument (30), but their expectation was similar to that of Gamow, Alpher, and Herman. It is curious that Zeldovich (32) employed similar reasoning at about the same time (just before the discovery of the radiation), but reached the conclusion that because we had not observed the radiation predicted, we did not live in a hot big-bang universe. Zeldovich's arguments were correct, but at that time the search for the radiation had not been pursued sufficiently.

Other types of observational evidence also support big-bang models. These include the result that counts of quasars seem to indicate that they are much more common at large redshifts (33), implying that they were more prevalent in the past. The same is true of most radio sources (25), indicating that galaxies were more likely to be strong radio sources in the (recent) past than now. This is clearly in disagreement with the steady-state model, in which no property of the universe can change. Other evidence came from measurements of the present deceleration parameter $q_0 \equiv -\ddot{R}R/\dot{R}^2$, where the universal scale factor $R(t)$ is proportional to the distance between galaxies. Neglect of galactic evolutionary effects (34, 35), as required by the steady-state model, allows the observations to be directly interpreted. The resulting values of q_0 (36, 37) are, however, larger than -1, the value predicted by the steady-state model. However, once we accept a big-bang model, the interpretation of measurements of q_0 is clouded by uncertain evolutionary corrections.

We show in what follows that the helium abundance itself provides strong evidence for a big bang, since no other method of production gives the amount observed in such a natural way. However, the primary reason for believing that our universe did emerge from such a state remains the 3-K background radiation.

2.2   Basic Assumptions

We consider element production in the early universe within the framework of two sets of assumptions. The first set of basic assumptions defines what we mean by big-bang models. They are discussed in this section, and are taken to be valid throughout this review. The second set of model assumptions defines a particular big-bang model, the "standard" model. They are discussed in the following section. After we explore nucleosynthesis within this model, we investigate the consequences of their individual abandonment in Section 5.

The basic assumptions are that (a) the principle of equivalence is valid; and (b) the universe was once at a temperature sufficient to impose statistical equilibrium among all particles present.

By the principle of equivalence we mean the requirement that the (nongravitational) laws of physics, as expressed in their usual special-relativistic form, hold locally in all freely falling (Lorentz) frames. For example, this requires all the nongravitational fundamental constants to remain constant during the expansion of the universe, which appears to be the case (134). This assumption restricts the theory of gravitation we employ to a metric theory (38). All presently viable gravitation theories effectively belong to this class.

The second basic assumption is a very powerful one, since it allows us to begin the calculation of nucleosynthesis at a time when most properties of the constituents of the universe are known, independent of the previous history of the universe. In most models, the minimum temperature required is about 10^{11} K. Since 1 MeV corresponds to a temperature of 1.16×10^{10} K, we see, moreover, that the particle energies are such that the laws governing their interactions are known at that time. For instance, the equilibrium of the weak reactions $n + e^{+} \rightleftharpoons p + \bar{\nu}_{e}$, etc at this temperature determine the initial number of neutrons per proton, so that this important quantity does not have to be specified as an initial condition.

It should be noted that the basic nature of gravitation in controlling the expansion of the universe guarantees that virtually all reasonable models of the universe that were dense enough to thermalize the 3-K background radiation must necessarily have been at a temperature and density sufficient for nucleosynthesis at an earlier time. In certain composite models of hadrons, analyses initiated by Hagedorn (39) and Veneziano (40) suggested that the number of hadrons grows so fast with mass that the universe could never have been at a temperature greater than $T \cong 2 \times 10^{12}$ K (41). In contrast, the elementary particle model (42) allows the initial temperature of the universe to be at least as high as the Planck temperature $T_{P} = (\hbar c^{5}/G)^{1/2}/k = 1.417 \times 10^{32}$ K (above which the space-time metric must be quantized). Note, however, that both of these models of hadrons, which are in a sense the two limiting cases of general particle theories, are consistent with assumptions (a) and (b).

2.3   Model Assumptions

The "standard" big-bang model is defined by the following model assumptions, in addition to the basic assumptions already discussed: (a) the lepton number of the universe is less than the photon number; (b) the baryon number of the universe is positive; (c) only those particles presently known were present during the epoch of nucleosynthesis; (d) the cosmological principle was valid; and (e) general relativity (including a possible cosmological constant) is the correct theory of gravity.

In effect, assumption (a) is equivalent to the requirement that all types of neutrinos be nondegenerate. Degeneracy of electron neutrinos or antineutrinos will shift the equilibrium ratio of neutrons or protons established by their weak reactions, as we demonstrate in Section 5. The other effect of degeneracy is to increase the expansion rate of the universe because of the increased density of neutrinos. The electrons can only become degenerate for large baryon densities, since they are related to the number of protons through the fact that the universe has no net charge.

Assumption (b) is equivalent to the statement that the universe is not symmetric with respect to matter and antimatter. Steigman (43) has presented strong arguments in support of this assumption. Nevertheless, we discuss nucleosynthesis in symmetric universes.

The third assumption concerns new particles with sufficiently long lifetimes to exist during element production, which began a few seconds after the "initial" state of much higher temperature and density. Possible violations of this assumption that we consider later include free quarks, superbaryons, and new types of neutrinos. In the standard model, we show that the only particles present at the beginning of nucleosynthesis were photons, electron and muon neutrinos, antineutrinos, electrons, positrons, neutrons, protons, and possibly gravitons.

The cosmological principle states that the universe presents the same aspect to all observers at any fixed epoch of cosmic time. An equivalent statement is that the universe is homogeneous and isotropic. The justification for this belief is based on the Copernican principle and the observations of isotropy. The Copernican principle states that we occupy a typical, not special, position in the universe. Beginning with Copernicus, we have learned that the earth is not at the center of the solar system, then that the solar system is not at the center of the galaxy, and finally, that our galaxy appears to be a typical one. Observations of the distribution of galaxies (44), radio sources (45), and the microwave background radiation (discussed in Section 2.1) are all consistent with there being $\lesssim 1\%$ anisotropy on large scales. If we accept the Copernican principle and the isotropy of the universe as viewed by us, then it follows that the universe is homogeneous and also appears isotropic to any observer moving with the expanding matter. This result is also consistent with the Hubble law discussed in Section 2.1, a law that would also be found to hold by an observer on any galaxy. Thus assumption (d) requires that there be no center, or any preferred position in the universe. This is an important restriction, since it allows us to infer global properties of the universe from observations of that portion visible to us. Although the present universe is not uniform within volumes containing only a few galaxies, the isotropy of the 3-K background radiation and the indications that galaxies condensed out of the gas that last scattered this radiation allows one to apply the cosmological principle to all length scales (including many particles) in the early universe.

The final model assumption concerns the nature of the gravitational interaction, which controls the large-scale dynamics of the universe. Its only effect on nucleosynthesis is through the expansion rate of the universe, which is related to the properties of the matter in different ways in different theories of gravity. This is discussed in detail in Section 3. All experiments carried out to date both on earth and within the solar system are completely consistent with the predictions of general relativity. In addition, most competing theories of gravitation have been shown to be nonviable when confronted with the increasing variety and accuracy of these observations (46). Nevertheless, we allow for other theories in Section 5 by considering expansion rates other than the one given by general relatively.

3   NUCLEOSYNTHESIS

3.1   Thermodynamic History of the Universe

The framework for understanding the process of nucleosynthesis is the evolution of the various constituents of the early universe. Their basic properties in the standard big-bang model are illustrated in Figure 1. Detailed discussions of this evolution can be found elsewhere (11, 26, 42, 47). It is indeed fortunate that we only need to know what occurs after the temperature has dropped below about 10^{11} K (10 MeV), since the electromagnetic, strong, and weak interactions keep all particles in statistical equilibrium above that temperature, making most of their properties independent of the previous history of the universe. In order to understand this result, as well as others, we must first know how the expansion rate of the universe is determined.

Consider a fixed number of baryons, which occupy an element of volume $V(t) \propto R^3$,

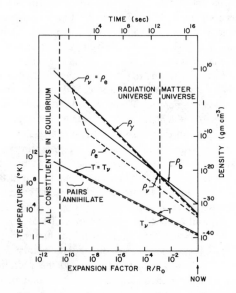

Figure 1   Evolution of the density and temperature of the constituents of the standard big-bang model (11). The temperatures of the baryons and electrons no longer equal the photon temperature T below 4000 K.

with $R(t)$ being the distance measure of the universe. This distance measure determines the evolution of the Robertson-Walker space-time metric

$$ds^2 = -c^2\,dt^2 + R^2(t)[(1-ku^2)^{-1}\,du^2 + u^2(d\theta^2 + \sin^2\theta\,d\phi^2)], \qquad 3.1$$

whose form follows from the cosmological principle (42). Here u is a radial co-moving coordinate (particle world lines are given by u, θ, ϕ = const.), and $k = 0$; ± 1 is the curvature constant. The Einstein field equations then give the expansion rate as

$$\frac{1}{V}\frac{dV}{dt} = \frac{3}{R}\frac{dR}{dt} = (24\pi G\rho + 3\Lambda c^2 - 9kc^2/R^2)^{1/2}, \qquad 3.2$$

where ρ is the total mass-energy density, G is the gravitational constant, and Λ is the cosmological constant. We know that $V^{-1}\,dV/dt \sim (24\pi G\rho)^{1/2}$ today, and we show that $\rho \propto R^{-4}$ in the early universe. Therefore, during nucleosynthesis we have, to very high accuracy,

$$V^{-1}\,dV/dt = (24\pi G\rho)^{1/2}. \qquad 3.3$$

Two other relations provided by general relativity (or most other metric theories) are conservation of energy:

$$\frac{d}{dt}(\rho V) + \frac{p}{c^2}\frac{dV}{dt} = 0, \qquad 3.4$$

where p is the total pressure, and conservation of baryons:

$$\frac{d}{dt}(\rho_b V) = 0, \qquad 3.5$$

where $\rho_b = M_u n_b$, with n_b the baryon number density and M_u the atomic mass unit. During nucleosynthesis the universe is radiation-dominated: $\rho_b \ll \rho$, but later, $\rho_b \cong \rho \propto R^{-3}$.

These relations allow one to describe quantitatively how the properties of the universe relevant to nucleosynthesis (density and temperature) evolve for each constituent. These are summarized in Table 1, where we introduce the useful quantity $T_9 \equiv T/(10^9\ \text{K})$, since nucleosynthesis occurs when $T_9 \sim 1$. Another property of interest is the mass of the visible universe. Because of the existence of a particle horizon (48) in these models, only a limited portion of the universe is causally related to any observer. One finds that the visible mass of any constituent k at time t is given by

$$M_k(t) \cong (4\pi/3)\rho_k(t)(2ct)^3. \qquad 3.6$$

For instance, the visible baryon mass (in units of the solar mass $M_\odot$) is given by

$$M_b/M_\odot = 1.58 \times 10^6\,hT_9^{-3} \quad (T_9 \gg 3) \qquad 3.7$$
$$= 3.14 \times 10^6\,hT_9^{-3} \quad (T_9 \ll 3), \qquad 3.8$$

in terms of the baryon density parameter h, defined in Table 1. Since we demonstrate

Table 1 Thermodynamic properties of the early universe ($10^3 \gtrsim T_9 \gtrsim 10^{-5}$)

Densities[a]	Temperatures
$\rho_\gamma = 8.42\, T_9^4$	$T \propto R^{-1}$ [b]
$\rho_v = \frac{7}{4}\rho_\gamma (T_v/T)^4$	$T_v \propto R^{-1}$
$\rho_e = \frac{7}{4}\rho_\gamma$ [c]	$T_v = T$ [c]
$\ll \rho_\gamma$ [d]	$= T/1.40$ [d]
$\rho_b = 2.75\, hT_9^3$ [c]	$T_9 = 10.4\, t^{-1/2}$ [c,e]
$= hT_9^3$ [d]	$= 13.8\, t^{-1/2}$ [c,e]

[a] In g cm^{-3}.
[b] Except when $T_9 \sim 3$.
[c] $T_9 \gg 3$.
[d] $T_9 \ll 3$.
[e] t in sec.

that this quantity has a value $10^{-6} \lesssim h \lesssim 10^{-3}$, only a small amount of surrounding matter could have affected nucleosynthesis within any volume element.

A particular particle will cease to be in statistical equilibrium when the time scale $\tau \sim \langle n\sigma v \rangle^{-1}$ of its interactions with particles of number density n becomes greater than the expansion time scale t, given as a function of temperature in Table 1. At 10^{11} K, only electron and muon neutrinos and antineutrinos, photons, nucleons, electrons, and positrons were present, since the age of the universe t was $\sim 10^{-4}$ sec, greater than the lifetimes of all other particles. In addition, the energy $kT \sim 10$ MeV had become too small to maintain a large equilibrium abundance of heavier particles. All particles present were in statistical equilibrium through the weak, strong, and electromagnetic interactions at the same temperature T. The total mass density of each type of particle is given in Table 1, with $\rho \cong \rho_\gamma + \rho_v + \rho_e$. Recall that the leptons are assumed to be nondegenerate.

As the temperature dropped to 10^{10} K, the neutrinos ceased to be in equilibrium, since the weak-interaction time scale $\tau \sim \langle n_e \sigma c \rangle^{-1}$ was proportional to T^{-5}, while the expansion time scale $t \propto T^{-2}$. The last weak reactions to drop out of equilibrium were ordinary electron-neutrino scattering and those involving neutral currents, such as $v_e + \bar{v}_e \rightleftharpoons e^+ + e^+$ and $v_\mu + \bar{v}_\mu \rightleftharpoons e^+ + e^-$, and the corresponding scattering processes. Thereafter, the neutrinos expanded freely, but maintained a Fermi-Dirac distribution corresponding to temperature $T_v(t)$ inversely proportional to the scale of the universe $R(t)$.

At about the same time, the reactions $n + e^+ \rightleftharpoons p + \bar{v}_e$, $n + v_e \rightleftharpoons p + e^-$, and $n \rightleftharpoons p + e^- + \bar{v}_e$ were no longer able to keep the neutrons and protons in their equilibrium abundance ratio

$$X_n/X_p = \exp[(M_p - M_n)c^2/kT] = \exp(-15.01/T_9). \qquad 3.9$$

The subsequent nuclear evolution is dealt with in the next section (3.2). Photodisintegrations prevented nuclear reactions from building up other nuclei until the temperature had dropped to $T_9 \sim 1$. Because the baryon density was negligible relative to the density of other particles, the energy released by nuclear reactions could not affect their evolution. There were (and are today) roughly 10^9 photons per baryon.

The next major change occurred at $T_9 \sim 3$, when the positrons (and most of the electrons) annihilated, adding energy to the photons relative to the neutrinos. Thereafter, the photon and neutrino temperatures remained in the ratio $(\frac{11}{4})^{1/3}$ given in Table 1. Until atoms could form at $T \sim 4 \times 10^3$ K, the electrons and nuclei remained in equilibrium with the photons at temperature T through electromagnetic interactions. The presently observed microwave background is the red-shifted photons from this epoch, because, once atoms formed and the free-electron density dropped, photons could freely propagate. At about the time that atoms formed, the universe switched from being radiation-dominated to being baryon-dominated.

We show below that the abundance of the elements produced in this standard big-bang model depends only on the baryon density parameter h. This quantity remains constant (except during electron pair annihilation), and is inversely proportional to the entropy per baryon, most of which is due to photons. Thus it is also inversely proportional to the number of photons per baryon

$$n_\gamma/n_b = 3.37 \times 10^4/h \qquad\qquad 3.10$$

present after pair annihilation. A more useful relationship involves the present photon temperature T_0 and baryon density ρ_0:

$$h = (\rho_0/2.44 \times 10^{-26})(2.9/T_0)^3. \qquad\qquad 3.11$$

Since T_0 is known to be about 2.9 K (27), an equivalent parameter is thus ρ_0. The present baryon density probably lies in the range

$$3 \times 10^{-32} \lesssim \rho_0 \lesssim 3 \times 10^{-29} \text{ g cm}^{-3}. \qquad\qquad 3.12$$

This range includes present uncertainties in the values of the Hubble constant H_0 and cosmological constant Λ, as well as the fact that there are observational limits on the matter (of all types) associated with galaxies (34, 59, 60).

3.2   Nuclear Reactions

The nuclear reactions that have actually been included in the computer calculations of Wagoner (13, 14) are indicated in Figure 2, while the ones of most importance for determining the abundances are presented in Table 2. Details of the computational method are given by Wagoner (49; see especially the appendices). In general, the fastest proton, neutron, and α-particle reaction with each nucleus is chosen, although some reactions of deuterium with light nuclei are also important.

We usually express the abundance of nucleus i in terms of its fraction X_i of the total baryon mass. All reactions of importance involve at most four different nuclei, and may be depicted as

$$N_i(^{A_i}Z_i) + N_j(^{A_j}Z_j) \rightleftharpoons N_k(^{A_k}Z_k) + N_l(^{A_l}Z_l), \qquad\qquad 3.13$$

where N_m is the number of nuclei with mass number A_m and atomic number Z_m involved. The rate equation that determines the evolution of the abundance of any nucleus i then has the form

$$\frac{dY_i}{dt} = \sum_{j,k,l} N_i \left(\frac{Y_l^{N_l} Y_k^{N_k}}{N_l! N_k!} [lk]_j - \frac{Y_i^{N_i} Y_j^{N_j}}{N_i! N_j!} [ij]_k \right),$$ 3.14

where

$$Y_i = X_i / A_i, \quad \sum_i X_i = 1$$ 3.15

and the sum includes all reactions involving nucleus i.

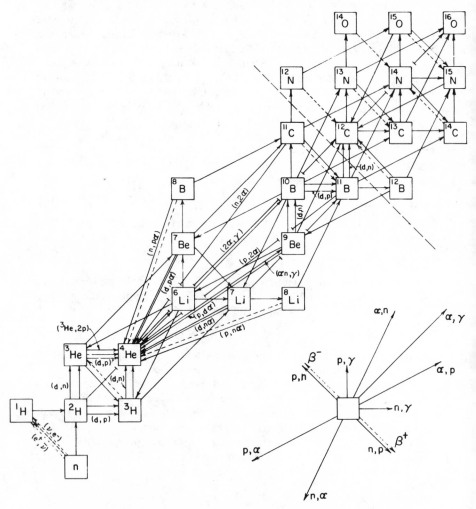

Figure 2   Network of nuclear reactions included in computer program (13). The exoergic directions are indicated by arrows, while the dashed line indicates that individual nuclear abundances were only computed for mass number $A < 12$.

Table 2   Important nuclear-reaction rates

1a. $n + v_e \to e^- + p, n + e^+ \to \bar{v}_e + p, n \to \bar{v}_e + e^- + p$:
$$\lambda_l(n) = 0.98\, \tau_n^{-1}(27.512\, Z^{-5} + 36.492\, Z^{-4} + 11.108\, Z^{-3}$$
$$- 6.382\, Z^{-2} + 0.565\, Z^{-1} + 1)\sec^{-1}$$

1b. $n + v_e \leftarrow e^- + p, n + e^+ \leftarrow \bar{v}_e + p, n \leftarrow \bar{v}_e + e^- + p$:
$$\lambda_l(p) = 0.98\, \tau_n^{-1}(27.617\, Z^{-5} + 34.181\, Z^{-4} + 18.059\, Z^{-3}$$
$$- 16.229\, Z^{-2} + 5.252\, Z^{-1})\exp(-2.531\, Z)\sec^{-1},$$

with $Z = m_e c^2/kT = 5.930/T_9$, $\tau_n = 926$ sec (free-neutron mean lifetime).
For the following reactions, the value of $N_A \langle \sigma v \rangle$ (in $cm^3\ sec^{-1}\ g^{-1}$) is given.
The number in parentheses is the value of Q_9.

2. $p + n \to d + \gamma\,(25.82)$:
$4.40 \times 10^4 (1 - 0.860\, T_9^{1/2} + 0.429\, T_9)$

3. $d + p \to {}^3\mathrm{He} + \gamma\,(63.75)$:
$2.65 \times 10^3\, T_9^{-2/3}(1 + 0.112\, T_9^{1/3} + 1.99\, T_9^{2/3} + 1.56\, T_9 + 0.162\, T_9^{4/3}$
$+ 0.324\, T_9^{5/3})\exp(-3.72\, T_9^{-1/3})$

4. $d + n \to {}^3\mathrm{H} + \gamma\,(72.62)$:
$66.2\,(1 + 18.9\, T_9)$

5. $d + d \to {}^3\mathrm{H} + p\,(46.80)$:
$4.17 \times 10^8\, T_9^{-2/3}(1 + 0.098\, T_9^{1/3} + 0.518\, T_9^{2/3} + 0.355\, T_9 - 0.0104\, T_9^{4/3}$
$- 0.0181\, T_9^{5/3})\exp(-4.26\, T_9^{-1/3})$

6. $d + d \to {}^3\mathrm{He} + n\,(37.94)$:
$3.97 \times 10^8\, T_9^{-2/3}(1 + 0.098\, T_9^{1/3} + 0.876\, T_9^{2/3} + 0.600\, T_9 - 0.0405\, T_9^{4/3}$
$- 0.0706\, T_9^{5/3})\exp(-4.26\, T_9^{-1/3})$

7. $n + {}^3\mathrm{He} \to {}^3\mathrm{H} + p\,(8.86)$:
$7.07 \times 10^8 (1 - 0.150\, T_9^{1/2} + 0.098\, T_9) + 1.29 \times 10^{11}\, T_9^{-3/2}\exp(-20.61\, T_9^{-1})$

8. $p + {}^3\mathrm{H} \to {}^4\mathrm{He} + \gamma\,(229.94)$:
$2.20 \times 10^4\, T_9^{-2/3}(1 + 1.68\, T_9^{2/3} + 1.26\, T_9 + 0.551\, T_9^{4/3}$
$+ 1.06\, T_9^{5/3})\exp(-3.87\, T_9^{-1/3})$

9. $n + {}^3\mathrm{He} \to {}^4\mathrm{He} + \gamma\,(238.81)$:
$6.62\,(1 + 905\, T_9)$

10. $d + {}^3\mathrm{H} \to {}^4\mathrm{He} + n\,(204.13)$:
$8.09 \times 10^{10}\, T_9^{-2/3}(1 + 0.092\, T_9^{1/3} + 1.80\, T_9^{2/3} + 1.16\, T_9 + 10.5\, T_9^{4/3}$
$+ 17.2\, T_9^{5/3})\exp[-4.52\, T_9^{-1/3} - (T_9/0.386)^2] + 1.21 \times 10^9\, T_9^{-3/2}\exp(-0.89\, T_9^{-1})$

11. $d + {}^3\mathrm{He} \to {}^4\mathrm{He} + p\,(213.00)$:
$6.67 \times 10^{10}\, T_9^{-2/3}(1 + 0.058\, T_9^{1/3} - 1.14\, T_9^{2/3} - 0.464\, T_9 + 3.08\, T_9^{4/3}$
$+ 3.18\, T_9^{5/3})\exp[-7.18\, T_9^{-1/3} - (T_9/1.373)^2] + 2.59 \times 10^9\, T_9^{-3/2}\exp(-3.30\, T_9^{-1})$

12. ${}^3\mathrm{He} + {}^3\mathrm{He} \to {}^4\mathrm{He} + 2p\,(149.24)$:
$5.96 \times 10^{10}\, T_9^{-2/3}(1 + 0.034\, T_9^{1/3} - 0.199\, T_9^{2/3} - 0.047\, T_9 + 0.0316\, T_9^{4/3}$
$+ 0.0191\, T_9^{5/3})\exp(-12.28\, T_9^{-1/3})$

13. $d + {}^4\mathrm{He} \to {}^6\mathrm{Li} + \gamma\,(17.08)$:
$480\, T_9^{-2/3}\exp(-7.44\, T_9^{-1/3})$ (uncertain)

14. $p + {}^6\mathrm{Li} \to {}^4\mathrm{He} + {}^3\mathrm{He}\,(46.67)$:
$3.46 \times 10^{10}\, T_9^{-2/3}(1 + 0.050\, T_9^{1/3} - 0.048\, T_9^{2/3} - 0.0165\, T_9 + 0.0016\, T_9^{4/3}$
$+ 0.0014\, T_9^{5/3})\exp[-8.41\, T_9^{-1/3} - (T_9/6.13)^2]$

15. $n + {}^6\mathrm{Li} \to {}^4\mathrm{He} + {}^3\mathrm{H}\,(55.53)$:
$1.25 \times 10^8 + 3.65 \times 10^9\, T_9^{-3/2}\exp(-2.53\, T_9^{-1})$

Table 2   *Continued*

16. $^3He + {}^4He \rightarrow {}^7Be + \gamma$ (18.42):
   $6.33 \times 10^6 \, T_9^{-2/3}(1 + 0.033 \, T_9^{1/3} - 0.350 \, T_9^{2/3} - 0.080 \, T_9 + 0.0563 \, T_9^{4/3}$
   $+ 0.0325 \, T_9^{5/3}) \exp(-12.83 \, T_9^{-1/3})$

17. $^3H + {}^4He \rightarrow {}^7Li + \gamma$ (28.63):
   $5.27 \times 10^5 \, T_9^{-2/3} \exp(-8.08 \, T_9^{-1/3})$

18. $n + {}^7Be \rightarrow {}^7Li + p$ (19.08):
   $6.77 \times 10^9 (1 - 0.903 \, T_9^{1/2} + 0.218 \, T_9)$

19. $n + {}^7Be \rightarrow {}^4He + {}^4He$ (220.39):
   $2.05 \times 10^4 (1 + 3760 \, T_9)$ (uncertain)

20. $d + {}^7Be \rightarrow {}^4He + {}^4He + p$ (194.59):
   $1.07 \times 10^{12} \, T_9^{-2/3} \exp(-12.43 \, T_9^{-1/3})$

21. $p + {}^7Li \rightarrow {}^4He + {}^4He$ (201.31):
   $7.66 \times 10^8 \, T_9^{-2/3}(1 + 0.049 \, T_9^{1/3} + 0.443 \, T_9^{2/3} + 0.152 \, T_9 - 0.149 \, T_9^{4/3}$
   $- 0.130 \, T_9^{5/3}) \exp[-8.47 \, T_9^{-1/3} - (T_9/30.07)^2] + 1.07 \times 10^{10} \, T_9^{-3/2}$
   $\exp(-30.44 \, T_9^{-1})$

22. $^7Be + p \rightarrow {}^8B + \gamma$ (1.56):
   $4.34 \times 10^5 \, T_9^{-2/3}(1 + 0.041 \, T_9^{1/3}) \exp(-10.26 \, T_9^{-1/3})$
   $+ 3.30 \times 10^3 \, T_9^{-3/2} \exp(-7.31 \, T_9^{-1})$

23. $^7Be + \alpha \rightarrow {}^{11}C + \gamma$ (87.56):
   $3.61 \times 10^7 \, T_9^{-2/3}(1 + 0.018 \, T_9^{-1/3} + 1.71 \, T_9^{2/3} + 0.215 \, T_9 + 2.88 \, T_9^{4/3}$
   $+ 0.919 \, T_9^{5/3}) \exp[-23.21 \, T_9^{-1/3} - (T_9/0.654)^2] + T_9^{-3/2}[7.27 \times 10^4 \exp(-6.50 \, T_9^{-1})$
   $+ 1.82 \times 10^5 \exp(-10.24 \, T_9^{-1})]$

24. $^7Li + n \rightarrow {}^8Li + \gamma$ (23.59):
   $4.90 \times 10^3 + 9.96 \times 10^3 \, T_9^{-3/2} \exp(-2.62 \, T_9^{-1})$

25. $^7Li + \alpha \rightarrow {}^{11}B + \gamma$ (100.55):
   $3.52 \times 10^8 \, T_9^{-2/3}(1 + 0.022 \, T_9^{1/3}) \exp[-19.16 \, T_9^{-1/3} - (T_9/0.268)^2]$
   $+ 1.51 \times 10^3 \, T_9^{-3/2} \exp(-2.62 \, T_9^{-1}) + 1.33 \times 10^4 \exp(-4.29 \, T_9^{-1})$

Because of the low baryon density ($\rho_b \sim 10^{-3} - 10^{-6}$ g cm^{-3}) during nucleosynthesis, all important reactions involve at most two nuclei in the initial state. Because of the high temperatures involved, most reverse reactions must also be included. There are three types of reactions involved:

1. Free β decays, for which $[i\beta]_k = \lambda_\beta(i)$, the inverse mean lifetime.
2. Reactions involving a lepton or photon ω with nucleus i, for which $[i\omega]_k = \lambda_\omega(i)_k = \langle n_\omega \sigma v \rangle$. Here v and n_ω are the velocity and number density of ω, σ is the cross section, and the average is taken over the appropriate distribution function of ω.
3. Reactions between two nuclei i and j (including the case $i = j$, $N_i = 2$, $N_j = 0$), for which $[ij]_k = \rho_b N_A \langle \sigma v \rangle$. Avogadro's number $N_A = M_u^{-1}$, v is the relative velocity, and the average is taken over the Maxwell-Boltzmann distribution.

The rates of the important reactions have also been included in Table 2. The energy release Q_9 (in units of $k T_9$) has also been indicated. The reverse rates $[\propto \exp(-Q_9/T_9)]$ are given in terms of the exoergic rates by Wagoner (49).

The procedure for converting experimental (or theoretical) cross sections into these rates is presented by Fowler, Caughlan & Zimmerman (50, 51) and Wagoner (49). The energy (in MeV) that is most effective at a given temperature T_9 is given by $E_0 \cong 0.086\,(l+\tfrac{1}{2})\,T_9$ for l-wave neutrons, $0.12\,(Z_i T_9)^{2/3}$ for protons, and $0.30\,(Z_i T_9)^{2/3}$ for α particles. Fortunately, big-bang nucleosynthesis occurs near $T_9 \sim 1$, requiring that we know the cross sections at energies much higher than effective in stars. Because of this fact, all the important nuclear cross sections have been measured in or near the relevant energy range. The weak-nucleon reactions (1) in Table 2 depend only on the free neutron lifetime τ_n. It appears that all these rates [except (13) and (19)] should be accurate to within a factor of two, with the rates of the reactions involved in ^{4}He production accurate to a few percentage points. The rates in Table 2 were the ones actually used in the calculation (13, 14) of abundances, although more recent compilations exist (51).

The evolution of the nuclear abundances and baryon density during the expansion of a typical standard big-bang model is illustrated in Figure 3. The equilibrium abundance of deuterium (and therefore all heavier nuclei) remains small above $T_9 \cong 1$ because of the many high-energy photons participating in the reactions $p+n \rightleftharpoons d+\gamma$. When the temperature drops to this value, mostly helium is produced because of its large binding energy and the stability gaps at $A = 5, 8$. Thus, virtually all neutrons present are converted into helium nuclei, making its abundance

$$X(^4\text{He}) \cong (2n/p)(1+n/p)^{-1}, \qquad\qquad 3.16$$

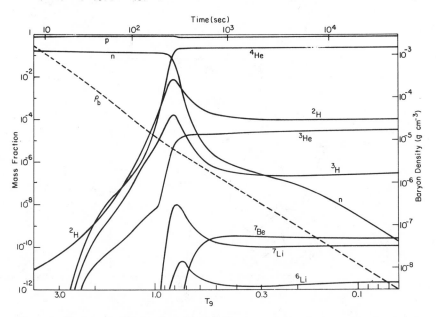

Figure 3   Nuclear abundances X_i and baryon density ρ_b during the expansion of a typical ($h = 1.15 \times 10^{-5}$) standard big-bang model (13).

where $n/p \equiv X_n/X_p$ when helium formation occurs. This ratio also affects the other abundances in less direct ways.

Thus it is important that we understand how the neutron-proton ratio is determined. Although the evolution of all the reactions is followed on the computer, a rough indication of the important factors involved can be gained by realizing that n/p is approximately equal to X_n/X_p when these reactions "freeze out" at $T_9 \sim 10$, minus the effects of free-neutron decay up to the time $t \cong 300$ sec, when nucleo-synthesis occurs. The major factor is the freeze-out temperature, which determines the neutron-proton ratio by Equation 3.9. This in turn is determined by the time when their weak-interaction rates λ became less than the expansion rate $V^{-1} dV/dt$ (due to the gravitational interaction). Thus it is essentially the balance between these two fundamental interactions that determines the helium abundance. This fact provides the basis for our later discussion (Section 5) of the sensitivity of the helium abundance to violations of our standard model assumptions.

The other nuclei are also produced rapidly once the photodisintegrations are reduced enough to allow other reactions to proceed. The Coulomb barriers and lack of neutrons below $T_9 \sim 1$ cause the final abundance levels to be reached quickly, except for subsequent β decays.

3.3 Resulting Abundances

The final abundances produced in the standard big-bang model are presented in Figure 4. These results can be found in tabular form as well (13, 14; but note that

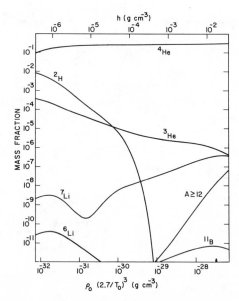

Figure 4   Final abundances (by mass) produced in standard big-bang models, whose only parameter is h, related to the present density and temperature by Equation 3.11 (13).

$h_0 = 2.75\ h$). Although a more detailed plot of the ^{4}He abundance is presented later (Figure 8), the analytic expression

$$X(^4\text{He}) = 0.333 + 0.0195 \log h \qquad\qquad 3.17$$

has been found (13) to be accurate for $h \geq 10^{-5}$. Since h is related to the present baryon density and photon temperature by Equation 3.11, the dependence on these quantities has also been included in Figure 4.

The insensitivity of the helium abundance to baryon density is consistent with our discussion of its production in the previous section. This follows from the fact that neither the weak-reaction rates nor expansion rate depended on ρ_b (recall that $\rho_b \ll \rho$). In fact, estimating the freeze-out temperature of the weak reactions and subsequent neutron decay gives a resulting neutron-proton ratio $n/p \cong \frac{1}{7}$, which, when used in Equation 3.16, gives $X(^4\text{He}) = \frac{1}{4}$, in good agreement with the full calculation.

The baryon density at $T_9 \sim 1$ strongly affects the abundance of the other nuclei. Recall that all the strong nuclear reaction rates $[ij]_k = \rho_b F(T_9)$, where $F = N_A \langle \sigma v \rangle$ is a sensitive function of temperature for charged-particle reactions, due to the Coulomb barrier. The reaction that initiates the buildup of the nuclei is $n + p \to d + \gamma$, which is much faster than the corresponding reaction $p + p \to d + e^+ + \nu_e$, which must operate in low-mass stars because of the lack of neutrons. The deuterium can be destroyed by many reactions involving neutrons, protons, and itself. In fact, these dominate the net effect in the sense that the higher the baryon density, the more deuterium is destroyed, as seen in Figure 4.

In the production of nuclei heavier than helium, the net effect becomes reversed, since the flow toward heavier nuclei is severely reduced by the increasing Coulomb barriers and the mass gaps at $A = 5, 8$. The heaviest nucleus produced in interesting amounts is in fact ^{7}Li. Its production peak in low-density universes is due mainly to the reaction $^4\text{He} + ^3\text{H} \to ^7\text{Li} + \gamma$, while in high-density universes, $^4\text{He} + ^3\text{He} \to ^7\text{Be} + \gamma$ dominates.

It should be noted that the rate of the reaction $d + ^4\text{He} \to ^6\text{Li} + \gamma$ is very uncertain. This reaction might produce interesting amounts of ^{6}Li if its actual rate were greater than the estimated value given in Table 2, but an independent estimate by Fowler gives a rate ~ 100 times lower at $T_9 \sim 1$.

4   COMPARISON WITH OBSERVATIONS

4.1   Abundance of ^{4}He

We begin our survey of observed abundances with ^{4}He, the major product of big-bang nucleosynthesis. A summary of the range of abundances (including the uncertainties) observed in various locations is presented in Table 3. Keep in mind that the probable range of values of the present baryon density given by Equation 3.12 corresponds to $10^{-6} \lesssim h \lesssim 10^{-3}$, giving a predicted primordial helium abundance of $0.17 \lesssim X(^4\text{He}) \lesssim 0.27$.

From observations of the emission lines of interstellar helium both in our galaxy and other galaxies, Peimbert (54) has concluded that its primordial abundance

Table 3   Helium-abundance determinations

Location	Mass fraction	Reference
Interstellar medium and young stars	0.26–0.32	52, 53, 61
Nearby normal galaxies	0.22–0.34	52, 53, 54
Large Magellanic cloud	0.24–0.27	55, 56
Small Magellanic cloud	0.21–0.28	56
Galaxy I Zw 18	0.21–0.29	54, 57
Galaxy II Zw 40	0.13–0.23	57

probably lies in the range $0.20 \leqq X(^4\text{He}) \leqq 0.25$. He sees evidence for a small increase in helium abundance with increasing abundance of heavy elements, indicating that stars do produce a small amount of the observed helium while they are producing all the elements heavier than boron. This would explain why helium abundances in "older" galaxies such as ours are somewhat higher than the "primordial" value. Probably the most significant recent results were those of Searle & Sargent (57), who investigated the two compact blue galaxies referred to at the end of Table 3. In the galaxy I Zw 18, for instance, the abundance of heavy elements is at least a factor of ten less than in normal galaxies, while the helium has almost the same abundance, indicating that it was present before the stars formed. Of course, there is still considerable controversy concerning the actual level of primeval helium abundance.

However, the fact that all these helium observations are consistent with the prediction of the standard big-bang model does provide strong support for that picture of the early universe. In addition, calculations indicate that local processes such as ordinary stellar evolution or supernovae could have produced at most one fourth of this amount, which is consistent with the range of helium observations (within their uncertainties). The resulting correlation between helium abundance and heavy-element abundance also agrees with the conclusion of Peimbert (54) that the "evolved" systems (2% of the mass in heavy elements) should have 4–6% more helium than the "primitive" systems ($\ll 1\%$ of the mass in heavy elements).

In fact, it can be seen from Equation 3.17 that if one could accurately determine the primordial helium abundance, then one could infer the actual value of h, or, equivalently, the present density of the universe. However, this is not possible at present, both because of uncertainties in the observations and their interpretation, and the weak dependence of $X(^4\text{He})$ on ρ_0. This very fact, that the helium-abundance prediction is relatively insensitive to the only degree of freedom in the standard big-bang model, on the other hand, adds strength to its agreement with observation.

It is extremely important to pursue these observations in order to determine whether there was a universal, pregalactic helium abundance. We indicate the full implications of such an abundance in Sections 5 and 6.

4.2   *Abundance of ^2H*

Another nucleus upon which much attention has been focused is deuterium. A summary of its various abundance determinations is presented in Table 4. Its

Table 4   Deuterium-abundance determinations

Location	$10^5 N(^2H)/N(^1H)$	Reference
Solar system:		
Earth (HDO)	15	135
meteorites (HDO)	13–20	136
Jupiter (CH_3D)	2.8–7.5	80
Jupiter (HD)	2.1 ± 0.4	81
present sun	< 0.4	137
primoridal sun		
from ^3He in gas-rich meteorites	1–3	64, 65
from ^3He in solar wind	< 5	63
from ^3He in solar prominences	< 6	138
Interstellar medium:		
Cassiopeia A (91.6-cm line)	< 7	89
Sagittarius A (91.6-cm line)	< 35	90, 91
Orion Nebula (DCN)	[a]	78, 79
nearby clouds (HD)	[b]	139
clouds (DCH^+)	[a]	83
galactic survey (DCN)	[c]	82
nearby (Lyman lines)	1.8 ± 0.4	85, 86

[a] Uncertain because of chemical fractionation.
[b] Uncertain because of chemical fractionation and self-shielding.
[c] Absolute ratio uncertain due to fractionation, but $DCN/H^{13}CN$ uniform except in direction of galactic center.

importance follows from the fact that its predicted abundance is a steeply decreasing function of the present baryon density, as seen in Figure 4. Of course, the strength of any conclusions we reach regarding ρ_0 from observations of $X(^2H)$ depends upon the strength of our belief that the helium observations indicate the validity of the standard big-bang model.

In principle, if one can determine what the abundance of deuterium was before galaxies formed, then one would be able to obtain the present density from Figure 4. Unfortunately, this approach is plagued with difficulties, one being the fact that deuterium is very easily destroyed within stars by the reaction $p + d \rightarrow {}^3He + \gamma$. This is, for example, the reason why deuterium has not been detected on the surface of the sun. Another problem arises from the possibility that the observed deuterium might have been produced by some process within galaxies. Although this appears unlikely, we consider this possibility in Section 6.2.

The revelation that deuterium could be an important tool in determining the density of the universe occurred after the discovery of the 3-K background radiation, particularly in the calculation of Wagoner, Fowler & Hoyle (11). They noted that the deuterium abundance derived from the isotopic ratio in sea water corresponded to a present-day density $\rho_b = 2 \times 10^{-31}$ g cm^{-3}, somewhat lower than that thought to be associated with galaxies. However, at that time many felt that deuterium could also be made within the solar system (62).

By the early 1970s, the situation had changed. One important factor was the realization that the hydrogen isotopic ratio in sea water must have been affected by chemical fractionation. In particular, in his measurements of the solar wind using the aluminum-foil window shade on Apollo 11, Geiss (63) found that the amount of ^{3}He was only a few parts in 10^5. Since all the deuterium in the sun had been burned to ^{3}He, and since ^{3}He has not been destroyed on the surface of the sun, the amount of ^{3}He in the solar wind provided an upper limit to the amount of deuterium present in the sun when it formed. This argument gave $X(^2H) \leqq 7 \times 10^{-5}$, a factor of three less than the terrestrially inferred value. Another argument was put forward by Black (64, 65), who showed that the amount of ^{3}He in the solar wind trapped by gas-rich meteorites implied a similar upper limit to the protosolar deuterium abundance. In addition, he noted that if one subtracts the amount of ^{3}He in the early solar system (as determined from various meteoritic analyses) from its abundance in the present solar wind, the difference should be due to deuterium burning, since the surface of the sun has not been contaminated with ^{3}He synthesized near the center. He thereby arrived at the abundance shown in Table 4.

In addition to these investigations, which lowered the abundance of deuterium by about an order of magnitude, other arguments changed the general viewpoint regarding its origin. Serious problems for the local theory (62) of deuterium production developed from careful analyses of meteorites (66, 67). In addition, any model of early solar-system production had difficulties on energetic grounds (68). By late 1971, the implications of these developments were recognized independently by a number of people (including Black, Schramm, Audouze, Reeves, and Fowler), and a general article summarizing the origin of all the light ($A < 12$) elements was published (69) in 1973. Reeves (103) has recently reviewed the situation.

The belief that the deuterium we observe was produced in the early universe has been strengthened because of the difficulties encountered by each new alternative process proposed for its origin (15, 70–77). This is discussed in more detail in Section 6.2. One of the basic difficulties with non-cosmological production is the tendency to overproduce the other rare light elements Li, Be, and B (74, 75).

The first detection of deuterium outside the solar system occurred in 1973 (78, 79), in the form of DCN in the Orion Nebula. Since then, there have been many observations of deuterium in both interstellar clouds and the solar system, as indicated in Table 4. However, chemical fractionation effects prevent one from deriving the total abundance of deuterium from the molecular observations. In particular, the abundance observed in Jupiter (80, 81) and future abundance determinations in Saturn, etc should be viewed with caution. Nevertheless, recent molecular observations (82, 83), which indicate a deficiency of deuterium in the center of our galaxy, are of great interest because such an effect is consistent with enhanced depletion of primordial deuterium due to the greater stellar processing there (84). On the other hand, if deuterium was produced within the galaxy, one might expect to find its abundance greater in the central regions (84, 143).

The most definitive observations of deuterium have been those from the *Copernicus* satellite (85–87). Absorption by the Lyman lines of interstellar atomic deuterium is observed in the spectra of nearby stars. The uncertainties in the derived abundance are small, giving a present mass fraction $X(^2H) = 2.5 \times 10^{-5}$ (within errors of about 25%) if the abundance is uniform. Although the observations are consistent with a uniform abundance within 200 parsecs of the sun, the abundance could vary by up to a factor of 4 smaller to a factor of 2 larger in the directions investigated (86).

The most relevant deuterium observations and their interpretation are indicated in Figure 5. The observations appear to be consistent with depletion rates due to stellar processing of the order of magnitude expected (88). The full implications of this picture are explored in Section 6.2. However, we might note that the present value of the deuterium abundance is a lower limit to the primordial value, and thus provides the upper limit to the present density of the universe indicated on Figure 5.

Future observations are vitally needed in this area. It is very important to look for variations in the deuterium abundance throughout the galaxy, in conjunction with more detailed models of the effects of galactic evolution in modifying the primordial abundance. In particular, an effort should be made to convert the upper limits provided by the observations of the 91.6-cm hyperfine transition (89–91) into observed values. The possibility of observing the radio recombination lines should also be pursued (84).

4.3   Abundance of Other Nuclei

Referring to Figure 4, let us compare the production of the other nuclei with their observed abundance. The mass fraction of heavy nuclei ($A \geq 12$) is about 1–2% (92), far greater than their production in the early universe. Even if one considers higher-density universes, their relative production does not resemble the observed distribution (11).

Among the other light nuclei, it appears as though only ^3He and ^7Li could have been made in the big bang. This is because the observed abundances (13) of ^6Li and

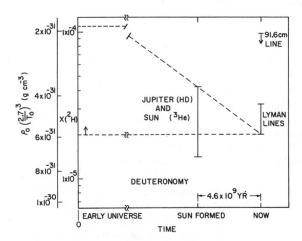

Figure 5   The evolution of the deuterium abundance during the history of our galaxy. Also indicated is the present density corresponding to the primordial value, from Figure 4.

^{11}B, $X(^6\text{Li}) \cong 4 \times 10^{-10}$ and $1.5 \times 10^{-9} \leq X(^{11}\text{B}) \leq 7 \times 10^{-8}$, are much greater than the amounts produced. On the other hand, the observed abundances (13) of ^{3}He and ^{7}Li are $X(^3\text{He}) \cong 3 \times 10^{-5}$ and $X(^7\text{Li}) \cong 6 \times 10^{-9}$. These numbers are probably uncertain within a factor of about two, but within the range of values indicated in Figure 4.

Unfortunately, the constraints provided by the abundances of ^{3}He and ^{7}Li are weakened by the fact that they can both be produced within stars. In particular, lithium is known to be enriched on the surface of many red-giant stars (93), as expected if it were being made within the star and transported to the surface by convection (94). Similarly, ^{3}He is produced during the normal course of hydrogen burning within stars on the main sequence, and convection followed by mass loss from the surface will also carry some of this ^{3}He into the interstellar medium (95).

Another source of light elements is the spallation of interstellar gas by the galactic cosmic rays. However, it has been shown (96–98) that the abundances of only ^{6}Li, ^{9}Be, ^{10}B, and perhaps ^{11}B can be produced in this way. It may be highly significant that these are precisely those light elements that cannot be made in the early universe. Theoretical analyses (96–98) and recent cosmic-ray measurements (99) indicate that twice as much ^{7}Li as ^{6}Li should be produced by galactic cosmic rays. Since meteoritic measurements (92) show that $X(^7\text{Li}) = 14.6\ X(^6\text{Li})$, this indicates that this process could not have produced the bulk of the ^{7}Li. The factors are even greater for ^{2}H, ^{3}He, and ^{4}He.

Although stars may have produced significant quantities of ^{3}He and ^{7}Li, it is also possible that they are primarily due to big-bang nucleosynthesis. For example, a universe with present density $\rho_0 = 6 \times 10^{-31}$ g cm^{-3} would have produced amounts of ^{2}H, ^{3}He, ^{4}He, and ^{7}Li within the uncertainties of their pregalactic abundances (93). The densities required to produce the most realistic abundances of ^{2}H and ^{3}He are too small to produce enough ^{7}Li, however.

An additional problem with ^{3}He is the lack of abundance data outside the solar system. The interstellar limit $X(^3\text{He}) < 10^{-4}$ comes from observations of the 3.46-cm hyperfine transition in singly ionized ^{3}He in HII regions (100). It has been pointed out (95) that even this upper limit could have been produced by stars during the evolution of the galaxy. This would imply some constraint on the amount that could have been produced by the big bang. In addition to more detailed models of stellar production of ^{3}He (and ^{7}Li), what is critically needed is a measurement of its present interstellar abundance. Only then can any cosmological information be extracted. We hope that work in progress by Rood, Steigman & Wilson will improve our knowledge of the interstellar abundance of ^{3}He.

Before considering other models in the next section, it is important to recall that no nuclei heavier than $A = 7$ are produced in significant quantities in the standard big bang. Thus we require other sources, such as stars, to provide the carbon, oxygen, iron, etc in the universe. For the oldest stars, we may require that processes other than those that occur during the normal course of stellar evolution produced the elements in the primordial gas out of which they condensed. Nevertheless, Gamow's original hope that such processes might occur in the early universe has not been fulfilled by any recent calculations. The existence of nuclei with half-lives much less than the age of the universe also dooms any such hope (101).

5   OTHER BIG-BANG MODELS

5.1   *Large Lepton Number*

We now consider the effects on element production of separately relaxing each of the five standard big-bang model assumptions listed in Section 2.3.

We first consider universes in which the lepton number is much greater than the number of photons. The electron-lepton number $N(l_e) = N(e^-) - N(e^+) + N(v_e) - N(\bar{v}_e)$ and the muon-lepton number $N(l_\mu) = N(v_\mu) - N(\bar{v}_\mu)$. Since $N(e^-) - N(e^+) = N(p) \ll N(\gamma)$, in practice $N(l) \gg N(\gamma)$ is equivalent to neutrino degeneracy. In fact, when any type of neutrino is degenerate $(|\Phi_v|/kT_v \sim |\Phi_v|/kT \gg 1)$, the number of leptons per photon becomes

$$N(l)/N(\gamma) = 0.069\,(\Phi_v/kT)^3, \quad (T_9 \ll 3) \qquad 5.1$$

where Φ_v is the Fermi level. The degeneracy parameter Φ_v/kT_v remains constant as the universe expands. The resulting neutrino mass density is

$$\rho_v = 2.92 \times 10^{-21}\,\Phi_v^4(\text{eV})\,\text{g cm}^{-3}, \qquad 5.2$$

if the Fermi level is measured in electron volts. It is seen from Equation 5.2 that the present limit $|\Phi_v| \lesssim 1$ eV imposed by measurements indicating a lack of proton-neutrino interactions in very high-energy cosmic rays (102) still allows degenerate neutrinos to dominate the mass of the universe.

Neutrino degeneracy affects element production in two ways. First, the increased density increases the expansion rate (Equation 3.3). This effect is discussed in more detail later. The second effect, modification of the rates of neutron-proton inter-conversion (Reactions 1, Table 2), is more important, however, if the electron-lepton number is large. For instance, if electron antineutrinos were degenerate (large negative electron-lepton number), the reactions $n + e^+ \rightleftharpoons \bar{v}_e + p$ would produce an equilibrium ratio of neutrons to protons much greater than unity. This would change greatly the subsequent production of helium and the other elements. On the other hand, electron-neutrino degeneracy would produce an equilibrium ratio of neutrons to protons much less than unity, resulting in much less production of helium.

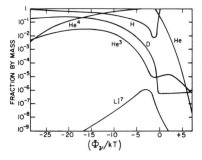

Figure 6   Abundances produced in typical $(h = 10^{-4})$ big-bang models containing equal numbers of degenerate electron- and muon-neutrinos $(\Phi_v > 0)$ or antineutrinos $(\Phi_v < 0)\,(11)$.

The effects of large lepton number were first investigated by Wagoner, Fowler & Hoyle (11). Their results are shown in Figure 6, for a particular value of the density parameter h. It is seen that the abundances, particularly that of ^{4}He, are strongly affected if the neutrinos or antineutrinos are degenerate ($|\Phi_\nu| \gtrsim kT$). The peak in the production of ^{4}He occurs for an antineutrino degeneracy such that the equilibrium ratio of neutrons to protons at freeze-out is slightly greater than unity. Neutron decay then reduces the value to unity by the time nucleosynthesis can proceed. For very large positive lepton number, the number of neutrons becomes so small that no significant synthesis occurs. For very large negative lepton numbers, the fast expansion rate becomes important in reducing the time available for nuclear reactions to proceed. In addition, the number of protons remains small until the Fermi level has dropped to about 1 MeV, at which point the neutrons can decay. Thus again, little synthesis occurs in this limit.

If only the muon neutrinos were degenerate, their sole effect would be to increase the expansion rate because of their higher density. A faster expansion rate means that the neutron-proton ratio will "freeze out" at a higher temperature, where it is

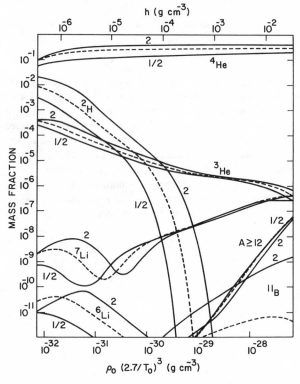

Figure 7 Abundances produced by big-bang models with expansion-rate factors of $\xi = \frac{1}{2}$ and 2 (*solid curves*), compared with abundances produced by the standard model (*dashed curves*) (13).

closer to unity. There is also less time for subsequent neutron decay, resulting in a neutron-proton ratio closer to unity during nucleosynthesis, and thus greater production of ^{4}He.

We show below that other factors also can affect the expansion rate. However, since nucleosynthesis only occurs during a short time interval in the evolution of the early universe, it is a good approximation to merely multiply the standard expression by a factor ξ to include such effects. That is, we consider variations of the parameter ξ in the more general expression (13)

$$V^{-1}\,dV/dt = \xi(24\pi G\rho)^{1/2},\qquad\qquad 5.3$$

where $\xi = 1$ in standard models. The abundances that result from faster ($\xi = 2$) and slower ($\xi = \frac{1}{2}$) expansions are shown in Figure 7 for all the elements, and in Figure 8 for ^{4}He. The results shown in Figure 8 can be represented by the generalization of Equation 3.17.

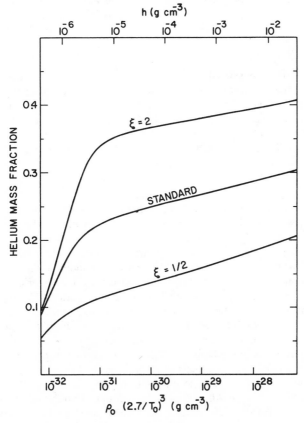

Figure 8   Same as Figure 7, but with an expanded scale showing only the ^{4}He mass fraction produced (13).

$$X(^4\text{He}) = 0.333 + 0.0195 \log h + 0.380 \log \xi, \qquad\qquad 5.4$$

for $h \geq 10^{-5}$ (13). In the case of muon-neutrino degeneracy, ξ is greater than unity if we take ρ to be the total density in the standard model. Thus for $|\Phi_v|/kT \gg 1$, $\xi \sim (\Phi_v/kT)^2$.

Yahil & Beaudet (104) have calculated element production when both the electron- and muon-lepton numbers are independently varied. Their results for variation of the electron-lepton number are similar to those indicated in Figure 6, while their results for variation of the muon-lepton number are similar to those indicated in Figures 7 and 8. It is true that by properly combining these two variations, better agreement with the observed abundances of ^3He and ^7Li can be obtained. However, since stars may well have produced these nuclei, there does not appear to be a compelling need to invoke such changes in the standard model, unless one also wishes to change the relation between the deuterium abundance and present density (104). Remember that in order to retain the observed abundance of ^4He while invoking large values of $N(l_e)$, one must also increase the expansion rate by invoking large values of $N(l_\mu)$ so that the correct n/p ratio is obtained.

5.2 Zero Baryon Number

Steigman (105) has considered nucleosynthesis within symmetric matter-antimatter cosmologies, such as those proposed by Omnes (106–108). He finds that annihilation is always very efficient, although the degree of efficiency depends on the particular model considered. However, in all cases the ratio of the number of nuclei remaining to the number of photons present can be calculated to be much less than the present value. Thus these symmetric models cannot produce significant quantities of any nuclei.

5.3 New Particles

Let us consider the effects of other particles that might have been present during the epoch of nucleosynthesis. Recall that the detailed properties of the early universe begin to affect nucleosynthesis at a time of about a second, when the temperature has dropped to 10^{10} K. Therefore, any particles present at that time must have either a lifetime $\tau \gtrsim 1$ sec or a mass $M \lesssim 1$ MeV. In addition, they should be at least weakly interacting, so that they would have been in equilibrium during the annihilation of the heavier hadrons and leptons at temperatures $T \gtrsim 10^{12}$ K. These requirements hold if the standard model is not modified greatly, so that the ^4He abundance agrees with observation. Particles that we specifically consider include gravitons, new neutrinos, quarks, and superbaryons.

We consider a graviton to be any particle that interacts only gravitationally. This includes the usual spin-2, zero-mass graviton in general relativity, as well as the spin-$\frac{3}{2}$, zero-mass particles that accompany them in a multiplet in the new super-gravity theories (109, 110). All such particles decouple from the other constituents very early in the history of the universe. Subsequent annihilation of heavy particles puts more energy into those light particles which interact weakly, electro-magnetically, or strongly. Their temperature and density are therefore increased relative to the gravitons. Since the gravitons then contribute negligibly to the total density, they cannot increase the expansion rate, which is the only way they could have affected nucleosynthesis.

One way in which the expansion rate would be changed is through the existence of new types of neutrinos. For example, new heavy leptons may exist (111), and have associated with them neutrinos. There would then be more types of neutrinos in the universe, contributing to the total density in the same way as electron- and muon-neutrinos. The density would be higher at a given temperature, thereby increasing the expansion rate during nucleosynthesis. In particular, the neutron-proton ratio would freeze out closer to unity, resulting in more ^{4}He production.

Shvartzman (112) and Wagoner (13) originally considered this effect of having additional particle species. Recently Steigman, Schramm, & Gunn (113) have investigated this question in more detail, with the goal of settling limits on the number of possible lepton types in theories where each heavy lepton would have a corresponding neutrino. They show that the increase in the ^{4}He abundance can be calculated in a straightforward manner, using Equations 5.3 and 5.4. The expansion rate factor ξ is given by

$$\xi^2 \cong \rho'/\rho = 1 + (\tfrac{7}{36})\Delta N_l, \qquad\qquad 5.5$$

where ρ' is the density with the new neutrinos, ρ is the density in the standard model, and ΔN_l is the number of additional massive leptons, assuming that each would have a characteristic neutrino and antineutrino. Although ξ does change during pair annihilation, the change is small and Equation 5.5 is a sufficiently good approximation. Upon substituting Equation 5.5 into equation 5.4, one finds a change in ^{4}He abundance given by

$$\Delta X(^4\text{He}) \cong 0.016 \, \Delta N_l, \qquad\qquad 5.6$$

provided that the change is small: $\Delta N_l \ll \tfrac{36}{7}$.

Since the primordial helium abundance is probably $X(^4\text{He}) < 0.29$, one can set a limit on the number of additional types of neutrinos. One can also set a limit on the number of heavy leptons in the universe, if the above relationship between heavy leptons and neutrinos holds. The limit is approximately 5 additional heavy-lepton types, corresponding to a total of 7 lepton types, including the electron and muon. It is also interesting that the maximum number of lepton types can only be accommodated in low-density ($h \sim 10^{-5}$) universes, since they produce less ^{4}He.

Another particle that might have been present in the early universe is the free quark. This possibility has been explored by Frautschi, Steigman & Bahcall (114), among others (115, 142). They showed that if free quarks can exist, their density would depend sensitively on the highest temperature reached in the big bang. Their high-temperature results were similar to those of Zeldovich (115), who showed that in the standard big bang, more free quarks would exist than the present limits allow. However, if the quark-confinement force is able to bind all quarks when the temperature has dropped to a sufficient value, then the Zeldovich (115) argument does not apply. An important question is whether the quark-confinement force has a finite range. In any case, it is unlikely that free quarks could have affected nucleosynthesis significantly, since their abundance relative to nucleons was probably the same then as it is today.

In statistical bootstrap theories of hadrons, superbaryons of initial mass $\sim 10^{38} \, m_\pi$ dominate the early universe (116). It is possible for the decay of these particles into ordinary particles to occur when the temperature has 'dropped below 10^9 K. In this case, nucleosynthesis proceeds quite differently because of the large Coulomb barriers. In fact, the main reaction is just $n + p \rightarrow d + \gamma$, with the result that most of the production consists of deuterium (14). In any case, such models should not be considered too seriously until their prediction of elementary hadrons with baryon number greater than one has been verified by accelerators.

5.4   Inhomogeneities and Anisotropies

Other possible big-bang models arise from a breakdown of the cosmological principle in the early universe. It is instructive to note, however, that in this case the element production within any co-moving volume element V depends on only three factors.

They are:

1. The expansion rate $V^{-1} \, \mathrm{d}V/\mathrm{d}t$, determined by the generalization of Equation 3.3 (49),

$$(V^{-1} \, \mathrm{d}V/\mathrm{d}t)^2 = 24\pi G \rho + 2V^{-2/3} c^2 \int \left[\dot{U}^u_{;u} + 2(\omega^2 - \sigma^2) \right] V^{-1/3} \, \mathrm{d}V. \qquad 5.7$$

The additional local properties of the matter that appear are the rotation scalar, ω, the shear scalar, σ, and $\dot{U}^u_{;u}$, the four-divergence of the acceleration (produced by pressure gradients).

2. $h = \rho_b T_9^{-3}$, the same density parameter that we have been using.

3. The neutrino distribution in phase space, since neutrinos are the only particles that are not coupled to form a perfect fluid. However, since this factor is less critical than the other two, we have neglected it in Equation 5.7 and neglect it in what follows.

Essentially three types of violation of the cosmological principle have been investigated. They are (a) homogeneous anisotropic models (117–120), (b) adiabatic inhomogeneities ($h = $ constant) (120, 121), and (c) isothermal inhomogeneities (T uniform) (13, 122). We consider each in turn.

In anisotropic universes that evolve to the present degree of isotropy, it is found that the presence of shear in the motion of the matter can strongly affect the expansion rate during nucleosynthesis. Equation 5.7 assumes the approximate form of Equation 5.3 at that time, with the expansion rate factor ξ influencing element production in the way indicated in Figures 7 and 8. In fact, the requirement that the observed ^{2}H and ^{4}He be made in the early universe strongly limits the amount of anisotropy possible at the epoch of nucleosynthesis and even more strongly (than even the 3-K background radiation) limits the amount of anistropy possible today (119).

Adiabatic fluctuations whose sizes are greater than that of the particle horizon

(Section 3.1) do not produce large differences from the standard-model abundances (121). This is because such fluctuations act like separate "universes," each controlled by essentially the same equations that govern the standard model. Adiabatic fluctuations of smaller size act like sound waves that can produce net changes, however (120).

The largest inhomogeneities are possible if they are isothermal, since the radiation density controls the expansion rate, which is therefore the same as in the standard model. Since the ^{4}He abundance is not very sensitive to changes in baryon density, the requirement that it agree with the observed amount still allows significant variation in h. One can thus apply the standard-model results in Figure 4 to each volume element characterized by its particular value of h.

If all the volume elements are subsequently mixed, so that the present interstellar abundances represent the average amounts produced, then it is possible to match the observed amounts of ^{2}H, ^{3}He, ^{4}He, and ^{7}Li. On the other hand, one may ask if such a model could alter greatly the limit on the present density provided by the observed deuterium. Wagoner (13) and Epstein & Petrosian (122) have shown that it could not, because the abundance of deuterium does not continue to rise sharply as one considers lower-density regions. Thus the attempt to add larger amounts of low-density material to compensate for the high-density regions (devoid of deuterium) does not add a correspondingly larger amount of deuterium. It is difficult to increase the upper limit on the present density by more than a factor of two.

On the other hand, it is possible that regions of higher h were not subsequently mixed, due to their entrapment in primordial stars (123) or black holes. This would weaken somewhat the constraint on the present density, since the interstellar medium would have a deuterium abundance higher than the average. However, since the present universe does not appear to be dominated by such objects, their effect should not be significant (58). It is interesting to note, though, that regions of very high h can produce heavy elements, although most will not have their observed ratios (11). In fact, Woosley (124) has carried out detailed computations that show that one can obtain interesting abundances of some elements heavier than iron.

5.5   Other Gravitation Theories

The sole effect on nucleosynthesis of considering metric theories other than general relativity will be through the expansion rate. The expansion rate predicted by general relativity, Equation 3.3, will in general assume a different form, but during nucleosynthesis can also be represented by Equation 5.3 in terms of the expansion rate factor ξ. Thus each theory of gravitation will predict a value for ξ during nucleosynthesis, which will affect the abundances, as indicated in Figures 7 and 8. The effects of much more rapid expansions have been considered by Peebles (10) and Wagoner (125). Although the ^{4}He abundance can return to the observed level for $\xi \sim 10^3$, the deuterium production is extremely large in such a universe.

Recent solar-system experiments have severely restricted the number of viable theories of gravitation (46). However, the consequences of such theories for the early universe have not been fully developed. Thus we cannot compare them in detail. However, it is clear that if one believes that the observed ^{4}He was produced in the early universe, only those theories that give a value of $\xi \cong 1$ will remain viable. It appears that most of the existing theories other than general relativity, such as those that predict a "changing gravitational constant," will fail this test.

6   REMARKABLE CONCORDANCE OF THE STANDARD MODEL

6.1   Implications of the Production of ^4He

The major feature that has emerged from the results discussed in the preceding section is the extreme sensitivity of the ^{4}He abundance to most types of deviation from the standard big-bang model. Any factor that affects either the expansion rate (controlled by the gravitational interaction) or the nucleon weak-interaction rate will affect the production of ^{4}He. In fact, we see from Figure 8 that changes in the standard-model expansion rate by more than a factor of two results in helium production outside the range of its observed abundance. It is truly amazing that these two rates happen to be balanced in precisely the correct way in the simplest model of the universe, the standard big bang.

However, we must also consider the possibility that the ^{4}He was not produced in the early universe. Of course ^{4}He is synthesized within stars, and in fact we mentioned in Section 4.1 that there are indications that perhaps as much as one fourth of its present-day abundance came from this source during the evolution of galaxies. This is not surprising, since presumably some helium would be ejected from stars during the latter stages of their evolution. The ejected material would probably be somewhat richer in helium than in heavier elements, since all stars evolve through a stage in which they produce a large amount of helium. Since we know that our galaxy has been enriched by this process by about 2% in heavy elements, a few percent increase in the helium abundance would not be surprising.

On the other hand, matter that has not been so enriched in heavy elements (in the atmospheres of most old stars and in the interstellar medium of some "young" compact galaxies) still seems to contain about 20% helium. This helium could not have been produced during the course of normal stellar evolution. What would be

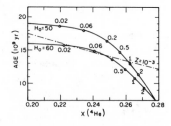

Figure 9   Comparisons of the age of the universe as a function of helium abundances, as determined by two independent methods. The solid curves show the relationship predicted by helium synthesis in the big bang, for $\Lambda = 0$ and two choices for the Hubble constant. The dashed line shows the result from globular cluster stars for a particular heavy-element mass fraction Z. Also shown are some values of Ω (129).

needed is other objects that existed before or during the formation of galaxies, which processed much of the matter in the universe into helium. For example, some "little bangs" that have been investigated (11, 49, 126) do produce a large amount of ^{4}He. However, all such sources are ad hoc, and made even less likely by the observation of the uniformity of the helium abundance from galaxy to galaxy (54).

Another point of interest is the fact that the abundance of helium produced in the standard model is consistent with the age of the globular clusters of stars in our galaxy. These clusters appear to have formed early in the evolution of the galaxy, so that their age (relative to the present) should be almost equal to the age of the universe. Their age is inferred by comparing the results of stellar evolution calculations with the observed brightness and color distribution of the stars in the cluster. It has been shown by Iben (127) and others that the most sensitive parameter in determining their age by this method is the primordial helium abundance. There is a direct relationship between their inferred age and the helium abundance in the gas from which they formed, shown in Figure 9.

On the other hand, we have seen that the helium produced in the big bang does depend to some extent on the present baryon density of the universe (Figure 8). It is convenient to define the dimensionless density parameter

$$\Omega = \frac{8\pi G \rho_0}{3H_0^2} = (50/H_0)^2 (\rho_0/4.7 \times 10^{-30}), \qquad 6.1$$

where H_0 is the Hubble constant and ρ_0 is the present total density of the universe (equal to the present baryon density in the standard model). Any method that employs the virial theorem to obtain the mass of the constituents of the universe will determine a value of Ω independent of the actual value of H_0 (42). Now within the standard model, the present age of the universe t_0 is related to these quantities and the cosmological constant by an expression of the form

$$t_0 = f(\Lambda c^2/H_0^2, \Omega)/H_0. \qquad 6.2$$

For fixed Λ and H_0, the helium abundance is therefore also related to t_0 by this equation. If we restrict our consideration to Friedmann models ($\Lambda = 0$), we can compare this relation for various values of H_0 with the relation from the globular clusters. This is shown in Figure 9, for two values of the Hubble constant close to its latest determination (128), $H_0 = 50 \pm 10$ km sec^{-1} Mpc^{-1}. It is impressive that these two independent relationships overlap to a large extent (129). [The dimensionless function $f(0, \Omega) \leq 1$; also, 1 Mpc = 3.086×10^{24} cm = 3.262×10^6 light-years.]

6.2   Implications of the Production of ^2H

If we adopt the standard big-bang model of the universe, as strongly indicated by the helium abundance, then the deuterium abundance can give us information about the present density of the universe. If other sources have not contributed to the present deuterium abundance, then the lower limit to its primordial (pregalactic) abundance is $X(^2H) \geq 2 \times 10^{-5}$, as indicated on Figure 5. Using the results in Figure 4, this corresponds to the upper limit $h \leq 3 \times 10^{-5}$, or

$$\rho_0(2.9/T_0)^3 \leq 7 \times 10^{-31} \text{ g cm}^{-3}. \qquad 6.3$$

Recall that it appears that the present temperature of the microwave background radiation is $T_0 = 2.9 \pm 0.1$ K (27). The corresponding upper limit on the density parameter is then

$$\Omega \lesssim 0.16\,(50/H_0)^2. \qquad 6.4$$

Let us compare this limit on Ω with the contribution to Ω due to the mass associated with galaxies. The most recent determination (60) gives a best value of 0.06, with a maximum value of about 0.3. With the most recent determinations of $H_0 \sim 50$, the agreement between these values of Ω and the deuterium limit, Equation 6.4, is certainly encouraging. They indeed seem to imply that $\Omega \lesssim 0.3$.

However, it should be stressed that this limit provides no information about the curvature constant k or cosmological constant Λ, both of which are needed to determine the future evolution of standard models of the universe. These constants can be determined in terms of the observational quantities H_0, Ω, and $q_0 = -(R\ddot{R}/\dot{R}^2)_0$ (deceleration parameter) by the relations (59)

$$\lambda = (3H_0^2/2c^2)(\Omega - 2q_0), \qquad 6.5$$

$$k = (R_0 H_0/c)^2(3\Omega/2 - q_0 - 1). \qquad 6.6$$

Gott, Gunn, Schramm & Tinsley (34) have investigated what various observations, including the abundance of deuterium, seem to indicate concerning the type of standard model we live in. However, they assumed that the cosmological constant $\Lambda = 0$. Gunn & Tinsley (59) have considered the general case (arbitrary Λ, to be determined by the observations). Unfortunately, it is at present difficult to draw any significant conclusions regarding the value of Λ or k because, although it appears that we may be approaching accurate determinations of H_0 and Ω, the measurements of q_0 are beset with fundamental difficulties (129, 130). These arise from the fact that a determination of q_0 by the best present method, the red-shift–apparent brightness relation, involves a knowledge of the evolution of the intrinsic luminosity of giant elliptical galaxies (42). This is due to the fact that we see these distant galaxies as they were billions of years ago. Models of the evolution of single galaxies tend to imply that galaxies were more luminous in the past (59). However, the interaction with other galaxies and gas in the cluster appears to result in a gain in mass of the giant ellipticals, producing a change in their luminosity in the opposite direction (35). What is clear is that the total evolutionary correction is very uncertain, and from the present observations (36, 37) all we can say is that probably $-2 \lesssim q_0 \lesssim 2$.

However, one may proceed by assuming $\Lambda = 0$, and see if the results are consistent, thus eliminating the need to know q_0. Gott, Gunn, Schramm, & Tinsley (34) found that when one combines all the observational constraints, one does indeed obtain a consistent value of $\Omega \sim 0.1$ and in addition predicts that the Hubble constant $H_0 \sim 60\,\mathrm{km\,sec^{-1}\,Mpc^{-1}}$, consistent with its directly measured values. This is shown in Figure 10. It may be significant that if one employs the best estimate for the pregalactic deuterium abundance (assuming that half of it was later destroyed by stellar processing), the implied density of the universe agrees almost exactly with the best estimate of the density of matter in galaxies if the most recent value of H_0 is used. In any case, the value of Ω is at least an order of magnitude below the value

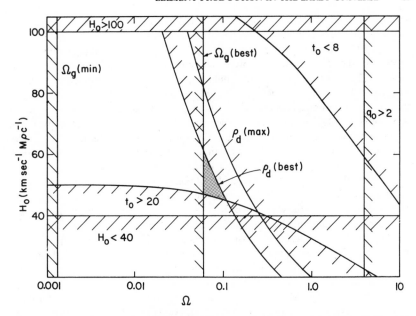

Figure 10   Constraints on the Hubble constant H_0 and density parameter Ω, which specify standard big-bang models with $\Lambda = 0$ (34). Included are limits from direct determinations of H_0; the age of the universe t_0 (in units of 10^9 yr); the deceleration parameter q_0; the minimum $[\Omega_g(\text{min})]$ and most probable $[\Omega_g(\text{best})]$ contribution of galactic matter to Ω; the maximum $[\rho_d(\text{max})]$ and most probable $[\rho_d(\text{best})]$ present density from standard big-bang deuterium synthesis.

($\Omega = 1$) required to close the universe in these Friedmann models. A value of $\Omega \sim 0.1$ also implies an age of the universe $t_0 \cong 16 \times 10^9$ years, from Figure 9, which is also quite consistent with the age estimated from the abundances of radioactive elements (131).

Before accepting these powerful implications of the deuterium abundance, alternative production processes should be investigated. If galactic sources produced the observed deuterium, then only a lower limit could be placed on Ω from the measurements of the mass within galaxies.

Proposed schemes for such production have been reviewed by Epstein, Lattimer & Schramm (74). Their basic problem seems to be that deuterium is extremely fragile, rapidly burning to ^{3}He in hot environments. Thus, in order to synthesize deuterium, conditions of relatively low temperature or density are required. At the same time, sufficient neutrons must be available to produce it by $p+n \rightarrow d+\gamma$ reactions or sufficient numbers of high-energy particles must be available to produce it from heavier nuclei by spallation. Such conditions are extremely rare in the universe, especially when one considers that deuterium is rather abundant compared to other light nuclei such as lithium, beryllium, and boron. In addition, spallation tends to overproduce these rare light nuclei if one requires conditions that produce sufficient deuterium. Thus it appears as if sources such as stellar flares and galactic cosmic rays cannot produce the deuterium we observe.

Colgate (70–72) and Hoyle & Fowler (73) proposed a way of making deuterium in supernova shock waves that involved a combination of spallation and neutron capture. The shock broke nuclei into nucleons, which later recombined to build up deuterium. However, Weaver & Chapline (77) showed that actual shocks do not seem to reach strengths sufficient to break up nuclei. In addition, Epstein, Arnett & Schramm (75, 76) showed that even if such strengths could be reached, the invariable outcome was an overproduction of Li, Be, and B relative to deuterium.

Another proposal put forward by Lattimer & Schramm (132) involved ejecting a large number of free neutrons at low temperature from a neutron star tidally disrupted by the gravitational field of a black hole. However, Lattimer et al (133) have subsequently pointed out that the bulk of such material would be built up to very heavy nuclei similar to those produced in the r-process. Since these elements are rare, such a process could only be a source for them, not deuterium.

It appears that in order to avoid the abovementioned problems of overproduction of heavier nuclei, one must consider pregalactic events involving primordial material devoid of nuclei other than helium to be spalled (74). The events must also be more energetic than supernovae if lithium is not to be overproduced by the reaction $\alpha + \alpha \rightarrow {}^7Li + p$ (73, 74). Although it is difficult to rule out such conditions in the early history of galaxies, it seems rather odd that such events would have occurred at a rate sufficient to synthesize all the deuterium, when the simple big bang does it in such a natural way. Nevertheless, the search for other sources of deuterium that are less ad hoc should be pursued, so that we can have more confidence in the implications of its abundance.

7   CONCLUSION

We have seen how the abundances of the lightest elements provide a nuclear probe of conditions in the early universe. In fact, virtually all the information we have concerning the nature of the primeval fireball a few seconds after its birth comes to us in this way.

At present, the nucleus which has given us the deepest insights is ${}^4He$. This is because its abundance is controlled by the neutron-proton ratio when their weak-interconversion rate becomes equal to the expansion rate of the universe. In the simplest big bang, the standard model, the balance is precisely that required to produce the observed abundance of helium. However, changes in either of these rates by a factor of two produces too much or too little helium. This places powerful constraints on fundamental properties of the universe such as its lepton and baryon numbers, homogeneity and anisotropy, number of possible new particles, and theory of gravitation.

The other nucleus that is an effective probe is deuterium, although ${}^3He$ or ${}^7Li$ may also turn out to be important in the future. Once the standard model is accepted on the basis of the ${}^4He$ abundance, the deuterium abundance depends only on the number of baryons per photon, or, equivalently, the present density ρ_0 of the universe.

This dependence is a very strong one, and indicates that $\rho_0 \leq 7 \times 10^{-31}$ g cm^{-3} This limit is consistent with measurements of the mass associated with galaxies, and has important implications for the future evolution of the universe, which can be realized when the cosmological constant is measured.

Of course, all these implications rest on the belief that alternative sources of these nuclei are less important than the big bang. For this reason, it is vitally important to continue the search for other processes of nucleosynthesis, and to investigate the uniformity of the abundance of these light nuclei.

Once we accept the validity of the standard big-bang model, we reach a deeper level of questions. These include

1. Which values of the curvature constant k and cosmological constant Λ describe our universe, and why?
2. What determined the number of photons per baryon?
3. Why was the early universe uniform?
4. How did galaxies form?

Behind all this lies the general question of why the universe is the way it is, which should keep us busy for the foreseeable future.

ACKNOWLEDGMENTS

We cannot individually thank all those who have helped and motivated us over the years. However, the controlling influence has been the inspiration provided by Willy Fowler, who has taught us that cosmology may be applied nuclear physics. One of us (Schramm) would also particularly like to thank Jim Gunn for many discussions of the consistency of the various cosmological constraints. In addition, the historical insights provided by Steven Weinberg were especially useful. Finally, we thank Victoria LaBrie for her usual expert typing, and Virginia Bonnici for help with the literature cited.

Literature Cited

1. Alpher, R. A., Bethe, H., Gamow, G. 1948. *Phys. Rev.* 73:803–4
2. Alpher, R. A., Herman, R. C. 1950. *Rev. Mod. Phys.* 22:153–212
3. Hayashi, C. 1950. *Prog. Theor. Phys.* 5:224–35
4. Alpher, R. A., Follin, J. W. Jr., Herman, R. C. 1953. *Phys. Rev.* 92:1347–61
5. Salpeter, E. E. 1952. *Ap. J.* 115:326–28
6. Hoyle, F. 1954. *Ap. J. Suppl.* 1(5):121–46
7. Gamow, G. 1955. Quoted by W. A. Fowler in *Orange Aid Prepr. 197*, Calif. Inst. Tech. (1970)
8. Hoyle, F., Tayler, R. J. 1964. *Nature* 203:1108–11
9. Peebles, P. J. E. 1966. *Phys. Rev. Lett.* 16:410–13
10. Peebles, P. J. E. 1966. *Ap. J.* 146:542–52
11. Wagoner, R. V., Fowler, W. A., Hoyle, F. 1967. *Ap. J.* 148:3–49
12. Penzias, A. A., Wilson, R. W. 1965. *Ap. J.* 142:419–21
13. Wagoner, R. V. 1973. *Ap. J.* 179:343–50
14. Wagoner, R. V. 1974. In *Confrontation of Cosmological Theories with Observational Data,* ed. M. S. Longair, p. pp. 195–210. Dordrecht, The Netherlands: Reidel
15. Schramm, D. N., Wagoner, R. V. 1974. *Phys. Today* 27:41–47
16. Fowler, W. A., Pasachoff, J. M. 1974. *Sci. Am.* 230:5, 108–18
17. Hubble, E. P. 1920. *Publ. Yerkes Obs.* 4:69–76
18. Hubble, E. P. 1925. *Ap. J.* 62:409–33
19. Hubble, E. P. 1929. *Proc. Natl. Acad. Sci. USA* 15:168–73

20. Friedmann, A. 1922. *Z. Phys.* 10:377–86
21. Lemaitre, G. 1927. *Ann. Soc. Sci. Brux. A* 47:49–57
22. Gamow, G. 1948. *Nature* 162:680–82
23. Bondi, H., Gold, T. 1948. *Mon. Not. R. Astron. Soc.* 108:252–70
24. Hoyle, F. 1948. *Mon. Not. R. Astron. Soc.* 108:372–82
25. Ryle, M. 1968. *Ann. Rev. Astron. Astrophys.* 6:249–66
26. Peebles, P. J. E. 1971. *Physical Cosmology.* Princeton, NJ: Princeton Univ. Press. 282 pp.
27. Woody, D. P., Mather, J. C., Nishioka, N. S., Richards, P. L. 1975. *Phys. Rev. Lett.* 34:1036–39
28. Thaddeus, P. 1972. *Ann. Rev. Astron. Astrophys.* 10:305–34
29. Weiss, R. 1977. *Proc. N.Y. Acad. Sci.* In press
30. Dicke, R. H., Peebles, P. J. E., Roll, P. G., Wilkinson, D. T. 1965. *Ap. J.* 142:414–19
31. Alpher, R. A., Herman, R. C. 1949. *Phys. Rev.* 75:1089–95
32. Zeldovich, Ya. B. 1966. *Usp. Fiz. Nauk.* 89:647–68
33. Schmidt, M. 1968. *Ap. J.* 151:393–409
34. Gott, J. R., Gunn, J. E., Schramm, D. N., Tinsley, B. M. 1974. *Ap. J.* 194:543–53
35. Ostriker, J. P., Tremaine, S. D. 1975. *Ap. J. Lett.* 202:L113–17
36. Sandage, A., Hardy, E. 1973. *Ap. J.* 183:743–57
37. Gunn, J. E., Oke, J. B. 1975. *Ap. J.* 195:255–68
38. Misner, C. W., Thorne, K. S., Wheeler, J. A. 1973. *Gravitation.* San Francisco: Freeman. 1279 pp.
39. Hagedorn, R. 1965. *Nuovo Cimento Suppl.* 3:147–86
40. Veneziano, G. 1968. *Nuovo Cimento A* 57:190–97
41. Huang, K., Weinberg, S. 1970. *Phys. Rev. Lett.* 25:895–97
42. Weinberg, S. 1972. *Gravitation and Cosmology: Principles and Applications of the General Theory of Relativity.* New York: Wiley. 657 pp.
43. Steigman, G. 1976. *Ann. Rev. Astron. Astrophys.* 14:339–72
44. Sandage, A., Tammann, G., Hardy, E. 1972. *Ap. J.* 172:253–63
45. Hughes, R. G., Longair, M. 1967. *Mon. Not. R. Astron. Soc.* 135:131–37
46. Will, C. M. 1972. *Phys. Today* 25(10):23–29
47. Harrison, E. R. 1973. *Ann. Rev. Astron. Astrophys.* 11:155–86
48. Rindler, W. 1956. *Mon. Not. R. Astron. Soc.* 116:662–77
49. Wagoner, R. V. 1969. *Ap. J.* 18: Suppl. 162, pp. 247–96
50. Fowler, W. A., Caughlan, G. R., Zimmerman, B. A. 1967. *Ann. Rev. Astron. Astrophys.* 5:525–70
51. Fowler, W. A., Caughlan, G. R., Zimmerman, B. A. 1975. *Ann. Rev. Astron. Astrophys.* 13:69–112
52. Danziger, I. J. 1970. *Ann. Rev. Astron. Astrophys.* 8:161–78
53. Searle, L., Sargent, W. L. W. 1972. *Comments Astrophys. Space Phys.* 4:59–63
54. Peimbert, M. 1975. *Ann. Rev. Astron. Astrophys.* 13:113–31
55. Peimbert, M., Torres-Peimbert, S. 1974. *Ap. J.* 193:327–33
56. Dufour, R. J. 1975. *Ap. J.* 195:315–32
57. Searle, L., Sargent, W. L. W. 1972. *Ap. J.* 173:25–33
58. Carr, B. J. 1975. *Ap. J.* 201:1–19
59. Gunn, J. E., Tinsley, B. M. 1975. *Nature* 257:454–57
60. Gott, J. R., Turner, E. L. 1977. *Ap. J.* In press
61. Churchwell, E., Mezger, P. G., Huchtmeier, W. 1974. *Astron. Astrophys.* 32:283–308
62. Fowler, W. A., Greenstein, J. L., Hoyle, F. 1962. *Geophys. J. R. Astron. Soc.* 6:148–220
63. Geiss, J., Reeves, H. 1972. *Astron. Astrophys.* 18:126–32
64. Black, D. C. 1971. *Nature Phys. Sci.* 234:148–49
65. Black, D. C. 1972. *Geochim. Cosmochim. Acta* 36:347–75
66. Burnett, D., Lippolt, H. J., Wasserburg, G. J. 1966. *J. Geophys. Res.* 71:1249–69
67. Eugster, O., Terg, F., Burnett, D., Wasserburg, G. J. 1970. *J. Geophys. Res.* 75:2753–68
68. Ryter, C., Reeves, H., Gradsztajn, E., Audouze, J. 1970. *Astron. Astrophys.* 8:389–97
69. Reeves, H., Audouze, J., Fowler, W. A., Schramm, D. N. 1973. *Ap. J.* 179:909–30
70. Colgate, S. A. 1973. *Ap. J. Lett.* 181:L53–54
71. Colgate, S. A. 1973. In *Explosive Nucleosynthesis,* ed. D. N. Schramm, W. D. Arnett, pp. 248–63. Austin: Univ. Tex. Press. 301 pp.
72. Colgate, S. A. 1974. *Ap. J.* 187:321–32
73. Hoyle, F., Fowler, W. A. 1973. *Nature* 241:384–86
74. Epstein, R., Lattimer, J., Schramm, D. N. 1976. *Nature* 263:198–202
75. Epstein, R., Arnett, W. D., Schramm, D. N. 1974. *Ap. J. Lett.* 190:L13–16
76. Epstein, R., Arnett, W. D., Schramm,

D. N. 1976. *Ap. J. Suppl.* 31:111–41
77. Weaver, T. A., Chapline, G. 1974. *Ap. J. Lett.* 192:L57–60
78. Jefferts, K. B., Penzias, A. A., Wilson, R. W. 1973. *Ap. J. Lett.* 179:L57–59
79. Wilson, R. W., Penzias, A. A., Jefferts, K. B., Solomon, P. R. 1973. *Ap. J. Lett.* 179:L107–10
80. Beer, R., Taylor, F. W. 1973. *Ap. J.* 179:309–27
81. Trauger, J. T., Roesler, F. L., Carleton, N. P., Traub, W. A. 1973. *Ap. J. Lett.* 184:L137–41
82. Penzias, A. A., Wannier, P. G., Wilson, R. W., Linke, R. A. 1977. *Ap. J.* In press
83. Snyder, L. et al 1977. *Ap. J.* In press
84. Ostriker, J. P., Tinsley, B. M. 1975. *Ap. J. Lett.* 201:L51–54
85. Rogerson, J. B. Jr., York, D. G. 1973. *Ap. J. Lett.* 186:L95–98
86. York, D. G., Rogerson, J. B. Jr. 1977. *Ap. J.* In press
87. Vidal-Madjar, A., Laurent, C., Bonnet, R. M., York, D. G. 1977. *Astron. Astrophys.* In press
88. Audouze, J., Tinsley, B. M. 1974. *Ap. J.* 192:487–500
89. Weinreb, S. 1962. *Nature* 195:367–68
90. Cesarsky, D. A., Moffet, A. T., Pasachoff, J. M. 1973. *Ap. J. Lett.* 180:L1–6
91. Pasachoff, J. M., Cesarsky, D. A. 1974. *Ap. J.* 193:65–67
92. Cameron, A. G. W. 1973. See Ref. 71, pp. 2–21
93. Boesgaard, A. M. 1976. *Publ. Astron. Soc. Pac.* 88:353–66
94. Cameron, A. G. W., Fowler, W. A. 1971. *Ap. J.* 164:111–14
95. Rood, R. T., Steigman, G., Tinsley, B. M. 1977. *Ap. J.* In press
96. Reeves, H., Fowler, W. A., Hoyle, F. 1970. *Nature* 226:727–29
97. Meneguzzi, M., Audouze, J., Reeves, H. 1971. *Astron. Astrophys.* 15:337–59
98. Mitler, H. E. 1972. *Astrophys. Space Sci.* 17:186–218
99. Simpson, J., Garcia-Munioz, M., Mason, G. 1975. *Proc. Int. Cosmic Ray Conf.*, Munich
100. Predmore, C. R., Goldwire, H. C. Jr., Walters, G. K. 1971. *Ap. J. Lett.* 168:L125–29
101. Schramm, D. N. 1974. *Ann. Rev. Astron. Astrophys.* 12:383–406
102. Cowsik, R., Pal, V., Tandon, S. V. 1964. *Phys. Lett.* 13:265–67
103. Reeves, H. 1974. *Ann. Rev. Astron. Astrophys.* 12:437–69
104. Yahil, A., Beaudet, G. 1976. *Ap. J.* 206:26–29
105. Steigman, G. 1976. *Ann. Rev. Astron.* *Astrophys.* 14:339–72
106. Omnès, R. 1971. *Astron. Astrophys.* 10:228–45
107. Omnès, R. 1971. *Astron. Astrophys.* 11:450–60
108. Omnès, R. 1971. *Astron. Astrophys.* 15:275–84
109. Deser, S., Zumino, B. 1976. *Phys. Lett.* B 62:335–37
110. Freedman, D. Z., van Nieuwenhuizen, P., Ferrara, S. 1976. *Phys. Rev. D* 13:3214–18
111. Perl, M. L. et al 1975. *Phys. Rev. Lett.* 35:1489–91
112. Shvartsman, V. F. 1969. *JETP Lett.* 9:184–86
113. Steigman, G., Schramm, D. N., Gunn, J. E. 1977. *Phys. Lett. B.* 66:202–4
114. Frautschi, S., Steigman, G., Bahcall, J. N. 1972. *Ap. J.* 175:307–22
115. Zeldovich, Ya B. 1970. *Comments Astrophys. Space Phys.* 2:12–17
116. Carlitz, R., Frautschi, S., Nahm, W. 1973. *Astron. Astrophys.* 26:171–89
117. Hawking, S. W., Tayler, R. J. 1966. *Nature* 209:1278–79
118. Thorne, K. S., 1967. *Ap. J.* 148:51–68
119. Barrow, J. 1976. *Mon. Not. R. Astron. Soc.* 175:359–70
120. Olson, D. W. 1977. *Ap. J.* In press
121. Gisler, G. R., Harrison, E. R., Rees, M. J. 1974. *Mon. Not. R. Astron. Soc.* 166:663–71
122. Epstein, R. I., Petrosian, V. 1975. *Ap. J.* 197:281–84
123. Hoyle, F. 1975. *Ap. J.* 196:661–70
124. Woosley, S. 1975. Presented at Symp. on Front. Astron. 1975, Venice
125. Wagoner, R. V. 1967. *Science* 155:1369–76
126. Wagoner, R. V. 1971. In *Highlights of Astronomy*, ed. C. de Jager, pp. 301–17. Dordrecht, The Netherlands: Reidel
127. Iben, I. 1974. *Ann. Rev. Astron. Astrophys.* 12:215–56
128. Sandage, A., Tammann, G. A. 1976. *Ap. J.* 210:7–24
129. Gunn, J. E. 1976. *Orange Aid Prepr. 464*, Caltech
130. Gunn, J. E., Tinsley, B. 1976. *Ap. J.* 210:1–6
131. Hainebach, K. L., Schramm, D. N. 1976. *Ap. J. Lett.* 207:L79–82
132. Lattimer, J., Schramm, D. N. 1974. *Ap. J. Lett.* 192:L145–47
133. Lattimer, J., MacKie, F., Ravenhall, G., Schramm, D. N. 1977 *Ap. J.* In press
134. Shlyakhter, A. I. 1976. *Nature* 264:340
135. Friedman, I., Redfield, A. C., Schoen, B., Harris, J. 1964. *Rev. Geophys.* 2:177–224

136. Boato, G. 1954. *Geochim. Cosmochim. Acta* 6:209–20
137. Grevesse, N. 1970. *Colloq. Liège* 19:251–56
138. Hall, D. N. B. 1975. *Ap. J.* 197:509–12
139. Spitzer, L., Drake, J. F., Jenkins, E. B., Morton, D. C. Rogerson, J. B., York, D. G. 1973. *Ap. J. Lett.* 181:L116–21
140. Weinberg, S. 1977. *The First Three Minutes.* New York: Basic Books. 188 pp.

141. Slipher, V. M. 1924. Table prepared for Eddington, A. S. 1924. *The Mathematical Theory of Relativity,* p. 162. London: Cambridge Univ. Press. 2nd ed.
142. Chapline, G. F. 1976. *Nature* 261:550–51
143. Audouze, J., Lequeux, J., Reeves, H., Vigroux, L. 1976. *Ap. J. Lett.* 208:L51–54

Ann. Rev. Nucl. Sci. 1977. 27: 75–138

HYBRID BUBBLE ✖5582
CHAMBER SYSTEMS[1]

J. Ballam and R. D. Watt

Stanford Linear Accelerator Center, Stanford University, Stanford, California 94305

CONTENTS

[1] Work supported by the US Energy Research and Development Administration.

1 INTRODUCTION

After nearly two decades of spectacular success, the bubble chamber (1), as a stand-alone detector of single events in high-energy particle collisions among hadrons, seems to be on the wane.[2] Its place is being taken by sophisticated wire–spark-chamber arrays; by more complicated combinations of a bubble-chamber and electronic detectors, usually called "hybrid chambers," or simply "hybrid systems;" and by the huge bubble chambers that are mainly used for neutrino physics at very high energies. Even these very large chambers are partially hybridized—that is, they have external associated electronic detectors that aid in the identification of certain particles that leave the chambers.

There are two main reasons for this change in emphasis: one is the need in strong-interaction physics for high-statistics experiments, at least an order of magnitude greater than the heretofore largest bubble-chamber exposures (100 events per μb of cross section). While these chambers could probably have produced the needed events, the required data analysis (the finding and measuring of the events) simply took too long. The early promises (2) concerning data-analysis capability failed by an order of magnitude;[3] as a result, a collaboration of institutions, with a joint measuring capability of 2 million events/year, would have required more than ten years to complete the analysis of a 1000 event/μb experiment in a 2-m hydrogen chamber (3) if they measured every event. Indeed, there are ways of cutting this down, such as selecting certain categories of events to measure, but even with such restrictions it would take three or four years of the entire collective resources of three very large institutions to do such an experiment, and during this period of time they would have completed only this one experiment.

The other reason is that the new very-high-energy accelerators (energy greater than 300 GeV) produce many events that have tracks with momenta too high to be

---

[2] Many of the productive chambers of medium size have been discontinued: the SLAC/LBL 82-in chamber in 1973; the BNL 80-in chamber in 1974; the CERN 80-cm chamber in about 1971. The CERN 2-m chamber will likely shut down in 1978.

[3] This calculation assumes a 2-m chamber with a useful length of 1.5 m and 10 tracks per picture. For such a chamber, an automatic measuring machine could measure about 250,000 events per year.

measurable in a bare bubble chamber. In such cases, the measured kinematic quantities, supplied by the bare chamber, are confined to the spatial angles and the momenta of the slower particles. Some assistance has to be provided by external means in order to be able to analyze an event more completely.

In spite of these difficulties, the bare chamber retains qualities that will continue to make it indispensible for a large class of experiments, provided the difficulties mentioned above can be overcome. Among these qualities are a very large detection efficiency; the ability to see very close to the origin of an event; almost 100% efficiency for observing the angular distributions of the charged products of an interaction; measurements of proper particle lifetimes between 10^{-11} and 10^{-7} sec; the ability to determine, by means of ionization and momentum, the masses of particles moving slowly ($\beta \leq 0.8$) in the laboratory; and good efficiency for observing the decays of short-lived unstable particles, both neutral and charged.

The Hybrid System (HS) is the result of an attempt to retain as many as possible of these advantages of the bare chamber, while at the same time extending its usefulness to experiments needing an order-of-magnitude increase in statistics, supplying a better method of particle identification, and providing better momentum measurements for the higher-energy particles.

This has been accomplished gradually over the past five years, first by the development of the rapid-cycling chamber, which operates reliably at a rate from 20 to 100 times faster than was previously possible; and second by the use of wire chambers placed both upstream and downstream of the bubble chamber, thus providing particle-by-particle measurement of the incoming beam momentum and angle as well as an accurate measurement of the momentum of the scattered particle. In addition, particle identifiers, such as large Cerenkov counters, can also be placed both in the beam and downstream of the chamber. The HS operate in either a triggered mode, in which the illuminating lights are flashed only when the downstream electronics have selected an event, or in an untriggered mode, in which every expansion is photographed and the downstream information is stored on tape for later use. Such systems are now in use in experiments at the Fermi National Accelerator Laboratory (FNAL) and the Stanford Linear Accelerator Center (SLAC), and are also being constructed for use at the Rutherford Laboratory and at the new Super Proton Synchrotron at the European Center for Nuclear Research (CERN).

In this article we describe some of the basic characteristics of these hybrid systems, and examine how they have been used in certain experiments in high-energy physics.

2 AREAS OF PARTICLE PHYSICS ESPECIALLY SUITED TO HYBRID SYSTEMS

Since the HS can no longer claim to own all of the desirable characteristics of the bare chamber, it is important to delineate those areas of investigation that in some way are best matched to the system. At least one of the particles produced in an interaction in the liquid has to leave the chamber and be detected by the downstream counters. The structure of the bubble chamber usually imposes restrictions on the momenta and angles of these emerging particles. For example, most of the area of the bubble-chamber walls is made of solid steel, between $\frac{1}{2}$ and 1 nuclear collision length and 5–10 radiation lengths thick. In addition, most chambers are surrounded by fairly thick-walled vacuum tanks and also, except for regions up- and downstream of the chamber, by steel magnets, thus making it virtually impossible to detect particles leaving the liquid at large angles with respect to the beam. In larger chambers, the best that can be done is to provide beam-exit windows in both the chamber and vacuum tank that are as large and as thin as can be made compatible with mechanical and safety requirements. These windows have been made as thin as $\frac{1}{100}$ radiation length of aluminum (2.3 mm) with areas up to 70 × 15 cm, but most are made of stainless steel roughly 1.5 mm thick. The design of such beam windows is discussed in detail in Section 3. An illustration of these geometrical restrictions is shown in Figure 1.

One should also design the downstream detectors so that they can accept the maximum number of particles leaving the beam-exit window.

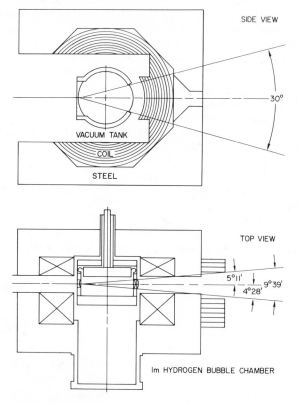

Figure 1   The SLAC 1-m bubble chamber, showing the limitations imposed on the exit aperture by the magnet steel and coils.

Finally, especially in the case of rapid-cycling chambers, it is important to design a trigger that will keep the number of pictures to be scanned and measured to a level such that the experimental group(s) can finish this process within, say, a year of their original production.

Bearing these restrictions in mind, two general modes of operation are employed, depending on whether the HS is triggered or untriggered. In the triggered case, one generally focuses on a specific interaction for which there exist the following conditions: (a) a calculable (and preferably large) acceptance; (b) a yield that is statistically significant; (c) a natural amenability to the design of a tight trigger so that the number of background pictures is not excessive; and (d) some aspect of the interaction that uses the bare bubble chamber's natural advantages.

Several classes of interactions fall naturally into these categories. The first to be considered is the class of single, fast-forward, charged-particle triggers in which the emerging particle takes up most of the momentum in the laboratory system and has a high probability of leaving through the beam-exit window. The momenta of the rest of the particles produced in the interaction are small and well-measured from the photographs themselves. Furthermore, these slow particles can be studied using all the advantages of the bare chamber. Some examples of this type of experiment are the following.

Nucleon Dissociation

In this process a nucleon is dissociated into a nucleon plus one or more pions, namely:

$$a + N \rightarrow a + N^*$$
$$ \hookrightarrow N + n\pi \, (n \geq 1), \qquad\qquad 1.$$

where $a = \pi^{\pm}$, $K^{\pm}$ or $p\bar{p}$; $N = p$ or n; and the emerging particle a carries a large fraction of the laboratory momentum in the forward direction. Such processes are very well suited to study by the HS. First, the N^* system is backward in the overall center-of-mass system and is therefore moving slowly in the laboratory. The mass of the N^* is well measured in the bubble chamber, and its production and especially its decay angular distribution can be studied with very little experimental bias. Since the trigger is a single particle emerging in the forward direction, the acceptance of the system as a function of the four-momentum transfer (t) from the beam particle to the triggering particle is easily calculable, as is the acceptance as a function of the mass recoiling against the triggering particle.

These processes have a fairly sharp dependence on t, and therefore a t acceptance that cuts off at about $t = 1$–$2 \, (\text{GeV}/c)^2$ (a typical range for most experimental arrangements); this eliminates a physical region in which the yield would, in any event, be very small. The momentum of the fast-triggering particle is measured in the downstream spectrometer with sufficient accuracy to allow the experimenter to determine whether or not any missing neutral mass consists of only one neutral or more than one neutral particle, and thus improves the mass resolution for the single-neutral case to a level comparable to that achieved when no neutral particles were produced in the interaction.

The diffractive reactions have been the most studied because their cross sections remain reasonably large even at high energies, and the background due to exchange processes is small even at energies as low as 15 GeV. For example, the quartet

$$\pi^+ p \to \pi^+ N^{*+} \qquad\qquad\qquad 2a.$$

$$\pi^- p \to \pi^- N^{*+} \qquad\qquad\qquad 2b.$$

$$\pi^- n \to \pi^- N^{*0} \qquad\qquad\qquad 2c.$$

$$\pi^+ n \to \pi^+ N^{*0} \qquad\qquad\qquad 2d.$$

can be used to study the I-spin dependence of diffractive dissociation for N^*'s produced in an energy-independent manner. The cross section for fast-emerging π's is about 5% of the total cross section, so that if one can eliminate elastic events in which the emerging particles are identical to the incoming ones, one can achieve good efficiency and a high ratio of useful pictures to total number of pictures taken.

Exchange Processes

These are described by the general process $a + b \to c + d$, in which the fast-incoming particle a exchanges some intrinsic property with the target particle b so that the fast-emerging particle c differs from a in electric charge, spin, strangeness, or rest mass, etc. Again, the triggering particle should be a single, charged, fast-forward particle so that the trigger criteria described above can be adhered to.

These processes have a much steeper energy dependence than do the diffractive ones, going roughly as p^{-2}. However, the power of a HS allows one to observe processes whose cross sections are as small as 1 μb with good statistics. Thus it will be possible to pursue the study of these interactions to energies as high as 70 GeV, where backgrounds due to competing processes become quite small.

Exchange-degenerate processes are good examples:

$$\begin{array}{l} \pi^+ p \to K^+ Y^* \\ \mapsto Y^* \to \Lambda + n\pi. \\ K^- p \to \pi^- Y^* \end{array} \qquad\qquad 3.$$

Here the K^+ or π^- is fast and forward-going. Again, the acceptance is easily calculable as a function of t and recoil mass. The Y^* is slow in the laboratory system, and its decay can be well studied in the bubble chamber itself. The polarization of the Y^* can be determined from the polarization of the Λ itself.

Another example is baryon exchange, where the incoming particle is a meson and the outgoing one is a baryon, e.g.

$$\pi^\pm p \to p(n\pi), \qquad\qquad\qquad 4.$$

where here the emerging fast particle is a proton.

There are many variations of the examples given here, including the case in which a fast meson is produced in the forward direction and decays into two charged particles, one of which is detected by the downstream chambers. Such a case is

$$\pi p \to \phi n$$
$$ \mathrel{\rightarrow} K^+ K^-.$$ 5.

When one of the K's is detected, the chamber lights are flashed.

Fast-Forward Neutral Particles

Another class of events suitable for a triggered HS are those that have a forward-going neutral particle. Provided the downstream neutral detector is efficient and has good energy resolution, these events do not have to be characterized by a fast or high-momentum trigger, since the neutral particles are unaffected by the magnetic field of the chamber. Examples of such reactions are

$$\pi^{\pm} p \to \pi^0 N^*$$
$$\phantom{\pi^{\pm} p \to} \mathrel{\rightarrow} p + n\pi.$$ 6.

Reaction 6 is an example of charge exchange. It is an important adjunct to Reaction 1 because it sorts out the nondiffractive background. It does have a steeper energy dependence than Reaction 1, and therefore is accessible only up to the 50–100-GeV energy ranges.

A further example is

$$K^{\pm} p \to K_L^0 p + n\pi.$$ 7.

Since the K_L^0 mean distance for decay is 1.55 m, many of the forward K_L^0's will leave the bubble chamber before decaying and can be made part of a neutral trigger. Similarly,

$$\pi^{\pm} p \to N + n\pi,$$ 8.

where the fast-forward N is detected in optical or wire spark chambers, and its energy is measured in a calorimeter. Such experiments examine backward production of ordinary mesons and also search for exotic mesons (which decay into two like charges).

Inelastic Muon Scattering

Since muons have a small nucleon-interaction cross section, it is possible to increase the incident beam flux by a factor of ten over what can be tolerated for hadron beams. The inelastically scattered muons can be separated from the elastics simply by demanding that the scattering angle be essentially nonzero before flashing the chamber lights. The advantage of the HS here is that of observing in the bubble chamber the hadrons associated with the scattered muon.

Since the total cross section for inelastic muon interactions is only a few μb, it would be virtually impossible to use a bare bubble chamber to study this process.

Reactions With More Than One Forward-Going Particle

The class of interactions that is most difficult to deal with in an HS is the class in which the forward particle decays into several particles before leaving the chamber.

An example is:

$$\pi^\pm p \to Bp$$
$$\qquad \qquad \hookrightarrow B \to \omega\pi$$
$$\qquad \qquad \qquad \qquad \hookrightarrow \pi^+\pi^-\pi^0. \qquad\qquad\qquad 9.$$

In Reaction 9, the acceptance of the downstream system is poor because only a portion of the decay products is detected. Reaction 9 is therefore best studied in large-acceptance wire-chamber spectrometers. However, there are special cases in which the decay angular distribution of the forward meson is well understood, and the object is to study a complicated N^* or Y^* state left behind in the bubble chamber. Examples of such reactions are

$$\pi^\pm p \to \rho N^*$$
$$\qquad \qquad \hookrightarrow p + n\pi$$
$$\qquad \hookrightarrow \pi^+\pi^- \qquad\qquad\qquad 10.$$

and

$$\pi^\pm p \to K^* Y^*$$
$$\qquad \qquad \hookrightarrow \Lambda + n\pi$$
$$\qquad \hookrightarrow K + n\pi. \qquad\qquad\qquad 11.$$

In both Reactions 10 and 11, the forward ρ and K^* mesons have been well studied. Therefore the effect of triggering on one of their decay products can be modeled, and the pseudo-two-body production process can be studied.

Interaction Triggers

This is a very general trigger that ensures that the interaction occurs in the useful volume of liquid in the chamber. The purpose is to eliminate reactions in the beam and vacuum windows, both upstream and downstream of the chamber, and to cut down on the number of photographs taken.

In the *untriggered* HS mode, the focus is on interactions for which (*a*) a number of the forward-going particles have momenta too high for measurement in the bare chamber, (*b*) it may be important to know the mass of the forward-going particle, and (*c*) the cross section is fairly large ($\geq$ several hundred μb).

Untriggered HS are mainly used to improve the accuracy of momentum measurement and for particle identification. The most successful systems are those set up at the FNAL, where the energy of beam particles reaches 400 GeV. The classes of experiments that have been done at FNAL so far are the following.

Elastic Processes

The HS is used to distinguish between elastic and inelastic scattering processes. Use of Cerenkov counters upstream of the chamber allows identification of the mass of particles in the beam, and therefore p-, $\bar{p}$-, K-, and π-initiated events can be differentiated in an unseparated beam. The reactions studied were:

$$\pi^- p \to \pi^- p \qquad \qquad \text{12a.}$$

$$\pi^- p \to \pi^- p + n\pi \qquad \qquad \text{12b.}$$

and

$$K^- p \to K^- p \qquad \qquad \text{13a.}$$

$$K^- p \to K^- p + n\pi. \qquad \qquad \text{13b}$$

Low-Multiplicity Nucleon Diffraction and Double Diffraction at Very High Energies

Interactions similar to those of Reactions of 2a and 2b have been studied for the case in which all the products are charged. In these cases, the fast-pion momentum is measured in the downstream system, and the momenta of the three remaining particles are measured in the bubble chamber. The slow proton is identified by ionization, and a four-constraint fit is made by balancing transverse momentum. Such experiments, as well as those that study the double-diffraction reaction

$$\pi^- p \to \rho N^*$$
$$\begin{array}{l} \quad\quad\; \big\lfloor \;\; \big\lfloor\!\!\to p\pi^+ \\ \quad\quad\; \big\lfloor\!\!\to \pi^+\pi^-, \end{array} \qquad \qquad \text{14.}$$

have been studied with the FNAL 75-cm HS. In these cases, the momenta of several forward-going particles are measured with the downstream system.

Inclusive Processes

These are dealt with in an HS because, in order to cover the fullest range of kinematic variables used in the subsequent physics analysis, it is necessary to measure the momenta of all particles of given electric charge. The variable most often used is called rapidity, which is symbolized by y and defined by

$$y = \tfrac{1}{2} \ln \frac{E + p_b}{E - p_b}, \qquad \qquad \text{15.}$$

where E is the energy of the particle, and p_b is its momentum along the beam axis. Specific inclusive reactions that have been studied are

$$\pi^- p \to \pi^{\mp} X$$
$$pp \to \pi^{\pm} X. \qquad \qquad \text{16.}$$

Up to the present, it has been assumed that all outgoing particles were pions (except those slow protons tagged by ionization, as seen in the chamber). These experiments are quite well suited to the HS because at very high energies, most fast particles leave the chamber through the beam-exit window and are seen in the downstream spark chambers. The ones remaining in the chamber are usually slow enough so that their momenta are directly measurable. Furthermore, since these results are mainly statistical and somewhat qualitative, a high degree of precision is not required. The incoming particle is again identified by the beam Cerenkov counters.

A further consequence of the increased acceptance at high energies is the ability to observe the inclusive interaction

$$\pi p \to \rho^0 X$$
$$\hookrightarrow \pi^+ \pi^-,$$

17.

again made possible by virtue of good downstream momentum measurements.

Muon Identification in Neutrino Experiments

A very successful use has been made of the HS in identifying muons emerging from neutrino interactions in the liquid of the very-large-volume chamber installed at FNAL. Since the muons penetrate the thick chamber and vacuum walls without interacting, they can be detected via wire chambers placed externally to the bubble chamber.

Reactions of the type

$$vp \to \mu X$$

$$vp \to \mu\mu X$$

18.

can be studied in various chamber materials (hydrogen, neon-hydrogen mixtures, and eventually deuterium).

Future extensions of triggered experiments, which will depend upon developments in fast on-line data processing, include multiparticle, neutral-kaon, and missing-mass triggers; whereas extensions of untriggered experiments depend on more sophisticated particle identification, especially the ability to distinguish among the masses of multiple forward-going fast particles.

3   CONSTRUCTING AND OPERATING FAST-CYCLING CHAMBERS

Because of the unique nature of its accelerator, SLAC has been a leader in the development of fast-cycling chambers. The accelerator can pulse as often as 360 times per second, and this makes the construction of rapid-cycling chambers a very desirable endeavor. In this chapter we discuss the SLAC chambers (4, 5) and also those in operation and being considered for construction at other laboratories.

Definition of a Rapid-Cycling Bubble Chamber (RCBC)

We have chosen to limit our discussion to those chambers that pulse 10 times per second or more steady-state, and those that can pulse 10 or more times during the beam-spill period of proton synchrotrons. Within the group so defined are the SLAC 38-cm chamber, the SLAC 1-m chamber, the Rutherford Rapid-Cycling Vertex Detector (C. Fisher, R. Newport, private communication), and the proposed CERN 1-m chamber (6).

Basic Requirements to be Met in the Design of an RCBC

SPATIAL RESOLUTION WITHIN THE CHAMBER   The prime purpose of an HS is the accurate location of the vertex in an event and the detection of all of the charged and short-lived neutral particles produced. It is very important that the liquid motion within the chamber be small so that minimum track displacement occurs between the time of beam injection and the flash of the camera lights. The best way to minimize liquid motion during expansion is to use the largest possible piston to produce a plane pressure wave in the liquid. The 1-m chamber at SLAC is a good example of this principle because it has a piston that needs to move only 4.6 mm to expand a chamber that is about 43 cm deep.

BUBBLE SIZE AND DENSITY ALONG A TRACK   In the bright-field–illuminated chambers, especially those using Scotchlite (7, 8), we have found that tracks having a mean bubble diameter of around 450 μm in chamber space and an average of 10–12 bubbles per centimeter are preferred by most experimenters. The bubble count per centimeter can be controlled by both the liquid temperature and pressure at the time of beam injection, but the size of the bubbles, although dependent on liquid temperature, is more easily controlled by the time delay between the beam injection and light flash. It is perhaps fortunate that not only does a delay of 2.5–3 m give the bubbles time to grow to the proper diameter, but it also gives the computer time to make the necessary calculations to decide whether or not to flash the lights.

SIGNAL-TO-NOISE RATIO OF TRACKS ON FILM   The film may be scanned by automatic scanning machines that are very limited in their ability to adapt to changes in signal-to-noise ratio, and this puts stringent requirements on the uniformity of illumination and the bubble diameter. Figure 2 shows a densitometer scan of typical SLAC 1-m RCBC tracks. It is possible under good conditions to maintain tracks of this quality for days at a time. In order to do this, the temperature must be maintained very accurately, and the expansion system must repeat its operation very precisely for millions of times. Both of these systems are described in detail later.

RELIABILITY AND REPRODUCIBILITY OF THE CHAMBER   Perhaps the greatest difference between a large slow chamber and an RCBC is the difference in the reliability requirements of parts. As an illustration, the 72-in chamber at Lawrence Berkeley Laboratory (LBL) (9) pulsed around 15,000 times a day, and it destroyed its expansion valve about every 60,000 pulses. In contrast, the SLAC 1-m RCBC pulses as often as 1 million times a day, and it needs a replacement for its expansion valve about every 10–15 million pulses. A great deal of effort has gone into developing that increase in reliability.

HEAT LOADS   It is very important that a fast-cycling chamber be made so that the heat load into the liquid is as small as possible. A sliding piston with rings not only leaks liquid from the chamber to the region behind the piston, thereby reducing expansion efficiency, but it also produces a large amount of heat as a result of sliding friction. Eckman (10) introduced the use of a single-convolution bellows on the LBL 25-in chamber, and most fast chambers now have such a bellows in one form or another. It is interesting to note that although one can obtain much more flexibility with multiconvolution bellows, it is possible that most of the motion is provided in the first convolution, and this of course would soon cause its failure.

An additional source of heat into the liquid results from eddy-current heating in the piston as it moves through a nonuniform magnetic field. Eddy-current heating is proportional to $N(VB_x)^2$, where B_x is the gradient of the magnetic field in the direction of the piston motion, V is the velocity of the piston during the stroke, and N is the number of pulses per second. Both N and V are determined by the design pulse rate and chamber depth. The field gradient can be reduced by shaping the iron pole pieces to a certain extent, but this is far from a satisfactory solution. The only other alternatives are the use of a piston made of nonconducting material, such as plastic or glass, as the Rutherford Laboratory group has done with the RCVD,

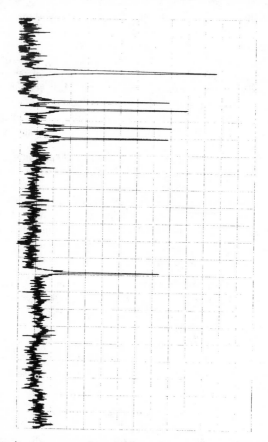

Figure 2   A densitometer scan of tracks from the SLAC 1-m bubble chamber. The bubble diameter is 400 μm. The scanning slit was 0.26 mm × 10 mm (parallel to track). The signal-to-noise ratio is 6:1.

or the incorporation of cooling loops within the piston in a manner similar to that used with the SLAC 1-m piston.

EXTRANEOUS BUBBLES   When pulsing at rates in excess of 10 pps, it is very important that bubbling from trapped volumes, leaking valves (4), glass-to-metal seals, and hot spots be kept as small as possible. The success of any fast-cycling chamber will be determined by how well these problems are solved.

It is possible to inject as many as 100 charged particles into the chamber and form bubbles whose volume will total 1.5 cc at the time when they are largest. If the chamber is operating well, this volume of gas will present no problem to either the expansion system or the refrigeration system. However, if extraneous bubbles from a poorly designed chamber collect in the top, it is easy to have 50–100 cc of gas continuously present, and this quite seriously affects both the piston stroke and the refrigeration requirements. The stroke, as calculated from the liquid-bulk modulus, should be much less than 1%. However, few small chambers are so free of local gas pockets and bubbles distributed throughout the liquid that they achieve sensitivity with a change in volume of less than $\frac{3}{4}\%$ during the stroke.

Description of Some RCBCs

THE SLAC 1-M RCBC   This chamber (5) was first operated in 1967 and has been gradually improved from its design pulse rate of 1 per sec to its present 10–12 pulses per sec. Since its origin, it has expanded a total of 300 million pulses and has taken 10.5 million pictures. The principal features of the chamber are described in Table 1, and Figure 3 shows an elevation view of the chamber. A detailed description of the various systems follows.

Illumination   The piston of this chamber is covered with high-quality Scotchlite (8) made by the Minnesota Mining and Manufacturing (3M) Company. Great care was taken in the selection of the glass beads so that they would be very uniform in

Table 1   Features of SLAC 1-m RCBC

Visible volume	360 l
Chamber depth	43 cm
Diameter	1 m
Glass thickness	19.5 cm
Expansion system	bellows-sealed piston/hydraulically controlled at 10–12 pulses per sec
Beam window	entrance 1.5 mm stainless steel exit 1.5 mm stainless steel
Magnetic field	26 kG
Magnet weight	285 tons
Refrigeration system	1-kW refrigerator
Illumination system	bright field (Scotchlite®)
Chamber camera control system	proportional wire chambers, Cerenkov counter, scintillators

diameter (about 30 μm) and thus all have about the same retrodirective characteristics. In addition, they have been spread on their support matrix so that their density is quite uniform. The Scotchlite comes in 30-cm-wide strips, and since the chamber piston is 102 cm in diameter, it requires four sections of Scotchlite to cover the surface. It is not feasible to have the Scotchlite overlap at the seams in an effort to provide continuous coverage, because this would cause bubbling from the volume trapped under the overlap. A compromise is made wherein the seams are left as narrow as possible without overlap, and the sections put on so that the seams are at right angles to the path of the beam as it traverses the chamber. This is shown in Figure 4.

Scotchlite is a composite made of beads held in an adhesive matrix and covered by a thin layer of Mylar® for protection. The Mylar is a very good insulator, and this attribute has, on occasion, been a source of trouble. During the piston motion, there is a small movement in the liquid hydrogen, and this motion causes an electrostatic charge to build up on the surface of the Mylar. If the chamber is not extremely clean and free of dust, lint, and other such debris, these contaminants will eventually end up deposited on the surface of the Scotchlite. The deposited material tends to clump

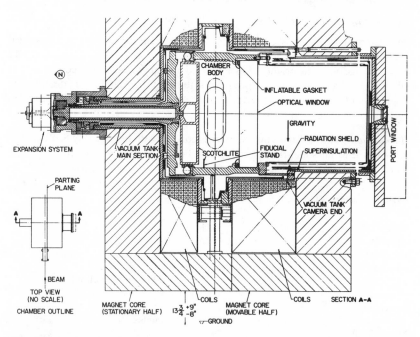

Figure 3   An elevation drawing of the SLAC 1-m bubble chamber. The outer shell is the 280-ton magnet. From right to left are shown the iron shield for the camera; the main glass window; the chamber body with its internal heat exchanger at the top; the hollow expansion piston, on which Scotchlite is glued and inside of which are refrigeration loops; the piston drive rod, which extends through the vacuum tank out to the room-temperature hydraulic-expansion system.

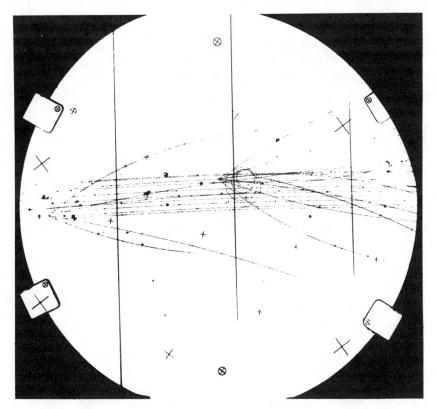

Figure 4   Photograph of the SLAC 1-m bubble chamber taken while the chamber was operating at 10 pulses per sec. Note the absence of bubbling along the Scotchlite seams.

together and eventually becomes large enough to be a source of small bubbles. The bubbles not only contribute to the dynamic heat load imbalance but also impair the retrodirectivity of the Scotchlite, causing a loss in picture quality. Very careful installation of the Scotchlite allows up to 50 million pulses or more of operation and several cool-down cycles before replacement is necessary.

Bellows   The stainless-steel omega-shaped bellows, which allows the piston to move, is an extremely critical part of a fast-cycling bubble chamber. The cross-sectional view shown in Figure 5 shows the method of construction used at SLAC. The bellows is essentially a tube bent into a circle of over 1 m in diameter, after which a segment of the inner portion is cut out. The inner edges of the circular tube are then welded to 2 flanges, which are then in turn welded, one to the piston and the other to the chamber body.

In the bellows design it is important to keep the bellows quite stiff so that there will be little tendency to excite oscillations in the mass of the bellows that may be harmonically related to the motion of the piston. During the early stages of the SLAC 1-m chamber development, a hydroformed bellows was installed whose cross section is shown in Figure 6. When the chamber was pulsed, this bellows failed very quickly (less than three million pulses) at the point where the metal rolls over the rounded nose of the flange. Tests ultimately proved that the "flopping mode," in which the

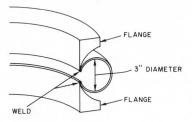

Figure 5   A cross section of the bellows used on the SLAC 1-m bubble chamber. The effective bellows diameter is 120 cm.

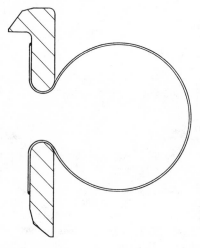

Figure 6   The hydroformed bellows first used with the SLAC 1-m bubble chamber. This bellows was later replaced by the type shown in Figure 5.

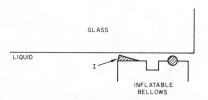

Figure 7   The special glass-to-metal seal used in SLAC bubble chambers.

whole mass of the bellows moves up and down, had a period quite close to that of the chamber pulse period, and this caused the bellows center to overshoot when the moving flange returned to the reset position as the chamber stroke was completed. If heavy liquids such as neon are used in a chamber, particular attention should be paid to these resonances, because the internal liquid mass can greatly affect the bellows' natural frequency. Even though the bellows contributes no significant heat to the liquid, it is necessary to have a cooling loop in the top internal section to condense any gas bubbles that may be formed and trapped there.

With proper attention to these problems, bellows can be made to last indefinitely. The 1-m chamber has 280 million pulses on its present bellows.

Pluming   In order to keep the expansion stroke as small as possible, it is important that the chamber be completely full of liquid—no large gas bubbles trapped any-where. This must be true both before the start of expansion and during the pressure drop caused by the piston motion. Any small pre-expansion bubbles can be taken care of by proper cooling of the periphery of the chamber, especially the top region. The bubbles that grow during the pressure drop and that are a result of hot spots, small cavities, or rough surface regions with sharp points are a continuing problem in fast chambers and must be kept to an absolute minimum.

It is probably safe to say that the major source of extraneous bubbles in SLAC's chambers is the seal between the chamber glass and the chamber body. During the evolution of the fast-cycling program at SLAC, it was necessary to develop a new type of seal before the 1-m RCBC could exceed 6 pulses per sec. Figure 7 shows the construction of this seal. It is made of pure indium that has been pooled over the indium-wetted surface of the stainless-steel rings that make up the inflatable gasket. After the indium has cooled, it is machined to a wedge shape that crushes into intimate contact with the glass when helium under pressure is introduced into the inflatable gasket.

The traditional glass-to-chamber seal (5) previously used was similar to that shown in Figure 8. The indium was pressed into round grooves and then later flattened when helium was introduced into the inflatable bellows region under pressure. This method of installation allowed small gas pockets to be trapped, both in the groove and under the indium, as it flattened down on the steel surface. When the chamber was later

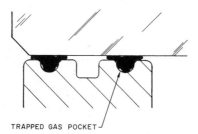

TRAPPED GAS POCKET

Figure 8   An inflatable-gasket glass-to-metal seal of the type conventionally used in bubble chambers. Note the voids around the indium seal.

filled, many of these gas pockets were continuous from the bottom of the round grooves out to the chamber liquid and thus caused pluming during the pulse. The new wedge gasket cannot have gas pockets between the indium and the steel, and is unlikely to have them between the glass and the indium.

Chamber refrigeration   When pulsing at high repetition rates, one finds that the time constants involved in changing the temperature of a local region by external means are very long compared to the time between two pulses. On the other hand, if a warm region develops within the chamber, it will probably cause a thermal runaway within four or five pulses. Since it is thus not possible to detect a thermal problem and correct it rapidly enough, great emphasis must be placed on preventing thermal instabilities from occurring at all.

One source of thermal instability is the development of gas pockets within the chamber at the top of the bellows and the chamber body. If the accumulation of gas is slow, it is possible to condense the bubbles and not cause any great perturbation in the local region. If it is fast and the refrigeration system is given the information to cool rapidly, then a situation can occur in which the chamber over-pressure is reduced to the point where it is too low to compress bubbles in the main body of the liquid, and a loss of control happens very quickly. If, in the normal course of operation, a heat exchanger is allowed to shut down for a period of several minutes because of lack of refrigeration demand for that loop, the pipes that supply the loop will warm up by radiation loss, etc. When next the loop is required to conduct liquid, the very first gas that comes through will be hot and the heat exchanger will have a hot spot in it that in turn heats the chamber liquid, causing a bubble and subsequent loss of control.

We have found that heat exchangers should be operated with some flow at all times, and the temperature of the liquid in the heat exchanger should be kept only very slightly below that of the chamber liquid. This small temperature difference between refrigerant and chamber liquid means that the heat-exchanger area needs to be large. The ideal situation would be one where the whole chamber body, especially the top, was a part of the heat-exchange system. One method is to suspend the chamber totally in a bath of liquid. The SLAC 1-m chamber approximates total immersion by having an internal upper heat exchanger welded to the chamber wall. In addition, the whole chamber casting is refrigerated externally, and the piston is hollow and contains refrigeration loops.

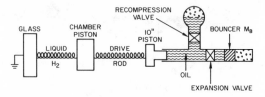

Figure 9   The spring-mass equivalent of the expansion system for the SLAC 1-m bubble chamber.

One additional benefit that accrues from the small temperature difference between the refrigerant and the chamber liquid is the relative absence of small volumes of cold chamber liquid near the heat exchangers. If present, these cold or hot volumes of liquid would cause "optical turbulence" because of the difference in the index of refraction between the main body of the liquid and the local perturbation. This results, in some cases, as apparent short-range deflections of the tracks.

Expansion system   Figure 9 is a schematic representation of the drive system used in the SLAC 1-m chamber. Examining this figure from left to right, one finds the chamber hydrogen, which can be treated as a massless spring. (When neon is the liquid, it is better treated as a distributed-mass spring.) Next is the chamber piston, whose weight is about 1200 lbs, followed by the rod, which extends from the chamber out through the vacuum tank to the 25.4-cm oil-drive piston. The drive rod, of course, is also treated as a spring, since it compresses by about 0.5 mm under the chamber pressure load. The oil system follows, with an expansion valve that releases the oil into the bouncer or energy-storage region that consists of an aluminum piston backed up against a gas volume. When the piston system has moved to the right (decompressed direction) as far as the bouncer gas will allow it to move, most of the potential energy that had been stored in the chamber liquid is now transferred to the gas in the bouncer. At this time the direction reverses and the piston system moves to the left (compressed direction) and "almost" travels back to its original starting point. However, it has lost some energy in sliding friction, oil friction, etc, and it resets about 80% of the way without help. At the proper time, the recompression valve is opened and a small burst of oil is added to make up for the losses. Figure 10 shows a typical trace of the motion of the warm end of the drive rod; the cold-end motion is inferred from the pressure change during the cycle. The phase lag between the ends can be seen, as well as the natural frequency of the drive-rod spring-mass system.

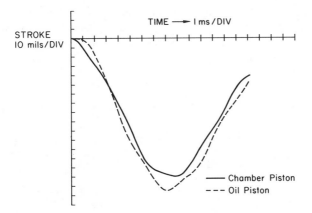

Figure 10   Displacement vs time for the piston system used in the SLAC 1-m bubble chamber.

Beam-window design   The beam-entrance window must be kept as thin as possible to minimize beam interactions within its material. On the other hand, it must be strong enough to withstand many millions of pulses without failure. From the standpoint of low mass and reasonable strength, aluminum is a good material to use. If, however, the main chamber body is made of stainless steel, then the aluminum window must be bolted in place with a gasket to provide the seal. For slow chambers this is a satisfactory solution, but RCBCs will not tolerate the additional pluming that such a seal would introduce, and one is forced to look for other solutions. Figure 11 illustrates the methods used at SLAC on the 1-m chamber. A section of stainless steel (2.5 cm thick) is milled so that it has a cutout of the same height and width as the beam window requires. The heavy portion (window frame) surrounding the thin portion will ultimately be welded into the chamber body in such a way as to produce a smooth, "plume-free" weld.

The flat milled piece is then bent into a cylinder whose radius is the same as the chamber radius. After it is properly bent, it is then placed in a hydroform jig, and the thin section is forced to assume the shape illustrated. Since the window is then welded into the chamber, particular attention must be paid to the transition from the thick window-frame portion to the thin window section. The thin section moves radially in response to the pressure change during the pulse, and any discontinuities will soon cause a crack and subsequent failure of the window. Therefore the milling cutters should not introduce any scratches or grooves that might cause high local stresses. If the window has been made with stainless steel, such a failure would give adequate warning because stainless steel is quite ductile, and the surrounding vacuum would thus deteriorate slowly over many pulses.

THE SLAC 38-CM RCBC   This chamber (4, 11), finished in 1973, was the first electro-magnetically driven bubble chamber. It receives its expansion force from a very large coil wrapped on a fiberglass form and suspended in a dc magnetic field, much like a large audio-speaker voice coil.

The chamber has achieved a pulse rate of 40 times per sec while taking pictures of acceptable quality. It has been pulsed a total of 90 million times and has taken over one million pictures. The principal features of the chamber are listed in Table 2, and the chamber itself is shown in Figure 12.

Figure 11   Cross section of the beam window used in the SLAC 1-m bubble chamber.

Table 2   Features of the SLAC 38-cm RCBC (11)

Shape	cylindrical
Beam plane dimensions	38-cm diameter
Depth	14 cm
Approximate visible volume	15 l
Orientation	window horizontal
Maximum pulse rate	40 pps (achieved)
Illumination	Scotchlite bright field
Entrance beam window	single-convolution omega bellows, 0.8 mm
Exit beam window	stainless steel
Magnet field	12 kG (superconducting coil)
Expansion	electromagnetic

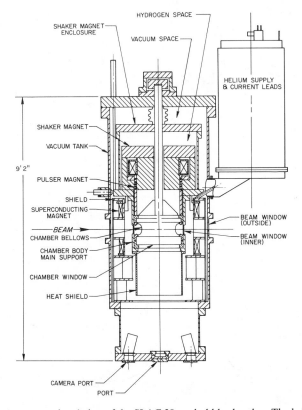

Figure 12   A cross-sectional view of the SLAC 38-cm bubble chamber. The bellows has an extended flattened central region in order to provide a thin window for particles through a full 360°. The chamber is immersed in liquid hydrogen for cooling purposes. However, most of the cooling occurs through the top plate. The piston motion (0.125 cm) comes from the curved portion of the bellows.

The bellows in this chamber performs three major functions. Figure 13 shows a cross-sectional view of the bellows, which is seen to differ from the usual omega-shaped bellows in the extended central section. This was done so that the beam particles could enter and exit through a section of thin material. In effect, it provides a beam window with a full 360-degree coverage. The whole chamber is immersed in a liquid hydrogen bath, and the thin bellows provides a large area for heat exchange between the chamber and the surrounding liquid. However, the major part of the refrigeration is done through the top plate of the chamber. The piston motion, about 1.27 mm, is made possible by the curved upper and lower sections of the bellows.

Perhaps the most unusual feature of this chamber is the method by which the drive force is obtained. Figure 14 shows the spring-mass system involved and the drive coil through which the pulsed current travels. The radial dc magnetic field through

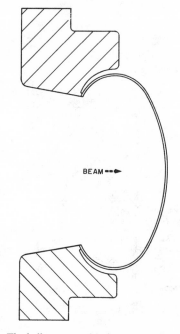

BEAM

Figure 13   The bellows assembly for the SLAC 38-cm RCBC.

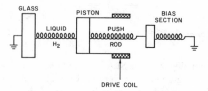

Figure 14   The spring-mass equivalent of the expansion system for the SLAC 38-cm RCBC.

which the pulsed coil moves is about 14 kG, and the pulsed current is 400–600 A. Figure 15 shows the control circuits for the pulsed current. Silicon-controlled rectifiers (SCRs) 1 and 2 discharge the energy stored in the capacitor through the moving coil. The electrical time constants are designed so that the coil current has gone through its peak and returned to zero at about the same time the piston reaches the most expanded part of its motion (see Figure 16). The charge on the energy-storage capacitor is now of the opposite sign and about 75% of the original value. When the piston starts to move back down in the direction of recompression, SCRs 3 and 4 are fired and cause a braking force to reduce the large piston overshoot that would otherwise occur. Proper phasing of the various SCRs can produce a pulse shape very close to the optimum, which would be expressed by $X = A(1 - \cos \omega t)$.

The problem of spurious bubbling or pluming from the glass-to-chamber seal was handled in the same manner as that for the SLAC 1-m chamber. However, one of the major obstacles to fast pulsing in this chamber was pluming from the fill valve between the chamber liquid and the reservoir outside the chamber. This problem was solved through the use of the special valve shown in Figure 17. An important feature of this valve is the conical seat, slightly flexible, which is pressed on by the

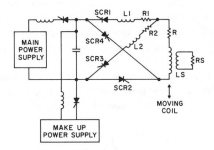

Figure 15   Control circuit for pulsing the SLAC 38-cm RCBC.

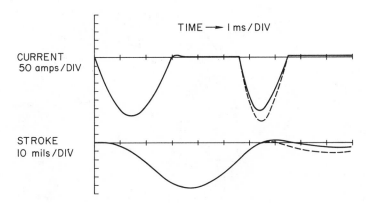

Figure 16   Current and stroke vs time for the SLAC 38-cm RCBC.

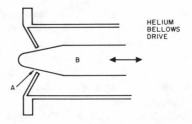

Figure 17   The chamber-liquid seal-off valve used with the SLAC 38-cm RCBC.

tapered center prod. A force of about 40 pounds on the prod will make a seal that is tight when tested by a mass spectrograph, which tolerates many operations without developing an appreciable leak-through rate.

THE RUTHERFORD RCVD   This chamber (12) first ran in July, 1976 and promises to be a very useful device. It achieved good sensitivity at pulse rates as high as 20 per sec, and it is planned to operate at 60 pps. Figure 18 shows the chamber in cross section. The chamber body is aluminum and the sidewalls are very thin so that the beam, which makes its entrance there, will cause few interactions in the walls. The cooling is done from the top by a pool of liquid hydrogen whose pressure and temperature are controlled by a condensing coil through which runs liquid hydrogen from the refrigerator. The chamber is filled through an aluminum valve in the top, of a design similar to that shown in Figure 17.

The expansion system is electromagnetic, and the force is derived from a large coil moving in a fixed magnetic field. The drive coil is at the bottom, at room temperature, and is separated from the cold chamber by an evacuated distance piece

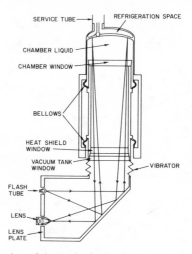

Figure 18   An elevation view of the Rutherford Laboratory rapid-cycling vertex detector (RCVD).

made of epoxy glass. The glass moves with respect to the chamber body and is sealed with adhesive to an annular ring that connects to the expansion bellows. In the final version, this chamber will be operated in an extremely high magnetic field, which would cause large eddy-current heating in moving metal parts. To avoid this heat load, the Rutherford group developed a unique bellows made of epoxy fiber-glass and molded over a form of the proper shape. As can be seen in Figure 18, the bellows does not close upon itself in the same way an omega bellows does, and this of course simplifies the construction process. It also precludes the trapping of bubbles within the bellows, a problem inherent in the omega form. A second bellows is used at the bottom of the distance piece to allow its free motion.

The optics system outside the chamber consists of a large diagonal mirror to direct the light to the cameras through a hole in the side of the distance piece. The inner upper part of the chamber body is polished and acts as a mirror to return the light from the light source.

The pulse mode, designed to make use of accelerator flat-top operation, will be one with bursts of about 20 pulses followed by a period of quiet. This mode has the advantage of allowing extraneous bubbles that are formed during the burst to condense in the next off-period.

THE CERN RCBC   At the present time the CERN RCBC proposal is the only known chamber of its size that is designed to pulse 30–40 times per sec. The possibility of accomplishing such a high pulse rate is enhanced by the mode of operation that will be used. Rather than a steady rate of 30 pps, it will be on for only 2 sec and off for 6 sec. The dead period will allow any residual bubbles to be condensed before the next pulse burst arrives.

Table 3   Main characteristics of the proposed CERN chamber

Diameter	80 cm
Exit angles for the chamber body:	
in the bending plane	$\pm 30°$
in the dip plane	$\pm 13.5°$
Operating fluids	hydrogen and deuterium
Depth	40 cm
Cycling rate with hydrogen	30 Hz for up to 2 sec every 8 sec
Material in the beam exit windows	$< 4\%$ of a collision length
over an angle of $\pm 12°$	$< 11.5\%$ of a radiation length
Beam-entry window size	± 5 cm horizontally
	± 12 cm vertically
Material in beam-entry windows	$< 4\%$ of a collision length
	$< 11\%$ of a radiation length
Precision with respect to the	50 μm in the bending plane
chamber fiducials	150 μm in the dip plane
Optical resolution	better than 300 μm
Precision of correlation with	once calibrated, to be stable within 100 μm in
downstream system	the bending plane and 200 μm in the dip plane

Although at this time the final design of the chamber and its expansion system is not complete, it is very likely that the usual CERN mechanism for expansion will be employed. A hydraulic drive system using a proportional-flow-control valve will force the chamber piston to conform to a prescribed displacement curve. The system frequency response will be high enough to do some pulse shaping within the pulse as it is demanded by the feedback-control system. A system such as this is versatile and will allow them to shape their pulse as they wish. A large overshoot in pressure in combination with low operating temperature (see bubble-growth description in Section 6) will be needed to achieve the pulse rate they need. Table 3 gives the important specifications for this chamber.

4   DESCRIPTION OF HYBRID SYSTEMS

General

As an introduction to a relatively sophisticated triggered HS, we consider the one associated with the SLAC 1-m chamber. This is shown in Figure 19. It consists of four main parts: the beam position, momentum- and mass-measuring system, the bubble chamber, the outgoing particle–measuring system, and the fast trigger. In addition, not shown in Figure 19, there is the on-line data-logging system. The fast trigger consists of a coincidence between the beam counter (S_1) and one of the hodoscope counters (H), together with a pulse $\leqq$ a small value in Cerenkov counter C. If this coincidence occurs, all of the information gathered by the beam multiwire proportional chambers (MWPC), the beam Cerenkovs (C_B), the downstream MWPCs, and the large Cerenkov counter (C) is dumped into the buffer. Since MWPCs are used, all of the above occurs within about 100 nsec. The system then has up to 3 msec (the time it takes the bubbles in the chamber to grow large enough to be photographed) to determine whether or not to flash the chamber lights and take a photograph.

This time to sort out unwanted events is extremely important in establishing a "tight" trigger, is unique to the triggered HS, and is one of the main reasons for its success. Without it, the number of photographs taken would be prohibitively large. The first plane of upstream MWPCs determines the deflection of the beam particle due to a bending magnet (not shown) and thus its momentum, and the two-beam MWPCs define the entry angle of the beam particle into the bubble chamber. The beam Cerenkov, used in a threshold mode, defines the mass of the beam particles. Several beam Cerenkovs in series may also be used. The three planes of downstream MWPCs are placed in the magnetic field common to the bubble chamber and determine the momenta of the outgoing tracks. Each of the planes, α, β, γ, consists of three sections in x, y, and u. All have 2-mm wire spacing, with u placed at 30° with respect to y.

If one of the outgoing tracks goes through C, its mass can be determined by pulse height or by the absence of a signal, and the hodoscope can be used later to make a proper matchup with a signal from one of the mirrors of C. In a typical experiment, such as $\pi^+ p \to K^\pm X$, it was desired that the lights be triggered by a K^+ of momentum greater than some p_{min}. The pressure in C was set so that at p_{beam} the K pulse was one

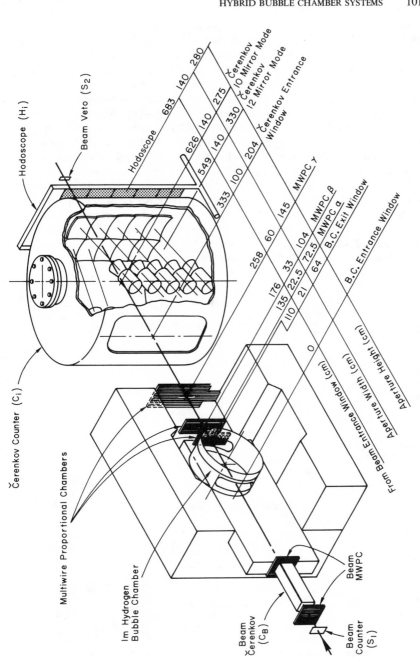

Figure 19 An isometric view of the present SLAC hybrid system.

half that of a π meson of momentum p_{beam}. After a fast trigger, $S_1 C_b H \overline{C}$, an algorithm using the information from the rest of the system was used to calculate the momentum and angle of the outgoing track(s). Then a check was made to see if the pulse from C corresponded to that for a K particle having the measured momentum, and also to see if the hodoscope counter corresponded to a particle having gone through the correct mirror in C. (The algorithm is described in more detail in a later section.) Thus, if the entering particle were truly a beam particle, if the interaction took place inside the fiducial volume of the bubble chamber (and not in external windows or other material), and if one of the outgoing particles was a K meson of the desired momentum, then the lights were flashed. The design goal of the system was to achieve an overall on-line efficiency of $\geq 50\%$.

In the untriggered case, the emphasis is on storing highly accurate information that is to be used later in off-line analysis. Figure 20 shows a typical setup designed for use with the FNAL 78-cm HS (13). The spectrometer is composed of MWPCs at locations A–G and, as in the previous case, makes use of the bubble-chamber magnetic field. The upstream portion utilizes MWPCs with 2-mm wire spacing. The wire chamber at each of the detector locations is again made up of three sections,

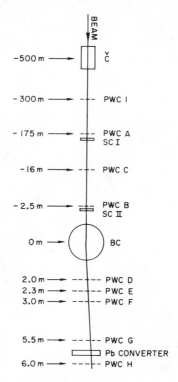

Figure 20   Schematic diagram of the FNAL hybrid system. The muon detector is not shown. The trajectories of a two-particle event are sketched in.

each of which is rotated 120° with respect to its neighbor. Planes D, F, and G have a fourth section for greater momentum precision. The upstream system includes a Cerenkov counter (14), and the downstream system includes a muon counter (not shown in the figure), followed by scintillators and a shower detector made up of plane H, immediately preceded by 1.27 mm of lead.

The system is triggered by a coincidence between scintillator counters SC I and II, and all other signals are strobed into memory. Since MWPCs are used, the time to strobe the data was less than 125 nsec, and therefore each beam particle entering the chamber during its sensitive time could be individually tagged and associated in time and space with tracks in the bubble chamber. This system did not have particle identification for outgoing tracks, so that its main function was to tag the beam and provide an accurate momentum measurement for the outgoing tracks. Particle tagging will be introduced in the near future (see Section 6).

Upstream Counters and Wire Chambers

These are usually quite conventional in composition. The object is to define the mass, angle of entry into the bubble chamber, and momentum of the beam particles: the emphasis is therefore on a high degree of spatial accuracy. This is usually achieved with MWPCs such as those shown in Figures 19 and 20, placed in such a way as to have a long lever arm between the first MWPC and the momentum-defining bending magnet. In general, 2-mm wire spacing gives sufficient spatial resolution to provide fractional percentage momentum resolution for beams whose momenta is as high as several hundred GeV/c. The beam angles are determined by two sets of MWPCs with no magnetic field between them. Beam Cerenkov counters are usually employed for mass determination. Since the beams enter these counters at a zero-degree or very small angle with respect to the mirror axis, they are highly efficient, and it is thus possible to know the mass composition, as well, to the level of a fraction of a percent (as is discussed in Section 5).

Counters Located in Vacuum Tank

There are special circumstances in which counters placed near the outside wall of the bubble chamber and inside its vacuum tank can be used to mark the occurrence of an interaction with a large acceptance relative to that available to a system such as shown in Figure 1. These have usually been scintillators, with light pipes coming through the outside wall of the vacuum and out to a region of small magnetic field. Such a system can also be used to mark the time of flight of a neutral beam. The chamber lights can be triggered on a rough time measurement, and a more exact determination can be made later by measuring the actual path length of the track in the bubble chamber that produced the pulse in the scintillator. Such a system of scintillators installed in the SLAC 1-m meter chamber has been used in a $K^0 p$ experiment (15). Because of the wall thickness of the bubble chambers, these counters cannot be used to get information on the precise location of a track after it leaves the illuminated volume.

An exception to this rule is that of the SLAC 38-cm fast-cycling chamber, which was constructed so that the expansion bellows also served as the cylindrical side wall and was only 1.2 mm thick. Proportional wire chambers (16) were placed in the vacuum space, both upstream and downstream. The beam used was mainly K_L^0 mesons, and the upstream chambers were used as an anticoincidence, while the downstream wire chambers were used as part of a positive-coincidence circuit that triggered the bubble-chamber lights when two or more forward charged particles left the bubble chamber.

The Downstream System

The most complicated part of the HS is the array of counters and spark chambers downstream of the bubble chamber. These must provide information on the momenta, angles, and masses of outgoing particles, as well as multiplicity and, on occasion, veto capability, either on- or off-line. For charged particles there are two classes of systems, depending on whether the analyzing magnet is the fringe field of the bubble chamber or a separate magnet, forming part of a downstream spectrometer. The former case is cheaper, but, more importantly, has a much greater acceptance, which is the reason that most HS use this method. The separate magnet has an inherently greater momentum accuracy, which it achieves at the expense of acceptance. However, this measurement of momentum can be done independently of any knowledge of the vertex point of the interaction and is especially useful if an accurate measurement of the mass recoiling against the out-going particle is desired. An example of such a system is shown in Figure 21, in which the momentum of the outgoing particle was known to a fraction of a percent.

This system was used to trigger on a missing mass greater than the mass of a proton, and it eliminated some 75% of the elastic pion-proton scatterings that otherwise would have swamped the experiment. This spectrometer, when used in off-line analysis of events with very-high-energy forward-going charged particles, is almost indispensible to the analysis of one-constraint (1C) events, namely, in separating events with one neutral particle from those in which there are more than one.

Another form of downstream trigger is that for neutral particles. An example (17)

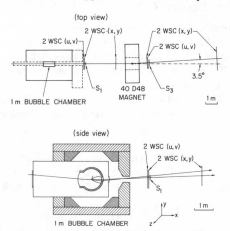

Figure 21   Top and side views of the first SLAC HS using a downstream spectrometer.

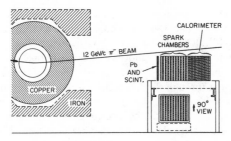

UCLA NEUTRON DOWNSTREAM TRIGGER

Figure 22   Top view of the neutron HS used with the SLAＣ 1-m chamber. The path of a noninteracting beam particle is sketched in.

is shown in Figure 22. The trigger system consisted of an array of 19 wide-gap optical spark chambers, preceded by lead plates and an anticoincidence counter S, and followed by a calorimeter consisting of 15 Fe/scintillator sandwiches. The bubble-chamber lights (and the optical spark chambers as well) were triggered whenever there was no pulse in S_1 (neutral particle and not a gamma ray) followed by a pulse height from the calorimeter corresponding to an energy deposited equal to or greater than half the energy of the incident π^- particle. A similar technique has been used with the SLAC fast-cycling 38-cm chamber (18) and with the 78-cm chamber located at the Argonne National Laboratory and designed by a group from the University of Wisconsin (19).

Another elaborate downstream spectrometer has been used with the SLAC 38-cm fast-cycling chamber in the scheme shown in Figure 23. The spectrometer (20)

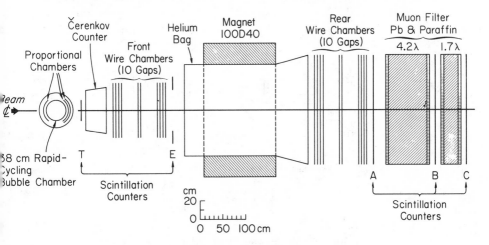

K° DOWNSTREAM TRIGGER

Figure 23   The hybrid system associated with the SLAC 38-cm bubble chamber. The downstream spectrometer has been used mainly for counter experiments.

information was fed into a computer whenever the T scintillation counter banks recorded two or more fast particles, the E counters did not fire (no slow accidental), and the A counters indicated that one or more particles had left the magnet. A picture was taken at the same time if two or more wires in the bubble-chamber PWCs fired. Off-line, the spectrometer calculated the mass of the pair in events in which two charged particles went through the spectrometer and determined if this was a K_S^0 mass. This was correlated with the bubble-chamber frame number, and that particular picture was subsequently scanned to see if the decay point of the K_S^0 was inside the chamber's fiducial volume. The chamber was then used to see if another particle (usually a slow proton) was associated with the interaction that produced the K_S^0.

Another downstream system that was used in an untriggered mode with the FNAL 78-cm chamber (21) is shown in Figure 24. The main elements are the four dual, wide-gap, optical spark chambers 1, 2, 3, and 4. The gaps are 20 cm long, the active transverse dimensions are 76×180 cm, and the two chambers are separated by 1 m. The spark-chamber cameras could be pulsed 5 times per sec. The spark chambers were triggered if an incoming beam track either produced two or more forward minimum-ionizing secondary tracks as detected in a set of three dE/dx counters, or

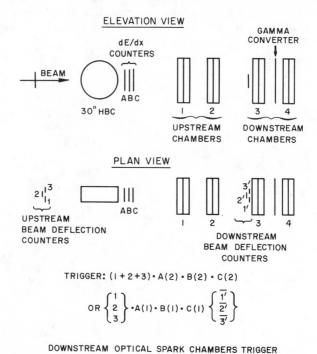

Figure 24 Schematic diagram of the first FNAL HS using optical spark chambers in the downstream system.

was deflected from a normal beam trajectory as detected by two sets of aligned counters. The dE/dx counters have high efficiency for high-multiplicity events, and the deflection counters are best for low-multiplicity events. Here again, the bubble-chamber magnet is used for momentum analysis of outgoing particles.

A downstream trigger for muons using the SLAC 1-m HS is shown in Figure 25. Muons with a scattering angle roughly between 1° and 6°, having an energy greater than 1.4 GeV, and penetrating 1 m of steel caused a coincidence between the scintillator trigger-counter hodoscopes, which provided a fast trigger for the spark chambers. The information from the hodoscope was then fed into a computer, which first checked that the hodoscope hit pattern could have been caused by a track coming from the bubble-chamber fiducial volume, and if so, caused the bubble-chamber lights to flash. The wire-chamber information was used off-line to improve the momentum measurement of the scattered muon after it was identified as a track that went through the requisite amount of iron. This trigger was used with incoming muon fluxes up to 100 per expansion.

The HS used in neutrino experiments with the FNAL 4.6-m chamber is called an External Muon Identifier or EMI (22), and the arrangement is shown in Figure 26. It consists of three vertical layers of proportional wire chambers each wrapped around the back third of the chamber (downstream of the coils) and behind a zinc absorber layer. Each chamber is 1.27×1.27 m square. The wires are spaced by 5 mm, and there are three coordinate sets x, y, and u. The readout is by delay line, so that while a single track is found with the usual wire-spacing resolution, two tracks have to be spaced by 5 cm to be resolved. However, this is a good match to the system, because of the combination of large distances and multiple scattering, and the electronics are relatively inexpensive.

Because there is only a single plane of counters, they do not by themselves determine a direction. The scheme is to record all resolved tracks in the MWPCs and then later to pick only the unambiguous cases where a single noninteracting particle is observed in the chamber and then predicted to go through a particular set of wires in the EMI. Those that satisfy the criteria are then classified as having a very high probability of being muons. Since the chamber is very large, the probability

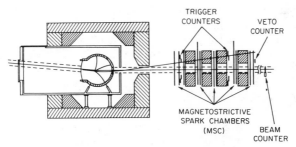

Figure 25   The HS used with the SLAC 1-m bubble chamber for triggering on scattered muons. A typical four-prong event is sketched in.

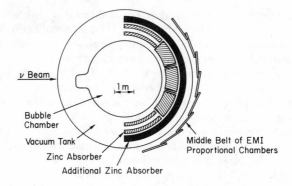

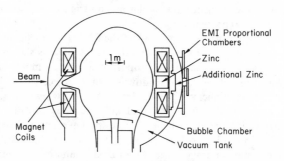

Figure 26   The External Muon Identifier (EMI) installed at the FNAL 4.6-m bubble chamber.

is large that the hadrons produced in the neutrino interactions will themselves interact, so this method of separation of muons has been quite successful.

Data-Gathering Systems

All HS have associated data-gathering systems. In the untriggered mode, their main function is data logging and monitoring. In this sense they are similar to the systems used in counter experiments. The information from wire chambers or scintillator hodoscopes is put into the computer on a signal from some coincidence circuit, and at some later time calculations are made and the results displayed. Their most important function is to generate position information accurately enough to do momentum measurements and to match the spark-chamber tracks with those measured in the bubble chamber. In addition, they perform the standard monitoring functions. It is of course very important to have a sample of the data from the downstream electronics analyzed a short time after pictures are taken so as to correlate tracks in test pictures of the bubble chamber with the spark-chamber information while the experiment is in progress.

A more complicated use of the data-logging system is used in the triggered mode. An algorithm is needed to decide whether or not to trigger the lights. Since the time for decision is small (several msec), this is usually written in machine language. As an example of such an algorithm we consider the one written for the experiment using the SLAC HS in order to study the reactions $\pi^+ p \rightarrow K^+ Y^*$ and $K^- p \rightarrow \pi^- Y^*$, with the subsequent decay $Y^* \rightarrow \Lambda + n\pi$, at an energy of 11.6 GeV/c. The trigger for the first reaction was a K^+ meson of momentum greater than 4.5 GeV/c. Referring to Figure 19, the algorithm had to

1. Find all possible tracks that went through all three MWPC stations.
2. Find all incoming beam tracks.
3. Calculate the intersection point between each MWPC track and the beam track to find which outgoing tracks came from the fiducial volume of the BC.
4. Calculate the momentum of all good candidates in step 3.
5. See which of the tracks of step 4 project back to the proper hodoscope counter.
6. Check to see whether the pulse height in the Cerenkov counter (24) corresponds to that for a K meson of the correct momentum. The Cerenkov counter has a 3-m path length, so that K mesons of momentum between 8 and 11 GeV/c can be distinguished from pions and protons by pulse height.

When conditions 1–6 are satisfied, the chamber lights are flashed. The time for the execution goes roughly as the third power of the number of tracks.

Because of the limited time available for computation, which is imposed by the thermodynamics of the bubble chamber (2–3 msec), it is instructive to examine in some detail an example of what current technology permits one to do during this time. The system in use with the apparatus of Figure 19 is shown in Figure 27.

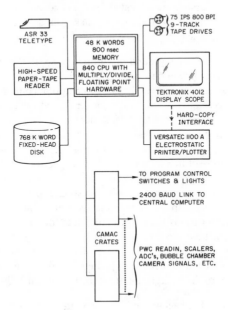

Figure 27   Block diagram of the computer hardware used in the SLAC HS.

HARDWARE   The central processor is a Data General Nova 840, which is a 16-bit, four-register machine with a memory cycle-time of 800 nsec and a 1.25-megaword/ sec direct memory-access channel. (Devices with controllers external to the Central Processor Unit (CPU) chassis are limited to a 500-kiloword/second transfer rate.) It is equipped with hardware floating point and multiply/divide instructions, and a memory mapping and protection unit using 1024-word pages. The total amount of memory is 48 kilowords.

The system has a typewriter master console and a high-speed paper-tape reader, used primarily for diagnostics. Two 9-track, 800 byte/in, 75 in/sec vacuum-column tape drives are used for data recording. A 786-kiloword fixed-head disk is used for nonresident system storage, program swap space, user files, and temporary data sets such as histogram displays. The main operating console is a graphics display storage scope, which is equipped with a hardware hard-copy interface to an electrostatic printer/plotter that also functions as the system printer. Two CAMAC crates provide a connection to the experimental apparatus. Proportional wire-chamber data, photomultiplier pulse heights, scalers, etc are read in through appropriate modules. The CAMAC crates also provide an interface to a set of 16 program-control-and-monitor switches and display lights (primarily for diagnostic purposes), and also to a 2400-baud link to SLAC's central computing facility.

SOFTWARE   The 840 system operates under the control of MRDOS (mapped real-time disk-operating system), a multitasking system supporting separate foreground and background partitions that communicate through a common array in memory. Twelve kilowords of memory are allocated for the resident system, 14 kilowords for the foreground program, 18 kilowords for the background program, and 4 kilowords for the communications array.

Foreground program   The foreground program handles all the "time-critical" tasks connected with data reading and recording, and generating a software trigger for the bubble-chamber camera. It is initiated by an interruption coming slightly before the beam spill. Upon receipt of a hardware trigger (based on counter information), Proportional Wire Chamber (PWC) data, scalers, pulse heights, etc are read in via CAMAC in 100 μsec or so. An experiment-dependent algorithm is then executed in roughly one msec to decide whether to take a picture of the triggering event. Provision is made to read in a second hardware trigger in the same 1.6 μsec beam spill to reduce software dead-time inefficiencies.

If the trigger algorithm generates a positive response to either of the two possible hardware triggers per beam spill, an appropriate signal is sent to the bubble-chamber camera, together with various human-readable data for recording on the film. The foreground program then reads in other data not needed for the picture-taking decision, and writes out a fixed-length record on magnetic tape. Various histogram tallies in the communications array are updated, and then the foreground program relinquishes control to the background until the next beam spill.

Background program   The background program can run whenever the foreground is idle, which is most of the time. When the background is entered, a "menu" is displayed listing the various routines which may be run. Among the functions available are the following:

1. Change parameters and logical switches in the foreground.
2. Display a picture of the current event on the display scope.
3. Dump the communications array on the scope or printer in octal or decimal format, and update the dump periodically, if it is displayed on the scope.
4. Start, stop, or clear one or more foreground histogram tallies.
5. Display one or more histograms on the scope or printer.
6. Add, delete, or list a histogram definition from the foreground.

Trigger algorithm   The object is to derive the interaction vertex and momentum of a (fast) outgoing track. Angles involved are required to be small, and the magnetic field is assumed to be uniform in Y and Z, i.e. $B = B(X)$. The trajectory of a track with momentum p and initial angle θ passing through the system from (X_A, Y_A) to (X_B, Y_B) can be parameterized:

$$\Delta Y_{BA} = Y_B - Y_A = \theta(X_B - X_A) + I_{AB}/p,$$

with   19.

$$I_{AB} = 3 \times 10^{-4} \int_{X_A}^{X_B} dX \int_{X_A}^{X_B} B(X') dX'.$$

We now utilize the fact that a typical noninteracting beam track calibrates the system, i.e. provides a reference momentum and relates the lateral offsets of one PWC plane with respect to another.

Assuming an interaction point at X_V, we can then derive several equations by considering the upstream and downstream parts of the system with respect to X_V. For the beam,

$$\Delta Y_{VA} = \theta_0 \cdot \Delta X_{VA} + I_{AV}/p,$$ 20.

and for the outgoing secondary track,

$$\Delta Y_{BV} = \theta_1 \cdot \Delta X_{BV} + I_{VB}/p'.$$ 21.

Defining $\delta Y = \Delta Y - \Delta Y^{\text{typical beam}} = \Delta Y - \Delta Y^{\text{b}}$, we find

$$\delta Y_{BV} = (\theta_1 - \theta_0^{\text{b}}) \cdot \Delta X_{BV} + (1/p' - 1/p) I_{VB}$$ 22.

and

$$\delta Y_{VA} = (\theta_0 - \theta_0^{\text{b}}) \cdot \Delta X_{VA}.$$ 23.

(Similar equations exist in Z without the momentum-dependent terms.)

Considering another point (X_C, Y_C) downstream of the interaction vertex (e.g. A, B, C could be PWCs 2, γ, and α), we also have

$$\delta Y_{BC} = (\theta_1 - \theta_0^{\text{b}}) \cdot \Delta X_{BC} + (1/p' - 1/p) I_{CB},$$ 24.

and we are now equipped to derive the vertex and momentum of the outgoing track.

In two experiments at SLAC, matters were simplified by choosing the beam to be parallel. Thus $\theta_0 = \theta_0^b$ always, and we arrive at the following expressions for X_V and p' that are convenient for computation:

$$\Delta X_{BV} = \frac{\Delta X_{BC} \cdot \delta Z_{BV}}{\delta Z_{BC}} = \frac{\Delta X_{BC} \cdot \delta Z_{BA}}{\delta Z_{BC}} \qquad 25.$$

and

$$(1/p' - 1/p) = \frac{(\delta Y_{BA} \cdot \Delta X_{BC} - \delta Y_{BC} \cdot \Delta X_{BV})}{F(X_V)}, \qquad 26.$$

where

$$F(X_V) = I_{VC} \cdot \Delta X_{BV} - I_{VB} \cdot \Delta X_{CV}. \qquad 27.$$

The main limitation on the programming of the algorithm is the requirement that the time spent in making a decision to take a picture should not exceed 3 msec. The program is therefore coded in NOVA assembler language using 16-bit integer arithmetic, with the intermediate results of the calculations being restricted in magnitude to avoid overflows. Lengthy instructions, such as multiply and divide, are avoided wherever possible. In order to complete the calculations in the allowed time, limits are placed on multiplicities, etc in the PWCs, and certain numerical approximations are made. For example, the quantity $\Delta X_{BC}(= 125 \text{ cm})$ in Equation 25 is replaced by the value 64, which serves the dual purpose of limiting the size of the result and avoiding a multiplication.

The following steps are executed in the algorithm used in the SLAC experiments:

1. Check that PWC 2 saw at least one, but no more than two, beam particles. With no information to link Y and Z in the beam planes, more than two hits in each plane leads to spurious triggers, as well as consuming too much time.

2. Multiplicity cuts are made on the number of hits in the downstream PWCs α, β, and γ.

3. Matched points are found in PWCs α, β, and γ and saved for later use. More than five matches in two PWCs causes the trigger to be rejected immediately.

4. It is now possible to decide which pair of downstream PWCs are to be used for the first try at the calculation. Clearly, it is desirable to select the two planes with the

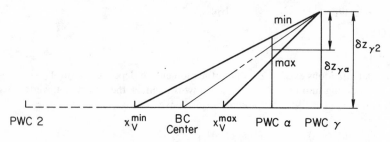

Figure 28   Sketch of the vertex calculation algorithm used with the SLAC HS.

largest separation to give greatest possible accuracy. The method using two PWCs was eventually preferred to making a fit using information from all three downstream planes simply because of the higher efficiency thereby obtained.

5. The beam tracks found in PWC 2 are projected into the downstream PWCs being used, and the predictions are compared with the matched points already found, to remove any noninteracting beam tracks for example, due to the spacing of the planes, predictions of Z_α, Z_γ can be made according to

$$Z_\gamma = Z_2 + \Delta Z^b_{\gamma 2} + \delta Z_{21}$$

$$Z_\alpha = Z_2 + \Delta Z^b_{\gamma 2} + \tfrac{3}{4}\delta Z_{21}$$

for a nonparallel beam. In the case of a parallel beam, the predictions are simplified even further by the absence of the δZ_{21} terms. Constants like $\Delta Z^b_{\gamma 2}$ are obtained by taking data triggering on incident beam tracks only.

6. Vertex calculation can now be attempted, as shown in Figure 28. The horizontal line defines the path of a noninteracting beam track, i.e. $\delta Z_{y2} = 0$. X_V^{max}, X_V^{min} are the limits of the fiducial volume.

To avoid wasting time on unnecessary divisions (Equation 25), we do the following: (a) Using δZ_{y2} and a table of values, we check that $\delta Z_{y\alpha}$ is within the fiducial volume, taking into account the possibility of multiple scattering; (b) if this is so and the angular deviation of the secondary from the beam track is sufficiently large, then the vertex is calculated. If the dip is too small, the vertex is imposed.

7. Once a vertex has been assigned, the momentum calculation is possible. Certain obvious checks (such as whether $\delta Y_{y\alpha}$ is small enough for the track to be within the momentum range of interest) are made, and then the substitution in Equation 26 is made. The value of the field function $F(X_V)$ (Equation 27) is obtained from a table constructed separately from a map of the magnetic field.

8. Finally, using the calculated momentum, the track is projected into the PWC so far unused, and predictions are made for Y, U, and Z. Confirmation is required from at least one of these before the trigger is accepted.

The steps described in Equations 23–26 are contained within a loop such that if at any point the calculation is unsuccessful, it is possible to continue with, for example, the next Z_α. It is to be noted, however, that considerable time can be gained by ordering the computation correctly. The wire numbers are arranged in ascending order, which can often prove useful in eliminating impossible combinations.

Should the two PWC calculations be unsuccessful, it is often possible to try another pair of PWCs, in which case the steps in Equations 23–26 are repeated.

5 PERFORMANCE OF HYBRID SYSTEMS

In this section we examine how well various operating HS have performed. For all systems, this includes the accuracy of momentum and missing-mass determination, as well as an evaluation of the losses due to limitations on acceptance imposed by the downstream system. In addition, on-line systems must be evaluated in terms of the efficiency and validity of their trigger systems.

Since track-matching and momentum measurements are intimately connected, we treat them together and begin with a general description of off-line track-matching procedure.

Off-Line Track Matching

Usually the track measured on the bubble-chamber film, together with its measurement errors, is projected into the space occupied by the wire chambers (either up- or downstream). Then the probability is calculated that the projected BC track and the observed sparks in the wire chambers do indeed match within their errors. This probability distribution is plotted in order to decide the conditions under which an event is acceptable. The position of the vertex of the event in the hydrogen is well known in all three coordinates, and this fact plays a major role in the procedure for rejection of spurious tracks formed in the spark chambers.

Usually, the space between the BC and WSC systems is filled only with the fringe magnetic field of the bubble-chamber magnet. Since this is not uniform, it must be measured to $\pm\frac{1}{2}\%$ at each point, so as not to contribute a significant error in the measured momentum.

In the following treatment (25), we evaluate the probability of a match at the first wire-chamber plane downstream (or upstream) of the bubble chamber. Other treatments (26, 27) use the vertex position in the chamber. We assume that all variables are transformed to the bubble-chamber space, and we define a five-vector called

$$ V = \begin{pmatrix} z \\ \theta \\ y \\ \phi \\ p \end{pmatrix}, $$

where z and y are the coordinates measured perpendicular to an x axis lying in the general direction of the beam, θ is the dip angle, ϕ is the azimuth angle, and p is the momentum of the particle to be matched. All variables are evaluated at an x somewhere along the bubble-chamber track, usually at the vertex or at the end of the track. This track, projected to the first spark-chamber plane, located at x_b, yields a four-vector

$$ V_b = \begin{pmatrix} z^b \\ \theta^b \\ y^b \\ \phi^b \end{pmatrix} $$

The error matrix associated with V_b is obtained by propagating the error at the vertex using a matrix $\mathbf{R} = (\mathrm{d}V_b/\mathrm{d}V)$.

Since there is a magnetic field between x and x_b, $\mathbf{R}$ is a 4×5 matrix that can be taken to be the same as that used in first-order beam transport theory, namely, the product of R(bend) $\times$ R(drift). To a good approximation, this beam transport consists of a bending magnet that turns the track through $\alpha = \phi^b - \phi$ and has an effective length L followed by a drift of length S. Then

$$
R(\text{drift}) = \begin{pmatrix} 1 & s & 0 & 0 & 0 \\ 0 & 1 & 0 & 0 & 0 \\ 0 & 0 & 1 & 0 & 0 \\ 0 & 0 & 0 & 1 & 0 \end{pmatrix};
$$

$$
R(\text{bend}) = \begin{pmatrix} 0 & L & 0 & 0 & 0 \\ 0 & 1 & 0 & 0 & 0 \\ 0 & 0 & 1 & L\cos\alpha/2 & \dfrac{-L\sin\alpha/2}{p} \\ 0 & 0 & 0 & 1 & \dfrac{-L\tan\alpha/2}{p} \end{pmatrix}
$$

Here σ_b, the 4×4 error matrix on V_b, is $\sigma_b = R^\sigma R^T$. Again, σ is evaluated from the bubble-chamber fitting program.

The above does not yet include multiple scattering in the region between the end of the measured track and the coordinate x_b. This includes Coulomb scattering in the hydrogen downstream of the measured track, the beam windows in the bubble chamber and its vacuum tank and in any counters or other material between the bubble chamber and the first spark-chamber plane.

To treat Coulomb scattering (28), each scatterer i is characterized by two independent variables in each of two perpendicular planes, namely α_i and β_i, the virtual displacement and scattering angle at its center. In the mean, we have

$$\langle \alpha_i^2 \rangle = \theta_s^2 L_i^3/24,$$

$$\langle \beta_i^2 \rangle = \theta_s^2 L_i/2, \qquad \qquad 27.$$

$$\langle \alpha_i \beta_i \rangle = 0.$$

When L_i is the length of the scatterer i and

$$\frac{\theta_s^2}{2} = \left(\frac{15\,\text{MeV}}{\rho\beta} \right)^2 \frac{1}{X_0}, \qquad \qquad 28.$$

then

$$
\sigma_i = \begin{pmatrix} \langle \alpha^2 \rangle & 0 & 0 & 0 \\ 0 & \langle \beta^2 \rangle & 0 & 0 \\ 0 & 0 & \langle \alpha^2 \rangle & 0 \\ 0 & 0 & 0 & \langle \beta^2 \rangle \end{pmatrix}
$$

Call l_i the drift distance between the center of the scatterer i and the plane of the first SC, and define the drift matrices as

$$
\mathbf{D}(l_i) = \begin{pmatrix} 1 & l_i & 0 & 0 \\ 0 & 1 & 0 & 0 \\ 0 & 0 & 1 & l_i \\ 0 & 0 & 0 & 1 \end{pmatrix}
$$

Then the total Coulomb scattering error at the first downstream SC plane is the sum

$$
\sigma_c = \sum_{i=1}^{N} \mathbf{D}(l_i)\,\sigma_i \mathbf{D}^{\mathrm{T}}(l_i), \qquad\qquad 29.
$$

where N is the total number of scatterers.

The spark-chamber four-vector is

$$
V_s = \begin{pmatrix} z^s \\ \theta^s \\ y^s \\ \phi^s \end{pmatrix}
$$

and its error matrix, σ_s, usually does not include any Coulomb scattering error.

The goodness of match between bubble chamber and wire chamber is now given by

$$
\chi^2 = (V_b - V_s)^{\mathrm{T}} (\sigma_b + \sigma_c + \sigma_s)^{-1}(V_b - V_s). \qquad\qquad 30.
$$

The results of using this procedure in an actual experiment, namely the study of 14-GeV/$c\,\pi^+ p$ interactions in the SLAC 1-m hybrid system (see Figure 21), are shown in Figures 29 through 33, which describe in detail how well the components of the five-vectors match up, and in Figure 34, which shows the χ^2 distribution for Equation 30. These show that the bubble-chamber and wire-chamber measurements agree for outgoing tracks to ± 1.5 mm and for angles to ± 1.5 mrad in ϕ and ± 2.5 mrad in θ.

For off-line purposes, the upstream and downstream wire-chamber values were used. According to Chadwick (26), a final, fitted error of ± 0.3 mrad in ϕ and ± 0.5 mrad in θ was achieved.

Missing-Mass and Momentum Accuracy

OFF-LINE   At FNAL energies, the error contribution of Coulomb scattering is relatively smaller, and more accurate results can be obtained. Bugg (29) reports results from matching beam tracks in the 78-cm FNAL chamber with the same tracks as determined by an upstream PWC spectrometer. A set of curves similar to those displayed for the 14 GeV/c case is shown in Figures 35 and 36.

Since, in these measurements the incident angle is known very well from the wire-chamber measurements, the σ's in Figures 35a and 35b are the angle errors introduced by the bubble-chamber measurements, and these must be used when calculating the momentum of secondary tracks produced in an interaction. Bugg (29) has provided

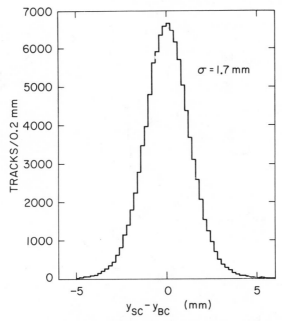

Figure 29   Track matching in the SLAC HS in the direction perpendicular to the magnetic field.

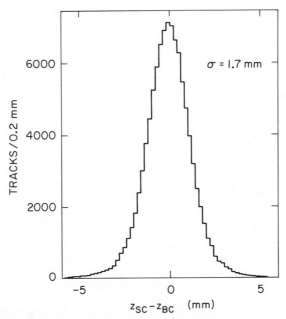

Figure 30   Track matching in the SLAC HS in the direction parallel to the magnetic field.

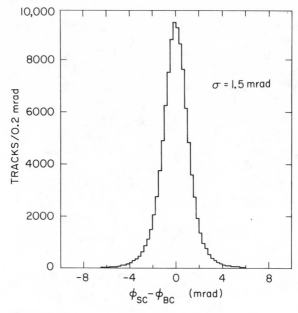

Figure 31   Track matching in the SLAC HS for θ, the angle of bend in the magnetic field.

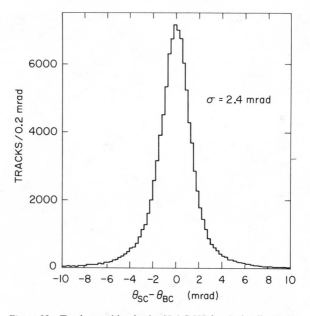

Figure 32   Track-matching in the SLAC HS for ϕ, the dip angle.

the information in Table 4, which summarizes the accuracies in space and momenta that have been achieved with the FNAL HS.

ON-LINE   The on-line momentum accuracy is important if (a) the trigger depends on the mass of the outgoing particles as determined by pulse height from a downstream Cerenkov counter, (b) the trigger depends on the mass recoiling against the outgoing

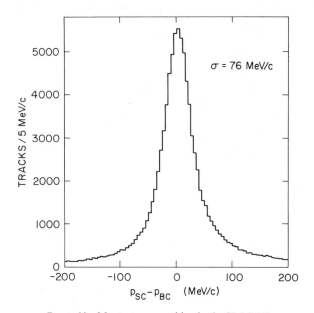

Figure 33   Momentum matching in the SLAC HS.

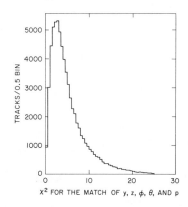

Figure 34 χ^2 distribution for the five variables (three are independent) in the track matching for the SLAC HS.

particle, (c) the trigger depends on the transverse momentum $p_\perp$ of the outgoing particle or on its total four-momentum transfer squared, t. The on-line accuracy depends, among other things, on whether there is a full-fledged spectrometer downstream of the bubble chamber, as in Figure 21, or whether only the fringe field of the bubble-chamber magnet is used as the analyzer, as in Figures 19 and 23.

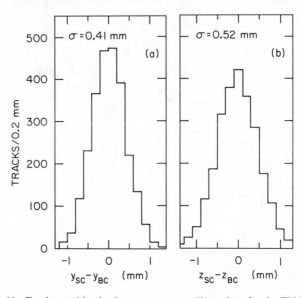

Figure 35   Track matching in the two transverse dimensions for the FNAL HS.

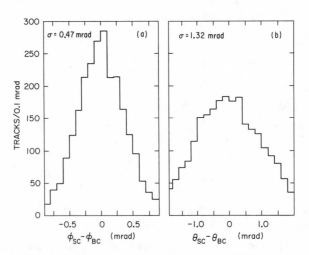

Figure 36   Track matching in θ and ϕ for the FNAL HS.

The on-line momentum measurement of the triggering track (i.e. the calculation made before bubbles have reached full size) has been used to determine the missing mass recoiling against the detected particle in one case and to identify the particle mass in conjunction with the Cerenkov counter in another. For the missing-mass calculation, we use

$$MM^2 = M_p^2 + 2M_p(E_{\text{beam}} - E_{\text{trig}}) + t, \qquad\qquad 31.$$

where t is the four-momentum transfer from beam to trigger particle, and M_p is the proton mass. The beam energy is well defined, and t is restricted to rather small values by the inherent aperture limitation, so that the error on MM^2 is dominated by the error on E_{trig}. Hence we discuss only the momentum error.

In all cases, the momentum error scales as the square of the trigger-track momentum, or

$$\frac{\Delta p}{p} = Ap.$$

Table 4   Resolution parameters of the FNAL PWC hybrid system

I   Angular accuracy

	Upstream system	Downstream PWC system	Bubble chamber (track length = 70 cm)
	$\sigma\phi = 0.002$ mrads	$\sigma\phi = 0.12$ mrads	$\sigma\phi = 0.45$ mrads
	$\sigma\lambda = 0.002$ mrads	$\sigma\lambda = 0.14$ mrads	$\sigma\lambda = 1.32$ mrads

II   Position accuracy at center of bubble chamber

PWC	Bubble chamber
$\sigma_y = 0.4$ mm	$\sigma_y = 0.12$ mm
$\sigma_z = 0.4$ mm	$\sigma_z = 0.34$ mm

III   Momentum
    A   Average beam momentum

        $p_{\text{beam}} = 146.75$ GeV/c Error = 0.4 GeV/c[a]

    B   Single-track accuracy on beam tracks (using upstream PWC only)

        $\Delta p = 2.1$ GeV/c

        $\dfrac{\Delta p}{p} = 0.011\ p/(\text{GeV}/c)$

    C   Secondary tracks in downstream PWC system

        $\dfrac{\Delta p}{p} = 0.053\ p/(\text{GeV}/c)$ (vertex close to entrance window)

        $\dfrac{\Delta p}{p} = 0.074\ p/(\text{GeV}/c)$ (vertex at center of chamber)

---

[a] All errors are standard deviations.

The most straightforward momentum determination uses only the downstream spectrometer measurements. In a diffractive πp-scattering experiment (see Figure 21), a 29 kG-m dipole magnet was used, surrounded by four wire-chamber stations. The two stations between the bubble chamber and the dipole were separated by 2.77 m, while behind the dipole the separation was 5.91 m. The bending angle could then be determined to 0.56 mrads (using 1 mm uncertainty at each wire plane), which resulted in a momentum error coefficient $A = 0.065\%$ per GeV/c. This implied an error in the missing mass (using Equation 31) of ~ 100 MeV at single-π threshold. In contrast, the missing-mass error for $MM = 3$ GeV was only 10 MeV.

In the later system, a large Cerenkov counter replaced the magnetic spectrometer, so that only the bending power of the bubble-chamber fringe field was available to use with this method (see Figure 19). Its accuracy was, however, quite adequate for particle identification, since approximately 15.3 kG-m of bending power was available between the PWC planes α, β, and γ. Each of these had three sets of wires z, y, and a diagonal set, u. The coordinate x is measured along the beam direction.

The momentum measurement comes from the displacement of the spark at y_β from a straight line joining the sparks at y_α and y_γ:

$$s = y_\beta - y_\alpha - 0.3464(y_\gamma - y_\alpha), \qquad\qquad 32.$$

where the numerical factor is the ratio of the α-β and α-γ separations. For a 10-GeV/c track, $s \approx 5$ mm, or only 2.5 wire spacings. However, a study of tracks found in the downstream system shows that an effective "setting error" of a single plane is ± 0.7 mm (we note that if two adjacent wires fired, we used the average location). From Equation 32 we find

$$\Delta s = 0.87 \text{ mm},$$

$$\frac{\Delta p}{p} = \frac{\Delta s}{s} = \frac{0.87 \text{ mm}}{5 \text{ mm}} \times \frac{p}{10(\text{GeV}/c)}, \qquad\qquad 33.$$

$$A = 1.7\%/\text{GeV}/c.$$

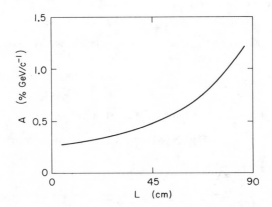

Figure 37   The coefficient A (see text) vs length of track in the bubble chamber.

The resolution coefficient A is 26 times worse than when the separate magnetic spectrometer was used. This makes the missing-mass measurement at single-particle threshold unusable: however, at $MM = 3$ GeV, the resolution would be ± 260 MeV.

In the experiment to trigger on the reaction $\pi^+ p \to K^+ X^+$ at 11.6 GeV/c, we used the momentum measurement only to exclude triggering particles below 4.5 GeV/c. At that momentum, we find $\Delta p/p \simeq 7.5\%$, $\Delta p = 345$ MeV/c, which was quite adequate for that purpose.

The above method of momentum measurement was not used in this particular experiment because the 1.5-μsec beam-pulse width of SLAC requires high detection efficiency in the planes used. The alternate procedure was to search for y, z, and u spark triplets in any 2 pairs of planes, project the z orbit (i.e. in the "unbent" plane parallel to the magnetic field) back into the bubble chamber from both downstream and upstream PWCs. The intersection of these orbits then locates the event vertex, x_v. The y plane orbits were then similarly projected, and the discrepancy of these projections at x_v measures the outgoing particle momentum.

The result of using this algorithm is a considerable improvement in on-line momentum accuracy, as well as improvement of the detection efficiency. The accuracy is downgraded by multiple scattering in the hydrogen and bubble-chamber entry and exit windows, but much more bending power is used. The error calculation is complicated (30) and resolution depends upon the location x_v of the event vertex within the bubble-chamber volume, and on the azimuth of the outgoing track with respect to the field direction. We found

$$A = A(x_v)/\cos \theta.$$

This dependence occurs because if the dip angle is small, the vertex x_v cannot be well located. However, if the production angle is small, the calculation becomes insensitive to x_v. Therefore, in practice, when small dip was observed, the value of x_v was taken to be at the center of the chamber.

The calculated values of $A(x_v)$, including the effect of multiple scattering, are shown in Figure 37 as a function of x_v corrected to BC track length L. There we see that for events at the beginning of the fiducial volume, we can achieve 2.5% momentum resolution for a 10-GeV/c track ($\Delta MM \approx 250$ MeV at single-particle threshold), but for events far downstream the precision becomes comparable to that for the first method.

Missing-Mass Errors

ON-LINE For some experiments it is convenient to include a missing-mass cutoff as part of the trigger. The accuracy of this trigger depends on the on-line determination of the production angle and momentum of the fast-triggering particle against which the missing mass is recoiling.

For the system shown in Figure 21, the error on the missing mass is ~ 150 MeV, small enough to eliminate most elastic events. If a downstream spectrometer is not used, but rather the fringe field of the bubble-chamber magnet, as in Figures 19 and 25, the error in MM becomes larger and in fact has not been used as part of an on-line trigger scheme in any actual experiment up to the date of this writing.

OFF-LINE   Here, missing-mass resolution is, of course, much improved, again for the main reason that the vertex is known. There are two cases, one in which the missing mass is calculated for a fitted event, e.g. one in which only one neutral particle is present, and the central values of the measured momenta and angles are displaced in order to make the best fit; and two, where there are no constraints and include those with more than one missing neutral. The two calculations can be made for the same sample of events in order to see how the fitted events improve the resolution. For the experiment of Figure 21, we compare the two cases $\pi^+ p \to \pi^+ N^{*+}$ (fitted events) and $\pi^+ p \to \pi^+ + MM$ (in this case any $n\pi^+$ or $p\pi^0$ combination is treated as unfitted). Figure 38 shows the error in MM against the missing mass for the fitted cases, and Figure 39 shows the same for the unfitted cases. One sees that in the fitted cases, a mass resolution of < 20 MeV is obtained for $n\pi^+$ masses < 2 GeV, while the unfitted cases range from 90 MeV at low masses to 3 MeV at 4 GeV. Thus a separation of the single neutral from multiple neutral events can be made with only a 5% residual contamination.

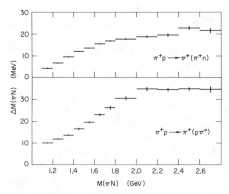

Figure 38   The error in missing mass vs missing mass for three-body events that had a three-constraint fit.

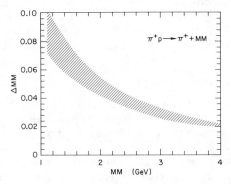

Figure 39   The error in missing mass vs missing mass for three-body events for which no fit was attempted.

Acceptances

GEOMETRICAL   This is determined in most cases (very-high-energy μ mesons may be excepted) by the fact that the walls of the bubble chamber have to withstand pressures up to 10 atmospheres. This results in walls that are several interaction lengths and many radiation lengths thick. In addition, the bubble-chamber magnet coils and iron also contribute a limitation on the geometrical acceptance for triggering particles (see Figure 1).

In order to maximize this acceptance, a thin exit window is installed in the downstream end of the chamber. The relative size and thickness of these windows is discussed in Section 3. In addition, the cryogenic chambers are surrounded by a vacuum chamber in which a matching exit window also has to be installed.

TOTAL ACCEPTANCE   This is determined by the product of the geometrical acceptance, the effect of the magnetic field, and the effect of the production and decay kinematics. Complicated cases have to be evaluated by the use of a Monte Carlo method. Here we discuss some simple cases in order to illustrate the limitations and possibilities.

First, we consider the case of a fast-forward single particle in a reaction $a + b \rightarrow c + x$ and define t as the four-momentum transfer between a and c, and M_x as the mass of particle x. The masses of a, b, and c are assumed to be known. One can now calculate for a given chamber the acceptance as a function of t and M_x for this forward particle c in a downstream detector whose position and dimensions are known. This has been done for a setup such as that shown in Figure 21. As can be seen from Figure 40, and as was mentioned earlier in this section, the downstream

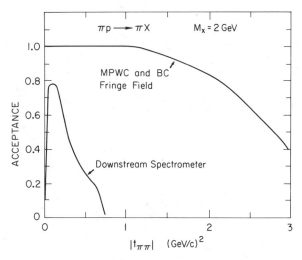

Figure 40   The t-acceptance of the SLAC HS for a fixed mass. Note how much this is reduced by the use of a downstream spectrometer.

spectrometer, while providing high momentum and mass resolution, does that with a sacrifice of acceptance. The same bubble chamber, but with the arrangement shown in Figure 19, has the t and M_x acceptance also shown in Figure 40.

In the case where particle c decays immediately into two or more particles, the acceptance depends on the decay angular distribution. Again, each individual case must be studied, but as an illustration we have calculated the acceptance of the system shown in Figure 19 for the following cases:

$$\pi^+ p \to A_2 p$$
$$\quad\quad\raisebox{0.2em}{$\llcorner$}\, K^+ K^0 \hspace{5cm} 35.$$
$$\pi^- p \to f^0 n$$
$$\quad\quad\raisebox{0.2em}{$\llcorner$}\, K^+ K^-,$$

with the results shown in Figures 41 and 42.

In general, the limitations of the exit-window size made the hybrid system a poor choice for the study of forward-going heavy mesons that decay into three or more particles, but an excellent choice for the study of recoil baryon resonances and backward-produced mesons. In cases where the forward meson decays into two particles, a choice has to be made on a case-by-case basis.

Performance of Algorithms for Triggering the Lights

The algorithms for doing the experiment $\pi^+ p \to K^+ X$, described in Section 2 with the apparatus of Figure 19, have been in operation in a routine manner. The times to do the calculations described in Section 4 are shown in Figure 43. As a result, the flash was set to $2\frac{1}{2}$ msec after the interaction occurred.

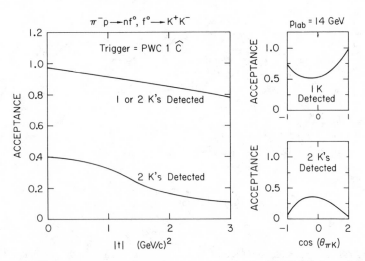

Figure 41 The t-acceptance and angular acceptance in the SLAC HS for a forward-produced f meson decaying into $K^+ K^-$.

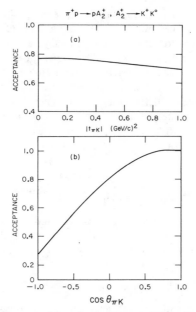

Figure 42 The *t*-acceptance and angular acceptance in the SLAC HS for a forward-produced A_2 meson decaying into $K^0 K^+$. It is quite evident that three-body decays would have a serious bias in angular acceptance.

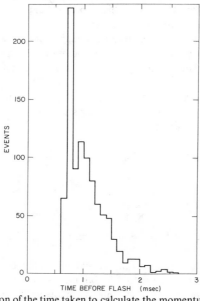

Figure 43 Distribution of the time taken to calculate the momentum of the scattered track and the vertex position of the event in the SLAC HS. The cutoff is made at 2.5 msec.

Approximately 40% of all pictures taken had an event in the fiducial volume. The overall efficiency was checked by taking data with low flux (2–3 particles per pulse) and using the fast counters to trigger the lights, namely triggering on any interactions in the chamber, recording, however, all other spark-chamber and Cerenkov information. These events were then processed off-line as though they were algorithm-triggered events. From those studies, the overall efficiency was found to be 80% for low flux. At high flux (8 particles/pulse), the efficiency drops to 65%. With these data it is possible to calculate how many expansions and pictures it would take to do a 500-eV/μb experiment with a flux of 8 particles/pulse and a $\frac{3}{4}$-m effective length of hydrogen when the triggering cross section is 500 μb.

This would be:

$$\left(\frac{127 \times 10^6 \times 1.33}{8}\right)\left(\frac{500}{27,000}\right)\left(\frac{1}{0.6}\right)\left(\frac{1}{0.65}\right) = 10^6 \text{ pictures.}$$

To do the same experiment in an untriggered mode would require 30 million pictures (because of a shorter usable length in hydrogen) in the same chamber, or perhaps 15 million in a 2-m chamber. This ratio of 15 to 1 changes the experiment from an "impossible" to a "possible" state.

Physics Results from Selected Experiments

The most extensive work with an untriggered system was done in the 78-cm chamber at FNAL with the two arrangements previously described and shown in Figures 20 and 24. Much of this work has been reviewed by Whitmore (31). The measurement of the individual particle momenta in the downstream system forms the basis for a study of the inclusive spectra at very high energies and allows comparison with lower energies. The results are shown in Figure 44.

The first thing to notice is the remarkable similarity of the spectra as one changes energy. Apart from diffractive effects, such as the bump at large X for $\pi^{\mp} p \to \pi^{\mp} X$, what matters most is the c.m. energy of the system, which seems to be responsible for the small differences.

A second interesting result comes from the four-constraint channels that work mainly when the forward particle is the same as the incident particle and the slow proton is identified by ionization:

$$\pi^- p \to \pi^- p \pi^+ \pi^-, \qquad\qquad\qquad\qquad 36a.$$

$$\pi^- p \to \pi^- p 2\pi^+ 2\pi^-, \qquad\qquad\qquad\qquad 36b.$$

$$pp \to pp\pi^+\pi^-, \qquad\qquad\qquad\qquad 36c.$$

$$pp \to pp 2\pi^+ 2\pi^-. \qquad\qquad\qquad\qquad 36d.$$

Reactions 36c and 36d are from counter experiments at the Intersecting Storage Rings (ISR) in CERN.

A comparison of Reactions 36a with 36c, and 36b with 36d, is made in Figure 45 and shows that after a rise from threshold, both decrease very slowly with incident energy, showing that they are processes mainly involving diffractive dissociation of the target nucleon.

Clearly, future work with the system depends a great deal on particle identification. The External Muon Identifier (EMI) installed behind the FNAL 4.6-m chamber (see Figure 26) has been an important part of an experiment (32) done by a Wisconsin,

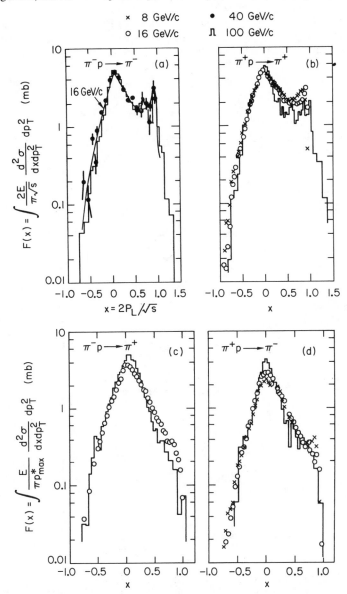

Figure 44 Inclusive distributions in the Feynman variable $x = p_l^*/p_{l(\max)}^*$ from data obtained with the FNAL HS compared with lower-energy data.

LBL, FNAL, and Hawaii group, in which either the decay of a new particle or the semileptonic decay of a charmed particle was observed for the first time in neutrino interactions. It is also being used in most neutrino experiments in hydrogen, the results of which are now being analyzed.

A series of reports and papers (33) on nucleon-diffraction dissociation, done by a SLAC, California Institute of Technology, and LBL group, has shown that, with the high statistics available from the HS (180 eV/μb), detailed interference effects may be studied. In particular, the interference between the nondiffractive Δ^{++} (1238) resonance and the diffractive component at the same mass has shown the need for nucleon exchange in diffractive dissociation.

The first look at the hadrons associated with inelastic muon scattering was also provided by an HS (34) that showed that even in the region where scaling begins to set in in e^-p inelastic scattering, the hadron component still looks very much the same as in photoproduction.

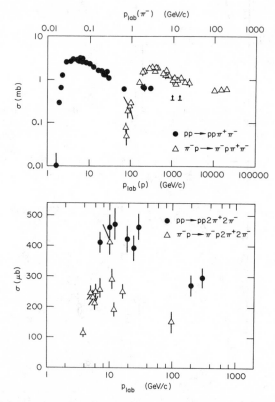

Figure 45 Cross sections for the reactions (a) $pp \rightarrow pp\pi'\pi$ and $\pi p \rightarrow p\pi'\pi$, and (b) $pp \rightarrow pp2\pi^+\pi^-$ and $\pi^-p \rightarrow \pi^-p2\pi^+2\pi^-$ as a function of incident beam momentum. For the ISR points, the mass of the $p\pi^+\pi^-$ system is chosen to be less than 2.5 GeV.

At the present time, the first very-high-statistics experiments are being run in the SLAC HS to look at the exchange-degenerate reactions (Reaction 3) described earlier. About 20% of a 500-eV/μb exposure has been completed, with the rest scheduled for completion in spring 1977.

Several experiments were performed with the SLAC 38-cm chamber. The first (18), using a trigger for a forward neutron, looked for exotic exchange in the reaction

$$\pi^+ p^+ \to n\pi^+ \pi^+ \qquad\qquad 38.$$

by analyzing the backward $\pi^+ \pi^+$ system for the presence of resonances. In a relatively short exposure on a new piece of apparatus, an upper limit of 2 μb was set by a group from Indiana, Purdue, and Vanderbilt.

Another group (20), using the same chamber, examined the reaction

$$K^0_L p \to K^{*0}(892)p, \qquad\qquad 39.$$

using the arrangement of Figure 23. The results give the differences in total and differential cross sections between $K^0 p \to K^{0*}(892)p$ and $\bar{K}^0 p \to \bar{K}^{0*}(892)p$ for momenta between 3 and 12 GeV/c.

A high-statistics experiment to study the reaction

$$\pi^- p \to nX^0$$

at 12 GeV/c incident momentum was carried out by a UCLA group using the arrangement of Figure 22 and required 6×10^7 chamber expansions. Some 240,000 pictures were taken, and the sensitivity of this experiment to backward ρ^0 production, after all corrections for efficiency, etc, was 187 ± 7 eV/μb. First comparison of the backward differential cross sections for ρ^0 and f mesons shows them to have the same slope.

6 FUTURE PLANS FOR HYBRID SYSTEMS

The future of hybrid systems depends on how well one can overcome the present limitations: low intrinsic data-gathering power, primitive particle identification at high momenta, and the biases associated with the downstream trigger system.

One way to improve the data rate is to make the bubble-chamber pulse faster. As pointed out in Section 3, small-volume chambers have cycled successfully at 36 per sec while running in a regular experiment. Other experiments (11) have shown that the thermodynamic properties of liquid hydrogen allow cycle rates up to 90 per sec. However, these chambers have been so small that more interactions take place in the surrounding wall than in the useful liquid volume, and they were thus used only for very specialized experiments. It is, however, practicable to think of a 2-m chamber operating at 30 per sec if one considers the following:

1. Keep the chamber as shallow as possible. A high pulse rate requires a short expansion pulse in order to provide enough time between expansions to recompress the bubbles that were formed. The short pulse generates higher harmonics, which can give rise to variations in sensitivity throughout the liquid volume. To minimize

this effect, it is necessary to keep the chamber depth as small as possible. In the SLAC 1-m chamber, the pulse width is 16 times the time taken by the pressure wave to go across the chamber and back again. A factor of at least 10 should be put into the design parameters.

2. Keep the bubble size as small as possible. Figure 46a shows a calculation (35) of the time it takes to grow and recompress a bubble of diameter 450 μm in the SLAC 1-m chamber, with operating conditions of 4.1 atmospheres of vapor pressure (26 K) and 7.15 atmospheres of overpressure. Measured times tend to be 20–30% longer (36). Changing the operating conditions by dropping the temperature by one K or raising the overpressure does not decrease the time significantly. However, as can be seen in Figure 46b, if one goes to a bubble diameter of 300 μm, the bubble-kill time is decreased enough to allow 30-per-sec operation.

3. Provide proper illumination. It is probably best to use a metal piston in order to get good heat conduction, and the most practical optics is to use some retrodirective material like Scotchlite to cover it with. Some effort should be put into getting a real point source (difficult) or a virtual one (using half-silvered mirrors). Such

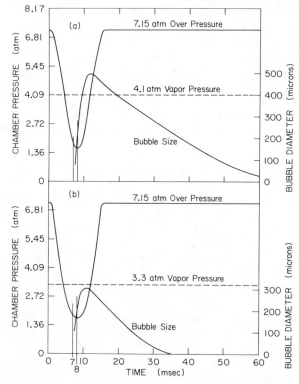

Figure 46   Bubble growth and decay time compared with the pressure trace for a maximum bubble size of (*a*) 450 μm and (*b*) 300 μm.

a source has been designed by Ferrie (37) at SLAC and is a great improvement over the usual ring sources used.

A most novel charged-particle identifier that has been proposed for use with an HS is ISIS (Identification of Secondaries by Ionization Sampling), developed by an Oxford group (38). This is a large enclosure filled with a mixture of 80% argon and 20% CO_2 at atmospheric pressure with 300 wires, each separated by 1 cm. A large electric field is established between the two side walls of the enclosure and the wires. A charged particle traveling along the 3-m length will create ions along its path that will be attracted to the wires and therefore the ionization of the particle will be sampled 300 times, allowing an accurate determination of the energy loss. This, together with a knowledge of the momentum of the particle, provides a determination of its mass. One interesting aspect of ISIS is that many particles can go through it at once and can be distinguished by the different drift times of the ions from their point of creation to the collecting wires. This is a device that is not particularly useful for counter spectrometers because of its relatively slow (10^{-4} sec) recovery time, but is eminently suitable for use in an HS. It can be made quite large so that it will not cut into the natural acceptance of the system.

Several new approaches are available to HS as electron or photon shower detectors with fairly good spatial resolution. One is the liquid argon ionization chamber (39). This consists of a cellular structure of lead walls, filled with liquid argon. The shower particles are detected by the pulse from the ionization in the liquid argon. The advantage is that a very large wall of this material can be placed downstream, 20 radiation lengths or so thick and with a 1 cm × 1 cm spatial resolution. Thus, unambiguous identification of π^0 mesons and single electrons is possible. Such a

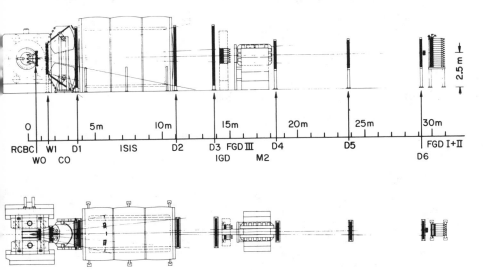

Figure 47   Side and top views of the proposed European HS. The Cerenkov counters, ISIS system, and γ detector will be built at a later time.

scheme provides neutral and multineutral triggers for the HS, opening up a little-known aspect of strong interactions to detailed exploration.

At very high energies, where most particles go forward, the full-fledged downstream spectrometer, using a separate magnet in addition to the fringe field of the bubble chamber, can be used without seriously cutting into the acceptance of the system. This is most important for missing-mass on-line triggers and for good off-line momentum analysis.

A system employing most of these new features (6) is being built for use at CERN, the European Center for Nuclear Research in Geneva, and will be installed in the experimental area surrounding the new 400-GeV accelerator there. The general outline of the system is shown in Figure 47. The bubble-chamber specifications were given in Section 3. The rest of the system consists of multiwire spark chambers

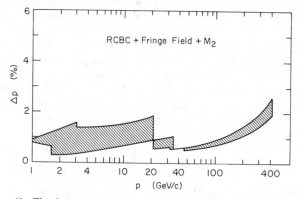

Figure 48 The design momentum resolution of the proposed European HS.

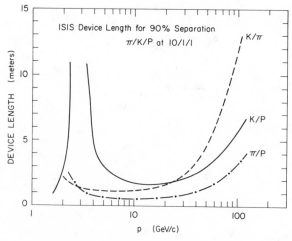

Figure 49 The design particle separation of the proposed European HS.

W_0 and W_1; drift chambers D_1, D_2, D_3; a dipole magnet M_2; and another set of drift chambers $D_4, D_5,$ and D_6. This part of the system, which consists of two momentum-measuring sections, one using the chamber field and the other a downstream spectro-meter, will be built first. A second stage is planned that will include two multicell Cerenkov counters and a three-cell ISIS system some 5.5 m long. A third stage will include two stages of γ detection labeled FGD and IGD, F for forward and I for intermediate. Note that this HS combine extends capabilities along directions mentioned previously, namely higher repetition rate, better particle identification, and superior momentum definition. Figure 48 shows the momentum resolution they hope to achieve. The particle identification as a function of momentum is shown in Figure 49.

A similar downstream system is being built for use in the FNAL HS (40). This will consist of an ISIS, a 5-m-long Cerenkov counter with eight mirrors, and a detector for forward γ rays and electrons.

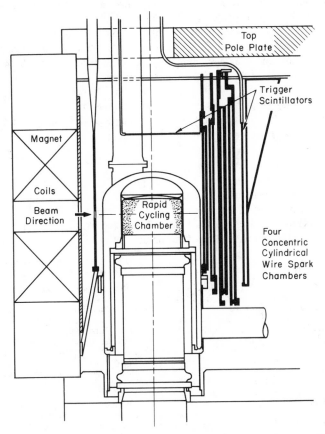

Figure 50   The chamber and trigger system of the Rutherford Laboratory RCVD.

Another HS is now undergoing initial tests at Rutherford. The bubble chamber was described in Section 3, and the HS is depicted schematically in Figure 50. The four cylindrical spark chambers have been designed to provide a high-multiplicity trigger, which is possible because the walls of this chamber are thin enough (3 mm of aluminum) to serve as beam-exit windows. The wire chambers will also be used to ensure that the event took place in the fiducial volume, a very important consideration with small chambers.

A further development, which may be a great help in providing a more unbiased trigger, is the use of solid-state light-sensitive modules that can be placed in a mosaic whose elements can be made as small as 13×13 μm and whose information content can be scanned electronically. With such a device, a low-level illumination may be applied to the bubble chamber after the bubbles have grown to almost full size (typically 1.5–2 msec with Scotchlite illumination). This light level can be made too low to sensitize photographic emulsion. There then remains another 1.5–2 msec of readout and calculation time before having to make a decision to flash full-intensity lights and take the photograph. The first application of such a system might be a multiplicity trigger that would be almost totally unbiased insofar as momentum and angle of outgoing tracks is concerned. This would be a marked improvement over previous systems. Encouraging preliminary trials of such devices have been made by Villa and Freytag of SLAC.

In summary, the HS represents a considerable advance in the use of bubble chambers in high-energy physics, especially with charged-particle beams. Its use has been extended to the very highest energies, using fixed-target accelerators, and it promises to be a very useful tool for many years to come.

ACKNOWLEDGMENTS

We wish to thank the following people for providing us with internal notes and information not easily found in the literature: I. Pless, V. Kistiakowsky, L. Montanet, C. Fisher, R. Newport, L. Stevenson, J. T. Carroll, G. Chadwick, R. C. Field, J. Brown, G. Hall, H. Ticho, D. Parker, R. J. Walker, W. Bugg, D. Freytag; and W. T. Kirk for his critical reading of the manuscript.

Literature Cited

1. Alvarez, L. W. 1969. *Science* 165:1071; Lasinski, T., Barbaro-Galtieri, A., Kelly, R. L., Rittenberg, A., Rosenfeld, A. H., Trippe, T. G., Barash-Schmidt, N., Bricman, C., Chaloupka, V., Söding, P., Roos, M. 1973. *Rev. Mod. Phys.* 45:S1
2. Watts, T. L., ed. 1970. *Proc. Int. Conf. PEPR, 2nd, Cambridge, Mass, 1970.* MIT Rep. 2098-60, p. 235
3. Davey, P. G., Hawes, B. M. 1974. *Oxford Conf. Comput. Scanning, Nucl. Ph . Lab., Univ. Oxford, 1974*, pp. 15, 17
4. Watt, R. D. 1973. In *Proc. Int. Conf. Instrum. High-Energy Phys. CNEN Frascati, 1973*, ed. S. Stipcich, p. 44
5. Ballam, J., Blumberg, R., Mark, J., Skarpaas, K., St. Lorant, S. J. 1966. In *Proc. Int. Conf. Instrum. High Energy Phys. SLAC, Stanford, Calif., 1966*, ed. D. M. Ritson, pp. 107–21
6. CERN, Geneva, Switzerland 1976. *CERN/SPSC/76-43*, Revision 2, May 1976
7. Powell, W. M., Oswald, L., Griffin, G., Swartz, F. 1963. *Rev. Sci. Instrum.* 34:1426–29
8. Minnesota Mining and Manufacturing Company, Minneapolis, Minn., registered trade name
9. Bradner, H. 1960. *Ann. Rev. Nucl. Sci.*

10:109–60
10. Barrera, F., Byrns, R. A., Eckman, G. J., Hernandez, H. P., Norgren, D. U., Shand, A. J., Watt, R. D. 1964. *Adv. Cryog. Eng.* 10:251–58
11. Rogers, A. 1970. In *Proc. Int. Conf. Bubble Chamber Tech., ANL, Lagrange, Ill., 1970*, ed. M. Derrick, pp. 346–73
12. Newport, R. W., David, D., Diplock, B. R., Edwards, B. W. H., Turner, W., Wheatley, J. D. 1973. In *Proc. Int. Conf. Instrum. High Energy Phys.*, CNEN Frascati, 1973, ed. S. Stipcich, pp. 50–59
13. Fong, D., Heller, M., Shapiro, A. M., Widgoff, M., Bruyant, F., Bogert, D., Johnson, M., Burnstein, R., Fu, C., Petersen, D., Robertson, M., Rubin, H., Sard, R., Snyder, A., Tortora, J., Alyea, D., Chien, C. Y., Lucas, P., Pevsner, A., Zdanis, R., Brau, J., Gunhaus, J., Hafen, E. S., Hulsizer, R. I., Karshon, U., Kistiakowsky, V., Levy, A., Napier, A., Pless, I. A., Silverman, J., Trepagnier, P. C., Wolfson, J., Yamamoto, R. K., Cohn, H., Ou, T. C., Plano, R., Watts, T., Brucker, E., Koller, E., Stamer, P., Taylor, S., Bugg, W., Condo, G., Handler, T., Hart, E., Kraybill. H., Jung, D. L., Ludlam, T., Taft, H. D. 1976. *Nucl. Phys. B* 102:386–404
14. Lach, J., Pruss, S. 1971. *FNALRep. TN-298*, Fermi Natl. Accel. Lab., Batavia, Ill.
15. Smart, W. M., Moriyasu, K., Leith, D. W. G. S., Johnson, W. B., Friday, R. G. 1973. *Rev. Sci. Instrum.* 44:1584
16. Coombes, R., Fryberger, D., Hitlin, D., Piccioni, R., Porat, D., Dorfan, D. 1972. *Nucl. Instrum. Methods* 98:317–28
17. Arenton, M. W., Bacino, W. J., Hauptman, J. M. 1976. Pap. No. 794. Submitted to *Int. Conf. High Energy Phys. 18th Tblisi, 1976*
18. Alam, M. S., Brabson, B. B., Galloway, K., Mercer, R., Baggett, N. N., Fowler, E. C., Huebschman, M. L., Kreymer, A. E., Rogers, A. H., Baglin, C., Hanlon, J., Kamat, R., Panvini, R., Petraske, E., Stone, S., Waters, J., Webster, M. 1974. *Phys. Lett. B* 53:207–11
19. Gunderson, B., Benvenuti, A., Brush, A., Erwin, A., Mistretta, C., Thompson, M. A., Walker, W. D. 1970. In *Proc. Int. Conf. Bubble Chamber Tech., ANL, Lagrange, Ill., 1970*, ed. M. Derrick, pp. 582–90
20. Dorfan, J. M. 1976. PhD thesis. Univ. Calif., Irvine
21. Smith, G. A. 1973. *A.I.P. Conf. Proc. DPF, Am. Inst. Phys., NY, 1973*
22. Cence, R. J., Harris, F. A., Parker, S. I.,

Peters, M. W., Peterson, V. Z., Stenger, V. J., Lynch, G., Marriner, J., Solmitz, F., Stevenson, M. L. 1976. *LBL-4816*, Lawrence Berkeley Lab., Berkeley, Calif.
23. Bashian, A., Finocchiaro, G., Good, M. L., Grannis, P. D., Guisan, O., Kirz, J., Lee, Y. Y., Pittman, R., Fischer, G. C., Reeder, D. D. 1971. *Phys. Rev. D.* 4:2667–79
24. Bowden, G. B., Field, R. C., Lewis, R., Hoard, C., Skarpaas, K., Baker, P. 1976. *Nucl. Instrum. Methods* 138:77
25. Carroll, J. T., Della Negra, M. 1972. *SLAC BC Note No. 16*, Stanford Linear Accel. Ctr., Stanford, Calif.
26. Chadwick, G. 1974. *SLAC BC Note No. 48*, Stanford Linear Accel. Ctr., Stanford, Calif.
27. Bugg, W., Condo, G., Hart, E., Pevsner, A., Sard, R., Snyder, A., Hulsizer, R., Kistiakowsky, V., Trepagnier, P., Cohn, H., McCulloch, R., Mills, M., Dauwe, D. 1973. *Bull. Am. Phys. Soc.* 18:564; Progr. Surv. Watts, T., Ou, T., Fong, D., Lucus, H., Pless, I., Trepagnier, P., Wolfson, J., McCulloch, R., Bugg, W., Ludlam, T. 1973. *Bull. Am. Phys. Soc.* 18:564, Progr. PWGP
28. Della Negra, M., 1972. *SLAC BC Note No. 15*, Stanford Linear Accel. Ctr., Stanford, Calif.
29. Bugg, W. M. 1974. *FNAL HS Note 30*, Fermi Natl. Accel. Lab., Batavia, Ill., and private communication
30. Chadwick, G. 1974. *SLAC BC Note No. 71*, Stanford Linear Accel. Ctr., Stanford, Calif.
31. Whitmore, J. 1976. *Phys. Rep. C* 27:187
32. vonKrogh, J., Fry, W., Camerini, U., Cline, D., Loveless, R. J., Mapp, J., March, R. H., Reeder, D. D., Barbaro-Galtieri, A., Bosetti, P., Lynch, G., Marriner, J., Solmitz, F., Stevenson, M. L., Haidt, D., Harigel, G., Wachsmuth, H., Cence, R., Harris, F., Parker, S., Peters, M., Peterson, V., Stenger, V. 1976. *Phys. Rev. Lett.* 36:710
33. Ochs, W., Davidson, V., Dzierba, A., Firestone, A., Ford, W., Gomez, R., Nagy, F., Peck, C., Rosenfeld, C., Ballam, J., Carroll, J., Chadwick, G., Linglin, D., Marcelja, F., Moffeit, K., Ely, R., Grether, D., Oddone, P. 1975. *Nucl. Phys. B* 86:253; Ochs, W., Davidson, V., Dzierba, A., Firestone, A., Ford, W., Gomez, R., Nagy, F., Peck, C., Rosenfeld, C., Ballam, J., Carroll, J., Chadwick, G., Linglin, D., Moffeit, K., Ely, R., Grether, D., Oddone, P. 1976. *Nucl. Phys. B* 102:405

34. Ballam, J., Bloom, E. D., Carroll, J. T., Chadwick, G. B., Cottrell, R. Leslie, Della Negra, M., DeStaebler, H. C., Gershwin, L. K., Keller, L. P., Mestayer, M. D., Moffeit, K. C., Prescott, C., Stein, S. 1974. *Phys. Rev. D* 10:765
35. Alexsandrov, Ya. A. 1963. *Prib. Tekh. Eksp.* 2
36. Harigel, G., Horlitz, G., Wolff, S. 1967. DESY Rep. No. 67/14, Deutsches Elektronen-Synchrotron, Hamburg, W. Germ.
37. Ferrie, J. Private communication

38. Allison, W. W. M., Bunch, J. N., Cobb, J. H. 1976. *Nucl. Instrum. Methods* 133: 315
39. Hitlin, D., Martin, J. F., Morehouse, C. C., Abrams, G. S., Briggs, D., Carithers, W., Cooper, S., Devoe, R., Friedberg, C. E., Marsh, D., Shammon, S., Vella, E., Whitaker, J. S. 1976. *Nucl. Instrum. Methods* 137:225
40. Kistiakowsky, V. 1976. Private communication and NALREP, Fermi Natl. Accel. Lab., Batavia, Ill.

Ann. Rev. Nucl. Sci. 1977. 27: 139–66
Copyright © 1977 by Annual Reviews Inc. All rights reserved

CHEMISTRY OF THE   ✺5583
TRANSACTINIDE ELEMENTS[1]

O. Lewin Keller, Jr.

Chemistry Division, Oak Ridge National Laboratory, Oak Ridge, Tennessee 37830

Glenn T. Seaborg

Lawrence Berkeley Laboratory and Department of Chemistry, Berkeley, California 94720

CONTENTS

1   INTRODUCTION

This review covers the atomic, chemical, and, where pertinent, nuclear properties of elements known and unknown beyond lawrencium, element 103, the last member of the actinide series.

Experimental and theoretical work, and their mutual interactions, are critically discussed and various approaches are compared. Gas chromatography and aqueous chemistry, as applied to one-atom-at-a-time chemistry, are described and evaluated. We also include the uses of modern atomic computer programs in obtaining ground-state and excited-electronic configurations, ionization energies, and radii. These

[1] Research sponsored by the US Energy Research and Development Administration under contract with Union Carbide Corporation.

calculated quantities can be of great use in predicting chemical and physical properties when combined with the correlative value of the periodic table and thermodynamic formulations.

Heavy-ion accelerators are the central tool for research in the heavy-element region. We hope they will be capable of propelling us into the most exciting prospective region beyond lawrencium—the superheavy elements. These elements are predicted by nuclear theorists to exist in an "island of stability" around $Z = 114$, $N = 184$, and possibly $Z = 164$ (1). Also, the heavy-ion accelerators are currently producing elements up to 106 for chemical study.

We proceed through the review of the transactinide elements more or less in numerical order, but make groupings that, we hope, will give some coherence to a field that has much of the confusion usually associated with a frontier. Thus, we begin with experiments concerning the chemistry of elements 104 and 105, rutherfordium and hahnium, and go on to mention the most recently discovered element, 106. These three known elements are of great importance to the field of chemistry because they occur at a critical juncture in the periodic system—the end of the actinide series and the beginning of a new series. The characteristics of this new series are yet to be determined in detail, but it has been shown that the chemistry of rutherfordium is distinctly different from that of neighboring trivalent lawrencium (2) and the other heavy actinides (3–5). This new series is thought to

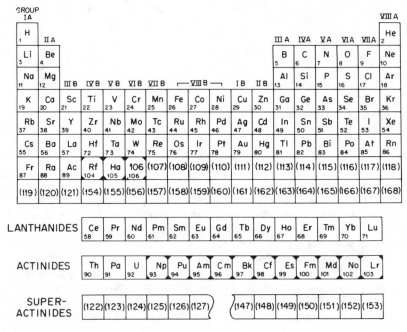

Figure 1 Simple form of periodic table, showing known transactinide and undiscovered elements in locations predicted with varying degrees of certainty.

extend from rutherfordium, eka-Hf, to 112, eka-Hg. It is thus thought to be a $6d$ series analogous to the $5d$ series, which extends from Hf to Hg. As shown in the accompanying periodic table (Figure 1), a new series is expected to begin following element 121, superactinium. This new series is called the superactinides (1). It will be longer and more complicated than the actinide series because $6f$, $5g$, and also $8p$ and $7d$ electrons are all added in mixed configurations. Nominally, a mixed $6f$, $5g$ series would end at 153, but current calculations stretch it to 156 (Table 1). For simplicity, however, we retain the symmetry of the periodic table in as straightforward a fashion as possible, even though the results of detailed electronic

Table 1 Dirac-Fock ground-state configurations of free neutral atoms of elements 104–168[a]

Element	Rn "core"$+5f^{14}+$								
	$5g$	$6d$	$6f$	$7s$	$7p_{1/2}$	$7p_{3/2}$	$7d$	$8s$	$8p_{1/2}$
Rf		2		2					
Ha		3		2					
106		4		2					
107		5		2					
108		6		2					
109		7		2					
110		8		2					
111		9		2					
112		10		2					
113		10		2	1				
114		10		2	2				
115		10		2	2	1			
116		10		2	2	2			
117		10		2	2	3			
118		10		2	2	4			
119		10		2	2	4		1	
120		10		2	2	4		2	
121		10		2	2	4		2	1
122		10		2	2	4	1	2	1
123		10	1	2	2	4	1	2	1
124		10	3	2	2	4		2	1
125	1	10	3	2	2	4		2	1
126	2	10	2	2	2	4	1	2	1
127	3	10	2	2	2	4		2	2
128	4	10	2	2	2	4		2	2
129	5	10	2	2	2	4		2	2
130	6	10	2	2	2	4		2	2
131	7	10	2	2	2	4		2	2
132	8	10	2	2	2	4		2	2
133	8	10	3	2	2	4		2	2
134	8	10	4	2	2	4		2	2
135	9	10	4	2	2	4		2	2
136	10	10	4	2	2	4		2	2

Element	118 "core"$+$						
	$5g$	$6f$	$7d$	$8s$	$8p_{1/2}$	$9s$	$9p_{1/2}$
137	11	3	1	2	2		
138	12	3	1	2	2		
139	13	2	2	2	2		
140	14	3	1	2	2		
141	15	2	2	2	2		
142	16	2	2	2	2		
143	17	2	2	2	2		
144	18	1	3	2	2		
145	18	3	2	2	2		
146	18	4	2	2	2		
147	18	5	2	2	2		
148	18	6	2	2	2		
149	18	6	3	2	2		
150	18	6	4	2	2		
151	18	8	3	2	2		
152	18	9	3	2	2		
153	18	11	2	2	2		
154	18	12	2	2	2		
155	18	13	2	2	2		
156	18	14	2	2	2		
157	18	14	3	2	2		
158	18	14	4	2	2		
159	18	14	4	2	2	1	
160	18	14	5	2	2	1	
161	18	14	6	2	2	1	
162	18	14	8	2	2		
163	18	14	9	2	2		
164	18	14	10	2	2		
165	18	14	10	2	2	1	
166	18	14	10	2	2	2	
167	18	14	10	2	2	2	1
168	18	14	10	2	2	2	2

[a] Reference 29

calculations, such as the configurations given in Table 1, show that the elements up to 168 will not be slavishly following this symmetry as set by the lighter known elements.

Fernelius, Loening & Adams (6) have described a systematic nomenclature, advocated by the IUPAC Commission on Nomenclature of Inorganic Chemistry, for temporary names to be applied to undiscovered elements for intermediate use until they are discovered and given their names in the traditional manner by the discoverers. They advocate using the "ium" ending preceded by the following roots: nil = 0, un = 1, bi = 2, tri = 3, quad = 4, pent = 5, hex = 6, sept = 7, oct = 8, and enn = 9. The corresponding chemical symbols would have three letters. Some examples from this system are: 107, Unnilseptium (Uns); 110, Unnnilium (Uun); 118, Ununoctium (Uno); 130, Untrinilium (Utn); 200, Binilnilium (Bnn); 900, Ennilnilium (Enn). We believe this system is unnecessarily cumbersome and would prefer to simply use the atomic number in parenthesis; thus the dioxide of element 108 becomes $(108)O_2$, the sodium salt of element 117 becomes Na(117), the trifluoride of element 126 becomes $(126)F_3$, etc.

A chart of the nuclides from Md through 106 is given for ready reference in Figure 2. All isotopes and elements that have been formally or informally reported are not included in this figure. In some cases this is because we have not had sufficient opportunity to evaluate the evidence at the time of writing.

Superheavy elements have been the subject of much theory and experiment over the last ten years. Other reviews for the interested reader would include those of Herrmann (7) and the Proceedings of the Welch Foundation Conference honoring the Mendeleev Centennial (8).

2 STATUS OF PRODUCTION AND CHEMICAL UNDERSTANDING OF RUTHERFORDIUM (Rf), HAHNIUM (Ha), AND 106

Currently, there are three known transactinide elements: 104, 105, and 106. The first two were named rutherfordium (Rf) and hahnium (Ha) by their Berkeley discoverers (9, 10); however, because of controversies between researchers at Berkeley and Dubna over the discoveries of these elements, both sides have refrained from naming 106. We do not engage here in a discussion of the relative merits of the Berkeley-Dubna arguments because such a discussion would obscure our purpose of describing the chemistry that has been done on Rf and Ha and developing an understanding of its meaning. We also want to build on past work and experience in giving suggestions as to what the future may hold in this pivotal area of the periodic table. We therefore simply refer to work carried out at Berkeley and Dubna on Rf and Ha as it befits this chemical framework.

Two overriding circumstances dominate studies of Rf, Ha, and 106—they have short half-lives (of the order of seconds) and they can be produced only at heavy-ion accelerators and in low yield (characteristically on the order of an atom per hour). Currently, the heavy-ion accelerators at Berkeley and Oak Ridge in the United

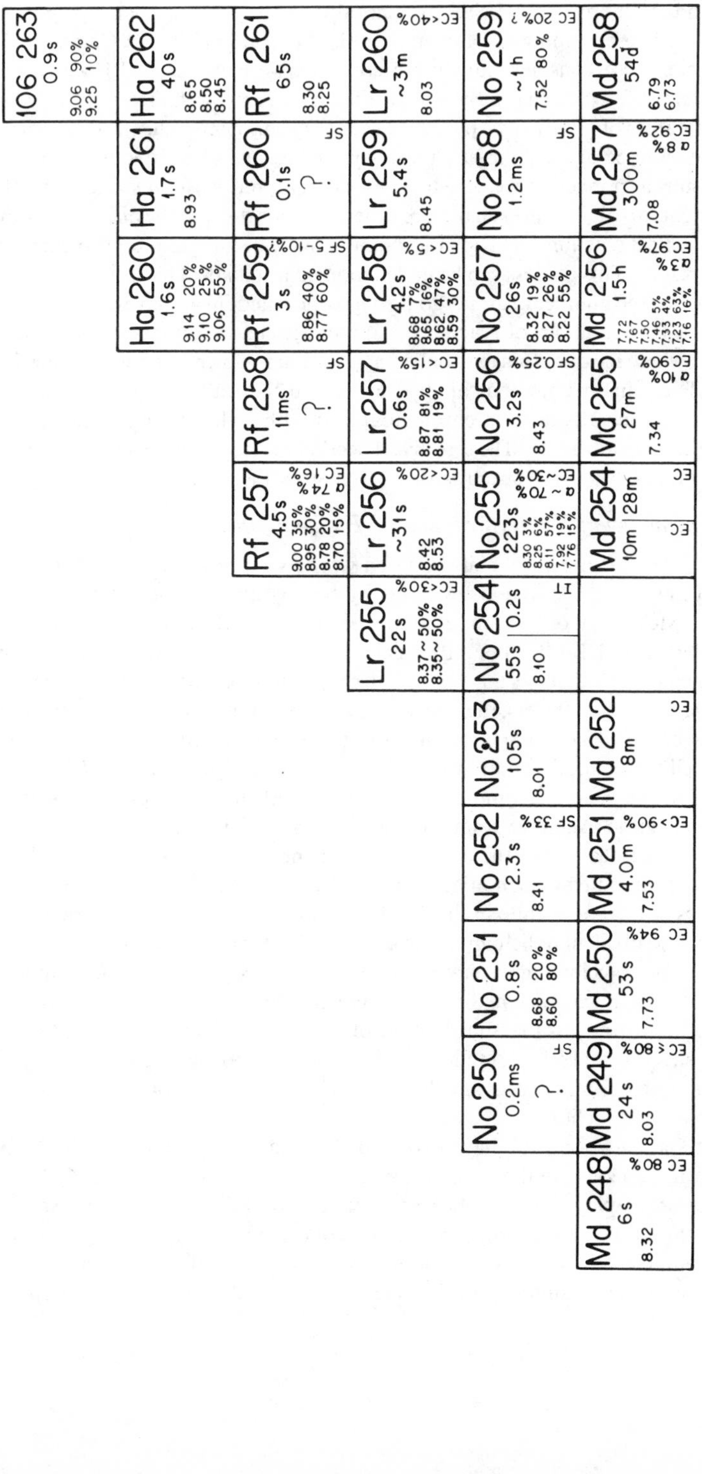

Figure 2   Chart of the nuclides for elements 101–106.

States and at Dubna in the USSR are partially devoted to this challenging field. A program is also beginning at Darmstadt in the Federal Republic of Germany. To make these considerations more concrete, we consider ^{261}Rf, which has a half-life, $t_{1/2}$, of ~ 65 sec (11). By bombarding a curium-248 target with oxygen-18, about one atom of ^{261}Rf can be produced every 4 experiments. Chemical experiments must therefore be of such a nature that they are fast and also capable of giving the same results as a macroscopic chemical experiment, but with only one atom of Rf. Experiments involving ion exchange, or other methods that put the one atom through many chemical reactions of an identical nature, have been found to yield chemically reliable results. This process can be understood on a statistical basis because, instead of having a large number of atoms go through one reaction, we have one atom going through a large number of reactions. Experiments based on co-precipitation, for example, where the one atom may only react once, are known to often give unreliable results. The principal methods applied to date to the transactinide elements have involved ion exchange and gas chromatography, although solvent extraction would often be equally attractive. The sensitive detection methods employed have involved high-resolution α-particle spectroscopy and fission-track counting.

2.1   Chemistry of Rutherfordium, Rf (Element 104)

2.1.1   AQUEOUS CHEMISTRY   Silva and co-workers (3) produced about 100 atoms of ^{261}Rf in several hundred experiments by bombarding a 47-μg target of $^{248}_{96}$Cm with a 92-MeV beam of $^{18}_{8}$O ions at a beam current of ~ 2 μamp (measured as O^{8+}) in the Berkeley HILAC. About 10 events were actually observed in the chemical experiments, the others being lost by decay, counting geometry, and chemical manipulations. Nonetheless, these investigators were able to carry out conventional aqueous chemical studies concerning the ion-exchange elution behaviour of Rf under well-understood conditions similar to those that had been much used for 25 years in the characterization of the chemistry of the actinides (elution from a Dowex 50 cation-exchange resin column using ammonium α-hydroxyisobutyrate) (12). In order to carry out the column experiments with the requisite speed, Silva et al set the experimental conditions such that hafnium and zirconium, the expected analogues of Rf, would elute in the first few column volumes, whereas the trivalent actinides and divalent nobelium would elute much later in terms of column volumes (and therefore precious time, as well). By having the elution early and automating much of the chemical system, Silva et al were able to make the time from beam-off to sample counting less than one half-life of ^{261}Rf, about 60 sec. The result of these experiments was that Rf was found to elute in the same position as Hf and Zr, rather than a hundred or more column volumes later, as did the 3+ ions Tm, Cf, and Cm, or the 2+ ion nobelium.

Hulet and co-workers (5) have extended the study of the chemistry of Rf to its chloride in aqueous solution, again using the 65-sec isotope of mass 261. These investigators employed a sophisticated computer-controlled automated system that allowed the chemical operations to be carried out rapidly and repeatedly. The actual experimental technique was similar to ion exchange except that a solvent extraction agent (trioctylmethylammonium chloride) was adsorbed on a fluorocarbon powder in a column. The HCl aqueous solution containing [261]Rf could then be passed through the column to study its extraction behavior for comparison with known elements. On the basis of 7 events, the elution characteristics of Rf were found to closely resemble those of a [181]Hf tracer used for comparison, but were markedly different from those of Cm and Fm tracers.

The results of Silva, Hulet, and co-workers thus confirm that Rf is distinctly different from the heavy actinides, as had been predicted. Also, the behavior of Rf in these experiments resembles that of Hf and Zr, thus opening up the possibility that Rf is a typical Group IVb element, as expected. However, this latter possibility is still a subject for experimental study because the species present and the oxidation state of Rf have not been definitely established in the experiments carried out so far.

2.1.2   GAS CHROMATOGRAPHY   In 1969 Ghiorso and co-workers reported (9) the discovery of [259]Rf, an α emitter with energies of 8.77 and 8.86 MeV and a half-life of 3 sec. A production rate at the Berkeley HILAC of ~ 150 atoms per hr is achieved using a 60-μg per 0.21 cm^2 target of [249]Cf and a bombarding beam of 69-MeV ^{13}C ions of 3×10^{12} particles per sec.

Studies of the spontaneous-fission decay of [259]Rf at Dubna (13) and Oak Ridge (14) indicate a possible branching of about 5%–10% for this mode of decay. A spontaneous-fission branching of [259]Rf is useful because it allows the application of methods developed by Zvara and his co-workers at Dubna, which involve chromatographic studies of gaseous species of extremely short-lived transactinide elements. Silva's use of ion-exchange techniques had the advantage that he could call on a broad technology built up over 25 years in the actinide and lanthanide field specifically and for most elements in general. Zvara set out on the difficult task of developing a new method based on gas chromatography. Zvara's basic chemical purpose is the determination of whether Rf forms a volatile chloride (analogous to a chloride species of hafnium), or whether it forms a nonvolatile chloride (analogous to LrCl$_3$). By 1972, Zvara and his associates had developed his methods to the point that he could determine whether Rf forms a volatile or nonvolatile chloride (4). Before analyzing these Dubna results, however, it may be helpful for future developments to delineate what such an experiment is capable of telling us. If it turns out that Rf actually has volatile chloride species, then it has been confirmed that the actinide series ends at Lr, as expected. The property of chloride volatility alone is too generally shared by groups in the periodic system, however, for us to learn much about the specific chemistry of Rf from this. There are also questions about the gas-chromatographic technique concerning the actual identity of species formed, possible aerosol formation that can lead to erroneous interpretations, and a general limitation on resolution among similar elements.

On the other hand, suppose Rf were found by gas-chromatographic methods to have a nonvolatile chloride. Then it would not be the first member of a new transition series with properties analogous to Hf, but it could still be the beginning of a new series with properties different from the actinides. Observation of nonvolatility of chlorides, for example, did not allow Zvara and co-workers to distinguish between trivalent actinides and divalent nobelium (15). So if Rf formed a nonvolatile chloride, it would neither confirm nor deny that the actinide series ends at Lr. As we show below, the experiments of the Dubna researchers indicate that Rf does form a volatile chloride; so the experiments of Zvara and associates augment those of Silva, Hulet, and co-workers to show that Lr marks the end of the actinide series, a most important result. Also, the likelihood that Rf will have a typical Group IVb behavior, as expected, is strengthened by the finding that it forms a volatile chloride analogous to the chlorides of zirconium and hafnium.

In the chemical experiments on Rf, Zvara et al (4) bombarded a 0.8 mg cm^{-2} target of 95% ^{242}Pu with 119-MeV ^{22}Ne ions from the Dubna heavy-ion cyclotron. The yield was quite low, since only 16 fission events were recorded in the final data. A small amount of samarium was also present in the plutonium target, to produce radioactive hafnium isotopes of masses 170 and 171 by interaction with the neon beam. Also, the target backing was aluminum, which led to the production of radioactive scandium. All of these products (Rf, Hf, and Sc) recoiled from the target because of the momentum introduced by the neon ions. They were therefore available for experimental purposes on a continuous basis, an advantage of the gas-chromato-graphic approach.

Figure 3 shows a schematic of the gas-chromatographic apparatus as developed by Zvara and co-workers, along with a histogram of their results on ^{259}Rf. The Rf, Hf, and Sc isotopes recoil from the ^{242}Pu target into a chamber through which nitrogen gas flows as a carrier. The recoil products are swept from the target chamber into a mixing chamber, where they meet the chlorinating agents titanium tetrachloride (TiCl$_4$) and thionyl chloride (SOCl$_2$). Here the chlorides of Rf, Hf, and Sc are formed so that each can play its role in the chromatographic column. The

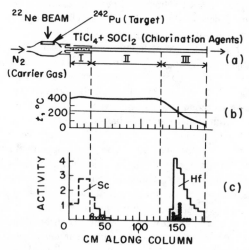

Figure 3 (a) Target chamber and gas chromatographic column used at Dubna for studies of the chlorine chemistry of ^{259}Rf; (b) temperature gradient in the column; (c) deposition position of ^{259}Rf (*black circles*) compared to deposition position of Sc and heavy actinides (*white circles*) and Hf.

hafnium, of course, is present to act as a marker in the column for the position where relatively volatile chloride species similar to those formed by Hf should appear. The scandium, which forms a relatively nonvolatile chloride, is a stand-in for the nonvolatile heavy-actinide chlorides. After chlorination, the products were swept into the glass gas-chromatographic column, which was 195 cm long and 4 mm in inside diameter. The glass tubing was smooth bore, except for the first 30 cm (section I). Turbulent flow was enhanced in this region through the use of numerous protuberances in order to enhance the deposition of nonvolatile chlorides. This section, as well as the second section (of 100 cm in length), were kept at $400 \pm 5°$ C. In the third section, whose length was 65 cm, an approximately linear temperature gradient from 400° to 50°C was maintained. Four-mm-wide mica plates were spaced along the axis of the glass tube to act as detectors for spontaneous fission of ^{259}Rf. The deposition positions of the ^{44m}Sc and the $^{170, 171}$Hf were detected by their γ activity. As can be seen in Figure 3, the scandium appears in section I of the column and at the beginning of section II. The ^{44m}Sc thus marks the position for deposition of elements that form nonvolatile chlorides. Also, some fission events appear at the beginning of section II. These events are quite reasonably attributed by Zvara and co-workers to spontaneously fissioning actinide nuclides such as ^{256}Fm. Most of the length of section II is free of activity so that a large buffer zone is maintained, giving confidence that all nonvolatile chloride molecules were stopped long before section III.

In section III, where the steep temperature gradient is maintained, Rf is deposited at the same position as Hf. The clear implication is that Rf forms volatile chloride species in a manner similar to Hf, and it is therefore clearly different in its chemistry from the heavy actinides.

Thus, both the aqueous chemistry experiments of Silva, Hulet, and co-workers and the gas-phase chemistry of Zvara and his associates substantiate that the actinide series ends at Lr and a new series begins at Rf. Additionally, the expectation is that Rf will have a chemistry similar to the Group IVb elements. The work of both the US and Soviet groups is consistent with this expectation, although neither gives detailed information on this point. Clearly, although some crucial questions have been answered, much remains to be done in developing the relationship of the chemistry of Rf to the members of Group IVb and to thorium.

2.2   Chemistry of Hahnium, Ha (Element 105)

Three isotopes of hahnium ($^{260, 261, 262}$Ha) are known for use in chemical studies as reported by Ghiorso and co-workers (10, 16). Also, ^{260}Ha and ^{261}Ha, as produced and characterized at Berkeley, have properties together that would closely fit the results of Flerov and co-workers on an ~ 1.5-sec activity, thought to be ^{261}105, that decays by spontaneous fission and by α decay with energies of 8.9 and 9.1 MeV (17, 18).

The production rates of these isotopes by currently employed reactions are in the few counts-per-hour range. Furthermore, the half-lives are very short for use in chemical studies, being in the range of 1–40 sec. These half-life and production limitations make meaningful experiments on the chemistry of Ha difficult in the extreme. There is hope for improvement, since new types of nuclear production reactions currently being investigated may increase yields substantially. Also, new automated chemical procedures are being conceived and developed that may allow reliable studies of important properties of elements with even these short half-lives (Figure 4).

The isotope with the longest half-life is ^{262}Ha ($t_{1/2} \cong 40$ sec). It can be produced by the reaction ^{249}Bk(^{18}O, $5n$)^{262}Ha. With an ^{18}O beam of 3 μA of fully stripped ions at 98 MeV and a 350-μg cm^{-2} ^{249}Bk target, the production rate is about 9 atoms per hr. Ghiorso and co-workers give the α energies as 8.45 (80%) and 8.66 (20%) MeV (16). In the chemical experiments, to eliminate confusion, it would be desirable to have α detection with sufficient resolution to distinguish between ^{262}Ha and possible interfering isotopes. Lawrencium-259 ($t_{\frac{1}{2}} = 5.4$ sec) and nobelium-256 ($t_{\frac{1}{2}} = 3$ sec), with α energies of 8.44 MeV, and ^{258}Lr ($t_{\frac{1}{2}} = 4.2$ sec), with an α energy of 8.61 MeV, are the closest interferences produced in the Berkeley experiments. Lawrencium-259 and nobelium-256 do not present potential difficulties, however, because of their short half-lives, if the experiments from beam-off to counting take roughly a half-life of ^{262}Ha (i.e. $\sim$40 sec). The daughter isotope ^{258}Lr can be resolved, since the energy of the main α group at 8.61 MeV is nearly 200 KeV away from that of the main α group of ^{262}Ha at 8.45 MeV.

The isotope ^{260}Ha ($t_{\frac{1}{2}} = 1.6 \pm 0.3$ sec) can be produced by the reaction ^{249}Cf(^{15}N,

Figure 4   G. T. Seaborg and R. J. Silva with an automatic chemical apparatus for studies of the aqueous chemistry of transactinide elements.

$4n$) ^{260}Ha with a cross section of 3×10^{-33} cm^2 (10). Ghiorso and co-workers produced about 6 counts per hr of ^{260}Ha using a 300-μg cm^{-2} target of ^{249}Cf.

In a continuation of their important and pioneering chemical studies of the heaviest elements, Zvara and co-workers (19, 20) have carried out gas-chromatographic experiments (Figures 5 and 6), intended to support discovery claims concerning element 105 made earlier by other members of the Dubna group (17, 18). Zvara and co-workers bombarded a ^{243}Am target with 119-MeV ^{22}Ne ions, a procedure described by Flerov and co-workers for producing ^{261}105. Although the Dubna nuclear results are compatible with the idea that they were producing a combination of ^{260}Ha and ^{261}Ha, as studied at Berkeley, their papers contained very little detail on isotopes of other elements produced in the americium-neon reaction that could possibly interfere in the detection methods used for studying the

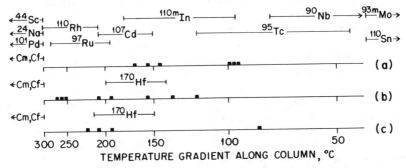

Figure 5   Location of fission tracks (squares), attributed to element 105, in the gas-chromatographic column compared to zones found for known elements using chlorine chemistry at Dubna.

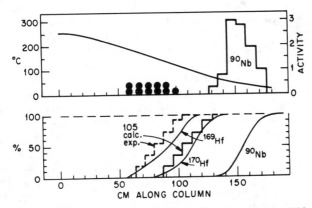

Figure 6   Locations of fission tracks (circles), attributed to element 105, in the gas-chromatographic column compared to zones found for ^{90}Nb($t_{1/2} = 14.7$ hr), ^{170}Hf($t_{1/2} = 12.2$ hr), and ^{169}Hf($t_{1/2} = 3.3$ min) using bromine chemistry at Dubna. The "calc" curve (marked 105) compensates for the relatively shorter half-life of element 105.

chemistry. Since spontaneous fission is nonspecific as an identification method, a thorough understanding of the nuclear reactions involved is especially important to be sure just which element or elements are entering into the chemical system being used. As things stand, some ambiguity must be said to exist because the fission activity or activities being studied by Zvara were not as clearly defined by the Dubna physicists (17, 18) as the α activities studied at Berkeley (10, 16).

Leaving aside, however, the nuclear-reaction and identification ambiguities, which can probably be cleared up through a more thorough study at Berkeley and at Dubna, let us assume for now that Zvara actually was studying element 105 and element 105 only. What should we expect if the chemistry is a straightforward extrapolation of V, Nb, and Ta chemistry? Since the stabilization of the oxidation state V in going to the heavier members of the series is clear, we would expect 105 to be most stable in the V state with little or no chemistry in lower oxidation states. Furthermore, the Dubna researchers have shown, as we have seen (Figure 3), that a well-defined peak is obtained in their chromatographic column for the chloride species of rutherfordium, and this peak occurs directly under the equally well-defined one for the chloride species of hahnium. The time scale, temperature gradient in the chromatographic column, and other parameters of the Rf experiment were evidently chosen such that no complexities appeared due to the widely differing half-lives of the isotopes employed (3 sec for ^{259}Rf and 12.2 hr for ^{170}Hf), although potentially this could happen. If we assume that the rutherfordium experiment is giving a correct chemical result, then it should be possible to carry out an analogous experiment with the ^{260,261}Ha isotopes (whose half-lives are near that of ^{259}Rf) using ^{90}Nb as the comparison isotope, since its 14.7-hr half-life is near that of ^{170}Hf. Thus the ambiguities found at Dubna (21, 22), due to half-life differences, in the experiments concerning the chemistry of element 105 (Figure 6) could presumably be eliminated if they were eliminated in the Rf case.

It is important to rule out, or thoroughly understand, all of these various ambiguities in the Dubna results, because, if their results are correct, we believe they show that Ha chemistry is more like that of vanadium than that of niobium and tantalum, an unexpected result. If this is true, then the trend in the periodic table for Group Vb is reversed between Ta and Ha, and Ha is not following the chemistry expected for eka-Ta as stated by the Dubna workers (19, 20). The implications for the chemistry of the heavier members of the series may be important as well.

Zvara and his associates have designed experiments for studying the chemistry of element 105 using both chlorinating and brominating agents (19, 20). The temperature gradient was made $\sim\frac{1}{3}$ that of the Rf experiments, presumably in an effort to separate Rf and Ha, as well as Hf and Nb. The fission tracks attributed to element 105 were broadly distributed along the temperature gradient in the column, in the case of the chlorine experiments, and did not form a peak. In Figure 5, we see that Hf, in the chlorine-chemistry experiments, is deposited over a range of $\sim75°C$, whereas the fission tracks attributed to element 105 appear over a range of $\sim200°C$. It is important to note also that the locations of the fission tracks occur in the position ranges of a number of elements, but none of the tracks appear in the range for niobium. Also, in Figure 6, the deposition range for the fission tracks attributed to element 105 in the bromine chemistry was found close to the range for Hf, but not Nb (20, 21), both before and after correction for the half-life effect.

If Zvara and co-workers are studying the chemistry of Ha, it appears plausible to suppose that their methods allow the formation of several species with different volatility characteristics, thus giving the broad distribution in the chlorine-chemistry case. It does appear, however, from the Dubna experimental results, that Ha is not following the expected trend set by V, Nb, and Ta in Group Vb toward stabilization of the oxidation state of five in the heavier members. Otherwise, some of the fission tracks should have appeared in the Nb region of the chromatographic column by analogy with the Rf-Hf experiments (Figure 3), at least after the half-life correction was made.

These Dubna experimental results thus suggest that hahnium does indeed have a chemistry more like vanadium than like niobium and tantalum. Whereas vanadium readily forms a tetrachloride, it does not form a pentachloride. Neither does it form a pentabromide. On the other hand, VF_5 is well known. So high covalencies are suppressed in vanadium chemistry and may well be in Ha chemistry as well. Also, $VOCl_3$ and $NbOCl_3$ are examples of volatile oxyhalides also possible with Ha. Our considerations of the Dubna experimental results suggest that the most clearly defined chemical conditions could be obtained using fluorine chemistry, since the oxidation state of five could more probably be stabilized under those conditions, oxyhalides would be less likely, and Ha would more probably behave like Nb and Ta.

The experimental results of the Dubna researchers are both important and tantalizing in their implications. Pressing forward under better-defined conditions is of the utmost importance in order to investigate the possible new trends in the periodic system.

The possible effect being seen at Dubna in Ha chemistry would not be detectable by them in the Rf case, since Ti is most stable in the IV state. So Ha would be the first case for them to observe that the transactinide-series elements may behave more like the lighter members of their groups than like the heavier ones.

2.3 Chemistry of 106 (Element Not Yet Named)

The synthesis and identification of two different isotopes of element 106 was announced in 1974 by Ghiorso and co-workers (23) and by Flerov, Oganessian, and co-workers (24). In view of the simultaneity of the experiments, and their very different natures, neither group has suggested a name for the element. The Berkeley group bombarded ^{249}Cf in the Super HILAC with 95-MeV ^{18}O ions at a beam current of 3×10^{12} particles sec^{-1} to produce $^{263}106$ ($t_{1/2} = 0.9 \pm 0.2$ sec, α particles of principal energy 9.06 ± 0.04 MeV) by the reaction $^{249}Cf(^{18}O, 4n)$. The definite identification consisted of the establishment of the genetic link between the previously known α particle-emitting daughter $^{259}104$ ($t_{1/2} = 3$ sec) and granddaughter $^{255}102$ ($t_{1/2} = 3$ min) nuclides; i.e. the demonstration of the decay sequence: $^{263}106 \xrightarrow{\alpha} {}^{259}Rf \xrightarrow{\alpha} {}^{255}No \xrightarrow{\alpha}$. The recoiling $^{263}106$ atoms were swept in a stream of helium gas to the surface of a wheel that could be rotated close to solid-state detectors in such a manner that the genetically related α particles of $^{263}106$, ^{259}Rf, and ^{255}No were recorded in their natural time sequence. The operation of the apparatus was controlled by a computer, which also recorded on magnetic tape all of the experimental data. A total of 73 ^{263}106 α particles and approximately the expected corresponding number of ^{259}Rf daughter and ^{255}No granddaughter α particles were recorded. The cross section was about 0.3 nb, corresponding to about one count every 2 hr.

The Dubna group (24) bombarded ^{207}Pb and ^{208}Pb with ^{54}Cr ions in their cyclotron to find a product that decays by the spontaneous fission mechanism with the very short half-life of 7 m-sec. They assign this to the isotope ^{259}106, suggesting reactions in which 2 or 3 neutrons are emitted: ^{207}Pb(^{54}Cr, 2n) and ^{208}Pb(^{54}Cr, 3n). They chose lead as the target because, they believe, its closed shells of protons and neutrons and consequent small relative mass lead to minimum excitation energy for the production of the desired product nuclide. The ^{54}Cr ions impinged on the lead target on the surface of a rotating disk. Dielectric (mica) detectors were placed in the position of a surrounding sleeve at a distance of 3 mm from the rotating lead target. The product nuclei, decaying by spontaneous fission, projected their recoiling fragments into the mica detectors, which were examined after the bombardment to determine the number and location, and hence the half-life, of these decaying nuclei. A total of 51 spontaneous-fission events were ascribed to ^{259}106. The identification depends on rather uncertain deductions concerning the nature of the new type of nuclear reactions that the Dubna scientists postulate and new correlations of the dependence of spontaneous fission half-lives on atomic and mass numbers. Such evidence does not meet criteria recently set forth for discovery of new elements by the present authors and others (25).

On the basis of its projected position in the periodic table, element 106 is expected to have chemical properties similar to those of tungsten. Zvara (21) has reported experiments with ^{175,176}W as a preparation for studies of element 106. With thionyl chloride, $SOCl_2$, as the chlorinating agent, he finds the tungsten deposition to be well separated from Hf and Nb, and believes that tungsten oxychlorides are involved. But the chemistry of Cr, Mo, and W is very complicated, and having well-defined nuclear reactions is clearly a prerequisite to the study of 106 chemistry. Preferably, fluorine chemistry should also be developed for the first experiments, since this would maximize the probability of obtaining a well-defined 106(VI) state.

It is interesting to speculate on the use of carbonyl chemistry in the case of 106, since $W(CO)_6$ is volatile, but Nb and Ta do not form carbonyls. It should be remembered, however, that $V(CO)_6$ is well known, so that the distinction between 105 and 106 on the basis of carbonyl chemistry may not be clear.

Penneman & Mann (26) have used equations developed by Jorgensen to predict that IV will be the most stable oxidation state of 106 in aqueous solution. Jorgensen's equations apply to the hydrated cation only, and are not intended to account for effects of stabilization of higher oxidation states by the binding of fluorides or the formation of oxyanions. Cunningham (27) used an extrapolation in the periodic table to obtain VI as the most stable state for element 106 as the oxide. If solution conditions are chosen so that oxyanions are formed, then the VI state of 106 will probably be stabilized over the IV, since W(VI) is stabilized in this way in the tungstate ion.

2.4   Element 107

The Dubna group has claimed discovery of element 107 on the basis of detection of a spontaneous fission activity (28). The evidence presented, however, does not meet criteria set forth recently for discovery of new elements by the present authors and others (25). Some expectations concerning the chemistry of 107, eka-Re, are discussed in Section 4.

3   THEORY AND SYSTEMATICS FOR PREDICTING PROPERTIES OF NEW ELEMENTS

Within a framework provided by the periodic table, modern quantum-mechanical calculations can provide a basis for predicting chemical and physical properties of undiscovered elements. Particular interest (1) has centered in recent years on elements in an "island of stability" around atomic number 114 because nuclear theorists have predicted an unusual stability in this region. Also, another island might exist around element 164. Various investigators have therefore applied the relativistic Hartree-Fock-Slater (rel HFS) and relativistic Hartree-Fock (rel HF) atomic calculational methods to elements in these areas. (29–32). The electronic configurations in Table 1 are one result of such calculations.

Since the connection between electronic properties and chemical properties is often complex, a number of theories must be used as an aid in the predictions in the superheavy region (33). Unfortunately, solutions to Schrödinger's equation for molecular systems are generally too crude to allow such predictive power. Theories such as valence-bond theory can be used with care, however. Also, the Born-Haber cycle, and Jorgensen's ingenious variation of it, have proven helpful (26, 33). Since the periodicity of chemical and physical properties of the elements results from the periodicity in their electronic properties, extrapolation within a given group in the periodic table is justified by theory. Although other types of extrapolations have been used, their theoretical justification has been obscure.

3.1   Atomic Calculations

The rel HF and rel HFS methods have proven very useful in calculating ground-state energy levels for neutral atoms of a number of superheavy elements (see Table 1, for example). These calculations allow a comparison of the electronic configuration of the superheavy element with the lighter members of its homologous group to see what cautions may be needed in making an extrapolation in the periodic table.

In the case of 113, eka-thallium, the ground-state configuration is found to be $7s^2 7p$, exactly analogous to the other members of Group IIIa. Furthermore, the $7p$ electron is expected to be the most active one chemically. These considerations give a strong indication that extrapolations in Group IIIa to 113 can be made with some confidence, although relativistic effects should be kept in mind.

Relativistic effects should be much more important in determining the chemical properties of 114, eka-Pb. In Group IVa, relativistic effects become noticeable (34) even at Si and increase in importance with atomic number. The relative importance of the relativistic effect can be gauged by the splitting of the $p_{1/2}$ and $p_{3/2}$ orbitals, as given by Fricke (29). The increased splitting between the $p_{1/2}$ and $p_{3/2}$ energies naturally decreases the interactions between electrons in these orbitals. The $p_{1/2}^2$ orbital then acts like a closed shell, and electrons in the $p_{3/2}$ orbital act like electrons being added outside of a closed shell. The effect of the $p_{1/2}^2$ "closed shell" will be of great importance in the chemistry of elements 114 through 117 (29, 33, 35).

The $p_{1/2}^2$ closed shell in 114 is analogous to an s^2 closed shell. This effect of the $p_{1/2}$ orbital is not so noticeable in Pb because the $p_{1/2}$ and $p_{3/2}$ orbitals are fairly close in energy in that case and large interactions between the orbitals can result. The $p_{1/2}$ orbital is simply an ordinary p orbital in that case, and the electrons in the $6p^2$ orbital of Pb can have parallel spins. In 114, the spins in the $7p_{1/2}^2$ orbital will be paired. Relativistic chemical effects appear, however, even in the lighter elements. A group-type extrapolation of well-behaved periodic properties of the elements in Group IVa can detect these relativistic effects on the p orbitals, which commence even at Si (34). Other methods of extrapolation have been used that are not so firmly rooted in the symmetry of the periodic table, and they naturally miss these more subtle effects in the lighter elements. Then the long extrapolation to the superheavy region allows a large margin for error.

Another interesting case is furnished by 115, eka-Bi. The ground state is not accurately characterized by saying it is $7s^2 7p^3$, the configuration expected on the basis of the configurations of the other group Va members. The reason is that in 115, the $p_{3/2}$ electron has a binding energy 5.6 eV lower than those in the $p_{1/2}^2$ closed shell (36), and 115 will be quite stable in the oxidation state of I. This extra stability in the I state arises from relativistic effects that have a small (though chemically detectable) effect in bismuth. These relativistic effects will have a profound effect on the chemistry of 115 (37), and we may one day (we hope) be studying relativity in the test tube.

Perhaps these illustrations will suffice to show some uses the atomic calculations can serve in evaluating the general applicability of the periodic table in the super-heavy region. Such calculations can also provide other useful data, such as energies of excited states, ionization potentials, radii, and X-ray energies. (29–31, 38, 39.)

3.2   Extrapolations and Correlations in the Periodic Table

Mendeleev and Meyer discovered independently the periodicity of properties of the elements as functions of their atomic weights. For the first time, then, it became possible to predict the chemical and physical properties of undiscovered elements. Mendeleev boldly presented the detailed properties of several undiscovered elements in his classic 1871 paper. Only four years later, in 1875, the discovery of gallium soon allowed a detailed verification of his predictions.

The modern atomic calculations support the concept that extrapolations into the superheavy region around 114 will be valid, although some care must be exercised. Certainly, interpolations (as used by Mendeleev) or extrapolations within a group are the method of choice. Indeed, if a group extrapolation cannot be made into the superheavy region because the points for the known elements are too scattered, this may be a warning that the property in question is not sufficiently periodic for a simple extrapolation of any kind. Thus the use of functions of unknown theoretical significance may yield a smooth curve that will extrapolate to an incorrect result. As an illustration, Keller and co-workers (40) found that the boiling points of Si, Ge, Sn, and Pb do not extrapolate smoothly to element 114. They found, however, that the heats of sublimation, S_M, fall on a smooth curve that can easily be extrapolated to yield a value of 10 kcal (g atom)$^{-1}$. They obtained a value for the heat of vaporization of 9 kcal (g atom)$^{-1}$ from the S_M of 114, and calculated a boiling point of 420 K using Trouton's rule. Trouton's constant is also a function of the group in the periodic table; a reasonable value of 22 was chosen for this constant. In this same way, the S_M for element 113 was found to be 34 kcal (g atom)$^{-1}$, the heat of vaporization, $\Delta H_v = 31$ kcal (g atom)$^{-1}$, and the boiling point = 1400 K. Keller and co-workers thus arrived at the result that the boiling point of 114 is much lower than the boiling point of 113, the difference, indeed, being about 1000 K. Also the difference in ΔH_v between 114 and 113 was obtained as 22 kcal (g atom)$^{-1}$. This large difference between the boiling points and heats of vaporization and sublimation between elements 113 and 114 is to be expected on the basis of relativistic effects, because the $7p_{1/2}^2$ orbital in element 114 is closed in essentially the same sense that the $6s^2$ orbital is closed in mercury. It is worthwhile to compare what we find in going from 113 to 114 with what is known about going from Au (which has a single s electron in its outer shell) to Hg (which has an s^2 closed outer shell). The difference in boiling points between Au and Hg is 2350 K and the difference in the heat of sublimation is 70 kcal (g atom)$^{-1}$. The comparison of the other group Ib and IIb elements with elements 113 and 114 is also instructive. The values of elements 113 and 114 in Table 2 are from Keller and co-workers (40). In the pairs Cu, Zn; Ag, Cd; and Au, Hg, an s electron is added to form a closed shell. In each case, the heats of sublimation and vaporization fall dramatically, along with the boiling point. Since the adding of a $7p_{1/2}$ electron in element 114 to form a closed shell is analogous to adding an s electron to form a closed shell in the lighter elements, a similar drop is expected in going from 113 to 114.

Table 2   Comparison of $7p_{1/2}$-shell closure to s-shell closure

	S_M kcal (g atom)$^{-1}$	Boiling Point (K)	ΔH_v kcal (g atom)$^{-1}$
Cu	81.1	2855	72.8
Zn	31.2	1181	27.6
Ag	68.4	2450	60.96
Cd	26.7	1038	23.9
Au	84.7	2980	77.5
Hg	14.6	630	13.6
113	(34)	(1400)	(31)
114	(10)	(420)	(9)

3.3   Oxidation-State Diagrams

The most important property to know about an element is the stable oxidation states it can assume. This is due to the fact that so many other chemical and physical properties are dependent upon the oxidation state. The second most important property to know is the relative stabilities of these oxidation states; that is to say, we need to know the standard electrode potential. The atomic computer codes allow calculations of ionization potentials and radii that can be used as input in oxidation-state diagrams to obtain the relative stabilities of oxidation states. These diagrams have been found useful in understanding much about inorganic chemistry (41).

Electrode potentials are by convention related to the standard $H^+/\frac{1}{2} H_2$ couple, whose potential is set equal to zero. We therefore consider the change in state for reduction of the aqueous metallic cation, M^{n+} (aq), to the metal, M(s)

$$M^{n+}(aq) + \tfrac{1}{2}n H_2(g) = M(s) + nH^+(aq). \qquad\qquad 1.$$

The change in Gibbs free energy is related to the standard reduction potential and to the enthalpy and entropy by the equation

$$-\Delta G^0 = nE^0 = -\Delta H^0 + T\Delta S^0. \qquad\qquad 2.$$

The ΔS^0 term is generally assumed to be small, or can be shown to be small (40); so we consider only ΔH^0, which can be obtained through the Born-Haber cycle. The heat of sublimation, S_M is obtained by extrapolation and the ionization potential, I_n is obtained through the use of a program such as the rel HF.

The next part of the Born-Haber cycle is most conveniently taken to be the single-ion hydration energy, $H_M n+$, although this quantity cannot be defined from a thermodynamic point of view. $H_M n+$ can be obtained by simple extrapolations or by calculation using various empirical modifications of the Born equation. For example, Keller et al, through the Born equation, obtained 72 kcal (g atom)$^{-1}$ for $H_{113}+$ (40).

Since we are not considering the entropy, we have for the change in state (1)

$$-\frac{1}{n}\Delta H^0 = E^0 = \frac{1}{n}\left[(I_n + S_M + H_M n+) - n(\tfrac{1}{2}D_{H_2} + I_H + H_H+)\right], \qquad 3.$$

where $\tfrac{1}{2}D_{H_2}$, half the dissociation energy of the hydrogen molecule, is 2.26 eV, I_H, the ionization energy of the hydrogen atom is 13.59 eV, and we accept the single-ion hydration energy of the proton as -11.3 eV, as derived by Halliwell & Nyburt (42). This yields a value of 4.5 eV for the energy released when one gram equivalent of hydrogen ions is combined with electrons.

Through the use of Equation 3, Keller et al calculated, for example, an E^0 (1 → 0) of $+0.6$ V for element 113 and E^0 (2 → 0) of $+0.9$ V for element 114 (40).

4   RESULTS OF CHEMICAL PREDICTIONS

Penneman & Mann (26) used equations developed by Jorgensen to predict the most stable states of elements 104–110 in aqueous solutions as follows: Rf(IV); Ha(V);

106(IV); 107(III); 108(II); 109(I); 110(O). In the case of some elements, several oxidation states appear likely. One suggestion is that Rf chemistry in solution may involve oxidation states of II and III, as well as IV. Jorgensen's approach is not intended to take into account the effects of oxyanion formation, such as that which stabilizes W(VI) in the tungstate ion. Since Re and Os are also stabilized by oxyanion formation, 106, 107, and 108 may be found in higher oxidation states in solution than postulated by Penneman & Mann.

Cunningham (27) used an extrapolation by group in the periodic table to obtain the prominent oxidation states for elements 104–110 as: 104(IV); 105(V); 106(VI); 107(VII); 108(VIII); 109(VI); 110(VI). Cunningham's predictions begin deviating from those of Penneman & Mann at 106. The higher oxidation states predicted by Cunningham are probably more appropriate for solutions where oxyanions, analogous to tungstate, may be formed.

For 111, eka-Au, Keller et al (43) predict the most stable oxidation state as III with the I state possible in the presence of highly polarizable ligands (such as CN^-). The II state is expected to be unstable. An unusual feature may arise in the $(111)^-$ ion, which would be analogous to the auride ion.

Pitzer (44) has combined relativistic quantum-mechanical calculations with a penetrating intuition to arrive at some most useful results on the chemical and physical properties of 112, eka-Hg. The relativistic effect on the closed $7s^2$ shell will make 112 more noble than Hg, so that the oxide, chloride, and bromide are expected to be unstable, although (112) Cl_4^{2-} and (112) Br_4^{2-} should be found in solution, and (112) F_2 should be stable. Pitzer also predicts that 112 will be a volatile liquid or even a gas, since the atoms will be bound together only by dispersion forces.

Table 3 summarizes a number of properties for Rf through 112. The standard electrode potentials of 110 and 112 were obtained by Hulet (personal communication), through the use of plots of the log of pertinent thermodynamic quantities versus the log of the atomic number, following the method of David (45, 46). The +5 states for Ha and 107 are formal designations taken from the literature (29, 46). Keller and Burnett (unpublished) derived the boiling and melting points of Rf in the same manner as for 113 (40).

According to Keller et al (40), the chemical behavior of 113 is expected to lie between those of Tl^+ and Ag^+. 113^+ is expected to bind anions more readily than Tl^+, so that (113)Cl will be rather soluble in excess HCl, whereas the solubility of TlCl is essentially unchanged. Similarly, (113)Cl is expected to be soluble in ammonia water, in contrast to the behavior of TlCl. The behavior of the 113^+ ion should tend toward Ag^+ in these respects. Also, although Tl(OH) is soluble and a strong base, the 113^+ ion should form a slightly soluble oxide that is soluble in ammonia water.

Pitzer has noted (44), for 114, that the effect of the $7p_{1/2}^2$ closed shell will cause eka-Pb to be a volatile liquid or even a gas like 112. His results for 114 are qualitatively similar to those of Keller et al (40). According to Pitzer, (114)Cl_2 and (114)F_2 will be stable, and probably (114)Br_2 will be stable as well. For both 112 and 114, Pitzer recommends using their great volatility and ease of reduction as separation methods in approaches to their discovery.

The chemistry of 115, 116, and 117 will be most interesting, since the $7p_{3/2}$ electrons

Table 3 Summary of properties predicted for Rf-112

Element	Rf	Ha	106	107	108	109	110	111	112
Stable oxidation states	III, IV	IV, V	IV, VI	III–VII	II–VIII	I–VI	I–VI	III,(−1)	I,II
First ionization energy (eV)[a]	5.6	6.6	7.6	6.9	7.8	8.7	9.6	10.5	11.4
Standard electrode potential (V)	(4→0) −1.8[b]	(5→0) −0.8[b]	(4→0) −0.6[b]	(5→0) +0.1[b]	(4→0) +0.4[b]	(3→0) +0.8[b]	(2→0) +1.7[c]	(3→0) +1.9[b]	(2→0) +2.1[c]
Ionic radius (Å)[d]	(+4) 0.71	(+5) 0.68	(+4) 0.86	(+5) 0.83	(+4) 0.80	(+3) 0.83	(+2) 0.80	(+3) 0.76	(+2) 0.75
Atomic radius (Å)[e]	1.50	1.39	1.32	1.28	1.26				
Density (g cm^{-3})[e]	23	29	35	37	41				
Heat of sublimation (kcal mol^{-1})[b]	166	190	205	180	150	142	115	80	
Boiling point (K)	5800[f]								
Melting point (K)	2400[f]								

111: electron affinity ~1.6 eV[g]

[a] Reference 30
[b] Reference 46
[c] E. K. Hulet, personal communication
[d] Reference 29

[e] Reference 26
[f] O. L. Keller, J. L. Burnett, unpublished
[g] Reference 43

will be added outside of the $7p^2_{1/2}$ closed shell. Keller, Nestor, & Fricke (37) therefore predict for 115, eka-Bi, a stable oxidation state of I, as well as III, with 115^+ being similar in its behavior to Tl^+. The group oxidation state of V will probably not be found. Although detailed predictions of the chemistry of 116, eka-Po, and 117, eka-At, have not been made, the II and III oxidation states will probably be of considerable importance, since two and three $7p_{3/2}$ electrons are present, respectively, outside of the $7p^2_{1/2}$ closed shell. The prominence of the -1 state for the halogen 117 has been questioned (29).

Grosse (47) has given detailed predictions of the properties of 118, eka-Rn. He expects 118 to be more reactive than Xe, forming compounds with chlorine as well as oxygen and fluorine. Pitzer (48) has made the important observation that the fluoride of 118 will most probably be ionic in form rather than molecular, as in the case of Xe. It is therefore predicted to be nonvolatile [as has already been observed for radon fluoride, in the exceptional work by Stein (49)].

Penneman & Mann (26) find, on the basis of rel HF calculations, that 119 will be the first alkali metal to have oxidation states higher than I. This result is in accord with the predictions of Grosse and Pitzer on the reactivity of 118, since it basically means reactivity of the rare-gas "core." The chemistry of 120 has not been examined in detail, but it is expected to follow alkaline earth chemistry in a vein similar to that in which 119 follows alkali metal chemistry.

Some of the more important quantities predicted for elements 113–120 are collected in Table 4.

Other reviewers (1, 29) have already treated the chemistry of the superactinides and still heavier elements (50). We do not pursue the subject further at this time, since we would only try to optimize in a sea of uncertainty.

5 SEARCHES FOR SUPERHEAVY ELEMENTS

Many investigators at laboratories throughout the world have searched in vain for superheavy elements in nature or have sought to synthesize them through heavy-ion bombardments (7, 51). Now the searches are concentrating at heavy-ion accelerator laboratories in Berkeley, Dubna, Darmstadt, and Orsay. If superheavy elements are found through the use of heavy-ion reactions, chemistry may very well be the key to their discovery because chemical separations allow the maximization of yields through the use of thick targets and the most intense beams.

We mention here a few attempts, then describe what may be the most hopeful approach, although even that has been unsuccessful so far.

Flerov and co-workers at Dubna have made a special study of "fusion-fission" reactions as a possible mechanism for producing superheavy elements (52). They postulate that perhaps projectile and target nuclei can fuse into some sort of intermediate nucleus that can then fission in such a way as to produce a superheavy element. As part of their program, they have bombarded uranium in their cyclotron with ^{136}Xe ions having an energy of ~ 900 MeV and separated the target chemically into an "actinide" fraction and a "heavy-metals–insoluble sulfide" fraction. About 40 spontaneous-fission events were detected, mostly in the "insoluble sulfides"

Table 4 Summary of properties predicted for 113–120

Element	113[a]	114[a]	115[b]	116	117	118[f]	119	120
Stable oxidation states	I	II	I,III	II,IV	(−1),I,III,V	II,IV,VI,VIII	I,others	II
Ionization energy (eV)								
I	7.4	8.5	5.2	6.6[c]	7.6[c]	8.7[c]	4.8[c]	6.0[d]
II		16.8	18.1					
Standard electrode potential (V)	(1→0) +0.6	(2→0) +0.9	(1→0) −1.5	(2→0) +0.1[e]			(1→0) −2.7[e]	(2→0) −3.0[e]
Ionic radius (Å)	(+1) 1.4	(+2) 1.2	(+1) 1.5 (+3) 1.0				(+1) 1.8[d]	(+2) 1.6[d]
Atomic radius (Å)	1.7[g]	1.6[g]	2.0					
Density (g cm⁻³)	18[g]	22[g]	11[g]					
Heat of sublimation (kcal/mole⁻¹)	34	10	34					
Boiling point (K)	1400	420	~1400			263	10[e]	
Melting point (K)	700	340	~700			258	33[e]	

[a] Reference 40
[b] Reference 37
[c] Reference 30
[d] Reference 29

[e] Reference 46
[f] Reference 47
[g] O. L. Keller, unpublished

fraction, corresponding to a production cross section of about 10^{-33} cm². A half-life of ~ 150 days was found for this spontaneous-fission activity and an average neutron-emission rate per fission of 1.5–3.5, a value much lower than expected for a superheavy element.

A separation scheme developed at Berkeley (53) has been used extensively at the SuperHILAC in searches for superheavy elements. The scheme, as shown in Figure 7, depends on the expected volatility of bromides and on the expected tendency of

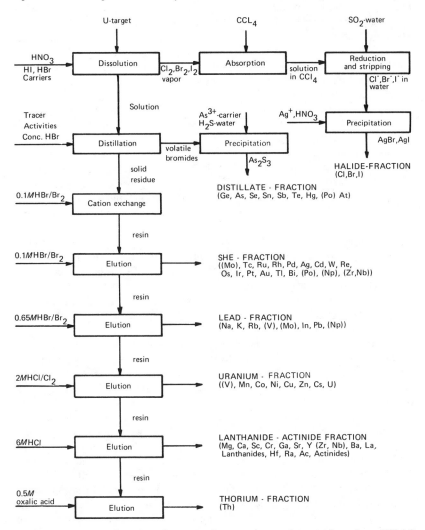

Figure 7 Processing flow-sheet for separating superheavy elements from SuperHILAC targets at Berkeley.

elements 108–116 to form strong complexes with bromide ions, as opposed to the much weaker complexes formed by the actinides. This difference in behavior is expected to allow a separation of superheavy elements from actinides. The scheme shown in Figure 7 has been tried on thick uranium targets that have been bombarded with 450–605-MeV ^{84}Kr (54) and 750–1150-MeV ^{136}Xe (55) ions. No evidence for superheavy elements was found in the experiments, which were designed to detect decay by spontaneous fission. The sensitivity of the ^{84}Kr experiments, covering a half-life range of a few days to 200 days, sets an upper limit for the formation cross section of about 10^{-35} cm^2, while for the ^{136}Xe experiments, covering a half-life range of about a day to 200 days, the upper limit is about 3×10^{-34} cm^2; very volatile super-heavy elements would have been detected with less efficiency, corresponding to some-what larger upper limits for the formation cross sections. The cross section for the fusion-fission reaction, as determined by the yield of fission products, was found to be too small to measure, less than 20 mb in each case. This suggests that the fusion-fission mechanism for the production of superheavy elements, which requires very heavy ions as projectiles, is probably not feasible. Looking in another direction, the transfer of nucleons between the heavy-ion projectile and the target nucleus by the quasi-elastic or by the quasi-fission (56) (deep-inelastic scattering) reaction leads to nuclei with atomic and mass numbers larger than those of the target nucleus. However, so massive an acquisition of nucleons by any available target nucleus is required in order to reach the region of superheavy elements, and the loss through the fission process is so severe that it appears unlikely that this, either, is a feasible route to superheavy elements.

Esterlund and co-workers (57) have bombarded 0.16-mm uranium metal foils with 620-MeV ^{63,65}Cu ions furnished by the University of Manchester LINAC. They made chemical separations based on co-precipitation of superheavy elements on cadmium sulfide from acid solution, on the hypothesis that they might be synthesized by the formation of a compound nucleus followed by a sequence of α- and β-decays. They observed 2 spontaneous-fission events in 178 days of counting, but drew no conclusions as to their significance.

Although the attempts to synthesize superheavy elements by bombardments with heavy ions have led to generally negative results, an analysis of heavy-ion reaction mechanisms provides some hope for success, through the compound-nucleus route, provided experiments with even greater sensitivity can be performed. Based on our present knowledge, the use of ^{48}Ca projectiles seems to be the most promising for several reasons. First, the ^{48}Ca nucleus is doubly magic, thus minimizing the excitation energy of the fused product nucleus and, one would hope, allowing some survival from the huge loss due to fission. Second, it is relatively neutron-excessive, which contributes to a closer approach to the closed-shell (magic) neutron number, 184. And third, its atomic number ($Z = 20$) is sufficiently small to allow the fusion reaction to occur with a relatively high probability. (Kratz et al (58) have measured the fusion-fission cross section of 200–288-MeV ^{40}Ar plus ^{238}U as ~ 600 mb). Thus, the bombardment of ^{248}Cm with ^{48}Ca ions with energy not too far above the Coulomb barrier height leads to the production of the compound nucleus ^{296}116 with an excitation energy of about 25 MeV; the emission of two neutrons, in competition with a tremendous probability for fission, might allow the detection, for example, of $^{294}_{178}116$, a nuclide only two protons and six neutrons removed from the doubly closed shell, $Z = 114$ and $N = 184$. Decay by electron capture might lead to nuclides closer to $Z = 114$ (or < 114) and $N = 184$. The bombardment of ^{244}Pu with ^{48}Ca holds forth similar promise. As another example, the bombardment of the tightly bound, doubly closed shell nucleus $^{208}_{82}$Pb with heavy ions such as $_{28}$Ni-$_{32}$Ge might lead to isotopes of elements 110–114 with sufficiently low excitation energy to survive loss by fission to the extent necessary to allow detection.

Hulet et al (personal communication), Oganessian et al (personal communication through J. M. Nitschke), and Ghiorso et al (personal communication) have bombarded ^{248}Cm with ^{48}Ca ions, and have looked for any resultant superheavy element nuclides using a variety of techniques.

In the work of Hulet et al, the most sensitive experiments were conducted with chemical separation of the sought-after superheavy-element products and related to half-lives in the range of one hour to years. Several different chemical schemes, which included removal of actinides by cation-exchange procedures, were used to isolate superheavy elements in the course of the experiments: (a) stabilization as bromide-complex ions based on the procedure of Kratz et al (53); (b) co-precipitation with copper sulfide from acid solution and from alkaline solution; and (c) electrodeposition from aqueous solution at variable controlled electrode potentials to cover expected reduction potentials for this region of elements. The bombardments, carried out at the Berkeley SuperHILAC, were conducted under the following conditions: duration of 10–45 hr, ^{48}Ca intensities of 2–4 $\times$ 10^{11} particles/sec^{-1}, and energy range of 245–265 MeV in a ^{248}Cm target (97% ^{248}Cm) of thickness about 2 mg cm^{-2} Cm (as CmF$_3$). Counting of spontaneous-fission events over a six-month period led to the observation of so small a number (no more than three in any experiment) that their assignment as background radiation cannot be ruled out. From the data presently available, the cross section for the production of superheavy elements within this range of half-lives is equal to or less than 10^{-34} cm^2. Future experiments, aided by more intense heavy-ion beams, with sensitivities increased 100-fold or more to detect products with formation cross sections of 10^{-36} cm^2 or less, might well yield positive results. Attempts to detect α particles, which might be due to superheavy elements, in the chemical fractions, were hampered by the presence of relatively high intensities of α particles due to the production of known, and perhaps new, isotopes of superheavy-element homologues (Bi, Po, At, etc) in these fractions. In supplementary experiments, recoiling nuclei were caught in aluminum foils, which were examined for spontaneous fission and α-particle activities. These experiments would have detected superheavy elements with half-lives as short as 10 min if they were produced with cross sections as large as 10^{-32} cm^2.

Oganessian et al used the Dubna 300-cm cyclotron as their source of ^{48}Ca ions for the bombardment of ^{248}Cm (8% ^{248}Cm). They performed chemical separations in which they searched for spontaneous fission and α particles, which might be due to superheavy elements, in volatile fractions and in insoluble sulfides precipitated from aqueous solution. Their negative results are consistent with the cross section limits set by Hulet et al for approximately the same range of half-lives.

Ghiorso et al also examined a broad range of possible shorter half-lives for recoiling superheavy-element nuclides that might be produced in bombardments of ^{248}Cm with ^{48}Ca ions. These experiments were less sensitive, in terms of limits on production cross sections, than those involving the chemical separations for longer-lived isotopes. They used the vertical wheel (VW) type of scheme employed in the first identification of isotopes of elements 104 (9), 105 (10), and 106 (23); in this system, the recoiling atoms are caught and swept in a stream of helium gas for deposition on the rim of a wheel that can be rotated to present the nuclides to a series of spontaneous-fission and alpha-particle detectors in measured time sequences. Sensitive to the half-life range of 0.1–10 sec, these experiments suggest that the production cross section for superheavy elements must be less than 10^{-32} cm^2. In experiments sensitive to half-lives as short as 10^{-8} sec, they used mica detectors to look for spontaneous-fission decay during the flight of recoiling atoms; here they could set an upper limit of about 10^{-31} cm^2 for the production cross section of superheavy elements.

The sensitivity for the detection of superheavy elements in the range of half-lives below that represented by the sensitive chemical experiments can be greatly increased by a number of techniques and by working with the much more intense heavy-ion beams that will eventually be available. It should be possible to detect superheavy elements, with half-lives down to nanoseconds or less, and production cross sections as low as 10^{-36} cm^2 or less, thus greatly increasing the probability for their detection.

6   CONCLUSION

The chemistry of the transactinide elements represents a continuing, challenging, and exciting field of investigation with much that can be, and undoubtedly will be, accomplished. The general form of the periodic table in this region seems to have been established. Of central importance has been the chemical evidence confirming that the actinide series of elements is completed with element 103, lawrencium—the demonstration that lawrencium has a relatively stable tripositive (2), and rutherfordium a relatively stable tetrapositive (3, 5) oxidation state in aqueous solution.

The chemical properties of all these elements must be studied on a one-atom-at-a-time (single-atom) basis. However, it is clear, on the basis of past accomplishments, that much can be learned despite these limitations. For the first few transactinide elements, it will be possible to study the chemistry of ions in aqueous solution. However, the half-lives decrease so fast with ascending atomic number that beyond these first few it will be impossible, at least for many years, to perform solution chemistry. The application of gas chromatography, gas-flow reactions, reactions with surfaces, and migration of gaseous atoms or ions should make it possible to get a limited amount of information about the chemical and physical properties of these elements down into the time span of milliseconds or less.

It is fervently hoped that it will be possible to enter a regime of longer half-lives in the region of the superheavy elements. We are optimistic that this will occur. The yields will apparently be so low, and, possibly, the half-lives so short as to impose formidable experimental difficulties. But the experimental capabilities, including increasing heavy-ion beam intensities and better experimental techniques, can be projected to improve tremendously in the future. It would add much to our understanding of basic chemical principles if we were able to test the predictions of the chemical properties of the superheavy elements—predictions based on relativistic considerations whose consequences are just becoming apparent in the atomic properties of the heaviest known elements. Experimental tests of these predictions would constitute observations on the concept of "relativity in the test tube" and would greatly expand our understanding of the Periodic Table of the Elements.

ACKNOWLEDGMENTS

The authors wish to express their gratitude to R. J. Silva and M. Nurmia for their aid in preparing this article.

Literature Cited

1. Seaborg, G. T. 1968. *Ann. Rev. Nucl. Sci.* 18:53–152
2. Silva, R., Sikkeland, T., Nurmia, M., Ghiorso, A. 1970. *Inorg. Nucl. Chem. Lett.* 6:733–39
3. Silva, R., Harris, J., Nurmia, M., Eskola, K., Ghiorso, A. 1970. *Inorg. Nucl. Chem. Lett.* 6:871–77
4. Zvara, I., Belov, V. Z., Domanov, V. P., Korotkin, Yu. S., Chelnokov, L. P., Shalaevskii, M. R., Shchegolev, V. A., Yussonnua, M. 1972. *Radiokhimiya* 14:119–122. (Engl. transl.) 1972 *Sov. Radiochem.* 14:115–18
5. Hulet, E. K., Nitschke, J. M., Lougheed, R. W., Wild, J. F., Landrum, J. H., Ghiorso, A. 1976. In *Transplutonium 1975. Proc. Symp. Baden-Baden,* ed. W. Müller, R. Linder, pp. 3–10. Amsterdam: North-Holland.
6. Fernelius, W. C., Loening, K., Adams, R. M. 1975. *J. Chem. Educ.* 52:583–84
7. Herrmann, G. 1975. In *MTP Int. Rev. Sci., Ser. 2, Inorg. Chem.,* ed. A. G. Maddock, 8:221–72. London: Butterworth
8. Milligan, W. O., ed. 1970. *Proc. Robert A. Welch Found. Conf. Chem. Res., 13th. The Transuranium Elements—The Mendeleev Centennial, Houston, 1969.* 494 pp.
9. Ghiorso, A., Nurmia, M., Harris, J., Eskola, K., Eskola, P. 1969. *Phys. Rev. Lett.* 22:1317–20
10. Ghiorso, A., Nurmia, M., Eskola, K., Harris, J., Eskola, P. 1970. *Phys. Rev. Lett.* 24:1498–1503
11. Ghiorso, A., Nurmia, M., Eskola, K., Eskola, P. 1970. *Phys. Lett. B* 32:95–98
12. Seaborg, G. T., 1963. *Man-Made Transuranium Elements.* Englewood Cliffs, NJ: Prentice-Hall. 120 pp.
13. Druin, V. A., Lobanov, Yu. V., Nadcarni, D. M., Kharitonov, Yu. P., Korotkin, Yu. S., Tret'yakova, S. P., Krashonkin, V. I. 1973. *At. Energ.* 35:279–80. (Engl. transl.) *Sov. At. Energy* 35:946–48
14. Dittner, P. F., Bemis, C. E., Ferguson, R. L., Plasil, F., Pleasonton, F., Hensley, D. C. 1974. *Oak Ridge Natl. Lab. Chem. Div. Ann. Prog. Rep.,* ORNL-4976, pp. 39–40
15. Zvara, I., Chuburkov, Yu. T., Belov. V. Z., Burklanov, G. V., Zakhvataev, B. B., Zvarova, T. S., Maslov, O. D., Caletka, R., Shalaevsky, M. R. 1970. *J. Inorg. Nucl. Chem.* 32:1885–94
16. Ghiorso, A., Nurmia, M., Eskola, K., Eskola, P. 1971. *Phys. Rev. C* 4(5):1850–55
17. Flerov, G. N., Oganesian, Yu. Ts., Lobanov, Yu. V., Lazarev, Yu. A., Tretiakova, S. P., Kolesov, I. V., Plotko, V. M. 1971. *Nucl. Phys. A* 160:181–92
18. Druin, V. A., Demin, A. G., Kharitonov, Yu. P., Akap'ev, G. N., Rud', V. I., Sung-Ching-Yang, G. Y., Chelnokov, L. P., Gavrilov, K. A. 1971. *Yad. Fiz.* 13:251–55. (Engl. transl.) 1971. *Sov. J. Nucl Phys.* 13:139–41
19. Zvara, I., Belov, V. Z., Korotkin, Yu. S., Shalayevsky, M. R., Shchegolev, V. A., Hussonnois, M., Zager, B. A. 1970. *Joint Inst. Nucl. Res. Dubna Rep.,* JINR P12-5120. 13 pp. (In Russian)
20. Zvara. I., Belov, V. Z., Domanov, V. P., Shalayevsky, M. R. 1975. *Joint Inst. Nucl. Res. Dubna Rep.,* JINR P6-8740. 15 pp. (In Russian)

166   KELLER & SEABORG

21. Zvara, I. 1973. *Proc. Int. IUPAC Congr.,* 24th Hamburg. 6:73–90
22. Zvara, I. 1976. In *Transplutonium 1975, Proc. Symp. Baden-Baden,* ed. W. Müller, R. Lindner, pp. 11–20. Amsterdam: North-Holland
23. Ghiorso, A., Nitschke, J. M., Alonso, J. R., Alonso, C. T., Nurmia, M., Seaborg, G. T., Hulet, E. K., Lougheed, R. W. 1974. *Phys. Rev. Lett.* 33:1490–93
24. Oganessian, Yu. Ts., Tretyakov, Yu. P., Iljinov, A. S., Demin, A. G., Pleve, A. A., Tretyakova, S. P., Plotko, V. M., Ivanov, M. P., Danilov, N. A., Korotkin, Yu. S., Flerov, G. N. 1974. *Joint Inst. Nucl. Res. Dubna Rep.,* JINR D7-8099. 13 pp.
25. Harvey, B. G., Herrmann, G., Hoff, R. W., Hoffman, D. C., Hyde, E. K., Katz, J. J., Keller, O. L., Lefort, M., Seaborg, G. T. 1976. *Science* 193:1271–72
26. Penneman, R. A., Mann, J. B. 1976. *Proc. Moscow Symp. Chem. Transuranium Elem., J. Inorg. Nucl. Chem., Suppl. 1976,* ed. V. I. Spitsyn, J. J. Katz, pp. 257–63. Oxford: Pergamon
27. Cunningham, B. B. 1970. See Ref. 8, pp. 307–22
28. Oganesyan, Yu. Ts., Demin, A. G., Danilov, N. A., Ivanov, M. P., Il'inov, A. S., Kolesnikov, N. N., Markov, B. N., Plotko, V. M., Tret'yakova, S. P., Flerov, G. N. 1976. *Pisma Zh. Eksp. Teor. Fiz.* 23:306–9. (Engl. transl.) 1976. *JETP Lett.* 23:277–79
29. Fricke, B. 1975. In *Structure and Bonding,* ed. J. D. Duntiz et al, 21:89–144. Berlin: Springer
30. Mann, J. B. 1976. In *Predictions in the Study of Periodicity,* ed. B. M. Kedrov, D. N. Trifonov, Acad. Sci., USSR, Inst. Hist. Sci. Technol. pp. 161–201. Moscow: Science. (In Russian)
31. Desclaux, J.-P. 1973. *At. Data Nucl. Data Tables* 12:311–406
32. Waber, J. T., Cromer, D. T., Liberman, D. 1969. *J. Chem. Phys.* 51:644–68
33. Keller, O. L. 1976. In *Predictions in the Study of Periodicity,* ed. B. M. Kedrov, D. N. Trifonov, Acad. Sci., USSR, Inst. Hist. Sci. Technol., pp. 202–23. Moscow: Science. (In Russian)
34. Desclaux, J.-P., 1972. *Int. J. Quantum Chem.,* Symp. No. 6, pp. 25–41
35. Fricke, B., Waber, J. T. 1971. *Actinides Rev.* 1:433–85
36. Lu, C. C., Carlson, T. A., Malik, F. B., Tucker, T. C., Nestor, C. W. Jr. 1971. *At. Data* 3:120
37. Keller, O. L., Nestor, C. W., Fricke, B. 1974. *J. Phys. Chem.* 78:1945–49
38. Carlson, T. A., Nestor, C. W., Malik, F. B., Tucker, T. C. 1969. *Nucl. Phys. A* 135:57–64
39. Lu, C. C., Malik, F. B., Carlson, T. A. 1971. *Nucl. Phys. A* 175:289–99
40. Keller, O. L., Burnett, J. L., Carlson, T. A., Nestor, C. W. 1970. *J. Phys. Chem.* 74:1127–34
41. Phillips, C. S. G., Williams, R. J. P. 1966. *Inorganic Chemistry,* Vols. 1, 2. Oxford: Oxford. 685 pp, 683 pp.
42. Halliwell, H. F., Nyburg, S. C. 1963. *Trans. Faraday Soc.* 59:1126–40
43. Keller, O. L., Nestor, C. W., Carlson, T. A., Fricke, B. 1973. *J. Phys. Chem.* 77:1806–09
44. Pitzer, K. S. 1975. *J. Chem. Phys.* 63:1032–33
45. David, F. 1971. *Inst. Phys. Nucl. Orsay Rep.* RC-71.05. 30 pp. (In French)
46. David, F. 1971. *Inst. Phys. Nucl. Orsay Rep.* RC-71-06. 27 pp. (In French)
47. Grosse, A. V. 1965. *J. Inorg. Nucl. Chem.* 27:509–19
48. Pitzer, K. S. 1975. *J. Chem. Soc. Chem. Comm.* 18:760–61
49. Stein, L. 1970. *Science* 168:362–64
50. Penneman, R. A., Mann, J. B., Jorgensen, C. K. 1971. *Chem. Phys. Lett.* 8:321–26
51. Silva, R. J. 1972. In *MTP Int. Rev. Sci., Ser. 1, Inorg. Chem.,* ed. A. G. Maddock, 8:71–105. London: Butterworth
52. Flerov, G. N., 1974. *Proc. Int. Conf. React. Complex Nucl.,* ed. R. L. Robinson, F. K. McGowan, J. B. Ball, 2:459–81. Amsterdam: North-Holland
53. Kratz, J. V., Liljenzin, J. O., Seaborg, G. T. 1974. *Inorg. Nucl. Chem. Lett.* 10:951–57
54. Kratz, J. V., Norris, A. E., Seaborg, G. T. 1974. *Phys. Rev. Lett.* 33:502–5 and additional work (To be published)
55. Otto, R. J., Fowler, M. M., Lee, D., Seaborg, G. T. 1976. *Phys. Rev. Lett.* 36:135–38 and additional work (To be published)
56. Hanappe, F., Lefort, M., Ngô, C., Péter, J., Tamain, B., 1974. *Phys. Rev. Lett.* 32:738–41 and references therein
57. Esterlund, R. A., Molzahn, D., Brandt, R., Patzelt, P., Vater, P., Boos, A. H., Chandratillake, M. R., Grant, I. S., Hemingway, J. D., Newton, G. W. A. 1976. *J. Radioanal. Chem.* 30:17–23
58. Kratz, J. V., Liljenzin, J. O., Norris, A. E., Seaborg, G. T. 1976. *Phys. Rev. C* 13:2347–65

Ann. Rev. Nucl. Sci. 1977. 27 : 167-207
Copyright © 1977 by Annual Reviews Inc. All rights reserved

THE WEAK NEUTRAL ×5584
CURRENT AND ITS EFFECTS IN
STELLAR COLLAPSE

Daniel Z. Freedman
Institute for Theoretical Physics, State University of New York at Stony Brook,
Stony Brook, New York 11790

David N. Schramm[1] and David L. Tubbs[2]
Enrico Fermi Institute (LASR), University of Chicago, Chicago, Illinois 60637

CONTENTS

1 INTRODUCTION

In the summer of 1973, experimental results indicating the presence of a neutral current in the weak interactions were announced, and the effect has since been confirmed in all other laboratories with neutrino-beam facilities (1–7). The conventional charged-current description of weak interactions goes back to Fermi (8), and the modern version evolved rapidly after the discovery of parity violation in β decay (9–12), the formulation of the conserved–vector current hypothesis (13–15),

[1] Also Department of Astronomy and Astrophysics.
[2] Also Department of Physics.

and the discovery that there were two distinct neutrino species, v_e and v_μ (plus corresponding antineutrinos), associated with the e^- and μ^- particles (16). The charged current requires that any particle reaction initiated in a v_μ neutrino beam, such as $v_\mu + n \to \mu^- + p$, must have a μ^- particle in the final state. The existence of a neutral current was inferred because of the observation of reactions whose final state did not contain the required charged lepton. Examples of neutral-current reactions that have been observed directly are $v_\mu + e \to v_\mu + e$ (1), $v_\mu + p \to v_\mu + p + \pi^0$ (5, 17), and $v_\mu + p \to v_\mu + p$ (18, 19). The final-state neutrino cannot be detected, but its presence is confirmed by kinematics, i.e. energy and momentum conservation. Originally, the existence of the neutrino was inferred from similar kinematic arguments in nuclear β decay.

The neutral current of the 1973–1974 experiments was not born into an unprepared physics world, because it had been predicted in theoretical work by Weinberg in 1967 (20) and Salam in 1968 (21). In this work, the neutral current is a natural feature of a unified treatment of weak and electromagnetic interactions that is based on non-Abelian gauge-field theory of a type first invented by Yang & Mills (22), together with a spontaneous symmetry-breaking mechanism studied by Higgs and others (23–25). An extraordinary synthesis of theoretical ideas is contained in the work of Weinberg and Salam, whose general correctness is confirmed by the experimental discovery. We refer readers to (26–29) for recent reviews of the theoretical framework.

It has been realized for many years (30) that neutrinos play an important role in stellar evolution. This occurs simply because their interaction with matter is so weak that under most circumstances a neutrino escapes instantaneously from a stellar interior, carrying away energy and momentum. As we discuss later, this energy-loss mechanism determines both the time scale and the qualitative features of the intermediate and late stages of stellar evolution (31).

Transparency of matter to neutrinos is not an absolute concept; it is a quantitative question of the size of the neutrino mean free path λ relative to the radius R of the system. Matter is transparent for $\lambda \gg R$ and opaque for $\lambda \ll R$. The mean free path depends on the neutrino energy and composition, density, and temperature of the material. For a 20-MeV v_e or $\bar{v}_e$ in matter consisting of protons and neutrons not bound in nuclei, plus electrons, the cross section is 10^{-41} cm^2, and one finds $\lambda \approx 10^{17}$ ρ^{-1} cm. In the hydrogen-burning stage of an ordinary star, we have $\rho \sim 1$ gm cm^{-3}, which means $\lambda \sim 10^{17}$, while $R \sim 10^{11}$ cm; therefore ordinary stars are transparent.

Although, in the late stages of evolution, the center of a large star consists of medium-weight elements, when stellar collapse occurs, a core composed predominantly of neutron matter begins to form at an approximate density of 10^{11}–10^{12} gm cm^{-3} and radius $\sim 10^7$ cm. The mean free path is then $\lambda \approx 10^5$–10^6 cm, so that the system is opaque or optically thick. Neutrinos diffuse out of the core via multiple microscopic processes of production, absorption, and scattering. Because neutrinos remain the chief transporters of the $\sim 10^{53}$ ergs of energy liberated in the stellar collapse to form a neutron star, the neutrino-transport problem is critical for determining the mechanism of supernova explosions and whether the dense compact remnant is, in fact, a neutron star or a black hole.

The key problem here is not the onset of the collapse that follows from general thermodynamic stability considerations and nuclear binding-energy systematics. The problem is to understand the reversal of the implosion after the central core has attained the mass $(1.4\ M_\odot)$ of the neutron-star pulsar remnant and the subsequent blowoff of the outer layers (containing most of the mass of the star), which constitutes the observed supernova explosion.

The weak-interaction neutral current affects stellar collapse simply because there are many more neutrino processes predicted than occur in the charged-current framework, and the implications of these new processes have received examinations (32–37). There are new processes, such as $v_e + n \to v_e + n$ and $v_e + p \to v_e + p$, which have cross sections in the collapsed core comparable to the traditional $v_e + n \to e^- + p$ and $v_e + e \to v_e + e$, and which do not suffer from the inhibition of the latter due to electron degeneracy. Similarly, v_μ and $\bar{v}_\mu$ can now be produced efficiently in the hot core even at temperatures much lower than $kT \simeq m_\mu c^2$. For example, production occurs via the reaction $e^- + e^+ \to v_\mu + \bar{v}_\mu$, and these neutrinos can subsequently scatter via $v_\mu + e \to v_\mu + e$ and $v_\mu + n \to v_\mu + n$. These processes must be taken into account in treatments of neutrino transport.

An extremely simple process predicted by the weak neutral current, namely elastic scattering from a nucleus (Z,N) containing protons and N neutrons (33), can be important for neutrinos emerging from the core and traversing the mantle that contains nuclei such as Fe^{56} and Ni^{56}. The principle of quantum-mechanical coherence applies to this process and for a 20-MeV neutrino, a scattering cross section about a factor of $(Z+N)^2$ larger than that for single-nucleon processes is predicted. Because of this large cross section, momentum transfer from the neutrinos to the nuclei, i.e. neutrino-radiation pressure, may be quantitatively important in blowing off the outer layers of the star.

The purpose of this article is to review the properties of a weak neutral current with emphasis in the major processes of astrophysical interest, and to discuss our current understanding of its role in gravitational collapse and explosions. Both subjects are in a state of uncertainty. First, the detailed properties of the weak neutral current and therefore cross-section values are still not well determined experimentally. Second, the collapse of the hot stellar core and subsequent blowoff of the outer layers are intrinsically so complicated that it might be said that although Nature has clearly arranged for supernova explosions, man has not clearly understood the arrangements. Because of these uncertainties, this is not a detailed review whose main function is bibliographic. Instead, the main qualitative ideas are stressed at a level consistent with the interdisciplinary character of the subject. An astrophysicist should be able to read this article and appreciate the relative sizes of cross sections for various important neutrino processes and how to calculate some of them, while a particle physicist should be able to learn the basic issues in the present description of stellar collapse, as well as the relation of supernova explosions to such general questions as cosmic rays and the abundance of the elements.

1.1   Neutrinos and the Latter Stages of Stellar Evolution

As mentioned above, neutrinos play a role throughout stellar evolution. For most of the star's life, they serve as an energy sink, completely removing some fraction of the energy from nuclear burning in the star's core. A well-known attempt at detecting these neutrinos from a standard main-sequence star—the sun—is being carried out by Davis and his collaborators using a tank of C_2Cl_4 in the Homestake gold mine (38). However, no definitive results have been obtained to date. These neutrinos from normal stellar evolution result from standard charged-current Fermi processes and thus are not of prime interest in this present review.

In the latter stages of a star's evolution, it is no longer just a small fraction of the nuclear energy that goes into neutrinos, but instead stellar energy loss becomes dominated by neutrinos rather than photons. Let us now briefly review the current ideas on stellar evolution [see also (39)], so that we can gain some perspective on the important role of neutrinos and see how they might even be the key to understanding how, at a massive star's death, it somehow manages to eject its outer material while leaving behind a dense neutron-star or black-hole remnant.

It is known that when stars first form from a gas of hydrogen and helium, the hydrogen burns in the core to helium ($4H \rightarrow {}^4He + 2e^+ + 2\nu_e +$ energy) as the star evolves along the main sequence. When the helium core from such burning becomes large enough, it will begin to gravitationally contract until it heats to a sufficiently high temperature that the helium can begin to burn. During this core contraction, the outer envelope of the star expands and the star becomes a red giant. Then the core is supported by the thermal pressure at the temperature of helium burning, with the energy being supplied by the helium burning. When the helium burns, it creates carbon and oxygen via the triple-α reaction ($3 \, {}^4He \rightarrow {}^{12}C + \gamma$) and the ${}^{12}C(\alpha,\gamma) \, {}^{16}O$ reaction. Following the carbon-core development, there is a branching point in the stellar evolution. If the carbon core is sufficiently massive, it too can reach a point where it will collapse, carbon will begin to burn and the star can proceed to the later stages of evolution, going successively to a neon core, an oxygen core, a silicon core, and finally to an iron-nickel core (40). However, if the core is not sufficiently massive, in particular, if it is of very low mass, the evolution could stop at this point, since sufficient pressure to support the entire star can come from degenerate electrons, in which case one would have a white dwarf (41). Thus, depending on the mass of the star, there is a branching point in the stellar evolution. For this review, it is important to note that for all the stages of evolution from carbon burning onward, the main energy-loss mechanism from the core is via neutrinos. This means that cores evolve on much more rapid time scales than if they had to wait for the photons to diffuse out. Thus, the cores evolve almost independently of the outer parts of the star.

STELLAR MASS RANGES   Let us divide the possible mass ranges in which a star might fall into four segments. Stars with a mass less than M_{MS} require times longer than the present age of universe to evolve off the main sequence. Although these stars will eventually end their lives as white dwarfs, the time required for them to reach that stage is excessively long. The next division is the range between M_{MS} and some mass M_{WD}, that range where stars will be able eventually to become white dwarfs. Now, M_{WD} is not the Chandrasekhar mass, 1.4 $M_\odot$, although 1.4 $M_\odot$ is the upper limit for the mass of a white dwarf, since M_{WD} can be significantly greater than 1.4

due to mass-loss processes during the star's lifetime. Between M_{WD} and M_{CD} is a region where the carbon in the core will ignite under degenerate conditions; that is, the core is being partially supported by degenerate electron pressure so there is no feedback on the pressure when the carbon begins to burn. When this happens, one may get the carbon-detonation supernova model (e.g. 42). There is currently some question as to whether a detonation wave or a deflagration occurs and what the results of stars igniting carbon under these conditions will be (cf 43, 44). However, for completeness, this range must be mentioned. It may be that M_{WD} is greater than M_{CD} in which case this range would not exist at all. The third mass range, between M_{CD} and M_{pair}, is where the carbon ignites nondegenerately and goes on to burn to neon, the neon burns to oxygen, the oxygen burns to silicon, and the silicon burns to iron and nickel, so that one obtains the onion-skin model (see Figure 1). This is the mass range that is of primary interest in this review. As we show below, stars in this range may be able to explain the origin of the bulk of the heavy elements, cosmic rays, pulsars (neutron stars), and maybe even black holes. There is also the range beyond M_{pair}, where the pair-formation supernova occurs (45–48). However, M_{pair} is probably around 60–70 $M_\odot$ and it is not at all clear that there are stars in this range, since none are seen. Of course, it may be that stars this massive always exist in clouds and thus it would be difficult to see them. In addition, it may be that such massive stars undergo such rapid mass loss that they never form a core large enough for the pair-formation supernova to take place (49). However, these very massive stars are believed to blow up via what is called the pair-supernova mechanism (45–47), assuming they would form and that they do not have as large a mass-loss rate as is implied by extrapolations up into that region. The range that

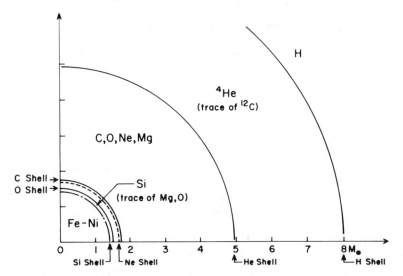

Figure 1   Cross section of an evolved 22-$M_\odot$ star showing the 1.4-$M_\odot$ Fe-Ni core with the surrounding shells of Si, O, Ne, C, He, and the H envelope.

is relevant for out current interest, though, is the range between M_{CD} and M_{pair}, where the star evolves into an onion-skin model with a very dense iron-nickel core, which then goes on to collapse gravitationally and produce many neutrinos.

Let us see what evidence there is concerning the characteristic masses M_{SD}, M_{WD}, M_{CD}, and M_{pair}. It is known that M_{MS} is somewhat less than 1 $M_\odot$, since we know the main-sequence evolution times. M_{WD} is a mass for which a star is able to shed enough mass, during its lifetime to get below the 1.4-$M_\odot$ Chandrasekhar limit. From the fact there is a white dwarf in the very young cluster, the Pleiades, and that all stars up to about 6 $M_\odot$ are still on the main sequence in the Pleiades, it seems clear that for that particular white dwarf, a star greater than 6 $M_\odot$ was able to shed enough mass to become a white dwarf. However, that does not mean that all stars from 6 $M_\odot$ down shed enough mass to become white dwarfs. There may be other variables that enter in, such as rotation and magnetic fields, which alter the rate of mass loss for different stars. For example, one finds that the number of white dwarfs in the Hyades is consistent more with a value for M_{WD} of around 3–4 $M_\odot$, than a value of 6 $M_\odot$. However, there may have been other factors contributing. For example, white dwarfs may have been lost from the Hyades, in which case the mass limit would go above 4 $M_\odot$. In view of the present uncertainties, it is probably best to estimate M_{WD} as somewhere between 4 and 8 $M_\odot$, with 6 being a reasonable best guess. The next mass limit is where carbon ignites under non-degenerate conditions, as contrasted with lower-mass cores having carbon that ignites under degenerate conditions. If we take the current estimates for the rate of the $^{12}C + ^{12}C$ reaction and the estimates of Paczynski (50) on the mass of the carbon core associated with the main-sequence mass, then we estimate the mass at which carbon burns nondegenerately to be approximately 8 $M_\odot$. However, because the mass of the carbon core associated with the main-sequence mass depends on convection theory, this mass is very uncertain. In fact, recent calculations of Barkat (48) show that this mass may go down to 5 or 6 $M_\odot$. Also, the rate of the $^{12}C + ^{12}C$ reaction is somewhat uncertain [cf the discussion in (39)], which also affects the location of this mass limit. Therefore it is best to just say that M_{CD} is somewhere between 4 $M_\odot$ and 10 $M_\odot$, with a "best" value of about 8. Note that there is an overlap between M_{WD} and M_{CD}. That is, there may be no stars in the range that undergo carbon detonation. In fact, it may very well be that the phenomenon of a degenerate core does contribute to the mass-loss phenomenon, in which case M_{WD} and M_{CD} may be the same number. This argument has been discussed by Paczynski, among others. It is actually very good that this mass region for the carbon-detonation model is small or nonexistent, in that current calculations of the carbon detonation, if the carbon really does detonate, indicate that the entire carbon core is converted to iron and ejected. Were ejection to happen for a very large-mass region, then there would be too much iron in the galaxy and we would have an iron catastrophe. Therefore, it is important either that this mass region be very small or that the final stage of life of these stars not be a carbon-detonation supernova.

The upper mass limit on the onion-skin model occurs when the carbon-oxygen core becomes sufficiently large that rapid pair formation causes a dynamic instability. Dearborn (49) has recently shown that any reasonable extrapolation

Table 1 Stellar mass ranges

Mass range	Consequence
$M < M_{MS} \lesssim 1\,M_\odot$	Main sequence life $>$ age of universe
$M_{MS} \lesssim M \lesssim M_{WD} \approx 6 \pm 2\,M_\odot$	Star loses sufficient mass to become a white dwarf of less than $1.4\,M_\odot$
$M_{WD} \lesssim M \lesssim M_{CD} \approx 8 \pm \frac{2}{3}\,M_\odot$	Carbon ignites under degenerate core conditions $\rightarrow$ Carbon-detonation supernova?
$M_{CD} \lesssim M \lesssim M_{pair} \approx 100 \pm 30\,M_\odot$	Onion-skin star develops finally, yielding Fe core surrounded by Si, O, Ne, C, He, and H shells $\rightarrow$ Core-collapse supernova
$M \gtrsim M_{pair}$	May not occur, due to such high mass loss that core never deviates from onion-skin scenario $\rightarrow$ Pair-formation supernova

of main-sequence mass-loss rates into the regime of massive stars would indicate that these massive stars lose mass at such a rapid rate that cores sufficiently massive for the pair-formation scenario may never develop. However, it is difficult to be certain of properties of these unseen massive stars. Thus, to be conservative, we put M_{pair} at $\sim 100\ M_\odot$. Table 1 summarizes the above comments [see also the review by Tinsley (51)].

SUPERNOVAE   There is common feeling in astrophysics that the observed supernovae may be related in name only to the theoretical supernovae. This misconception occurs to a large degree because, until recently, theoreticians had not tried to make detailed connections between explosion mechanisms that originate in the interior of the star and observational characteristics of supernovae that are determined by the expansion of the hot surface material.

Recently, this has changed, as several groups (52–54) have begun to calculate what happens when a great deal of energy is released in the center of a big gas cloud. A supernova is in fact just that, since it has a very-large-radius, low-density envelope with a very small, compact stellar core from which the energy comes. The calculations of Lasher (53), Chevalier (54), and Falk & Arnett (52) seem to be able to fit many of the observed characteristics of supernovae. In particular, they find that less than 1% of the neutron-star binding energy is needed to produce an observed supernova. However, one thing they are not able to resolve is how this energy is to be deposited, since by the time the supernova light is seen, almost all memory of how the "bomb" was set off in the middle of the cloud disappears. (Perhaps neutrino astronomy will one day resolve this problem.)

In the rest of this review, we focus on the details of this energy release. First, we recall some of the observed properties of supernovae. One of the most interesting is the rate of appearance of supernovae. Tammann (55) has done a systematic study and points out that, contrary to folklore, the supernova rate in our galaxy seems to be about one every $15 \pm \frac{15}{5}$ years. This high rate comes from the fact that we can only see supernovae in a narrow segment of our galaxy, so the fact that only 5 have been observed in the past 1000 years should not be discouraging. Tammann goes on to show that this is consistent with observed supernovae rates in other galaxies similar to our own. It has also been noted (56) that pulsar-formation rates may be much higher than previously thought. Since pulsars are presumably neutron stars left by supernovae, then pulsar-formation rates should be very closely related to supernovae rates.

Another observational point that tends to confuse physicists is the term Type I or Type II supernovae, which refers to the observational characteristics of the light curve, i.e. light intensity as a function of time after the explosion. Type I supernovae have light curves of a special shape with a peak and then an exponential drop [cf (57)], and Type II supernovae are a more loosely defined grouping of everything else (50). [Zwicky (58) originally employed many more supernovae classifications, but Type I and Type II are the only ones in frequent usage.] Type I's also show very little or no hydrogen, whereas Type II's show high levels of hydrogen. Recently Tammann (55) has argued that the light curves and energetic characteristics of one type seem to blend into those of the other type. Thus, the only clear distinction may be one of the spectrum. The location of young stars near the inner edge of galactic spiral arms suggests (51) that Type II supernovae probably have massive stars as their progenitors. The situation for Type I's is not so clear. At one time it was assumed that since Type I's occur in elliptical galaxies that normally are thought to have old stellar populations, then Type I's must be low-mass objects. However, objections have been raised to that argument (51, 55). It may be that a Type I results whenever the hydrogen envelope is stripped from the core. This may occur in the course of binary star evolution or by some other means. It may also be that Type I's are related to carbon-detonation models and Type II's to more massive stars. (Carbon detonation might occur by accretion in binary evolution, even if it is bypassed by mass loss in single-star evolution.)

In 1054 A.D., Chinese observers recorded the appearance of the most famous supernova, which was bright enough to be seen during the day. It is not easy to tell from the remnant we observe today, the Crab Nebula, whether the supernova was Type I or Type II. Of particular importance is the fact that the Crab's spectrum is dominated by helium. Arnett (59) has used this point to argue that the progenitor for the Crab was a $10–12\ M_\odot$ star, since such stars should produce large amounts of helium. The mass of the Crab remnant is very uncertain (60). It is greater than a solar mass, and could be as large as 8 or $10\ M_\odot$. It is also conceivable that the progenitor of the Crab lost large amounts of mass (its hydrogen) prior to the explosion. The most exciting aspect of the Crab is the existence of the pulsar at its center. This observation is the prime reason why one wants a supernova mechanism that not only blows off the outer part of a star, but also allows the central region to collapse into a neutron star.

Another interesting supernova type is the "invisible supernova," of which the remnant Cas A seems to be an example (61). It went off ~ 300 years ago, but its appearance was never recorded on earth. This is all the more interesting when one notes that Cas A shows evidence for nucleosynthesis of heavy elements, and Chevalier argues that it is a good cosmic-ray source. Arnett (59) has argued that the progenitor for Cas A was a massive star of $\sim 24\ M_{\odot}$. Perhaps some mass-loss mechanism so stripped Cas A that it had no envelope to produce a bright optical light curve when it went off.

Supernovae observations are rapidly improving and more calculations are being done to relate core explosions to exterior light. Thus, the relation between observed and theoretical supernovae seems to be improving.

NUCLEOSYNTHESIS   The stars in which we are most interested are those that do evolve to an onion-skin model with an iron-nickel core surrounded by a silicon shell, an oxygen shell, neon shell, carbon shell, helium shell, and a hydrogen envelope, and that may become Type II supernovae. One particularly attractive feature about these more massive stars is that they produce supernovae and they seem to be able to provide us with the synthesis sites for the bulk of the heavy elements (40, 62). In fact, they also seem to be able to provide good sources for the galactic cosmic rays (62, 63). In particular, with regard to the production of the heavy elements, it is interesting that the conditions in these massive stars may very well be close to those conditions achieved in the explosive nucleosynthesis calculations.

In the late 1960s and early 1970s, much work on explosive nucleosynthesis was carried out. This work has been reviewed recently by Arnett (64), and discussed in the book *Explosive Nucleosynthesis* (42), and so is not described in detail here. What was done in these explosive nucleosynthesis calculations was to set up a nuclear-reaction network that coupled all of the reactions thought to be important in a particular region of the chart of the nuclides. The reaction rates were dependent on temperature and density. The temperature and density in these calculations were made time-varying, following a simple exponential expansion law $\rho = \rho_0 \exp(-t/\tau)$, where the expansion rate was a multiple of the free-fall time $1/(24\pi\ G\rho)^{\frac{1}{2}}$. These calculations also tended to assume that the temperature varied in an adiabatic manner, with ρ proportional to T^3. It was found that good agreement with the observed abundances was obtained only for a certain set of conditions—values of T_0 and ρ_0 and the composition variable η_0 that was just a measure of the neutron enrichment [η is equal to $(n-p)/(n+p)$]. However, if ρ, T, and η were away from these values, then agreement was not possible. In particular, Table 2A shows the temperature and density fits that were obtained for the conditions of explosive carbon burning, explosive oxygen burning, and the equilibrium process.

The fits for these three processes enable one to explain the origin of all the elements from neon through the iron peak if these processes really occurred in a supernova. It was found that in the standard course of stellar evolution, the main α-particle nuclei were produced in roughly the right abundances in the onion-skin models (40, 62). However, the non–α-particle nuclei were not produced in the main-line course of stellar evolution. Table 2B shows the temperatures and densities associated

Table 2a   Explosive nucleosynthesis conditions

Process	Explosive C burning	Explosive O burning	e-Process
Fits	Ne-Si	Si-Ca	Fe peak
T_0(K)	2×10^9	3.1–3.9×10^9	$\geq 10^9$
ρ_0 (g cm^{-3})	10^{4-7}	10^{5-7}	$\geq 5 \times 10^7$
η	0.001–0.004	0.002–0.004	0.002–0.004

2b   Mantle of 22-$M_\odot$ star (8-$M_\odot$ evolved He core)

Zone	C	Ne	O	Si
T	1.1×10^9	1.1×10^9	2.8×10^9	4.0×10^9
"T"	1.8×10^9	1.8×10^9	4.0×10^9	7.0×10^9
ρ	9.4×10^9	1.1×10^5	2.8×10^6	5.9×10^9
η	≥ 0.0011	0.0019–0.0027	0.0019–0.0027	> 0.0019–0.0027

with the different zones in a 22-$M_\odot$ star. Notice that these temperatures are not the same as those needed in the explosive burning conditions to fit the non–α-particle nuclei. However, the densities are roughly right. It is interesting to note that if a shock were propagated through the star to turn it into a supernova, then the temperatures in those zones would be changed. In fact, if we make the shock necessary to turn one of these massive stars into an observed Type II supernova, then we can calculate what the effective temperature would be in each of these zones, and we find that the temperatures obtained, labeled "T" in Table 2B, are very close to those temperatures that one estimates from the explosive nucleosynthesis calculations. Thus it seems that if these massive stars do make the observed supernovae, then the ejecta from these stars will produce the elements from carbon to iron in their correct abundances, not just the α-particle nuclei, but all the nuclei, since the conditions would be those of explosive nucleosynthesis.

The above argument used a 22-$M_\odot$ star; however, it should be remembered that each different-mass star produces a slightly different composition. Figure 2 shows the density profiles of a 22-$M_\odot$ and a 15-$M_\odot$ star. Notice that the central core regions of both of these stars have almost identical iron-nickel cores. This core convergence results from the fact that neutrinos are the dominant energy-loss mechanisms for the cores, which thus evolve independently of the rest of the star at each neutrino-dominated stage. This tends to mean a 1.4-$M_\odot$ core is formed. When additional material is added to the core, it is pushed over the Chandrasekhar limit and collapse occurs. Since each composition zone in the star is roughly at the same density regardless of the mass of the star, the more massive star will have much more carbon and oxygen than the lower-mass star, whereas the lower-mass star will tend to have much more helium relative to the other material in the star. Therefore, if one blows up the 15-$M_\odot$ star, one would get a different ejecta than if one blows up the 22-$M_\odot$ star. We have already shown that the 22-$M_\odot$ star, if it becomes an observable supernova, will yield ejecta that produce an abundance of elements that is very similar to the observed abundance in the solar system. In fact, from the Dearborn (49) mass-loss estimates, it can be argued that most stars more massive than this will end up with a very similar internal structure due to their large main-sequence mass loss.

If one does not weight by the amount of mass ejected, then the mass of the typical star that blows up is 12–15 $M_\odot$. Thus a typical supernova would not produce as many heavy elements, but would eject relatively more helium than the 22-$M_\odot$ star. If you look at the Crab Nebula, it has a large amount of helium; thus it may be an example of a lower-mass star rather than of the more massive star.

It has also been demonstrated (62, 63) that the cosmic rays may be more like the typical supernova than like the mass-weighted supernova, and some of the differences in composition between the solar-system abundances and the cosmic rays may be due to just this different way in which the cosmic rays weight their selection of supernovae.

CORE EVOLUTION   We have seen that all these massive stars have a very similar core structure; thus, if we can understand what happens in the core, we might be able to understand how these stars blow up. Let us look at the evolution of the core once it reaches the point of being an iron-nickel core of about 1.4 $M_\odot$. All of the cores tend to pass through a point with the temperature around 10^{10} °K and the density around

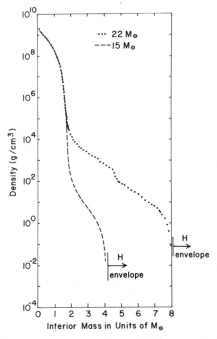

Figure 2   The density profile of a $\sim$22-$M_\odot$ and a $\sim$15-$M_\odot$ star.

10^{10} g cm^{-3}. They are supported by degenerate electrons with an adiabatic index $\gamma = 4/3$. Around this point, four things begin to happen:

1. Electron captures occur, electrons being captured due to their high chemical potential. Each electron capture causes a neutrino to be emitted; capturing electrons remove the electrons that are supporting the core.
2. Photodisintegration occurs, breaking the iron and nickel down into alphas, and then eventually down into free neutrons and protons.
3. The surrounding silicon shell continues to burn and thus adds more iron and nickel to the core, increasing the core mass to greater than 1.4 $M_\odot$.
4. The core is extremely hot, so neutrino pair emission takes place, carrying more energy out of the core.

All four of these processes yield the net effect of the core collapsing. Eventually the collapsing core will switch over from its adiabatic index γ being dominated by the electron-degeneracy pressure to the pressure being dominated by a thermal neutron gas with a γ of 5/3. This transition from a γ of 4/3 to a stiffer γ of 5/3 will cause a hydrodynamic bounce (65). It is interesting to note that the mass of the core inside of the bounce region always seems to be greater than about 1.1 $M_\odot$. This implies that one can never eject that inner 1.1 $M_\odot$, which would always go to form a remnant. Thus it seems that the mass of any remnant must be greater than 1.1 $M_\odot$. It may be significantly greater and go on to become a black hole, but in any event there should be no neutron stars with masses less than 1.1 $M_\odot$ as a result of the core collapse of these massive stars. This is in good agreement with the observations of the masses of neutron stars (66).

What happens next in the collapse is a matter about which there is currently much uncertainty. It is clear that the manner in which the core releases the bulk of its binding energy is via neutrinos. Thus, in order to form a neutron star, the star must release $\sim 10^{53}$ ergs of neutrinos. Whether or not these neutrinos are what cause the explosion, the ejection of the outer part of the star is a matter of some debate. However, it is important to realize that the neutrinos invariably seem to carry the bulk of the binding energy regardless of whether or not they cause the explosion. It seems clear that if one wants to observe gravitational collapse to the formation of neutron stars and black holes, one should look where the bulk of the energy is, in the neutrinos.

Rather than go into the complete collapse scenario now, let us just stop and examine the central-core configuration as this final gravitational collapse occurs. What we have is a hot, dense central region of neutron gas surrounded by a shell of alphas and neutrons, which in turn is surrounded by a shell of iron and nickel. This comprises the ~ 1.4-$M_\odot$ core. The surrounding Si, O, Ne, C, He, and H shells are at progressively lower densities and are not considered part of the core. One great possibility that weak neutral currents introduce is coherent neutrino scattering with a cross section approximately proportional to the square of the atomic weight (33). It is apparent that such a dependence might help in getting high neutrino fluxes through the $A = 1$ neutronized region to deposit momentum into the $A \sim 60$ iron-nickel shell. In principle, this could be the mechanism that enables the central core to collapse and form a neutron star while the outer material, containing the newly synthesized elements from carbon to iron, is ejected.

In the remainder of this review, we discuss the microscopic weak-interaction physics that is so important in understanding the collapse, and may even be important in understanding the ejection of the outer material. Following this discussion of weak-interaction physics, we discuss the gravitational-collapse scenario of the core.

The study of core collapse is an extremely active field at the present time. Thus the point of this review is not to emphasize any particular set of detailed core-collapse calculations that may be supplanted tomorrow, but rather to provide an overview of the important weak-interaction physics, particularly that concerning the neutral current.

1.2   Coherent Nuclear Scattering

Situations in which simple quantum-mechanical principles have macroscopic effects have a special appeal because they illustrate the unity of science. Electron-degeneracy pressure, which led to Chandrasekhar's prediction of the properties of white-dwarf stars, is one well-established astrophysical example. Stellar collapse may contain another example if, as current ideas indicate, nuclear elastic scattering $v + (Z, N) \rightarrow v + (Z, N)$ does indeed play a special role. The enhanced cross section for this process is due to the basic principle of quantum-mechanical coherence; it is worth discussing the coherent-scattering phenomenon to illustrate its generality.

In its simplest form, the phenomenon occurs whenever an elementary projectile (in our case the neutrino) scatters elastically from a composite system (in our case the nucleus), assumed to consist of A individual constituents (the nucleons) at positions $\mathbf{x}_i$, $i = 1, 2, \ldots A$. Due to the superposition principle, the amplitude $F(\mathbf{k}', \mathbf{k})$ for scattering from an incident momentum $\mathbf{k}$ to an outgoing momentum $\mathbf{k}'$ can be written as the sum of the sum of contributions from each constitutent:

$$F(\mathbf{k}',\mathbf{k}) = \sum_{j=1}^{A} f_j(\mathbf{k}',\mathbf{k}) \exp i(\mathbf{k}' - \mathbf{k}) \cdot \mathbf{x}_j, \qquad 1.1$$

where the individual amplitudes $f_j(\mathbf{k}', \mathbf{k})$ are added with a phase factor that takes into account the relative phase of the wave scattering at $\mathbf{x}_j$. This expression is valid under the simplifying assumption that individual amplitudes are small enough to neglect multiple scattering. This is extremely well satisfied for neutrinos!

The differential cross section is

$$\frac{d\sigma}{d\Omega} = \left| F(\mathbf{k}', \mathbf{k}) \right|^2, \qquad 1.2$$

and its size depends on the relation between the momentum transfer $\mathbf{q} = \mathbf{k}' - \mathbf{k}$, ($q = |\mathbf{q}| = 2k \sin \theta/2$, where k is the beam momentum and θ the scattering angle) and the size R of the composite system ($R = \max_{ij} |\mathbf{x}_i - \mathbf{x}_j|$).

If the dimensionless quantity qR is of order 1 or greater, then the relative phase factors are important, and contributions from individual scatterers tend to cancel (in the absence of special spatial relations among the position vectors $\mathbf{x}_i$ such as occur in a crystal). The elastic scattering cross section is therefore expected to be small.

For momentum transfer small compared to inverse target size, i.e. $qR \ll 1$, the relative phase factors are inconsequential and contributions from individual scatters add coherently, yielding

$$\frac{d\sigma}{d\Omega} \simeq A^2 |\overline{f}(\mathbf{k}', \mathbf{k})|^2, \qquad\qquad 1.3$$

where $\overline{f}(\mathbf{k}', \mathbf{k})$ is an amplitude averaged over all constituents,

$$\overline{f}(\mathbf{k}', \mathbf{k}) = \frac{1}{A}\sum_j f_j(\mathbf{k}', \mathbf{k}). \qquad\qquad 1.4$$

If there is only one type of constituent, then $\overline{f}(\mathbf{k}', \mathbf{k}) = f_j(\mathbf{k}', \mathbf{k})$, and the coherent scattering cross section is A^2 times the cross section for scattering from a single constituent.

If individual constituent scattering amplitudes differ, as is generally expected for the protons and neutrons of different spin projections in a nucleus, there can be cancellations among the contributions to Equation 1.4. No rigorous statement, therefore, can be made about the size of the coherent cross section without knowledge of the detailed scattering amplitudes involved. However, barring extreme cancellation, it is expected that the coherent cross section is approximately A^2 times a typical single constituent cross section.

To clarify the physical ideas involved, we assumed in the previous discussion that individual constituents were located at definite points $\mathbf{X}_j$ in the composite target. This restriction is inessential and the discussion can be generalized to the case where the target is treated as a quantum-mechanical system. There is one point of principle concerning the various quantum states, such as charge and spin projection, of the individual constituents of the composite system: namely, the instruction to sum over constituent amplitudes in Equation 1.1 follows from the superposition principle only if there is no change in quantum states. Equation 1.1 does not apply to spin-flip and charge-exchange processes, and these cross sections are not amplified by a factor of A^2 in a composite target.

Further insight may be gained by introducing the spatial density of scatterers $\rho(\mathbf{x}) = \Sigma_j \delta(\mathbf{x} - \mathbf{x}_j)$. In the case of one type of constituent, we can drop the subscript j and Equation 1.1 becomes

$$F(\mathbf{k}', \mathbf{k}) = f(\mathbf{k}', \mathbf{k})\int d^3 \times \exp{(i\mathbf{q} \cdot \mathbf{x})}\rho(\mathbf{x}). \qquad\qquad 1.5$$

Coherent scattering therefore measures the form factor of the composite target, i.e. the Fourier transform of its density. This interpretation is very general and applicable to neutrino-nucleus coherent scattering, which we calculate in Section 2.4.

The spherically symmetric Gaussian density $\rho(r) = A(4\pi R^2)^{-3/2} \exp{(-\frac{1}{4}R^{-2}r^2)}$, with $R = r_0 A^{1/3}$ and $r_0 \simeq 10^{-13}$ cm illustrates the systematics of nuclei of $A = Z + N$ constituents as mass number varies. The form factor corresponding to this density is $\rho(q) = A \exp{(-q^2 R^2)}$. We now discuss the qualitative behavior of coherent scattering cross sections for such nuclei.

For fixed beam momentum k, the maximum momentum transfer is $q_{max} = 2k$. For moderate or large energies ($2kR \gtrsim 1$), one finds that $d\sigma/d\Omega$ has a forward peak and falls off sharply at large angles or momentum transfer because of the form factor. Under these conditions the integrated elastic cross section

$$\sigma_{el} = \int d\Omega \, \frac{d\sigma}{d\Omega} \qquad\qquad 1.6$$

has the undramatic A dependence $A^{4/3}$.

However, at low energy, $2kR \ll 1$, the condition of coherence is satisfied for all angles, and the angular dependence is determined by the constituent amplitude $f(\mathbf{k}', \mathbf{k})$ in Equation 1.5 because the form factor is very nearly constant. The full benefits of coherence are therefore realized at low energies and $\sigma_{el} \propto A^2$.

These conclusions are largely independent of the nature of the projectile, whether pion, electron, or neutrino. In Fe^{56}, the radius is $R^{-1} \simeq 80$ MeV ($c = \hbar = 1$), so the condition for full coherence in the scattering of neutrinos in the mantle surrounding a collapsed stellar core is $E_\nu \ll 40$ MeV, which is reasonably well satisfied for the 20-MeV neutrinos expected in stellar collapse.

Finally, we note that removal of some of the simplifying assumptions of our discussion actually leads to broader application for the coherent scattering concept. When individual scattering amplitudes are large, terms describing multiple scattering in the target must be added in Equation 1.1. The multiple-scattering formalism describes π-nucleus scattering very well and has been reviewed recently (67, 68). In crystals and large molecules, maxima can occur in Equation 1.1 at nonzero momentum transfer because of the regular spacing of constituents. This is just the Bragg scattering phenomenon that has been fruitfully applied to determine the structure of biological molecules by X-ray scattering. Electron scattering in nuclei [for a recent review see (69)] is a very-well-studied application of the coherent scattering formalism. It is similar to the neutrino-scattering case, both because the single-scattering approximation is usually valid and because both electrons and neutrinos couple to nucleons through vector (and axial vector) currents.

2   WEAK INTERACTIONS

2.1   Charged-Current Processes

The formalism conventionally used to describe charged-current weak interactions involves a phenomenological or effective Lagrangian that is a product of local current operators. Amplitudes for weak decays and scattering processes are calculated in the lowest order of perturbation theory and are in good agreement with experimental data. However, the theory cannot be fundamental because it is non-renormalizable, and higher-order corrections to physical amplitudes are infinite. An alternate hypothesis is that the charged-current weak interactions are mediated by a charged intermediate vector boson of mass m_w. For momentum transfers small compared to m_w, the experimental predictions are the same as for the current-current theory. The intermediate vector-boson theory is also non-renormalizable.

Renormalizability problems of the weak interactions are solved in the non-Abelian gauge theories (26–29), such as those first proposed by Weinberg (20) and Salam (21) and many more recent models. In these theories there are one or more massive charged vector bosons (W's) that mediate charged-current weak interactions. The electromagnetic interactions mediated by the massless photon A_μ are naturally included, and there are generally one or more massive neutral-vector particles (Z's) that mediate neutral-current processes. The W and Z bosons are predicted to have masses on the order of 50 GeV or more, which is consistent with experimental evidence that suggests that the W boson, if it exists, has a mass greater than 20 GeV. In addition to vector particles, there are spinless Higgs mesons that are important for the mathematical structure of gauge theories, but have more subtle experimental effects.

The fundamental fermion fields of the gauge theories are those that correspond to leptons such as e, v_e, μ, and v_μ, and quarks, which are presumed to be the constituents of hadrons such as p and n. Currents J_λ^W and J_λ^Z, which are bilinear in the fermion fields, determine the observed properties of weak processes. The properties of processes of astrophysical interest and even most processes measured at present accelerators are insensitive to the full details of the gauge theories and may be described by current-current effective Lagrangians, which we use in this paper.

For charged-current processes, the effective Lagrangian is

$$\mathcal{L} = (2)^{-1/2} G J_\lambda^W (J_\lambda^W)^\dagger, \qquad\qquad 2.1$$

$$G = (1.015 \pm 0.03) \times 10^{-5} m_p^{-2},$$

and the current is decomposed into leptonic and hadronic parts:

$$J_\lambda^W = l_\lambda^W + h_\lambda^W; \qquad\qquad 2.2$$

with

$$l_\lambda^W = e_e \gamma_\lambda (1 - \gamma_5) v_e + \bar{\mu} \gamma_\lambda (1 - \gamma_5) v_\mu, \qquad\qquad 2.3$$

where we use the γ matrices

$$\gamma^0 = \begin{pmatrix} 1 & 0 \\ 0 & 1 \end{pmatrix}, \qquad \gamma = \begin{pmatrix} 0 & \sigma \\ -\sigma & 0 \end{pmatrix}, \qquad \gamma_5 = \begin{pmatrix} 0 & 1 \\ 1 & 0 \end{pmatrix},$$

and σ_1, σ_2, and σ_3 are the 2×2 Pauli matrices. Because of the extremely low experimental limits on the neutrino mass,

$$m_{v_l} < 60 \text{ eV} \qquad m_{v_\mu} < 0.65 \text{ MeV},$$

the neutrinos are assumed to be massless. The matrix $(1 - \gamma_5)$ projects out the left-handed chirality components of fermion fields. For the massless neutrino, this has the implication that the neutrinos that participate in charged-current processes are purely left-handed, while antineutrinos are right-handed.

A particle physicist would write the hadronic current in terms of quark fields, and there is strong theoretical and experimental support for four quarks u, d, s, and c, whose quantum numbers are listed in Table 3. The weak hadronic charged current is then

$$h_\lambda^W = d_c \gamma_\lambda (1 - \gamma_5) u + \lambda^c \gamma_\lambda (1 - \gamma_5) c. \qquad\qquad 2.4$$

Table 3   Quantum numbers of the four quarks: u, d, s, and c

Quarks	Electric charge (Q)	Baryon number (B)	Strangeness (S)	Charm (C)
u (up)	$+\frac{2}{3}$	$\frac{1}{3}$	0	0
d (down)	$-\frac{1}{3}$	$\frac{1}{3}$	0	0
s (strange)	$-\frac{1}{3}$	$\frac{1}{3}$	-1	0
c (charm)	$+\frac{2}{3}$	$\frac{1}{3}$	0	$+1$

There is a striking analogy with the form of l_λ^W, in accord with the idea that all fundamental fermion fields participate in the weak interaction with universal strength. Universality is in agreement with experiment if the fields d_c and λ_c are rotated by an angle θ_c with respect to d and λ:

$$d_c = \cos\theta_c d + \sin\theta_c \lambda,$$

$$\lambda_c = \sin\theta_c d + \cos\theta_c,$$

which is the quark-model form of a formulation of universality due to Cabibbo (70). Experiment gives the result $\cos\theta = 0.976$, $\sin\theta = 0.218$.

Ordinary β decay and the related reactions

$$n \rightarrow p + e^- + \bar{\nu}_e,$$

$$e^- + p \leftrightarrows \nu_e + n, \qquad\qquad 2.5$$

$$e^+ + n \leftrightarrows \bar{\nu}_e + p$$

are important for stellar collapse. To describe them, it is sufficient to use a phenomenological term in the current h_λ^W of the form (26)

$$\gamma_\mu(g_V - g_A\gamma_5)p, \qquad\qquad 2.6$$

with $g_V = 1$, $g_A \approx 1.21$. This form is consistent with Equation 2.4 and was suggested by the conserved vector-current (CVC) hypothesis (13–15). According to the CVC idea, the strangeness-conserving ($\Delta S = 0$) part of h_λ^W is a component of the isotopic spin current.

In this review we are only concerned with the $\Delta S = 0$ part of the hadronic current, parameterized as in Equation 2.6. Other $\Delta S = 1$ and $\Delta C = 1$ processes predicted by Equation 2.4 are irrelevant in stellar collapse because the temperatures ($kT < 30\,\text{MeV}$) and densities ($\rho < 10^{14}\,\text{gm cm}^{-3}$) are insufficient to produce strange and charmed particles. At the higher densities encountered in the collapsed state of the neuteron star, the nucleon Fermi energies are high enough so that strange hyperons appear and are stable. However, the weak interaction is less important in determining the structure of the star at this point.

Reactions that play an important role in stellar collapse, for both charged and neutral currents, are listed in Table 4. Feynman diagrams for the processes in Equation 2.5 are given in Figure 3 as they would occur in a theory with intermediate

Table 4   Astrophysical weak processes

$e^- + p \leftrightarrows \nu_e + n$	
$e^+ + n \leftrightarrows \bar{\nu}_e + p$	unchanged by neutral currents
$e^- + \nu \to e^- + \nu$	
$e^+ + e^- \to \nu + \bar{\nu}$	
$\Gamma \to \nu + \bar{\nu}$	altered by neutral currents
$e^\pm + \gamma \to e^\pm + \nu + \bar{\nu}$	
$e^- + Z \to e^- + Z + \nu + \bar{\nu}$	
$n + n \to n + n + \nu + \bar{\nu}$	
$A^* \to A + \nu + \bar{\nu}$	
$\nu + \nu \to \nu + \nu$	
$\nu + p \to \nu + p$	allowed only by neutral currents
$\nu + n \to \nu + n$	
$\nu + A \to \nu + A$	

vector-boson exchange. At low energy, the amplitudes take the form suggested by the local-current effective Lagrangian.

2.2 Neutral-Current Processes

Neutral current processes can be divided into $\Delta S = 0$ and $\Delta S = 1$ reactions and decays. Prior to the 1973–74 experiments, upper limits on the $\Delta S = 0$ reactions were not very stringent, and experiments (1–7, 17–19) now suggest a strength of about 20% of corresponding charged-current reactions. The absence of weak decays such as $\Sigma^+ \to p + e^+ + e^-$, $K^+ \to \pi^+ + \nu + \bar{\nu}$ and $K^0 \to \mu^+ + \mu^-$ has long indicated that $\Delta S = 1$ neutral-current processes are strongly suppressed.

The Weinberg-Salam model was originally proposed as a theory of purely leptonic processes, but extensions to include hadrons have been given (71, 72). Both $\Delta S = 0$ and $\Delta S = 1$ neutral-current processes would occur in models based on the three quarks u, d, s. The $\Delta S = 1$ neutral current is eliminated very naturally if a fourth charmed quark c is added, as was suggested by Glashow, Maiani & Iliopoulos (73). The recent experimental discovery of charmed particles is a strong confirmation of this idea (74).

The Weinberg-Salam model is based on the gauge group SU(2) × U(1). It contains an arbitrary parameter $\tan \theta_W$, which is the ratio of the gauge coupling constant of the U(1) and SU(2) factors. The masses of the vector bosons W and Z which mediate

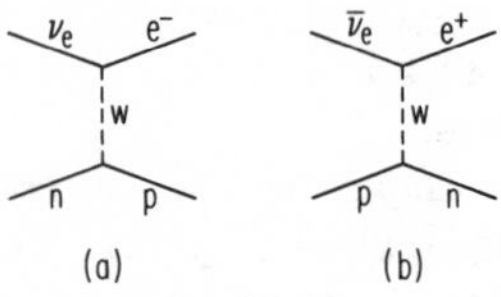

Figure 3   Feynman diagrams for neutrino absorption.

charged- and neutral-current processes are, respectively,

$$m_W^2 = \frac{e^2 (2)^{1/2}}{8G \sin^2 \theta_W},$$

$$m_Z^2 = \frac{e^2 (2)^{1/2}}{2G \sin^2 2\theta_W},$$

where $e^2/4\pi = \frac{1}{137}$ is the fine-structure constant.

The form of the effective Lagrangian for neutral-current processes is not very well determined experimentally. For this reason, we present first a reasonably general expression in terms of vector and axial vector currents that is consistent with most gauge-theory models, and we then give the special choice of parameters corresponding to the Weinberg-Salam model. It is worth noting that even the V-A structure is not fully confirmed experimentally, and that theories with scalar, pseudoscalar, and tensor currents (75, 76) are difficult to rule out. Some astrophysical implications of S, P, and T neutral currents are discussed later.

The effective Lagrangian is

$$\mathcal{L} = (2)^{-1/2} GJ_\lambda^Z J^{Z\lambda},$$

$$J_\lambda^Z = l_\lambda^Z + h_\lambda^Z. \tag{2.7}$$

The leptonic current may be parameterized as

$$l_\lambda^Z = \bar{v}_e \gamma_\lambda \tfrac{1}{2}(1 - \gamma_5) v_e + \bar{e} \gamma_\lambda (a_e + b_e \gamma_5) e$$

$$+ c_\mu \bar{v}_\mu \tfrac{1}{2}(1 - \gamma_5) v_\mu + \bar{\mu} \gamma_\lambda (a_\mu + b_\mu \gamma_5) \mu \tag{2.8}$$

$$+ \sum_L c_L \bar{v}_L \gamma_\mu (1 \pm \gamma_5) v_L + \bar{L} \gamma_\lambda (a_L + b_L \gamma_5 L).$$

Note that $\mu - e$ universality is not assumed. There is experimental evidence (108) for a charged heavy lepton and a possible associated neutrino. We have therefore included terms for heavy leptons and either left- or right-handed neutrinos in Equation 2.8. Possible astrophysical effects of such particles are discussed briefly below.

For neutral-current processes involving nucleons as targets, it is sufficiently general to write the hadronic current as a sum of isoscalar and isovector terms

$$h_\lambda^Z = a_0 V_\lambda^{I=0} + b_0 A_\lambda^{I=0} + a_1 V_\lambda^{I=1}, \quad \text{and} \quad b_1 A_\lambda^{I=1}, \tag{2.9}$$

for which the phenomenological parameterization is

$$V_\lambda^{I=0} = \bar{p}\gamma_\lambda p + \bar{u}\gamma_\lambda u,$$

$$A_\lambda^{I=0} = \bar{p}\gamma_\lambda \lambda_5 p + \bar{n}\gamma_\lambda \gamma_5 n,$$

$$V_\lambda^{I=1} = \tfrac{1}{2}(\bar{p}\gamma_\lambda p - \bar{n}\gamma_\lambda n), \tag{2.10}$$

$$A_\lambda^{I=1} = \tfrac{1}{2}g_A(\bar{p}\gamma_\lambda \gamma_5 p - \bar{n}\gamma_\lambda \gamma_5 n).$$

The general expressions for l_λ^Z and h_λ^Z contain nine free parameters. Although in principle these can be determined from neutral-current experiments, it is much too difficult to do this at present. One virtue of the Weinberg-Salam model is that there is only one free parameter, so that experimental determination is much simpler. The various general parameters take the values

$$c_\mu = 1,$$

$$a_e = a_\mu = -\tfrac{1}{2}(1 - 4\sin^2\theta_W),$$

$$b_e = b_\mu = \tfrac{1}{2},$$

$$a_0 = -\sin^2\theta_W,$$ 2.11

$$b_0 = 0,$$

$$a_1 = 1 - 2\sin^2\theta_W,$$

$$b_1 = -1.$$

Note that $\mu - e$ universality is assumed in the W-S model and that h^Z of Equation 2.9 can also be written as

$$h^Z_\lambda = \mathscr{I}^3_\lambda - 2\sin^2\theta_W J^{\rm em}_\lambda$$ 2.12.

where $\mathscr{I}^3_\lambda$ is the V-A isospin current in the same multiplet as h^W_λ in Equation 2.4 or 2.6, and $J^{\rm em}_\lambda$ is the electromagnetic current. Note that baryon number B, strangeness S, hypercharge Y, electric charge Q, and third component of isospin I_3 are related according to $Y = S + B$ and $Q = I_3 + \tfrac{1}{2}Y$. These relations facilitate the phenomenological parameterization of h^Z_λ in Equation 2.12, leading to expressions analogous to Equations 2.9 and 2.10, with the particular parameters of the Weinberg-Salam model.

The neutrino-electron-scattering process $v_e + e^- \rightarrow v_e + e^-$ (and processes such as pair annihilation $e^+ + e^- \rightarrow v_e + \bar{v}_e$) receive contributions from both charged and neutral currents, as is indicated in the Feynman diagrams of Figure 4. The amplitude is usually parameterized as

$$\frac{iG}{(2)^{3/2}\pi^2} \bar{u}(v')\gamma^\lambda(1 - \gamma_5)u(v)\bar{u}(e')\gamma_\lambda[C_V - \gamma_5 C_A]u(e).$$

Then W exchange alone (i.e. pure charged current) gives

$$C_V = C_A = 1,$$

while the Weinberg-Salam model predicts

$$C_V = 2\sin^2\theta_W + \tfrac{1}{2} \qquad C_A = \tfrac{1}{2},$$

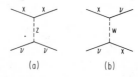

(a)   (b)

Figure 4   General Feynman diagrams for neutrino scattering: for $e^- + v \rightarrow e^- + v$, diagrams (a) and (b) are added. For $v + p$ (nor A) vp (nor A), only diagram (a) contributes.

and in the general parameterization of Equation 2.8,

$$C_V = a_e + 1,$$

$$C_A = 1 - b_e.$$

We now discuss coherent nuclear scattering of neutrinos, which was calculated by Freedman (33) for spinless $(J = 0)$ nuclei with $Z = N$. In this case, only the $I = 0$ vector current contributes, and the amplitude of Section 1.2 is

$$F(k', k) \sim G \exp(-q^2 R^2) \, A a_0 \bar{u}_v(v') \gamma^0 u_v(v), \qquad 2.13.$$

where an exponential nuclear form factor is assumed, as discussed in Section 1.2. For nuclei with non-zero spin $(J \neq 0)$ and $Z \neq N$, one must sum over nucleons, and one then finds, from Equations 2.9 and 2.10,

$$F(k', k) \sim G \exp(-q^2 R^2) [a_0(Z+N) + a_1 \tfrac{1}{2}(Z-N) \qquad 2.14.$$

$$+ (b_0 + \tfrac{1}{2} b_1 g_v)(Z_\uparrow - Z_\downarrow) + (b_0 - \tfrac{1}{2} b_1 g_A)(N_\uparrow - N_\downarrow)] \, \bar{u}(v') \gamma^0 u(v),$$

where the arrow indicates spin-up and spin-down nucleons that scatter via the axial neutral current with opposite sign. For all except very light nuclei, the most important term in Equation 2.14 is the term involving $a_0(Z+N)$, so that Equation 2.13 is generally approximately valid.

Nuclear coherent scattering is therefore mainly dependent on the coefficient a_0 of the isoscalar polar-vector part of the weak neutral current.

Nuclear coherent scattering of neutrinos has been considered by Adler (78) in a model (called model A) in which the weak neutral current is a spatial scalar. The effective Lagrangian is simply

$$\mathscr{L}_A = (2)^{1/2} G \bar{v} v J_s,$$

where J_s is a postulated scalar hadronic neutral current. Astrophysical consequences are discussed below.

2.3   Implications of Present Experiments

All neutrino experiments are notoriously difficult because of the low interaction rate of neutrinos, and associated background and systematic problems. Neutral-current experiments are particularly difficult because the final-state neutrino cannot be detected directly. As a result, present experimental support for any particular model of the weak neutral current is inconclusive. Nevertheless, high-energy accelerator experiments are generally consistent with the Weinberg-Salam model for a value of $\sin^2 \theta_W$ between 0.3 and 0.4. This is true of deep inelastic inclusive reactions at CERN (1–2) and FNAL (3, 4, 6). Furthermore (18), the momentum-transfer distributions in elastic scattering $v_\mu + p \to v_\mu + p$ are in good agreement and yield $\sin^2 \theta_W < 0.35$.

An interesting experimental measurement of antineutrino scattering from electrons $\bar{v}_e + e \to \bar{v}_e + e$ by Reines and coworkers has been recently reported (109). This experiment is performed with very-low-energy antineutrinos at the Savannah River reactor. The total cross section is consistent both with the pure V-A charged-current theory and with the Weinberg-Salam model for $\sin^2 \theta_W = 0.29 \pm 0.05$. A more striking comparison emerges when the cross section is plotted against recoil electron energy. The distribution differs from pure V-A and is in qualitative agreement with the W-S model. It is worth emphasizing that this is the first laboratory experimental evidence for neutral-current effects involving electron-type neutrinos, and thus is particularly significant for astrophysics. Accelerator experiments involve only v_μ and $\bar{v}_\mu$, and the untested assumption of universality ($C_\mu = 1$) is generally made in order to calculate v_e and $\bar{v}_e$ processes of astrophysical interest.

In principle, parity-violating effects in atomic physics occur because of the intrinsic parity violation in weak interactions. It was pointed out (112) that with a neutral current, such effects would be enhanced in heavy atoms, due essentially to the nuclear coherence, and that there would be amplitudes of order $Gb_e[a_0(Z+N)+ \frac{1}{2}a_1(Z-N)]$ for the mixing of atomic levels of opposite parity. Preliminary results (95) of experiments at Oxford University and the University of Washington, on two separate atomic transitions in bismuth, indicate no parity-violating effect with an upper limit of a factor of 3 or 4 below the Weinberg-Salam model prediction. This result provides the first serious challenge for the W-S model, and confirmation by experiments in other atoms and further study is awaited. The results are also relevant for the astrophysical effectiveness of nuclear coherent scattering because a strong limit on the product $b_e a_0$ is implied by the present negative results on parity violation. It is too early to conclude that a_0 is small. Indeed, theoretical models (113, 114) have been constructed that predict no atomic parity violation, but appreciable coherent scattering.

Uncertainty about the nature of the weak neutral current poses a dilemma for the astrophysicist. The weak interaction is considered fundamental in determining the outcome of the late stages of stellar evolution. The weak neutral current seems to be particularly important for stellar collapse and supernova explosions. Which model should be used? The Weinberg-Salam model is simple and elegant, but it may be incomplete at best. Much more experimental data is needed before the parameters of the weak neutral current can be determined. Information on the space-time structure and, in particular, on the isoscalar component of the neutral current are very much needed and anxiously awaited.

2.4   Neutrino Interactions in Stellar Matter

A physicist working at CERN, FNAL, or SLAC would rapidly develop ulcers were the laboratory conditions those of the stellar interior. In STAR, neither the target nor the projectile are confined to beams; instead, the local motions of the particles are for the most part random, with a wide variation of kinetic energies. Matter densities are so high that the less massive particles become packed nearly one atop the other, and particles are continuously created and destroyed by the thermal bath.

Nature quite successfully handles the situation. You and I are living proof—the stuff of the stars. Astrophysicists, though not as clever as Nature, are nevertheless stubborn and curious. They persist in the effort to understand how all of those precious elements like carbon and oxygen are released into a universe full of hydrogen. With many neutrinos and several theories, several of which are discussed in

Section 2.2 and none of which have been so well tested that we should stake our lives on them, the astrophysicist might come close to the answer.

The discussion that follows involves general results of theories that include only vector and axial vector neutral currents (e.g. gauge theories). When the calculations of pertinent processes have been done in other weak-interaction models, those results are compared with the predictions of the Weinberg-Salam gauge theory.

Electrons, and positrons if the temperature is not too low or the density too high, are in thermal equilibrium. (Temperatures in the region of interest are generally insufficient to thermally produce muons.) The electromagnetic interactions are able to maintain this equilibrium on time scales much shorter than those of the weak interactions. Thus the electrons and positrons that enter into the neutrino processes may be described by Fermi-Dirac distributions. In neutral matter, the proton-number density, matter density, average number of protons per amu (Y_e), and electron- and positron-number densities are related [cf (81)] by

$$n_p = N_A \rho Y_e = n_{e^-} - n_{e^+}. \qquad 2.15.$$

When the electrons are degenerate ($\mu_{e^-}/kT \gg 1$), $n_{e^+} \ll n_{e^-}$. By the time nuclear photodisintegration and collapse have been initiated, the electrons will be degenerate; the degree of degeneracy depends upon how advanced the collapse is (47, 77, 82).

The nucleons, on the other hand, will become noticeably degenerate only at advanced stages of collapse. The chemical potential of the electrons will be sufficiently high that the average electron will be relativistic ($\mathbf{p}^2 \simeq E^2$), whereas the nucleons can be described by a nonrelativistic ($\mathbf{p}^2 \simeq 2mE$) Maxwellian distribution. The lepton energies ($\sim$ tens of MeV) will still be low enough compared with the nucleon and nuclear rest masses that, in neutrino processes involving nucleons and nuclei, the calculation may be performed in the target rest frame (which is approximately the fluid rest frame). Yueh & Buchler (83) have included the Maxwellian distribution in their treatment of the transport equation.

One effect of the stellar matter on the single-particle processes is thus to inhibit those interactions with electrons in the final state. A factor of

$$1 - f_{e^-}(E) = 1 - [\exp(E - \mu_{e^-})/kT + 1]^{-1}$$

appears in the differential interaction rate for final-state electrons, and a factor of $f_{e^-}(E)$ appears for initial-state electrons. The single-particle interactions become statistical averages, the interaction probabilities being weighted by the product of available and filled momentum space.

Consider the β decay processes, which only proceed by way of charged currents. In the absorption of ν_e, the proton is essentially produced at rest. The cross section for $\nu_e + n \rightarrow p + e^-$ in stellar matter is just the single-particle cross section multiplied by the inhibition factor $1 - f_{e^-}(\omega + \Delta m)$ (84, 34).

$$\sigma_{\nu_e n \rightarrow p e^-} = \sigma_0 [\tfrac{1}{2}(\alpha^2 + 1) + \tfrac{1}{4}(\alpha^2 - 1)] \omega^2(\omega + \Delta m)$$

$$\times [\omega^2 + 2\omega\Delta m + \Delta m^2 - 1]^{1/2} [1 - f_{e^-}(\omega + \Delta m)], \qquad 2.16.$$

$$\sigma_0 = 1.721 \times 10^{-44} \text{ cm}^2.$$

Absorption of $\bar{\nu}_e$ is uninhibited, the final state containing a positron:

$$\sigma_{\bar{\nu}_e p} \to ne^+ = \sigma_0[\tfrac{1}{2}(\alpha^2+1)+\tfrac{1}{4}(\alpha^2-1)]\omega^2 \cdot (\omega-\Delta m)[\omega^2-2\omega\Delta m+\Delta m^2-1]. \quad 2.17.$$

A simple argument of detailed balance applied to the single-particle rates, or direct calculation from the matrix element (83, 85) yields the differential rate of production of neutrinos of energy ω', for the electron-capture process:

$$dR_{ep} \to n\nu_e = R_0\left(\frac{2n_p^f}{\pi}\right)[\tfrac{1}{2}(\alpha^2+1)+\tfrac{1}{4}(\alpha^2-1)]\omega'^2(\omega'+\Delta m)$$

$$\times [\omega'^2+2\Delta m\omega'+\Delta m^2-1]^{1/2}f_{e^-}(\omega'+\Delta m)\,d\omega'\,d\Omega',$$

$$\quad 2.18a.$$

$$R_0 = \frac{2G^2m_e^5c^2}{(2\pi)^3\hbar} = 5.6739 \times 10^{-5}\ \mathrm{sec}^{-1}. \quad 2.18b.$$

The dimensions of $R_{ep} \to n\nu_e$ are cm^{-3} sec^{-1}; and n_p^f is the density of free protons, not to be confused with Equation 2.15. Note that Equation 2.18 yields the energy spectrum of the electron-capture neutrino, and that the emissivity $d\mathscr{E}_{ep} \to n\nu_e$ is obtained by multiplying Equation 2.18 by $\omega(m_ec^2)$. Approximate expressions for $R_{ep} \to n\nu_e$ and $\mathscr{E}_{pe} \to n\nu_e$ when Δm can be neglected are given by Bludman & Van Riper (86).

If the electron gas has a dielectric constant less than unity, the photon can acquire a small mass, and thereby decay. In fact, the plasmon process, $\Gamma \to \nu + \bar{\nu}$, is a stellar parasite, robbing the star of energy, but unable to survive without its host.

The plasmon decay, pair annihilation, and photoproduction neutrino-pair emissivities (erg sec^{-1} cm^{-3}) have been calculated by Dicus (32), in the context of the Weinberg-Salam model. For present values of the Weinberg angle, the ratios of emissivities in the W-S model to the W-meson theory are $\lesssim 2$. Using arbitrary vector and axial vector couplings, Dicus's results may be extended to include heavy-lepton neutrino production. General expressions for the emissivities are:

$$\mathscr{E}_{\text{pair}} = [7(C_V^{e2}+C_V^{\mu2}+C_V^{L2})-2(C_A^{e2}+C_A^{\mu2}+C_A^{L2})]P_1$$

$$+9(C_V^{e2}+C_V^{\mu2}+C_V^{L2})P_2 \qquad\qquad 2.19a.$$

$$+(C_V^{e2}+C_V^{\mu2}+C_V^{L2}+C_A^{e2}+C_A^{\mu2}+C_A^{L2})P_3,$$

$$\mathscr{E}_\Gamma = (C_V^{e2}+C_V^{\mu2}+C_V^{L2})F, \qquad\qquad 2.19b.$$

$$\mathscr{E}_\gamma = (C_V^{e2}+C_V^{\mu2}+C_V^{L2}+C_A^{e2}+C_A^{\mu2}+C_A^{L2})G_1$$

$$+(C_V^{e2}+C_V^{\mu2}+C_V^{L2}-C_A^{e2}-C_A^{\mu2}-C_A^{L2})G_2. \qquad 2.19c.$$

The emission rates, $R(\mathrm{sec}^{-1}\ \mathrm{cm}^{-3})$, for these interactions follow analogous formulae. Both $\mathscr{E}$ and R have been numerically determined for general C_V^i and $C_A^i(i = e, \mu, L)$, over a wide range of densities and for temperatures up to several tens of MeV (87). If the arbitrary couplings are interpreted in the WS model, and the additional assumption of universality is made, then

$$C_V^e = \tfrac{1}{2}+2\sin^2\Theta_w \simeq 1.3, \qquad C_A^e = \tfrac{1}{2},$$

$$C_V^\mu = C_V^L = (1 - C_V^e),$$

and

$$C_A^\mu = C_A^L = (1 - C_A^e).$$

The ratios of the rates in this model of neutral currents to the charged-current theory, as functions of the number of μ and L neutrino types, J ($J = 2$ for 1 heavy-lepton neutrino), are approximately $(0.845 + 0.145\ J)$ for the pair and photo rates, and $(1.44 + 0.04\ J)$ for the plasma rates. The branching ratios for $v_J \bar{v}_J$ to $v_e \bar{v}_e$ in the theory with neutral currents are

$$\left(\frac{v_J \bar{v}_J}{v_e \bar{v}_e}\right)_{\text{pair},\gamma} \simeq 0.172,$$

$$\left(\frac{v_J \bar{v}_J}{v_e \bar{v}_e}\right)_\Gamma \simeq 0.028.$$

The electron bremsstrahlung process, $e^- + Z \rightarrow e^- + Z + v + \bar{v}$, will only be important when the electron degeneracy is extreme and the thermal processes above have shut off. The result of a calculation that includes general neutral currents (37) is that the charged-current emissivity is enhanced by only a factor of ~ 1.31 when v_e and v_μ pairs are produced, and by a factor of ~ 1.52 when a single v_L pair is also included ($C_V^e \simeq 1.4$, $C_A^e = \frac{1}{2}$). The branching ratio $v_J \bar{v}_J / v_e \bar{v}_e$ is ~ 0.186.

Neutral currents thus seem to have little effect on the charged-current neutrino-production mechanisms, and electron-neutrino pairs are the dominant neutrino type produced. In actual collapse situations, comparison of the emissivities shows (82, 88, 89) that electron capture is the dominant neutrino source; at later stages of the collapse, antineutrinos can become important, but still considerably less so than neutrinos (77); however, at late times, when the core is quite opaque to neutrinos, Wilson (77) has found that muon neutrinos may become the dominant luminosity source, due to their smaller electron-scattering cross sections.

Tensor, scalar, and pseudoscalar neutral currents have also been included in the basic neutrino-luminosity sources, and we briefly note the results. Based upon the $\Gamma \rightarrow v + \bar{v}$ process and a comparison of degenerate-dwarf cooling times with observations of hot white dwarfs, Sutherland et al (90) have deduced that an upper limit on the neutral-tensor coupling constant is $C_T \lesssim 0.06$. Dicus & Kolb (91), in a study of the pair annihilation, photoproduction, bremsstrahlung, and $\gamma + \gamma \rightarrow v + \bar{v}$ processes, have found less restrictive bounds, based upon a similar comparison with observed stellar lifetimes: $C_T \lesssim 3.6$, $C_S \lesssim 16$, $C_P \lesssim 14$.

The general conclusions of the effect of neutral currents on the emissivities hold likewise for the neutrino-electron scattering interaction (34). In terms of arbitrary vector and axial vector couplings, the reaction rates and cross sections follow the general form (for electron-type neutrinos)

$$R_{ev} \rightarrow ev = (C_V + C_A)^2 R_1 + (C_V - C_A)^2 R_2 + (C_A^2 - C_V^2) R_3. \qquad 2.20.$$

For relativistic electrons, $R_3 \rightarrow 0$. Muon neutrinos can now scatter with electrons, by the exchange of a neutral meson, but the rates are strongly suppressed, being at best

$\sim 17\%$ of the v_e scattering rate. The overall effect of the neutral current is to decrease the $R_e v_e$ rate by $\sim 20\%$ from the rate for charged currents. Table 5 gives some values for $(v_e e)$ scattering rates.

The great importance of neutral currents to astrophysics is not that they alter existing charged current processes, but that neutral currents permit new interactions that may have dramatic effects in the stellar matter.

Within the neutron star, neutron bremsstrahlung $(n + n \rightarrow n + n + v + \bar{v})$ may play as important a role in cooling by neutrino loss (118) as does the modified URCA process. While in white dwarfs with central temperature $\lesssim 10^8 \,^\circ K$ (at the high temperatures of gravitational collapse, the purely leptonic processes will dominate), semileptonic cooling by nuclear de-excitation to neutrinos $(A^* \rightarrow A + v + \bar{v})$ may compete with the purely leptonic cooling mechanisms (36). Both processes involve V and A neutral currents only, and with the assumption of universality, both produce e, μ, and L neutrinos with equal probability.

In the collapsing stellar core, purely neutral-current processes significantly increase the opacity of the matter to neutrinos—not only in the mantle of heavy nuclei at the exterior of the core, but deep within the core as well, where the nuclei are nearly fully dissociated into free nucleons. The possible implications of these neutrino processes is drawn in Section 3. The discussion that follows is concerned with providing estimates and comparison of the sources of neutrino opacity. Because the degeneracy (if it is present at all) of the nucleons and nuclei may be neglected, target-inhibition factors are absent in the cross sections. Furthermore, because $\omega \ll m_n, m_p, m_A$ is a good approximation for neutrinos produced in the star, the cross sections are calculated in the c.m. frame (fluid rest frame), and thus are conservative $(\omega' \simeq \omega)$ and proportional to ω^2.

We have already seen how neutrino scattering at small momentum transfer on spinless, isoscalar nuclei is greatly enhanced, due to the simple principle of coherent superposition of amplitudes. In the model of Freedman (33), the formulae for the cross sections (with z the cosine of the scattering angle) are:

$$\frac{d\sigma_A}{dz} \simeq \tfrac{1}{8} a_0^2 \sigma_0 A^2 \omega^2 (1 + z), \qquad\qquad 2.21a.$$

$$\sigma_A \simeq \tfrac{1}{4} a_0^2 \sigma_0 A^2 \omega^2, \qquad\qquad 2.21b.$$

where it is assumed that $\omega \ll m_A$ (the nucleus is pointlike). The application of this model to $Z \neq N$ spinless nuclei (34) introduces a coupling to the nuclear isospin

$[I_3 = \tfrac{1}{2}(Z - N)]$:

$$\frac{d\sigma_{Z,N}^{(J=0)}}{dz} \simeq \frac{\sigma_0}{8} a_0^2 A^2 \left[1 + \tfrac{1}{2}\frac{a_1}{a_0}\frac{Z-N}{A} \right]^2 \omega^2 (1 + z), \qquad\qquad 2.22a.$$

$$\sigma_{Z,N}^{(J=0)} \simeq \frac{\sigma_0}{4} a_0^2 A^2 \left[1 + \tfrac{1}{2}\frac{a_1}{a_0}\frac{Z-N}{A} \right]^2 \omega^2. \qquad\qquad 2.22b.$$

In the W-S theory, $a_1 = 1 + 2a_0 = 1 - 2\sin^2 \Theta_w$.

Table 5 Neutrino-electron scattering rates (sec^{-1} per neutrino)[a]

$T/10^9$	$\dfrac{\mu_e}{m_e c^2}$ [a]	ρY_e (g cm^{-3})	$\Omega = 4$ MeV			$\Omega = 8$ MeV			$\Omega = 16$ MeV			$\Omega = 24$ MeV		
			R_1	R_2	R_3	R_1	R_2	R_3	R_1	R_2	R_3	R_1	R_2	R_3
10	10	1.2+9	1.9+0	7.4−1	8.0−3	6.6+0	1.9+0	1.8−2	1.6+1	4.7+0	2.2−2	2.6+1	7.7+0	2.3−2
	25	1.6+10	9.1+0	5.8+0	9.2−3	4.3+1	1.9+1	4.0−2	2.8+2	7.0+1	1.9−1	5.5+2	1.4+2	2.5−1
	50	1.2+11	3.2+1	2.6+1	9.2−3	1.4+2	9.5+1	4.0−2	9.9+2	4.4+2	2.5−1	3.6+3	1.1+3	8.1−1
25	10	2.7+9	1.1+1	4.1+0	2.6−2	2.6+1	8.9+0	3.8−2	6.2+1	1.9+1	3.1−1	9.7+1	3.0+1	5.0−2
	25	1.9+10	4.2+1	2.3+1	4.4−2	1.2+2	5.4+1	9.2−2	4.5+2	1.4+2	3.1−1	8.5+2	2.4+2	3.1−1
	50	1.3+11	1.4+2	9.9+1	4.6−2	4.0+2	2.5+2	1.0−1	1.7+3	7.5+2	4.5−1	4.8+3	1.5+3	9.3−1
50	10	7.8+9	7.2+1	2.8+1	1.0−1	1.6+2	5.7+1	1.2−1	3.6+2	1.2+2	3.3−1	5.6+2	1.8+2	1.6−1
	25	3.2+10	2.0+2	9.0+1	1.8−1	4.7+2	1.9+2	2.5−1	1.2+3	4.2+2	4.0−1	2.1+3	6.7+2	5.0−1
	50	1.6+11	5.6+2	3.3+2	2.4−1	1.4+3	7.3+2	3.5−1	4.0+3	1.7+3	7.4−1	8.5+3	3.0+3	1.3+0
100	10	2.8+10	7.2+2	2.7+2	5.3−1	1.5+3	5.5+2	5.8−1	3.2+3	1.1+3	6.6−1	5.1+3	1.7+3	7.2−1
	25	8.3+10	1.3+3	5.4+2	8.2−1	2.8+3	1.1+3	9.3−1	6.3+3	2.3+3	1.1+0	1.0+4	3.5+3	1.3+0
	50	2.6+11	2.9+3	1.4+3	1.2+0	6.3+3	2.9+3	1.5+0	1.5+4	6.1+3	2.0+0	2.6+4	9.6+3	2.6+0

[a] The first number is the mantissa; the second is the power of ten.

The scalar neutral current model of Adler (model A) produces a different angular dependence in the differential cross section, as well as a different interaction strength:

$$\frac{d\sigma_A}{dz} = \frac{\sigma_0}{8} \mathscr{A}_s^2 A^2 \omega^2 (1 - z)$$ 2.23a.

$$\sigma_A = \frac{\sigma_0}{4} \mathscr{A}_s^2 A^2 \omega^2.$$ 2.23b.

The angular difference is understood as follows: in coherent transitions when the nucleus angular momentum remains unchanged, the scalar interaction flips the neutrino helicity and must vanish in the forward direction, whereas the vector interaction does not alter the helicity and must vanish in the backward direction. An upper limit to the magnitude of the scalar coupling is $|\mathscr{A}_0|^2 \lesssim 4$ (78). The main effect of this different angular dependence is to increase the neutrino radiation pressure.

The speculative model of De Rujula, Georgi & Glashow (DG^2) (79), which includes an additional $V + A$ neutral current, has been discussed and compared with the WS theory by Bernabeu (35). The results have been given in terms of the cross section for momentum transfer,

$$\Sigma = \int d\Omega (1 - z) \frac{d\sigma}{d\Omega},$$

for general nuclei with nonzero spin. Coherent scattering is modified by an isospin term T (which agrees with Equation 2.16), and a correction, $\mathscr{S}$, which contains both spin and isospin effects. In DG^2, the $\mathscr{S}$ correction is absent, and in WS it depends upon the particular nuclear model chosen. Bernabeu has expressed $\mathscr{S}^2$ in terms of the ft value for the β^+ decay of an isospin state $(I, I) \rightarrow (I, I - 1)$. The resulting formulae are

$$\Sigma = \tfrac{1}{6}\sigma_0 \eta_0^2 A^2 \omega^2 [(1 + T)^2 + \mathscr{S}^2],$$ 2.24a.

$$T = \frac{Z - N}{\eta_0 A} \begin{cases} \tfrac{1}{2}(1 + 2a_0) & \text{WS}, \\ 1 + \alpha_0 & DG^2 \end{cases}$$ 2.24b.

$$\mathscr{S}^2 = \frac{5}{12a_0^2 A^2}(N - Z)\left[\frac{K}{ft} + (Z - N)\right],$$ 2.24c.

$$K = 2\pi^3 \ln\left(\frac{2h}{G^2 m_e^5 c^2}\right),$$ 2.24d.

where η_0 is equal to a_0 for WS, and α_0 for DG^2. Some of Bernabeu's results for nuclei near Fe are presented in Table 6. The first row under each species is for the WS model $(a_0^2 = 0.16)$; row two is DG^2 $(\alpha_0^2 = 0.49)$. General conclusions to be drawn are that for $A \gtrsim 10$, modifications to coherent scattering $\lesssim 10\%$, independent of the model. For $A > 10$, the ratio of cross sections $\Sigma(DG^2)/\Sigma(WS) \simeq 3.1$, while the same ratio for neutrons is 0.45. Thus, the DG^2 model is a more effective means of transferring momentum from the core to the mantle than is the WS theory; that is, in the core,

Table 6   Spin and isospin corrections to coherent scattering

Nucleus	$(1+T)^2$	S^2
Mn^{55}	1.05	0
	1.08	
Fe^{56}	1.04	0
	1.06	
Ni^{58}	1.02	0
	1.03	
Co^{59}	1.04	0
	1.07	
Ni^{60}	1.03	0
	1.06	

λ_{eff} is larger in DG^2 than in WS, while in the mantle, λ_{eff} is less in DG^2 than in WS. Observe that for equal coupling strengths, $|a_0| = |\mathscr{A}_0|$, the scalar neutral current is twice as effective as the V-A neutral current in terms of momentum transfer in the $A > 1$ mantle. Results of Wilson's collapse calculations tend to reflect this model-dependent gradient of effectiveness in momentum transport.

Within the center of the star, the principal semileptonic neutrino opacities arise from elastic scattering with nucleons. With the added contribution from the axial vector current, the cross sections are significantly greater than those of coherent scattering for $A = 1$. At low momentum transfer, the interactions are (34):

$$\sigma_{vp} = \tfrac{1}{4}\sigma_0[(C_V - 1)^2 + 3(C_A - 1)^2]\,\omega^2, \qquad\qquad 2.25.$$

$$\sigma_{vn} = \tfrac{1}{4}\sigma_0\omega^2. \qquad\qquad 2.26.$$

Due to the assumption of a neutral current of the form of Equation 2.5a, the cross section for neutron scattering is independent of the Weinberg angle when neutron-structure effects are ignored. The contribution to scattering comes solely from the isovector V-A current (Equation 2.12). The proton interaction had previously been calculated by Weinberg (92), and the general result (2.24) of Bernabeu, which includes both nucleon cross sections, shows that

$$\Sigma_{A=p}(\text{WS})/\Sigma_{A=p}(DG^2) \simeq 13 \quad (\mathscr{S}^2 = 12.18).$$

A further correction to coherent scattering arises from the ion-ion correlations of nuclei in the stellar core (cf Ref. 93). The effective cross section is obtained from the liquid structure factor $S(\Delta a)$:

$$\frac{d\sigma_{\text{eff}}}{d\Omega} = \frac{d\sigma}{d\Omega} S(\Delta a),$$

where

$$\Delta = \frac{\omega}{c}[2(1-z)]^{1/2}$$

is the momentum transfer, and a is a measure of the mean spacing of the nuclei,

$$a = \left(\frac{4\pi}{3} \frac{N_A\rho}{A}\right)^{-1/3}$$

However, for conditions within the collapsing core, and for values of the typical neutrino energies encountered (~ 20 MeV), the suppression by the structure factor is estimated to be only ~ 10–15%. At neutron-star densities, when the neutrino wavelength encompasses several neutrons, the resulting multiple scattering would be like an index-of-refraction effect by the continuous neutron media, rather than coherent scattering from point-scattering centers. Correlation effects from such high-density scattering are also unimportant.

Finally, we must mention a bugaboo that we have tactfully ignored up to this point. What happens when densities become high enough that all of these wonderful cross sections cause a significant buildup of the neutrino number? At some time in the evolution of stellar collapse, in some locations of the core, and at some neutrino energies, the filling of neutrino phase space must be included in all of the above neutrino processes (cf Ref. 94). That is, a factor of

$$1 - f_\nu(\mathbf{r}, \mathbf{q}, t)$$

must be inserted into each differential interaction rate that has a neutrino in the final state. The general neutrino-distribution function depends upon seven parameters, although symmetries of the transport problem make some coordinates ignorable.

In addition to requiring inhibition factors in the rates, when the neutrino number becomes sufficiently high that they begin to scatter elastically with each other, they may do so with cross sections that are comparable to those of neutrino-electron scattering (80). This process, like the semileptonic interactions that have just been discussed, depends solely on the exchange of neutral currents.

In order to provide some feel for the size of typical mean free paths and the relative importance of the neutrino opacities, we have included in Table 7 the approximate data for ν_e in the Weinberg-Salam model ($C_V = 1.3$, $C_A = 0.5$). The table is loosely based upon an actual collapse configuration (88, 89). The local physical conditions should only be considered roughly representative of two stages in a stellar collapse. The situation is not simply a homogeneous mixture of electrons, free nucleons, and a single nuclear species as heavy as iron. The number densities are those for unbound particles; that is, absorption of neutrinos within a nucleus has been ignored. At the temperatures and densities indicated in Table 7, the number of positrons is negligible with respect to the number of electrons; therefore, $n_{e^-} \simeq N_A\rho Y_e$, the total number density of protons, both free and bound. For a given interaction, the mean free path can vary considerably—over several orders of magnitude—with the incident neutrino energy. The mean free paths show similar variation in magnitude from interaction to interaction.

Before leaving the weak-interaction details, let us reiterate that the weak neutral current can only cause dramatic effects when it allows a process to go that would otherwise be forbidden, as in the case of coherent scattering. Otherwise, the effects tend to be small ($\lesssim$ factor of 2 change in the rate).

Table 7 Electron-neutrino mean free paths[a]

Local physical conditions	mfp (cm)	$\omega = 4$ MeV	$\omega = 8$ MeV	$\omega = 16$ MeV	$\omega = 24$ MeV
$\rho/10^9 = 10$ g cm^{-3} $T/10^9 = 10$ K	λ_e	$2.8+8$	$6.5+7$	$9.6+6$	$2.7+6$
$X_A = 0.80$ $X_p = 0.093$	λ_n	$5.9+9$	$1.5+9$	$3.7+8$	$1.6+8$
$X_n = 0.107$ $Y_e = 0.464$	λ_p	$8.1+9$	$2.0+9$	$5.0+8$	$2.2+8$
$n_A/10^{31} = 8.60$ cm^{-3} $n_p/10^{31} = 55.93$ cm^{-3} $n_n/10^{31} = 64.53$ cm^{-3}	λ_A	$8.5+7$	$2.1+7$	$5.3+6$	$2.4+6$
$n_e/10^{31} = 279.47$ cm^{-3} $\mu_e/m_e c^2 = 46.54$ $A = Fe^{56}$	λ_{ab}	$4.2+9$	$1.0+9$	$2.6+8$	$5.6+7$
$\rho/10^9 = 200$ g cm^{-3} $T/10^9 = 27.14$ K	λ_e	$7.8+7$	$2.9+7$	$7.9+6$	$3.2+6$
$X_A = 0.50$ $X_p = 0.050$	λ_n	$7.0+7$	$1.8+7$	$4.4+6$	$1.9+6$
$X_n = 0.450$ $Y_e = 0.282$	λ_p	$7.5+8$	$1.9+8$	$4.7+8$	$2.1+7$
$n_A/10^{31} = 107.55$ cm^{-3} $n_p/10^{31} = 602.30$ cm^{-3} $n_n/10^{31} = 5420.7$ cm^{-3}	λ_A	$6.8+6$	$1.7+6$	$4.2+5$	$1.9+5$
$n_e/10^{31} = 3397.0$ cm^{-3} $\mu_e/m_e c^2 = 35.41$ $A = Fe^{56}$	λ_{ab}	$5.0+7$	$1.2+7$	$2.2+6$	$6.5+5$

[a] Same representation as Table 4. Mean free paths for v_e in the Weinberg-Salam model ($\sin^2 \theta_w = 0.40$, $C_A = \frac{1}{2}$), as functions of incident neutrino energy (ω in MeV). Subscripts on λ refer to scattering with electrons, neutrons, protons, and nucleus A; and absorption on neutrons. Local physical conditions are approximate. Effects of neutrino degeneracy have been ignored.

3 COLLAPSE AND SUPERNOVA EXPLOSIONS

Let us now examine how these weak processes effect the supernova–gravitational collapse scenario. Other applications, such as the cooling of neutron stars and the effects on stellar evolution, are also important, but are not as potentially dramatic as supernovae.

3.1 Collapse Scenario

As we mentioned in Section 1, all massive stars tend to evolve to the same core configuration—a 1.4 $M_\odot$ Fe-Ni core surrounded by burning shells of Si, O, Ne, C, He, and H. This Fe-Ni core begins collapsing on a ρ-T trajectory that passes through $\rho = 10^{10}$ g cm^{-3} and $T \sim 10^{10}$ K, as shown in Figure 5 (88). As the core collapses through a density of around 10^{11} g cm^{-3}, electron captures go at extremely high rates. Thus, at this point the neutrinos from electron captures dominate, and are regular electron neutrinos. As the collapse continues and the temperature continues to rise, the neutrino pair processes may become important.

Then, because the pair processes go both by a charged and a neutral current, they not only produce electron-neutrino pairs, but also produce muon-neutrino pairs and, if heavy leptons exist, heavy lepton-neutrino pairs in comparable numbers to the muon neutrino pairs. It has been pointed out by Mikaelian (119) and Freedman (97) that this could be extremely important, because it could increase the total neutrino flux.

It seems very clear that neutrinos are produced and are the main way in which the collapsing star loses its energy. However, the question of whether or not these neutrinos are capable of causing mass ejection is much more complicated. The primary neutrino-opacity sources are described in Section 2. The opacity sources divide themselves into those that go via a charged current, or a charged and a neutral current, such as neutrino absorption by neutrons and neutrino-electron scattering, and the pure neutral-current process, which is neutrino scattering from nucleons and nuclei. We have seen that the cross section for neutrino scattering from complex nuclei scales like atomic weight-squared at low energies and thus, it has been

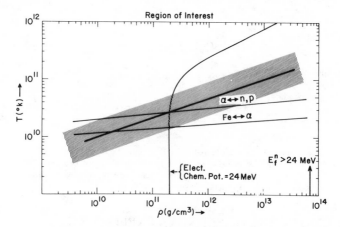

Figure 5   The temperature-density $(T-\rho)$ plane shows the probable region of interest in stellar-collapse calculations. A trajectory with $\rho \propto T$ going through the point $\rho = 10^{10}$ is shown. Cores for stars with $M \gtrsim 8 \, M_\odot$ follow similar trajectories. The region of interest may include the shaded region that surrounds this line. Also illustrated in the figure are the photo-disintegration contours, where 50% of the Fe-Ni has broken up into alphas and where 50% of the alphas have broken up into n and p. In fact, each of these photodisintegration lines turns down at high densities, but the exact location of the turndown depends on poorly known nuclear physics and thus has not been shown. The contour for the chemical potential $\mu = 24$ MeV is shown. For points to the right of this contour, electron capture will occur, even on the most neutron-rich nuclei. Therefore, to the right of $\alpha \sim 2 \times 10^{11}$ is a neutrino source and the average atomic weight $A \sim 1$, whereas to the left of $\sim 2 \times 10^{11}$ and below the $\alpha \to n, p$ line, $A > 1$ and not as many neutrinos are being produced.

noted, this mechanism may provide a way to eject the heavy element–rich mantle (iron and beyond), while allowing the central, neutronized core to collapse to make a neutron star. This would explain why the outer part of the star is ejected and the inner part implodes to make a very dense remnant. However, there have been a number of complications in this seemingly nice scenario. One important one is the effect of degeneracy (82, 88, 94), both the degeneracy of electrons and that of neutrinos in the very central core. The electron degeneracy causes inhibitions in all reactions that have electrons in the final state, such as the neutrino-electron scattering and the neutron absorption by neutrons going to a proton plus an electron.

The neutrino degeneracy also causes an inhibition in reactions that have neutrinos in the final state, but in completely conservative scattering processes, the effect can cancel (98, 99). That is, the enhancement in the scattering caused by the initial neutrino state driving the reaction in a forward direction balances the inhibition caused by neutrinos in the final state. This cancellation of the inhibition effects, due to the neutrino being both in the forward and backward channel, occurs only when one has conservative scattering; that is, scattering where there is no energy change, or only a very small one, for the neutrino. This is the case for the neutrino scattering off nuclei and nucleons. However, the neutrino-electron scattering is very non-conservative (34). Therefore the cancellation of the degeneracy effects does not occur and the situation is much more complex. In the central-core regions of the collapsing stars, the neutrino scattering from nucleons dominates and the neutrino-degeneracy effects for this region tend to cancel. This cancellation and noncancellation is extremely important because it determines the energy spectrum of the neutrinos emerging from the core, as well as their ability to emerge from the core. Since neutrino cross sections go roughly like E^2, the energy of the neutrinos coming out of the core is extremely important.

The effect of all of these processes is to make the following of the detailed neutrino transport in the core a very difficult problem. It is currently being attacked by a variety of methods from simple diffusion to flux-limited diffusion, to S_n diffusion by Yueh & Buchler (100, 115), to the Monte Carlo treatment by Tubbs (85) and Tubbs, Schramm, Weaver & Wilson (116). It does seem that the central core has extremely large neutrino opacities, which would make some form of diffusion a very good approximation. However, in the region just outside the central core, where the neutrino mean free paths are comparable to the size of the region, and where the neutrinos must act to produce the supernova (if they do so at all), the transport is extremely complicated. The importance of this region was pointed out by Schramm & Arnett (88). In the past few years, it has been noted that an equally important question is what is the luminosity of the neutrinos coming out of the central core into this mantle region, where the action might take place. It is in this central-core region that neutrino degeneracy seems to be very important, and it is here that the neutrino luminosity is determined.

It has been shown by Bruenn, Arnett & Schramm (39, 65) that the critical question of whether or not the neutrinos are able to drive off the outer material via momentum transport is that of whether or not the neutrino luminosity emerging from the central-core region is greater than the Eddington limit for neutrinos. We can estimate the

Eddington limit luminosity in the following way (39):

$$L_v^{Edd} \simeq \frac{4\pi GMc}{\kappa_v \langle 1 - \cos\theta \rangle},$$

where M is the interior mass, κ_v is the neutrino opacity in the shell, and $\langle 1 - \cos\theta \rangle$ is the average scattering angle.

The neutrino opacity in the shell is dominated by the coherent scattering process. If the shell has a composition of iron or heavier metals, it is very easy to estimate its opacity and thus to calculate the neutrino luminosity necessary to push it off. This luminosity is approximately 10^{54} ergs sec^{-1}. The number of neutrinos coming from electron capture alone is of the order of 10^{57} neutrinos, and the typical energy of these neutrinos is of the order of 15–20 MeV. If they were emitted within the free-fall time for a core at 2×10^{11} g cm^{-3}, then the neutrino luminosity would indeed be greater than the Eddington limit. However, we know that they are not emitted over a free-fall time since a core is collapsing somewhat slower than free-fall. In fact, due to the finite diffusion time out of the core, the time scale for which the neutrinos are escaping is quite a bit longer than free-fall. The whole collapse is regulated by the rate at which neutrinos carry away the binding energy. The question is, what is this time scale over which it takes the neutrinos to leave the core? If the neutrinos can get out more

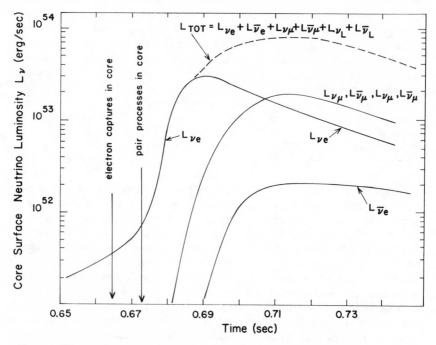

Figure 6   The neutrino luminosity versus time for a typical (though by no means conclusive) gravitational-collapse calculation. This is based on the work of Wilson (101).

rapidly, then in fact we will have a luminosity near the Eddington limit and the neutrinos can cause ejection. However, it should be remembered that even if the neutrinos could stream out of the core, the collapse would not be free-fall, because the electron degeneracy pressure and thermal nucleon pressures are not zero.

In simple hydrodynamic calculations by Wilson (101), it is found that using reasonable, though by no means certain, estimates for cross sections and emission rates, the luminosity of the core falls below the Eddington limit by a factor of approximately 3, thus meaning in these particular calculations that the neutrinos are unable to cause the ejection. One should remember, though, that all of these calculations have large numbers of uncertainties so that the factor of 3 is certainly not fatal at the present stages of the calculations. Thus the questions as to whether or not neutrinos are able to cause an ejection via momentum transport come down to whether or not the neutrino transport is sufficiently rapid to get the neutrinos out of the core with a luminosity greater than the Eddington limit. Another way to augment the luminosity is via neutrinos from heavy leptons, since these neutrinos are produced in equal numbers with the muon neutrinos in the pair processes. The more types of neutrinos there are, the greater the neutrino luminosity would be.

Figure 6 shows the neutrino time spectrum of typical gravitational-collapse calculations. Note that initially there is a burst of the electron neutrinos that gradually diffuse out of the core, followed by the emission of the pair neutrinos, the muon neutrinos, and the heavy-lepton neutrinos. The electron antineutrinos are suppressed for a while, due to the fact that the electron-neutrino phase space in the core has been filled by the neutrinos from electron capture. Only after these neutrinos from electron capture have been able to diffuse out can the pair processes be successful in producing electron antineutrinos. Eventually, as the electron-capture neutrinos have all diffused out, these electron antineutrinos have a flux equal to that of the electron neutrinos; however, that is not in the initial collapse, but only after several tenths of a second.

Figure 7 shows the trajectories of mass zones as a function of time for a particular model of Wilson's (101), which caused mass ejection. Notice how the zones undergo a hydrodynamic bounce. In this particular model, which had the opacities in the mantle turned up by a factor of 3 to allow ejection, as the bounce material reached its peak and was about to fall back in, it was hit by the neutrino flux and was thus, due to this factor of 3 in the opacity, at the Eddington limit and able to drive off the material before it fell back onto the core. Thus ejection, and a supernova, occurred. If, for some other reason, the neutrino flux could be increased or if the location of the bounce could be slightly changed relative to the neutrino flux, one might also be able to obtain an ejection. Due to the large uncertainties in the equation of state, which determines the strength and location of the bounce, and due to some uncertainties still present in the neutrino transport out of the core, it is clear that the calculations are by no means definitive at the present time. It has been found (65, 101) that by slight changes in the equation of state, well within the range of uncertainty, the hydrodynamic bounce can become sufficiently large that it alone is capable of causing mass ejection. Thus, depending on the equation of state, the bounce may be sufficient to eject matter regardless of the neutrino effects; it may be marginal, so that the neutrinos are crucial; or it may be so weak that the neutrinos cannot save the day. These equation-of-state uncertainties revolve around the photodisintegration and electron-capture rates for extremely neutron-rich nuclei between the iron-nickel region and free nucleons.

In view of the present status of uncertainty on all of these points, a strong statement cannot be made with regard to whether or not ejection occurs via neutrinos, a bounce, or some other mechanism in the core. However, it does seem reasonable to say that in view of the fact that these stars can produce the heavy elements and seem to be able to produce the cosmic rays and, apparently, definitely have very dense cores that would lend themselves to producing dense remnants like neutron stars or black holes, that this mass range of stars ($M_{CD} \lesssim M \lesssim M_{pair}$) is the one of key importance to the supernova problem. However, the actual mechanism whereby the ejection occurs is still a matter of great uncertainty. The weak interaction is obviously of great importance to the whole scenario; however, whether it is the key element in causing an ejection or just an important factor depends on the actual equation of state.

In order to try to observe gravitational collapse, it is important to remember that the neutrinos coming out of the core, whether or not they are able to cause ejection of matter, are still the dominant mechanism by which the energy is lost from the core and they are going to be emitted on a time scale somewhat greater than the collapse time, which in turn is somewhat greater than the free-fall time. In fact, calculations to date seem to indicate that the collapse time is $\gtrsim 5$ times greater than the free-fall time. Thus the time scale over which one would detect the neutrino pulse from gravitational collapse would be of the order of tens of milliseconds, with the duration

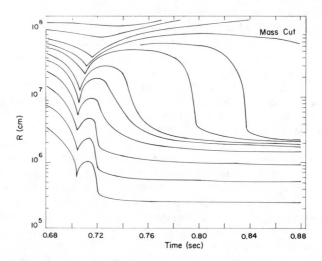

Figure 7   The trajectories of various mass zones in a calculation of Wilson's (101), with the coherent scattering opacity arbitrarily increased by a factor of ~ 3. The mass ejection occurs when the shocked mantle material is hit by the neutrino flux (compare times with figure 6) before it can fall back on the core.

of the radiation lasting for the amount of time required for the core to settle down to a neutron star, which is the order of tenths of seconds.

Another important point for possible detection is the energy of these neutrinos. The electron-capture neutrinos are going to be emitted at roughly 15–20 MeV. [The bulk of the electron captures take place when the Fermi energy is of the order of 24 MeV (88), and the electron-capture neutrinos have on the average an energy of $\frac{5}{6}$ of the Fermi energy (88).] This makes the neutrinos very low in energy, less than ~ 20 MeV, thus making it very difficult for detectors such as the Deep Underwater Muon and Neutrino Detector (DUMAND) (102) to see extragalactic supernovae. The thermal neutrinos continue to escape after the electron captures cease. The thermal neutrinos are the only ones that produce the easier-to-detect electron antineutrinos, which are inhibited, as we mentioned before, by a factor of 10–100 over the electron neutrinos until well after the initial burst of neutrinos. The thermal neutrinos have an energy characteristic of the temperature at which they are emitted, so as the core continues to heat up, the energy of the thermal neutrinos continues to increase. However, it never seems to reach an extremely high temperature in the core; that is, the energy of the thermal neutrinos never seems to get greater than a few tens of MeVs, again indicating that the detection of these neutrinos would be extremely difficult.

3.2   Summary of Uncertainties Regarding the Collapse

Understanding core collapse has proved difficult in the past; one factor is the problem of understanding the presupernova evolution; that is, the relative composition of the different zones in the star and the density profile of the core and the material just outside of the core. This can be strongly influenced by convection calculations.

We can summarize the problems in understanding core collapse as follows:

1. The equation of state, including the composition as the core collapses, in particular taking into account the rate of electron captures on complex nuclei and the rate of photodisintegration of these complex nuclei. This can make very large changes since, if many heavy nuclei are still present in the central core, they will significantly increase the neutrino opacity in the central core and thus significantly change the equation of state and the pressure balance in the central core. A very important point here with regard to the equation of state is the density at which the core bounce takes place and the temperature of the material at the core bounce. This is extremely sensitive to the equation of state and the composition.

2. The second problem is the weak-interaction theory—not just the strength of the isoscalar component of the neutral current that determines the strength of the coherent scattering opacity, but also the number of types of neutrinos that exist; that is, the number of heavy leptons and neutrinos associated with them, since we have seen that the more heavy leptons with their associated neutrinos, the greater the flux of neutrinos coming out of the core that can add to exceed the Eddington limit. As for the coherent scattering, the type of neutrino does not matter, so long as it interacts via a neutral current.

3. The third problem involves the neutrino transport. This must include the effect of both neutrino and electron degeneracy and the effect of conservative and non-

conservative scattering to be able to follow the neutrinos coming out of the core in both their flux and energy. As we have noted, the energy spectrum is extremely important due to the sensitive dependence of neutrino cross sections upon the energy of the neutrino.

4. The fourth problem is that of the effects of rotation and magnetic fields. The calculations mentioned so far have been with a spherically symmetric, nonrotating collapsing star. No rotation or magnetic fields have been mentioned. It seems clear that if rotation and magnetic fields are added, one could change the nature of the problem. In fact, in calculations by LeBlanc & Wilson (103) and further work by Meier et al (104), an ejection of a very different nature was obtained due to the magnetic field winding up and ejecting jets from the poles during a collapse. The effects of rotation and magnetic fields on the pressure balance, etc have not been explored in detail in any of the present calculations.

So we see that there are many problems in the present core collapse; however, these problems, although they affect the final stages of the collapse and whether or not mass is ejected, as far as neutrino detection is concerned, are of relatively little importance since the main fact that neutrinos are emitted in copious numbers from collapse and that the neutrinos emitted are relatively low in energy seems to be relatively insensitive to these details.

Let us summarize. The present models for a nonrotating star always yield a collapsing core greater than $1.1 \, M_\odot$. If it is necessary to eject $\sim \frac{1}{10} M_\odot$ of iron in order to explain the iron abundances that are in the galaxy, and it seems that neutrinos may eject that sort of mass, we would end up with a collapsing core mass of about $1.3 \, M_\odot$, which would be quite consistent with the observed masses of neutron stars (66). We also see that if we have a mass average of the ejecta of all of these massive objects, we would get a good fit to the carbon-to-iron abundances in the solar system and we also see that if we have a slightly different way of averaging the supernovae, we could get a reasonable fit to the cosmic-ray abundances and could even make predictions with regard to the cosmic-ray isotopic composition, which has yet to be determined.

We have seen that the bulk of the binding energy is released as neutrinos in these stars. In fact, it has been shown that even nonspherical perturbations are more rapidly damped out via neutrino radiation than via gravitational radiation (105, 106). However, it seems to be difficult to carry away significant angular momentum from the core via neutrino radiation (107, 117). Although neutrinos do remove some angular momentum (96, 107), they also remove energy, and thus the net change in the angular momentum per unit mass is small (117).

We do have some problems with the core. One very serious one is the conditions at the mass cut, that is, the boundary between the matter that implodes to make the neutron star or black hole and the matter that is ejected. Are these conditions at the mass cut similar to those that give the r-process? The r-process is a nucleosynthetic process thought to be responsible for the neutron-rich heavy elements. The standard scenario places the r-process at this boundary between the matter that falls in and the matter that is ejected (39), but until we understand the conditions at the mass cut, we cannot verify this scenario. Another very important question is whether the location of the mass cut is sensitive to the mass of the star. For example, is it possible that a more massive star, say 40 or $50_\odot$, might have a collapsing core that would be somewhat greater in mass than the lower-mass stars? In fact, in the stellar-evolution calculations, these extremely massive stars tend to have cores slightly over the Chandrasekhar limit. More massive stars still have the silicon shell burning very rapidly and at very high density, so that it, too, could form part of the collapsing core, whereas the lower-mass stars near 10–15 $M_\odot$ tend to have a core much closer to the 1.4-$M_\odot$ core and evolve in such a way that there is a very thin silicon shell so that it could not significantly increase the mass of the core. Thus the more massive cores may be over the Oppenheimer-Volkoff limit and would collapse to form black holes, whereas the less massive ones would simply make neutron stars. Thus, this question of which-mass stars make black holes and which make neutron stars is one of great uncertainty at the present time and one that will depend on understanding how the mass cut varies with the mass of the star. Another possibility may be that all of these nonrotating, noninteracting stars would yield neutron stars, and that only by putting some other property into the collapse, such as rotation, magnetic fields, or binary interaction, could one get a black hole. The only black holes noted today are in binary systems. (Of course, it would also be very difficult to observe a black hole that was not in a binary system.) On this point, one other comment that should be noted is that, although it seems to be an extremely delicate matter as to whether or not one gets mass ejection in the collapsing core, this is not as bad as it might first seem, since it may mean that a very delicate change one way or another could affect whether a collapse would yield a black hole or a neutron star and ejection. Since we probably see both neutron stars and black holes, the need for some sort of delicate shift from one to the other may be correct.

Let us reiterate that one problem that really needs to be explored is the dependence of the mass of the core on rotation. Where does rotation enter into this thing? How is angular momentum lost from a collapsing star? We know that massive stars on the main sequence are rotating relatively rapidly. Do the cores form while rapidly rotating or have they lost their angular momentum prior to their formation? We do know that all white dwarfs for which rotational velocities are known are slowly rotating. We also know that all pulsars are observed to be slowly rotating. However, that does not answer the questions of whether or not the pulsars formed are rapidly rotating and what the initial velocity of rotation is of a forming neutron star.

In conclusion, we have seen that weak-interaction processes are extremely important to the later stages of stellar evolution, in particular the final gravitational collapse to form a neutron star or a black hole where the bulk of the binding energy is emitted as neutrinos. In fact, the weak neutral current may even be the mechanism that enables the core to collapse while ejecting the outer regions in a supernova explosion. However, until a better understanding of the equation of state of the collapsing matter is obtained, no strong conclusion can be made.

ACKNOWLEDGMENTS

We would like to thank Dave Arnett, Jim Wilson, and Duane Dicus for much thoughtful discussion, and Demosthenes Kazanas and Martin Murphy for their valuable assistance throughout the course of this work. This research was supported in part by NSF grant AST 76-21707 at the University of Chicago, and in part by the Enrico Fermi Institute Research Fund and the Shirley Farr Fund.

Literature Cited

1. Hasert, F. J. et al 1973. *Phys. Lett. B* 46:121
2. Hasert, F. J. et al 1973. *Phys. Lett. B* 46:138
3. Benvenuti, A. et al 1974. *Phys. Rev. Lett.* 32:800
4. Aubert, B. et al 1974. *Phys. Rev. Lett.* 32:1454, 1457
5. Barish, S. J. et al 1974. *Phys. Rev. Lett.* 33:448
6. Barish, B. C. et al 1974. *Phys. Rev. Lett.* 34:538
7. Lee, W. et al 1974. In *Proc. International Conf. High Energy Phys., 17th*, ed. J. R. Smith, pp. IV–127. Chilton, Didcat, Engl: Rutherford Lab.
8. Fermi, E. 1934. *Z. Phys.* 88:161
9. Lee, T. D., Yang, C. N. 1956. *Phys. Rev.* 104:254
10. Wu, C. S., Ambler, E., Hayward, R. W., Hoppes, D. D., Hudson, R. P. 1957. *Phys. Rev.* 105:1413
11. Garwin, R. L., Lederman, L. M., Weinrich, M. 1957. *Phys. Rev.* 105:1415
12. Friedman, J. J., Telegdi, V. L. 1957. *Phys. Rev.* 105:1681
13. Feynman, R., Gell-Mann, M. 1958. *Phys. Rev.* 109:193
14. Marshak, R. E., Sudarshan, E. C. G. 1958. *Phys. Rev.* 109:1860
15. Sakurai, J. J. 1958. *Nuovo Cimento* 7:649
16. Dauby, G., Gaillar, J. M., Goulianos, K., Lederman, L. M., Mistry, N., Schwarts, M., Steinberger, J. 1962. *Phys. Rev. Lett.* 9:36
17. Lee, W., et al 1977. *Phys. Rev. Lett.* 38:202
18. Lee, W. et al 1976. *Phys. Rev. Lett.* 37:186
19. Cline, D. et al 1976. *Phys. Rev. Lett.* 37:252
20. Weinberg, S. 1967. *Phys. Rev. Lett.* 19:1264
21. Salam, A., 1968. In *Proc. Nobel Symp., 8th.* New York: Wiley
22. Yang, C. N., Mills, R. L. 1954. *Phys. Rev.* 95:191
23. Higgs, P. W. 1964. *Phys. Lett.* 12:132; 1964. *Phys. Rev. Lett.* 13:508; 1966. *Phys. Rev.* 145:1156
24. Englert, F., Brout, R. 1964. *Phys. Rev. Lett* 13:321
25. Garalnik, G. S., Hagen, C. R., Kibble T. W. B. 1964. *Phys. Rev. Lett.* 13:585
26. Abers, E. S., Lee, B. W. 1973. *Phys. Rep. C.* 9:1
27. Weinberg, S. 1974. *Rev. Mod. Phys.* 46:255; *Sci. Am.* 200:50
28. Bēg, M. A. B., Sirlin, A. 1974. *Ann. Rev. Nucl. Sci.* 24:379
29. Bernstein, J. 1974. *Rev. Mod. Phys.* 46:7
30. Gamow, G., Schönberg, M. 1941. *Phys. Rev.* 59:539
31. Fowler, W. A., Hoyle, F. 1964. *Astrophys. J. Suppl.* 9:201
32. Dicus, D. A. 1972. *Phys. Rev. D.* 6:941
33. Freedman, D. Z. 1974. *Phys. Rev. D.* 9:1389
34. Tubbs, D., Schramm, D. N. 1975. *Astrophys. J.* 201:467
35. Bernabeu, J. 1976. CERN Rep.
36. Bahcall, J., Trieman, S., Zee, A. 1977. *Phys. Lett* 52:275
37. Dicus, D., Kolb, E., Schramm, D., Tubbs, D. 1976. *Astrophys. J.* 210:481
38. Davis, R., Bahcall, J. 1976. *Science*
39. Schramm, D. N. 1976. *Nuckleonika* 21:727
40. Schramm, D. N., Arnett, W. D. 1975. *Mercury.* 4:16
41. Chandrasekar, S. 1931. *Astrophys. J.* 24:81
42. Schramm, D. N., Arnett, W. D., ed. 1973. *Explosive Nucleosynthesis*, Chap. 4. Austin, Tex: Univ. Tex. Press
43. Checketkin, V. M. et al 1977. In *Supernovae*, ed. D. N. Schramm. Dortrecht, West Germ: Riedel
44. Mazurek, T., Meier, D. L., Wheeler, J. C. 1977. *Astrophys. J.* 213:518
45. Barkat, Z., Rakauy, G., Sack, N. 1967. *Phys. Rev. Lett.* 18:379
46. Fraley, G. S. 1968. *Astrophys. Space Sci.* 2:96
47. Arnett, W. D. 1973. See Ref. 42
48. Barkat, Z. 1977. See Ref. 43, p. 131
49. Dearborn, D. 1977. *Astrophys. J.* In press
50. Paczynski, B. 1971. *Acta Astron.* 21:1
51. Tinsley, B. M. 1977. See Ref. 43, p. 117
52. Falk. S., Arnett, W. D. 1977. *Astrophys. J. Suppl.* 33:515
53. Lasher, G. 1977. See Ref. 43, p. 13
54. Chevalier, R. 1976. *Astrophys. J.* 208:826
55. Tammann, G. 1977. See Ref. 43, p. 95
56. Taylor, J. 1977. *Ann. NY Acad. Sci.*
57. Rossino, L. 1977. See Ref. 43, p. 1

58. Zwicky, F. 1965. *Stellar Structure*, ed. Dvuiper, G., Middlehurs, B. Chicago: Univ. Chicago Press
59. Arnett, W. D. 1975. *Astrophys. J.* 195:727
60. Kazanas, D., Schramm, D. 1975. *Bull. Am. Astron. Soc.* 7:4
61. Chevalier, R. 1977. See Ref. 43, p. 53
62. Arnett, W. D., Schramm, D. N. 1973. *Astrophys. J.* 184:L47
63. Hainebach, K., Norman, E. B., Schramm, D. N. 1976. *Astrophys. J.* 203:245
64. Arnett, W. D. 1973. *Ann. Rev. Astron. Astrophys.* 11:73
65. Bruenn, S. W., Arnett, W. D., Schramm, D. N., 1977. *Astrophys. J.* 213:213
66. Rappaport, A. S., Joss, P. 1977. *Nature.* In press
67. Sternheim, M. M., Silbar, R. R. 1974. *Ann. Rev. Nucl. Sci.* 24:249
68. Saudinos, J., Wilkin, C. 1974. *Ann. Rev. Nucl. Sci.* 24:341
69. Donnelly, T. W., Walecka, J. D. 1975. *Ann. Rev. Nucl. Sci.* 25:329
70. Cabibbo, N. 1963. *Phys. Rev. Lett.* 10:531
71. Weinberg, S. 1971. *Phys. Rev. D.* 55:1412
72. Weinberg, S. 1971. *Phys. Rev. Lett.* 27:1688
73. Glashow, S., Iliopoulos, J., Maiani, L. 1970. *Phys. Rev. D.* 2:1285
74. Goldhaber, G. et al 1976. *Phys. Rev. Lett.* 37:255
75. Kayser, B., Garvey, G. T., Fischbach, E., Rosen, S. P. 1974. *Phys. Lett. B* 52:385
76. Kingsley, R. L., Wilczek, F., Zee, A. 1974. *Phys. Rev. D.* 10:2216
77. Wilson, J. R., Couch, R., Cochran, S., Le Blanc, J. 1975. *Ann. NY Acad. Sci.* 262:54
78. Adler, S. L. 1975. *Phys. Rev. D.* 9:1954
79. De Rujula, A., Georgi, H., Glashow, S. L. 1975. *Phys. Rev. Lett.* 35:69
80. Flowers, E., Sutherland, P. 1976. *Astrophys. J.* 208:L19
81. Clayton, D. D. 1968. *Principles of Stellar Evolution and Nucleosythesis.* New York: McGraw-Hill
82. Schramm, D., Arnett, W. D. 1975. *Phys. Rev. Lett.* 34:113
83. Yueh, W., Buchler, J. R. 1976. *Astrophys. Space Sci.* 41:221
84. Euwema, R. N. 1964. *Phys. Rev. B.* 133:1046
85. Tubbs, D. L. 1977. PhD thesis, Univ. Chicago, Chicago.
86. Bludman, S., Van Riper, K. 1976. *Astrophys. J.* 212:859
87. Wiita, P. J., Tubbs, D. L., Schramm, D. N. 1977. In preparation
88. Schramm, D. N., Arnett, W. D. 1975.

Astrophys. J. 198:629
89. Arnett, W. D. 1975. *Ann. NY Acad Sci.* 262:47
90. Sutherland, P., Ng, J. N., Flowers, E., Ruderman, M., Inman, C. 1976. Columbia Univ. Press
91. Dicus, D., Kolb, E. 1977. *Phys. Rev. D.* In press
92. Weinberg, S. 1972. *Phys. Rev. D.* 5:1412
93. Itoh, N. 1975. *Prog. Theor. Phys.* 54:1580
94. Mazurek, T. 1975. *Astrophys. Space Sci.* 35:117
95. Baird, P. E. G. et al 1976. *Nature* 264:528
96. Mikaelian, K. 1977. *Astrophys. J. Lett.* 209:L77
97. Freedman, D. Z. 1976. Talk at Chicago, Neutrino Workshop, April
98. Bludman, S., 1976. *Proc. DUMAND Workshop*, publ. Fermilab
99. Pethick, C., Lamb, D. 1976. Prepr.
100. Yueh, W., Buchler, J. R. 1977. Univ. Fla. Prepr.
101. Wilson, J. R. 1976, private communication
102. *Proc. DUMAND Workshop Univ. Hawaii*, publ. Fermilab, 1976
103. Le Blanc, J., Wilson, J. 1970. *Astrophys. J.* 161:541
104. Meier, D., Epstein, R., Arnett, W. D., Schramm, D. N. 1976. *Astrophys. J.* 204:869
105. Kazanas, D., Schramm, D. 1976. *Nature* 267:671
106. Kazanas, D., Schramm, D. 1977. *Astrophys. J.* 214:819
107. Kazanas, D. 1977. *Bull. Am. Phys. Soc.* 22:88
108. Perl, M. L. et al 1975. *Phys. Rev. Lett.* 35:1489
109. Baltay, C. 1977. Joint Meet. APS-AAPT, Chicago
110. Perl, M. 1976. *Proc. 1976 Aachen Neutrino Conf.*, West Germany
111. Weinberg, S. 1977. Joint Meet. APS-AAPT, Chicago
112. Bouchiat, M. A., Bouchiat, C. C. 1974. *Phys. Lett. B* 48:1974
113. Mohapatra, R. N., Sidhu, D. P. 1977. *Phys. Rev. Lett.* 38:667
114. Cheng, T. P., Li, L. F. 1977. *Phys. Rev. Lett.* 38:318
115. Yueh, W., Buchler, J. R. 1977. *Astrophys. J. Lett.* 211:L121
116. Tubbs, D. L., Schramm, D. N., Weaver, T. A., Wilson, J. R. 1977. In preparation
117. Kazanas, D. 1977. *Nature.* In press
118. Flowers, E. G., Sutherland, P. G., Bond, J. R. 1975. *Phys. Rev. D* 12:315
119. Mikaelian, K. O. 1975. *Astrophys. J.* 200:336

Ann. Rev. Nucl. Sci. 1977. 27:209–78
Copyright © 1977 by Annual Reviews Inc. All rights reserved

NEUTRINO SCATTERING AND NEW-PARTICLE PRODUCTION

✕5585

D. Cline and W. F. Fry

Department of Physics, University of Wisconsin, Madison, Wisconsin 53706

CONTENTS

1 INTRODUCTION[1]

The neutrino has played a unique role in the development of subatomic physics for the past 40 or more years. Initially, the very existence of the neutrino and its intrinsic properties were of central importance. The neutrino interaction with matter was feeble and thus of secondary interest. In the early 1960s, high-energy particle accelerators were used to produce copious beams of high-energy neutrinos (1, 2). After the completion of Fermilab in 1973, very intense high-energy neutrino beams became available. At the same time, a new generation of neutrino detectors ranging in target mass from 1 to 2000 tons were developed to operate at Fermilab and the CERN SPS (3, 4). The coupling of intense neutrino beams and high-mass targets allows the accumulation of large numbers of neutrino interactions and thus detailed study of the interaction of the neutrinos with matter. Again, the unique property of the neutrino, which to first order interacts only through the weak interaction, plays a crucial role. There are three general areas in which neutrino scattering allows a sensitive probe of nature: (a) the study of the properties of the semileptonic and leptonic weak interaction, (b) the study of the locality of the weak interaction itself, and (c) the study of new forms of matter that are stable to strong and electromagnetic transitions. These three areas of study have already shown impressive payoff. In the first case, the discovery of weak neutral currents and the initial determination of the space-time properties of the interaction have come about (5); in the second, the limits of nonlocality have been pushed beyond an

[1] The following abbreviations and acronyms are used throughout the article:

FNAL	Fermi National Accelerator Laboratory (at Batavia, Illinois)
SPS	The 400-GeV accelerator at CERN, Geneva, Switzerland
GGM	The Gargamelle heavy-liquid bubble-chamber collaboration (at CERN)
GPWF	The Harvard-Pennsylvania-Wisconsin-Fermilab neutrino collaboration (at Fermilab)
CITF	The California Institute of Technology–Fermilab neutrino collaboration (at Fermilab)
ACMP	The Argonne Laboratory–Carnegie-Mellon–Purdue neutrino bubble-chamber collaboration (at Fermilab)
FIMS	The Fermilab-ITEP (USSR)-Michigan-Serpukhov neutrino heavy-liquid bubble-chamber collaboration (at Fermilab)
A-P	The Aachen-Padone neutrino collaboration (at CERN)
CHYM Model	The model developed at Caltech, Harvard, Yale, and Maryland
G	The weak coupling constant
M	The proton mass
E_ν	The neutrino energy
E_H	The hadronic energy released in a neutrino collision
F_2, xF_3, F_L	The three structure functions that enter into the differential-scattering equations for both neutrino and antineutrino scattering in the scaling limit. See (9) for more detail.
$\sigma_{Q.E.}^\nu, \sigma_{Q.E.}^{\bar\nu}$	The cross sections for quasi-elastic scattering of neutrinos and antineutrinos from protons and neutrons, respectively

effective mass of 20 GeV. This implies that the intermediate vector boson is very likely too massive to be produced with the current generation of proton sychrontrons or electron-positron machines. The third area of study is just beginning, and already the existence of at least one new hadronic quantum number has been discovered (6, 7). Further study indicates that one of the quantum numbers is charm (8).

The experimental search for new elementary particle families with lifetimes in the region 10^{-12}–10^{-18} sec requires three ingredients to be successful: (a) sufficient energy in the center of mass of any reaction to make the new particles, (b) sufficient cross section to permit the new-particle production to stand above background, and (c) a unique experimental signature to insure detection and identification of the new particles. Factor (a) needs no explanation but (b) and (c) require additional comment. A particle with a new internal quantum number that decays weakly, primarily to hadrons, is expected to be difficult to detect through its hadronic decay modes, if it is relatively massive. This is because, in strong and electromagnetic production processes, such particles will presumably be produced in particle-antiparticle pairs or by associated production. In addition, they will have, in general, several multiparticle hadronic decay modes. The relatively large hadron multiplicity, including neutrals, in the final state, as well as the multiplicity of final states and the limited effective mass resolution of most particle detectors now in use, all contribute to the difficulty of recognizing two particles of unique mass as the parents of the hadronic decay products.

The production of massive particles with new quantum numbers in high-energy neutrino interactions, and the subsequent decay of those particles into a neutrino, a charged lepton, and possibly hadrons, provide both the means of production of new particles and also the signal of their existence. Any new hadron, for example, that decays semileptonically or leptonically can be produced in the inverse process, in which an incident neutrino interacts with a hadron and gives rise to a muon, other hadrons, and the new particle. This is simply the equivalent of β decay and its inverse. In general, such new-particle production and decay would lead to the experimental observation of two muons (dimuons) in the final state of a neutrino- (or antineutrino-) induced reaction. The empirical description of the overall process would be

$$\nu_\mu + (A, Z) \rightarrow \mu^- + l^\pm + \nu_l + \text{hadrons}, \hspace{2cm} 1.$$

because lepton conservation requires that the unobserved ν_μ be present in the final state.

In this review article, we focus on the development in high-energy (> 10 GeV) neutrino physics during the past three years. The data included in the review were collected prior to November 1976 (9).

2   NEUTRINO AND ANTINEUTRINO BEAMS

A number of neutrino beams have been devised and operated at various accelerators. Table 1 gives a partial list of these beams along with some of their virtues and

Table 1 Muon neutrino and antineutrino beams

Type	Virtues	Weakness	Primarily used by
Bare target	"natural" ratios of $\bar{v}$ to v flux	low intensity	counter experiments
Pulsed magnetic horn beam	sign-selected high intensity at low energies all neutrino energies observed at some time in the detector	hard to compare v and $\bar{v}$ fluxes falling neutrino spectrum	counters and bubble chambers
Quadrupole-triplet beam	"natural" ratio of $\bar{v}$ to v flux hardened neutrino spectrum direct flux measurement all neutrino energies observed at some time in the detector	low intensity at low energy	counter experiments
Dichromatic beams	v and $\bar{v}$ separated	low intensity	counters and bubble chambers
pulsed	large flux acceptance at low energy—suitable for bubble chambers	different neutrino energies measured at different times—requires good monitor	counters (possibly heavy-liquid bubble chambers)
static	v and $\bar{v}$ very well separated, direct flux measurement		

weaknesses. In one case, the "pulsed dichromatic beam" is still in the planning stage at Fermilab. The use of a variety of neutrino and antineutrino beams is a powerful tool for providing additional redundancy for the measurements that are carried out, even with the most sophisticated detector. This is especially true of the massive counter experiments, since the high accelerator energy and high proton-beam intensity make it possible to carry out certain neutrino experiments with good statistical significance in relatively short periods of time. For example, a precision measurement of the ratio of $\sigma^{\bar{v}}/\sigma^{v}$, the total cross sections, can be carried out in a natural quadrupole triplet beam where the flux of $\bar{v}$ to v is directly related to the π^{-}/π^{+} and K^{-}/K^{+} ratios. An important part of the understanding and use

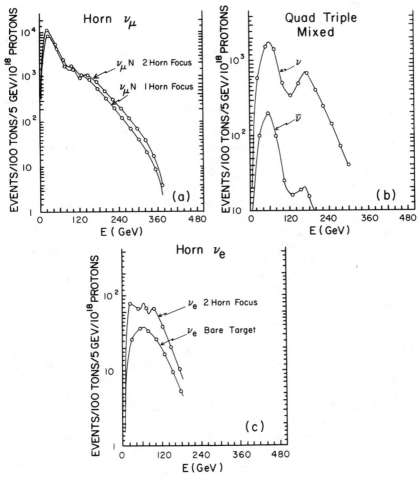

Figure 1 Neutrino event rates for various neutrino beams at Fermilab.

of any neutrino beam involves a knowledge of the hadronic spectrum produced from thick targets at nearly zero mrads. The uncertainty in this hadron yield contributes a corresponding uncertainty in total cross-section measurements. Two different yield-measurement experiments have been carried out at Fermilab (10).

As an illustration, the event rates for various beams described in Table 1 are given in Figure 1 for a 100-ton target mass and for 10^{18} protons on target. Included in Figure 1 are the yields of electron-neutrino events.

3 EXPERIMENTAL TECHNIQUES—NEUTRINO DETECTORS

The principal experimental problems in studying charged-current neutrino inter-actions in broad-band beams are

1. The large diameter of neutrino beams.
2. The low rate of neutrino interaction, requiring a massive target-detector.
3. Lack of knowledge about the incident neutrino energy, which requires a measure-ment of the total energy after the collision $(E_H + E_\mu)$.
4. Identification and momentum measurement of the outgoing lepton (p_μ), requiring a large-aperture lepton spectrometer.
5. The use of heavy nuclear targets for increased target mass.

A number of neutrino detectors and neutrino beams have been devised that over-come these problems in various ways. Figure 2 shows 4 different types of neutrino detectors that are used at various accelerators.

Figure 2A shows a bubble chamber with an external muon identifier (11). Figures $2C_1$, and $2C_2$ show massive calorimeter–magnetic spectrometer detectors of the HPWF and CITF groups (3, 4). The variables that are directly measured are the muon four-momentum p_μ, the muon angle θ_μ, the total hadronic energy E_H, and the angle of the hadronic system θ_H. Since the neutrino direction is assumed to be well known, the measurement of any three of these four variables permits a one-constraint (1C) determination of the kinematic variable of the inclusive collision. Of course, this assumes that the collision was on a stationary target and the effects of Fermi motion must be estimated. If all four variables are measured, then a 2C fit is obtained. From these kinematic variables it is possible to obtain several sets of variables that are presumed to facilitate interpretation of the neutrino data, such as Q^2, the momentum transfer from the neutrino to the muon; W, the invariant mass of the hadronic system; x and y, the Bjorken-scaling variables; or any of the other scaling variables that have been devised to interpolate between the expected scaling and nonscaling kinematic regions. The resolution function for the scaling variables is directly dependent on the kinematic variables. For example, in the Gaussian approximation, the Bjorken variables are related as

$$\left(\frac{\delta x}{x}\right)^2 = 4\left(\frac{\delta \theta_\mu}{\theta_\mu}\right)^2 + (2-y)^2(\delta p_\mu/p_\mu)^2 + (1-y)^2\left(\frac{\delta E_H}{E_H}\right)^2, \qquad 2.$$

$$(\delta y/y)^2 = (1-y)^2(\delta E_H/E_H)^2 + (1-y)^2(\delta p_\mu/p_\mu)^2. \qquad 3.$$

For any given experimental detector, the effects of resolution can be assessed by comparing the resultant distortion of a test x or y distribution.

Another type of calorimeter detector has been operated at the Brookhaven National Laboratory to study neutrino, antineutrino elastic scattering, as shown in Fig. 2B (12). This detector is a segmented scintillation counter capable of recording a proton or π recoil at low momentum. Finally, large emulsion stacks have now

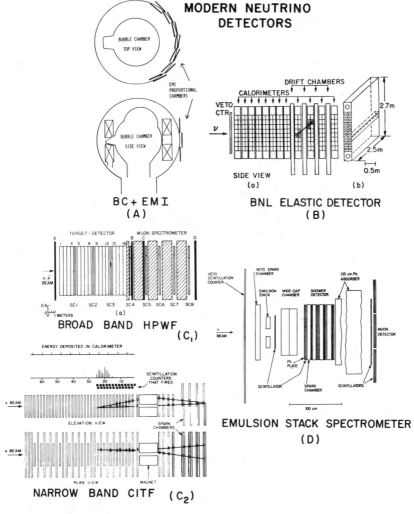

Figure 2   Schematic of neutrino detectors used at Fermilab and Brookhaven National Laboratory: (A) bubble chamber with external muon identities, (B) liquid-scintillation calorimeter detector used at BNL for neutrino elastic scattering measurements, (C_1) calorimeter-iron spectrometer detector used by the HPWF group at Fermilab, (C_2) iron calorimeter-iron spectrometer detector used by the CITF group, and (D) emulsion-stack spectrometer used to make first direct observation of a charmed-particle decay in flight.

operated in high-energy neutrino beams (13). One such detector is shown in Figure 2D. This detector has been used to record the first actual charm-particle track.

The variety of detectors is well matched to the variety of neutrino beams and to the expected and observed richness in the new physics of high-energy neutrinos.

4   CHARGED-CURRENT PROCESSES. I. DIFFERENTIAL CROSS SECTIONS

4.1   Deep Inelastic Scattering Formalism and Data

A deep inelastic charged-current collision of a neutrino or antineutrino with a nuclear target is described by two scaling variables (measured in the laboratory system):

$$x = \frac{Q^2}{2E_H}, \quad Q^2 = 2P_\mu E_\nu(1 - \cos^2 \theta_\mu),$$

$$y = \frac{E_H}{E_\nu},$$

where P_μ is the muon momentum, E_H the hadronic energy, θ_μ the μ-ν angle, and E_ν the neutrino energy. Three structure functions, F_2, xF_3, and F_L, are required to describe the differential cross section. For the three structure functions, the positive definite nature of the cross section requires the relations

$$F_2 \geqq 0, \, F_L \geqq 0, \, F_2 - F_L \geqq x\,|F_3|. \tag{4.}$$

For future reference, we define the ratios

$$B^{\nu,\bar{\nu}}(x) = -\frac{xF_3^{\nu,\bar{\nu}}(x)}{F_2^{\nu,\bar{\nu}}(x)},$$

$$C(x) = \frac{F_2^{\bar{\nu}}(x)}{F_2^{\nu}(x)} \tag{5.}$$

If charge-symmetry invariance applies to high-energy neutrino collisions, on an isoscalar target we expect

$$B^{\nu}(x) = B^{\bar{\nu}}(x);\, C(x) = 1 \tag{6.}$$

for all x. Such relations are expected to hold if the quark-parton model is valid in the absence of the production of single strange particles or new hadrons with new quantum numbers. The production of strange particles or new hadrons can result in an apparent violation of charge-symmetry invariance well above the physical threshold.

In the quark-parton model with no new-particle production and neglecting antiparton contribution, it is expected that

$$F_i^{\nu} = F_i^{\bar{\nu}}, \, F_L = 0; \, B \doteq B^{\bar{\nu}} = 1,$$

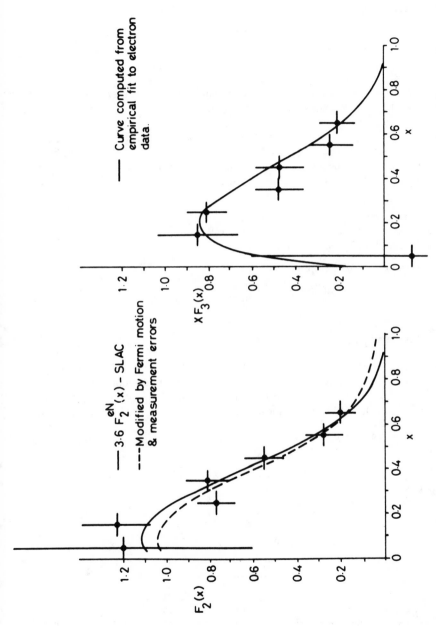

Figure 3 Separated values of $F_2(x)$ and $x[F_3(x)]$ extracted from CERN-Gargamelle data of $q^2 > 1$, $W^2 > 4$ GeV^2.

leading to the distributions

$$\frac{d^2\sigma^\nu}{dx\,dy} = \frac{G^2 mE_\nu}{\pi} F_2(x),$$
7.

$$\frac{d^2\sigma^{\bar\nu}}{dx\,dy} = \frac{G^2 mE_\nu}{\pi} (1-y)^2 F_2(x).$$
8.

Note that $xF_3 = -F_2$ in this model, corresponding to the scattering of neutrinos from left-handed fermions. For scattering from antiparticles, the opposite is found: $xF_3 = +F_2$.

At low energy, x and y distributions have been reported for approximately isoscalar nuclear targets by the GGM group (14). Figure 3 shows the most recent determination of the $xF_3(x)$ distributions by this group. These distributions are obtained from the corrected x and y distributions for $I = -$ (targets), using three assumptions.

1. Scale invariance [i.e. $w(q^2, v) \to F_2(x)$].
2. Charge-symmetry invariance (CSI) [i.e. $F_i^\nu(x) = F_i^{\bar\nu}(x)$].
3. The relation $2xF_1(x) = F_2(x)$.

$F_2(x)$ and $xF_3(x)$ are empirically defined by

$$xF_3(x) = \frac{3\sigma^\nu}{2C}\left[\frac{1}{N^\nu}\frac{dN^\nu}{dx} - R\frac{1}{N^{\bar\nu}}\frac{dN^{\bar\nu}}{dx}\right],$$
9.

$$F_2(x) = \frac{3\sigma}{4}\left[\frac{1}{N^\nu}\frac{dN^\nu}{dx} + R\frac{1}{N^{\bar\nu}}\frac{dN^{\bar\nu}}{dx}\right],$$
10.

where $C = 2mE_\nu G^2/\pi$ and $R = \sigma^{\bar\nu}/\sigma^\nu$. If CSI fails, then $F_2(x)$ and $xF_3(x)$ must be determined separately for neutrino and antineutrino interactions using the full $d^2\sigma/dx\,dy$ distributions. The corresponding $F_2(x)$ and $xF_3(x)$ distributions have been obtained by the HPWF group for neutrino energies below 30 GeV and are shown in Figure 4.

The $F_2(x)$ and $xF_3(x)$ distributions below 30 GeV are in reasonable agreement

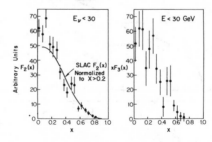

Figure 4   (A) $F_2(x)$ obtained from the neutrino and antineutrino x distributions for $E_\nu < 30$ GeV and assuming $\sigma^\nu/\sigma^\nu = 0.37$. (B) $xF_3(x)$ obtained in the same way $F_2(x)$ was determined in (A).

with the expectations of the parton model and the predictions made using electro-production data.

The HPWF group has obtained scaling distributions for neutrino and anti-neutrino interactions (15). The target is primarily carbon for these data. The distributions have been fully corrected for acceptance and resolution. In Figure 5, the x distributions for neutrinos and antineutrinos for three separate energy intervals are shown. For comparison, the $F_2(x)$ obtained from the SLAC-MIT data is shown normalized to the data in the internal $x > 0.2$. It appears that the electroproduction distribution does not adequately describe the shape of the neutrino x distribution, especially at the higher energies.

Antineutrino x distributions from an exposure of the FNAL 15-ft bubble chamber

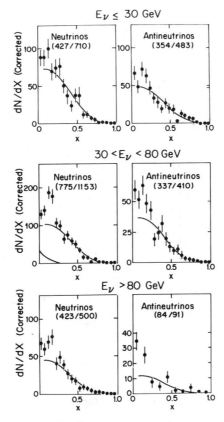

Figure 5   Corrected neutrino x distribution from HPWF group in three energy bins. The mean energy of the events in each bin is 20, 50, and 140 GeV, respectively (c, d, e). Same as (a, b, c) for antineutrino data. The numbers in the right-hand corner denote uncorrected and corrected numbers of events, respectively. The curves are obtained from electroproduction data and normalized for $x > 0.2$.

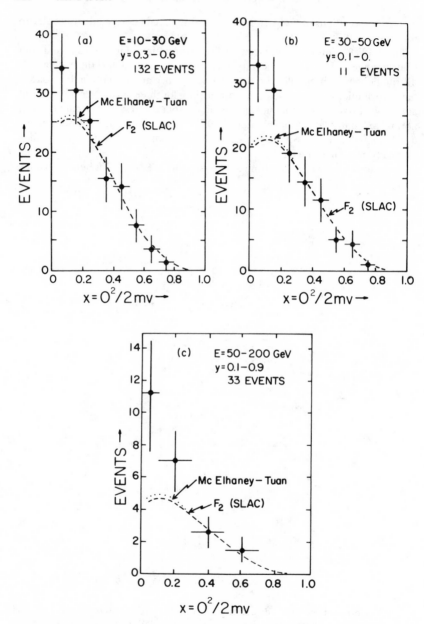

Figure 6 x distribution for v̄+neon experiment of the FIMS group in different energy bins and compared with the predictions obtained from the parton model and the SLAC electroproduction data.

are shown in Figure 6 for various neutrino energies and are compared with the expected distribution based on the SLAC electroproduction data (16, 17). As shown in both Figures 5 and 6, neither the neutrino nor antineutrino x distributions agree well with the theoretically expected distributions. The origin of the difference is not yet known, but may arise in part from new-particle production and in part from scaling violation from asymptotic-freedom corrections.

4.2 Scaling Tests

Three types of tasks of Bjorken scaling have been carried out to date: (a) the energy dependence of moments of the scaling distributions, (b) the shape of the scaling distribution with energy, and (c) the total cross section with energy. We address the first two experimental measurements in this section.

A test of Bjorken scaling can be carried out by showing that the neutrino and antineutrino distributions are independent of neutrino energy. A precise comparison requires very extensive statistics data and is not yet feasible. An intermediate test of scaling can be made by comparing the first moment of the distributions as a function of beam energy. Such a test has been carried out for the two moments $\langle Q^2 \rangle$ and $\langle y \rangle$. The $\langle Q^2 \rangle$ should be proportional to E_v if scaling holds. In the low-energy region, the GGM group finds

$$\langle Q^2 \rangle = 0.23 \pm 0.02 E_v.$$

Two $\langle Q^2 \rangle$ vs E_v measurements have been carried out at higher energy. The analysis of the first exposure of the Fermilab H_2 bubble chamber to a neutrino beam has been carried out by a Fermi-Michigan group. The muon identification was made on the basis of a $P_\perp$ requirement and only events with $E_{visible} > 10$ GeV were accepted in order to eliminate the background from incident neutrons. The data have been fit to the form $\langle Q^2 \rangle = kE_v$. Several $\langle Q^2 \rangle$ vs E measurements have been made at higher energy (17). For example, the analysis of an exposure of the Fermilab neon-H_2 bubble chamber to a neutrino beam has been carried out at Fermi-Michigan-USSR groups (16). The muon identification was made on the basis of an

Table 2 Determination of the values of k in various experiments on H_2 and heavy-liquid targets

	k	Energy	Group	Ref.
$\bar{v}$	0.14 ± 0.02	< 60 GeV	FIMS	17
	0.17 ± 0.011	< 60 GeV	ACMP ($\bar{v}p$)	17
	0.11 ± 0.011	52 GeV	CITF	17
	0.14 ± 0.03	< 10 GeV	GGM	14
v	0.17 ± 0.01	< 60 GeV	BFHM (vp)	17
	0.24 ± 0.01	52 GeV	CITF	17
	0.22 ± 0.01	149 GeV	CITF	17
	0.21 ± 0.01	< 10 GeV	GGM	14

identification in the electromagnetic interaction and only events with $E_{visible} > 10$ GeV were accepted in order to eliminate the background from incident neutral hadrons. The neutrino energy was obtained from an approximate measurement of the angle of the hadrons (Θ_H), the Θ_μ and p_μ for the muon. Data from the CITF, ACMP, and FIMS groups are summarized in Table 2. Within statistical error, the values of k for neutrino data taken at several energies are in agreement. Thus the neutrino data is consistent with Bjorken scaling over a large range of energies.

A comparison of the $\langle y \rangle$ moment with energy is deferred until Section 4.4. The shapes of the neutrino and antineutrino x distributions, as shown in Figures 5 and 6, do not change appreciably over the energy range ~ 30–100 GeV. This also constitutes a positive test for Bjorken scaling.

To first order, the high-energy neutrino data follow Bjorken scaling. There is a distinct deviation from scaling in the antineutrino y distribution, which has been attributed to new-particle production or to expected corrections to the predictions of Bjorken scaling obtained from the postulate of asymotopic freedom. A much smaller deviation in the neutrino data has been observed and is discussed later.

4.3 Locality Test

There are two general ways to search for nonlocality in high-energy neutrino collisions: (a) measurement of σ_{total} (σ_+) as a function of energy, which requires a good knowledge of the neutrino flux; and (b) measurement of $d\sigma/dQ^2$ or $\langle Q^2 \rangle$ vs E_ν, which is flux-independent. The present σ_T measurements are inadequate to set restrictive limits on nonlocality due to neutrino-flux uncertainty, and therefore flux-independent measurements are necessary. A very sensitive and stable parameter

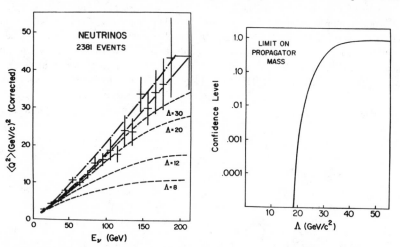

Figure 7 (A) Mean Q^2 as a function of neutrino energy and Λ. The —·—· is obtained from the SLAC $F_2(x)$ and the ——— curve is obtained from the HPWF x distribution. (B) Limit on the propagator mass using the data from (A). The — — — curves in (A) are obtained from varying Λ.

to measure is

$$\langle v \rangle = \langle Q^2/s \rangle = \langle xy \rangle,$$

where $S = 2mEv$. This can be written as a function of scaling variables

$$\langle v \rangle^v = \frac{\int v(d\sigma^v/dv)[1 + (vs/\Lambda^2)]^{-2} \, dv}{\int (d\sigma^v/dv)[1 + vs/\Lambda^2)]^{-2} \, dv}$$

as a function of the propagator mass Λ. For $\Lambda \to \infty$, $\langle v \rangle$ is constant. In order to evaluate the expectation for $\langle v \rangle$ in the equation for finite Λ, it is necessary to obtain the scaling functions. Figure 5 showed the x distributions from HPWF compared with the SLAC $F_2(x)$ in the scaling region. Because of the discrepancy between the two x distributions, an empirical fit to the neutrino x distribution was used to evaluate $\langle v \rangle$. Figures 7 and 8 show the HPWF and CITF data compared with the expected $\langle Q^2 \rangle$ or Q^2 distribution for various values of Λ (18, 19). The data are clearly consistent with $\Lambda \to \infty$. Figures 7 and 8 also give the confidence limits for Λ from the two groups. To 95% certainty, $\Lambda > 24$ GeV is obtained. In order to refine this limit, it will be necessary to obtain more precise x distributions and to take into account the effects of new hadron production on the distributions. The $\langle v \rangle$ variable is probably least sensitive to these unknown factors, whereas a direct comparison of data with $d\sigma/dQ^2$ predictions is likely misleading until the structure functions are better known.

4.4   The y Anomaly

The y distributions of high-energy neutrino and antineutrino collisions provide a sensitive test of the weak-current, hadron constituent models and for new-particle production. Since y distributions are directly connected to helicity states of the current, they are less sensitive to small scaling violations, as perhaps have been observed in inelastic μ-scattering experiments at Fermilab.

Experimentally, the neutrino y distribution is found to follow roughly the predictions of the quark-parton model, whereas a notable departure from the expectations of this model has been recorded for the antineutrino y distribution in the HPWF data. In order to establish a deviation from the prediction of the model, it is important to determine whether antineutrino scattering follows the simple quark-parton model predictions in any antineutrino-energy interval. Present data do not allow a precise test of the $(1-y)^2$ behavior at low energy, but do indicate that the antineutrino scattering follows this form closely. There are three pieces of evidence.

1. The GGM data follows the distribution for $\overline{B} \sim 0.8$ that is very nearly $(1-y)^2$.
2. The low-energy data ($\overline{E}_v \sim 25$ GeV) from the 15-ft bubble chamber analyzed by several groups at Fermilab indicates close agreement with $(1-y)^2$, but is based on relatively few statistics (16, 17, 20).
3. The data from HPWF follow very closely the $(1-y)^2$ form below 30 GeV, as shown in Figure 9 (21).

A significant breakdown of scale invariance in the $\bar{v}$ data has been observed by the HPWF group (21). This is clearly shown by the energy dependence of the first

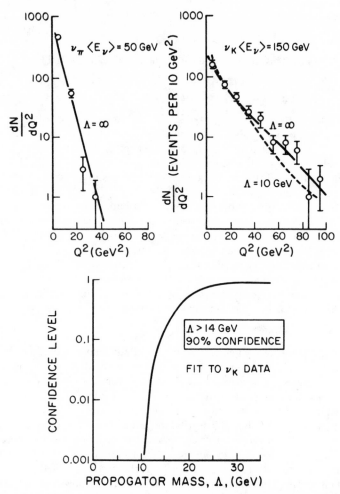

Figure 8 Q^2 distribution for neutrino scattering in the CITF experiment for two different energies and the χ^2 confidence level for a fit to the data with different propagator masses.

moment of the y distribution, $\langle y \rangle$, given in Figures 10A and B for v and $\bar{v}$, respectively. Also included in Figure 10 are the Monte Carlo predictions for $B = 0.9$ and $B = 0.4$, which include correction for the finite acceptance of the apparatus and the experimental resolution. The v data are not sensitive to the particular value of B chosen and can be described throughout the entire energy region by either B value. In contrast, the $\bar{v}$ data show a sharp rise of $\langle y \rangle$ with energy from a value corresponding to $B = 0.9$ below $E \simeq 30$ GeV to a plateau value corresponding to $B = 0.4$ above $E \simeq 60$ GeV.

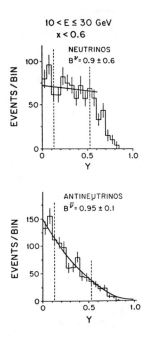

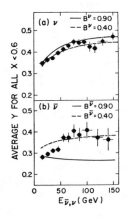

Figure 9   A y distribution for v and $\bar{v}$ for events with $10 < E_{v,\bar{v}} \leq 30$ GeV and $0 \leq x < 0.6$. The dotted lines indicate the region of the y distribution used in the fit for B given in the text.

Figure 10   Plot of the average value of y for v and $\bar{v}$ single-muon events in a function of $E_{v,\bar{v}}$.

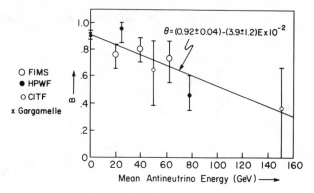

Figure 11   Compilation of fitted B values for existing $\bar{v}$ scattering data.

Independent fits to the y distributions for v and $\bar{v}$, made for two energy regions, yielded the following results (21):

$$B^v = 0.9^{+0.1}_{-0.6}, \ B^{\bar{v}} = 0.95^{+0.05}_{-0.1} \text{ for } 10 \text{ GeV} < E < 30 \text{ GeV},$$

$$B^v = 0.75^{+0.2}_{-0.1}, \ B^{\bar{v}} = 0.45^{+0.15}_{-0.1} \text{ for } E > 50 \text{ GeV},$$

indicating again a breakdown of scaling in the $\bar{v}$ data (B depends on the $\bar{v}$ energy) and an apparent violation of charge-symmetry invariance, ($B^v \neq B^{\bar{v}}$). This effect, the large y anomaly, was first recognized in a considerably smaller sample of events and its experimental validity has been recently corroborated in a detailed study of the systematics of the HPWF apparatus.

Data from the CITF group have also been analyzed and are consistent with a changing y distribution as a function of energy, as shown in Figure 11, where $B^{\bar{v}}$ values are reported from the various groups (16).

5   CHARGED-CURRENT PROCESSES. II. TOTAL CROSS SECTIONS

5.1 $v_\mu + N$ Total Cross-Section Measurement

Since neutrino beams are neutral, it is always necessary to devise an indirect monitor for the neutrino flux and hence for the total cross section. In addition, neutrino detectors rarely have full acceptance for identified leptons and therefore sometimes large corrections are necessary to obtain the total number of interactions. For these and other reasons, an accurate measurement of the neutrino total cross section is extremely difficult. Since the energy dependence of the total cross section is a very important facet of neutrino scattering, some experiments concentrate on the shape of the cross section independent of the absolute value. Other experiments attempt to measure both simultaneously. In most cases, it is essential to know the hadron-

Table 3 Total cross-section measurement techniques

Experiment	Relative E_ν flux	Absolute E_ν flux	Relative $\bar\nu$ to ν flux	Comments
GGM	Hadron spectrum and calculated flux compared with detectors in shield	Muon monitors in the shield—quasi-elastic events used at low energy	Calculation and monitor in the shield	High-energy neutrino flux uncertain due to K-meson yield uncertainly
HPWF	Hadron spectrum measured off-line—flux calculated—crude proton and beam on target monitor	Quasi-elastic events used for monitor	Neutrino and antineutrino data taken simultaneously. Ratio of π^+/π^- and K^+/K^- measured and flux ratio used	The shape of σ_T is obtained from flux calculation—absolute σ_T relative to quasi-elastic and $\sigma_T^\nu/\sigma_T^{\bar\nu}$ from "natural" ratio
CF	Hadron spectrum measured on-line with experiment—neutrino flux calculated	On-line monitor of the proton and meson flux at target	Calculations and use off-on-line monitor for π^+, K^+, and π^-, K^- to proton ratios on target	Careful flux monitor is required to obtain σ_T vs E_ν and $\sigma_T^\nu/\sigma_T^{\bar\nu}$. This technique has best potential for precise σ_T at a fixed E_ν.

production spectra and the effects of the focusing device (horn, dichromatic beam, etc.) on the neutrino spectrum. A technique for internal normalization of the total cross section to the quasi-elastic cross section has been devised and used. This technique assumes that the quasi-elastic cross section is constant with neutrino energy well above the energy where threshold effects are important. Table 3 presents

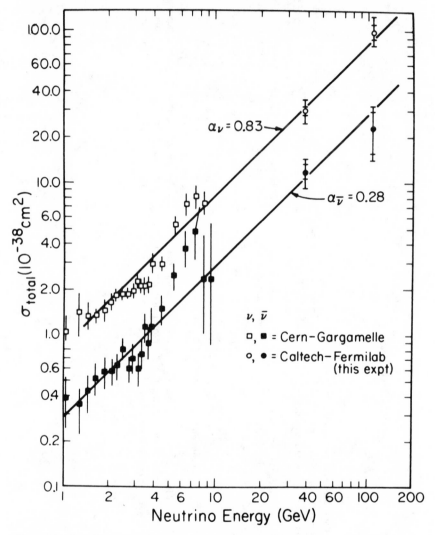

Figure 12 Neutrino and antineutrino total cross sections vs E_v for the GGM and CITF experiments.

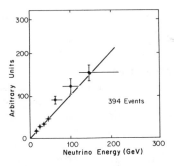

Figure 13   Relative neutrino total cross section obtained by HPWF group using a quadrupole-triplet neutrino beam at Fermilab, reported at the Palermo conference in 1975.

a summary of the methods used to measure total cross sections for the experiments that have published or reported such data.

The first demonstration of a linearly rising neutrino cross section was obtained by using the small CERN bubble chamber about 10 years ago. These measurements were greatly refined using Gargamelle and by now extended up to approximately 15 GeV. The total cross section at much higher energy has been measured at Fermilab by the HPWF and CITF groups using the techniques discussed in Table 3. Figure 12 shows the GGM and CITF total cross-section measurements (22, 23). The relative cross section has also been measured using the Fermilab quadrupole triplet beam tuned to 140 GeV and hadron-flux measurements. These results, which were reported at the Palermo Conference, are shown in Figure 13 (15); within the limited statistics they are consistent with a linear rise in σ_T^{ν}.

By now the approximate linear rise of the neutrino cross section has been well established up to ~ 150 GeV. However, it should be noted that departures from linearity by $\pm 15\%$ at the higher energies would not yet have been detected due to the limited statistics of analyzed events at the higher energy.

There is some evidence for a departure from a linear rise in the neutrino cross section that has been obtained in an indirect manner by using relative normalization of W distributions in the HPWF experiments. The sample of events used for the W distributions is the same as the one in (21) and (24), where the means of acquisition and the selection criteria are described. In Figure 14 the observed W distributions for ν and $\bar{\nu}$ in the energy interval 10 GeV $< E_{\nu,\bar{\nu}} <$ 100 GeV, for $0 \leq x \leq 1$ and $0 \leq y \leq 1$ are shown. These are compared with the continuum distributions obtained from a Monte Carlo calculation, which includes the detection efficiency and resolution of the apparatus. The observed and calculated W distributions in Figure 14 are normalized to have the same area above 2 GeV; below that W value, quasi-elastic scattering and N^* production make substantial contributions to ν and $\bar{\nu}$ scattering at energies less than 30 GeV, as is evident from Figure 14. Once the W distributions are normalized in this way, the relative normalization of the calculated and observed W distributions in Figure 14 is completely determined by the equation

$$\frac{A^{\bar{v}}(E)}{A^{v}(E)} = \frac{\sigma_c^{\bar{v}}(E)}{\sigma_c^{v}(E)} \frac{f^{\bar{v}}(E)}{f^{v}(E)} \frac{\varepsilon^{\bar{v}}(E)}{\varepsilon^{v}(E)},$$

where $A^{v}(E)$ and $A^{\bar{v}}(E)$ are the number of events.

The W distributions in two higher-energy intervals are shown in Figure 14. The calculated continuum distributions are obtained, complete with the normalization previously specified from the neutrino data and theory in the interval 10 GeV < E_v < 30 GeV. Since scale invariance is assumed then $\sigma_c^{v}(E_1)/\sigma_c^{v}(E_2) = E_1/E_2$. There are no free parameters.

There are three significant features of the W distributions shown in Figure 14.

1. Below 30 GeV, there is good agreement between the observed and calculated distributions for $W > 2$ GeV, i.e. above the resonance (here the normalization) region.
2. In the energy interval 30 GeV < E_v < 100 GeV, which is concentrated at large x, as indicated by the crosshatched areas in Figure 14, the events comprising the excess have y values between 0 and 0.2, x values between 0.3 and 0.6, and $Q^2 < 4$ GeV^2.
3. Apart from the excess mentioned in the second feature, the neutrino data do in fact scale linearly with E_v between 10 and 100 GeV, as the agreement within error of the observed and calculated distributions in Figure 1 indicates. There is a suggestion of a departure from linear rise in these data.

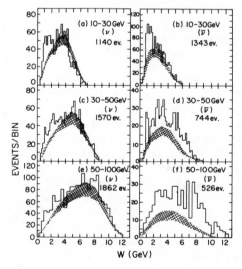

Figure 14 (a, b): W distributions for v and $\bar{v}$ in the energy interval 10–30 Gev; (c, d): 30–50 GeV; (e, f): 50–100 GeV. The shaded regions are calculated assuming charge symmetry and scale invariance and exhibit the total estimated error in the calculation.

The CITF group has obtained normalized total cross-section data. The measurement technique is given in (23). Figure 13 shows the normalized v data of the CITF group, as well as the data of the GGM group. The data are consistent with a linear rise within about 15%. The expected production of charmed particles and the relative departure from a linear cross section in the HPWF data suggest that refined measurements of σ_T versus E_v will be very valuable.

5.2 Measurement of $\sigma^{\bar{v}}/\sigma^{v}$

The neutrino and antineutrino total cross sections are related by the relative contribution of three structure functions, W_1, W_2, and W_3, or in the scaling approximation by the structure functions $F_1(x)$, $F_2(x)$, and $F_3(x)$. In the extreme case that the scaling functions obey

$$xF_3(x) = -F_2(x) \text{ and } 2xF_1(x) = F_2(x),$$

a ratio of

$$\frac{\sigma_t^{\bar{v}}}{\sigma_t^{v}} = \tfrac{1}{3}$$

for $\Delta S = 0$ process is obtained. This choice of structure function coincides with the expectations for the scattering of neutrinos and antineutrinos from elementary spin-$\tfrac{1}{2}$ particles by a V-A weak current. The measurement of the ratio $R = \sigma_t^{\bar{v}}/\sigma_t^{v}$ should be carried out at sufficiently high energy that quasi-elastic and N^* production processes are negligible because these cross sections give

$$\sigma_{Q.E.}^{\bar{v}}/\sigma_{Q.E.}^{v} \simeq \tfrac{1}{3} \quad \text{at low energy and}$$

$$\sigma_{Q.E.}^{\bar{v}}/\sigma_{Q.E.}^{v} \to 1 \quad \text{at high energy,}$$

providing an energy-dependent contamination to R. High-energy measurements of R have been carried out by the HPWF and CITF groups at Fermilab (23, 25). Although these data are statistically limited, they do indeed suggest that R is considerably smaller than 1 below 50 GeV (22, 23, 25). At much lower energy, the ratio of 0.39 ± 0.02 for $E_v > 1$ GeV has been obtained by the GGM group. However, the contribution from the quasi-elastic channel, which makes a statistically significant contribution at the lower energies, has not been subtracted from this ratio. We note that the production of new particles, including heavy leptons and new hadrons, can cause an energy-dependent change in R.

The ratio of antineutrino to neutrino charged-current cross sections on $T = 0$ targets and at higher energies is an important parameter in neutrino physics. A measurement of this ratio at higher energies has now been carried out by the HPWF group (25).

Neutrino and antineutrino events were collected in the HPWF detector. The cross-section ratio was determined by two independent techniques using two samples of data.

1. A sample of 2900 neutrino and 570 antineutrino events that were obtained from a run in the quadrupole triplet-focused neutrino beam in which both neutrino and antineutrino events were detected simultaneously.
2. The full sample of 4994 neutrino and 2408 antineutrino events, which were obtained using both quadrupole triplet- and horn-focused beams.

Sample 1 is used for a measurement of the ratio that depends on the knowledge of neutrino and antineutrino fluxes, whereas sample 2 is used for a flux-independent determination of the ratio. The measured cross-section ratio, using the quadrupole triplet data (technique 1) is shown in Figure 15A.

The ratio of cross sections has also been determined in a flux-independent way, using the full sample of neutrino and antineutrino data. The method is based on

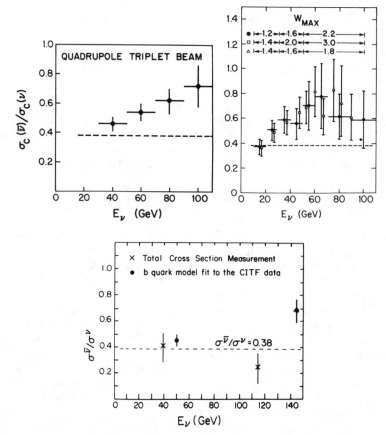

Figure 15 (A) Ratio of cross-section measurements made by the CITF group as explained in the text. (B) Energy dependence of the ratio of the charged-current cross sections $\sigma_c^{\bar{\nu}}/\sigma_c^{\nu}$ using the quadrupole-triplet beam. These data were obtained by the HPWF group.

charge-symmetry invariance and the energy independence of quasi-elastic and N^*-production cross sections at high energy; namely,

$$\sigma(\nu_\mu + T \to \mu^- + N \text{ or } N^*) = \sigma(\bar{\nu}_\mu + T \to \mu^+ + N \text{ or } N^*) \approx \text{const.},$$

where T is an isoscalar target. The small number of quasi-elastic and N^* events above 60 GeV prohibits a measurement of the ratio above this energy by this technique. We note that the lowest-energy measurement (below 30 GeV) is in very good agreement with previous measurements (0.38 ± 0.06).

In order to extend the flux-independent measurement above 60 GeV and to obtain better statistical accuracy, a technique prescribed by Sakurai was used. This technique extends the quasi-elastic and N^* flux normalization to include events with higher W, where W is the recoiling hadronic mass in the inelastic collision. In this case, the technique is fundamentally related to the Lee & Yang theorem, which is based on charge-symmetry invariance. The theorem asserts that

$$\lim_{E \to \infty} \frac{d\sigma}{dW^2} \left[\nu_\mu + \text{isoscalar target} \to \mu^- + \text{hadrons} \right]$$

$$= \lim_{E \to \infty} \frac{d\sigma}{dW^2} \left[\bar{\nu}_\mu + \text{isoscalar target} \to \mu^+ + \text{hadrons} \right].$$

Sakurai has indicated the range of W values for which this relation is expected to hold to better than 15%. Applying this technique to the data gives the ratio of cross sections shown later in Figure 18. Further variations of the W interval in an arbitrary way lead to consistent results with the Sakurai prescription as shown in Figure 15B. The effects of resolution smearing on the cross-section ratio have been investigated and found to be insensitive to these effects.

The ratio of $\sigma_c(\bar{\nu})/\sigma_c(\nu)$, as measured by these independent techniques, is in agreement (Figure 15A, B); a growing ratio with energy is indicated. It is remarkable that the V-A form of the weak interaction, as manifested in this cross-section ratio, is observed up to a center of mass (c.m.) energy of ~ 8–10 GeV $(E_\nu \sim 30$–50 GeV), down to the very low energies of K decay and neutron decay ($\sim \text{MeV}/c$ in the c.m.), and then an abrupt change suggestive of V+A currents is observed. Certainly, some aspect of the new physics must differ fundamentally from the old physics.

The CITF group has carried out measurements of $\sigma^{\bar{\nu}}/\sigma^\nu$ (23). Two different interpretations of these results have been given, as shown in Figure 15C (26). In one case, the b-quark model was used to fit the y distribution and the $\sigma^{\bar{\nu}}/\sigma^\nu$ ratio was obtained. In the other case, a normalized cross-section ratio was obtained. Although there appears to be a difference between the two results, it is important to realize that the high-energy measurement is based on only 11 $\bar{\nu}$ events.

6   NEUTRAL-CURRENT PROCESSES

6.1   Discovery of Weak Neutral Currents

Historically, weak interactions were detected through charged-current semileptonic nuclear decays, i.e. $A \to B + e^- + \bar{\nu}_e$. The observation of weak decays with two charged leptons in the final state would have provided evidence for neutral weak currents (i.e. $A \rightarrow A' + e^+ + e^-$ or $A \rightarrow A' + v_e + \bar{v}_e$). The experimental search for neutral currents with two charged leptons in the final state is effectively impossible in nuclear decays because the probability for a neutral-current transition is many orders of magnitude smaller than that of a corresponding electromagnetic transition. The search for weak nuclear decays with two neutrinos is equally difficult. Thus the question of weak neutral currents was never put to the experimental test in the study of nuclear decays.

The earliest definitive search for weak neutral currents focused on the decays of strange particles (27). The semileptonic weak neutral-current decays of K mesons

$$K^+ \rightarrow \pi^+ + e^+ + e^- \qquad\qquad 11.$$

$$K^+ \rightarrow \pi^+ + v + \bar{v} \qquad\qquad 12.$$

$$K^\circ \rightarrow \mu^+ + \mu^- \qquad\qquad 13.$$

are forbidden for electromagnetic interactions and therefore provide excellent tests for strangeness-changing weak neutral currents. Over the past twelve years, intensive experimental searches for these and similar decays have been carried out, and no example of a first-order weak neutral-current decay has been found.

With the advent of a new generation of heavy-liquid bubble chambers and with higher-energy, higher-intensity neutrino beams and massive target-calorimeter detectors, the search for weak neutral currents was reopened (5). There are two serious backgrounds that can simulate neutrino-induced events without final-state charged leptons. The most important background comes from neutral hadrons that are incident on the experimental detector. Proton accelerators produce copious quantities of neutrons and neutral kaons. The interaction of these particles in neutrino detectors is easy to distinguish from high-energy, charged-current neutrino events because of the presence of a charged lepton in the final state. However, neutral-current events, i.e. events without the charged lepton in the final state, are very difficult to separate from neutron and K° interactions. Furthermore, neutrons and K° mesons are also produced in quantity by the neutrino beam itself in the shielding that is used to protect the neutrino detector from muons. A second source of background for neutral-current events is provided by charged-current events in which the outgoing charged lepton either escapes from the detector before identification as a lepton or is misidentified as a hadron.

Since muon neutrinos are the principal source of high-energy neutrino interactions, events with undetected muons provide the largest source of background from charged-current interactions. Thus, in order to search for neutral currents, good muon identification is required, and interactions due to incident neutral hadrons must either be suppressed or subtracted out.

One of the initial observations of events without muons in the final state was carried at CERN using the large heavy-liquid bubble chamber Gargamelle (5). In this experiment, a semiempirical estimate of the neutrino flux and a background subtraction are used to eliminate neutron and K° interactions. The charged-current background is eliminated by selecting events in which all particles are identified as hadrons through scattering or decay.

The large-mass electronic target-detector experiments cope with the background in another way (5). It is experimentally observed that the neutral hadron interactions are eliminated by choosing a target volume that requires the hadrons to traverse many protective interaction lengths of material in which the neutrons and K° mesons are attenuated (an active shield). Identification and rejection of charged-current events is carried out by placing a hadron filter near the detector through which the muons pass and are detected. The primary uncertainty is related to the semiempirical estimate of the efficiency for muon detection. In the HPWF experiments at Fermilab, the detection efficiency was obtained directly from the muon angular distribution, which was measured out to 500 mrad, and a muonless-event signal was observed after correcting for detection efficiency. It was demonstrated that the existence of this signal does not depend on the muon angular distribution outside of the measured angular region, as can be shown by a simple argument (28, 29). The kinematics of high-energy neutrino collisons require that, in events with muons produced at angles greater than 500 mrad, the hadronic energy carried away in the collision be nearly equal to the incident-neutrino energy. This is a direct consequence of the conservation of transverse momentum in the collision. Thus, if the observed muonless signal were actually due to misidentified wide-angle muon charged-current events, then the spectrum of visible energy for the muonless events should be approximately the same as the incident-neutrino spectrum. The comparison of the spectra for the muonless events and for the charged-current events indicated a substantial difference. The muonless events were also shown to have the uniform spatial distribution in the detector, expected from neutrino interactions. From this evidence it was concluded that neutrino-induced muonless events had been observed.

The combined positive results of the bubble-chamber and counter experiments, working in different energy ranges and with vastly different experimental bias, provided dramatic evidence for the existence of muonless neutrino interactions. All present evidence points to these events as being due to neutral weak-current interactions.

6.2   Status of Inelastic Scattering Experiments

Three experiments have reported improved measurements of the deep inelastic scattering distributions that are considerably more reliable than the early discovery measurements discussed in Section 6.1. The HPWF, CITF, and GGM groups have all reported a comparison of the total cross section for neutrino and anti-neutrino neutral-current scattering and all groups agree that $\sigma_{NC}^{\bar{\nu}}/\sigma$ is less than 1 (30–32). We describe the HPWF measurements in some detail, since several precautions to avoid systematic errors were followed and a technique to correct for the inherent bias against low–hadronic energy events was used (30), and it is the only result published at the writing of this report.

3600 neutrino and antineutrino events were recorded in this experiment. The data were acquired during different running conditions, making possible important tests for systematic errors. Values of R^{ν} and $R^{\bar{\nu}}$, the ratios of neutral to charged-current cross sections for ν and $\bar{\nu}$, which are used to obtain values of the neutral-current cross sections were obtained. The observed inequality of the ν and $\bar{\nu}$ neutral-current cross sections leads directly to the conclusion that a parity-nonconserving component is present in that current, in the standard V + A models.

The apparatus was shown in Figure $2C_1$. The spectra of the incident v and $\bar{v}$ beams are shown in Figure 2B. Neutrino interactions that produced a hadronic cascade with energy $E_H > 4$ GeV triggered the apparatus with an efficiency greater than 99%. Counter A and the first seven calorimeters were in anticoincidence. Pulse-height information from the last eight calorimeter modules yielded the value of E_H. In addition, the wide-gap spark chambers (WGSC) were photographed in two $\pm 7.5°$ stereo views and in a 90° stereo view. Muons were identified by their presence in detectors 1 or 2 after passing through the iron hadron absorbers.

Candidates for both charged- and neutral-current interactions were selected from a scan of the film. The scanning efficiency was greater than 95% and was the same for NC and CC events because the scanning criteria involved only properties of event vertices. The vertex was measured in three views and its location was checked for consistency with the electronic information. The small fiducial region was chosen to insure high muon-detection efficiency and good shower containment, and to reject neutron backgrounds. To obtain $\sigma_N^{\bar{v}}/\sigma_N^{v}$, however, the effect of the experimental requirement $E_H > 4$ GeV on the measured values of R^v and $R^{\bar{v}}$ must be considered. More generally, for any linear combination of V and A, $R(E_H > 0) \geqq R^v(E_H > 4)$, while $R^{\bar{v}}(E_H > 0) \leqq R^{\bar{v}}(E_H > 4)$, so that $[\sigma_N^{\bar{v}}/\sigma_N^{v}]_{V-A}$. The numerical upper limit on $\sigma_N^{\bar{v}}/\sigma_N^{v}$ at $(E_{v,\bar{v}}) = 41$ was found to be

$$\sigma_N^{\bar{v}}/\sigma_N^{v} \leqq (R^{\bar{v}}/R^v)(\sigma_c^{\bar{v}}/\sigma_c^{v}) = [(0.39 \pm 0.10)/(0.29 \pm 0.04)](0.45 \pm 0.08)$$
$$= 0.61 \pm 0.25,$$

where R^v and $R^{\bar{v}}$ are obtained from Table 3A, and $\sigma_c^{\bar{v}}/\sigma_c^{v}$ at 41 GeV is taken from (24). Note that the values of R^v in Table 3A are approximately constant over the average energy interval from 53 to 85 GeV. This largely justifies the extrapolation of constant R^v to 41 GeV, and implies, within experimental error, a linear rise with energy of the total neutral-current cross section for neutrinos.

The ratio $\sigma_N^{\bar{v}}/\sigma_N^{v}$, rather than its upper limit, can be obtained by correcting for unobserved events with $E_H < 4$ GeV according to various forms of the neutral-current interaction. The high energy of the neutrino beams, particularly the 1975

Table 3A   The values of R^v or $R^{\bar{v}}$ for $E_H > 4$ GeV

Beam type	R^v	$R^{\bar{v}}$	Comment
single-horn	0.31 ± 0.06	—	Pure v beam; $(E_v) = 53$ GeV
1974 quadrupole triplet	0.24 ± 0.06	—	Mixed beam; $(E_v) = 78$ GeV
1975 quadrupole triplet	0.29 ± 0.04	—	Mixed beam; $(E_v) = 85$ GeV
double-horn with plug	—	0.39 ± 0.10	Pure $\bar{v}$ beam; $(E_{\bar{v}}) = 41$ GeV

Table 3B   Determination of the corrected value of $\sigma_N^{\bar{v}}/\sigma_N^{v}$ for various V, A combinations[a]

Form of the weak neutral current	Corrected experimental value	Expected value
V − A	0.61 ± 0.25	0.38
V or A	0.40 ± 0.17	1.00
V + A	0.37 ± 0.16	2.67

[a] The measured values of $\sigma_N^{\bar{v}}/\sigma_N^{v}$, after correction for the loss of events with $E_H < 4$ GeV according to the form of the weak neutral current in the first column. The corresponding values of $\sigma_N^{\bar{v}}/\sigma_N^{v}$, expected from theory, are given in the third column. An antiquark contribution of 5% has been assumed.

quadrupole triplet beam, gives rise to few events with $E_H < 4$ GeV, so that for any combination of V and A, the difference between $R^v(E_H > 0)$ and the measured values of $R(E_H > 4)$ is negligible. The ratio $R^{\bar{v}}$ is more sensitive to the exact form of the interaction because of the lower $(E_{\bar{v}})$. Thus for V − A, $R^{\bar{v}}(E_H > 0) = R^{\bar{v}}(E_H > 4)$, while for pure V or A, $R^{\bar{v}}(E_H > 0) - 0.71 R^{\bar{v}}(E_H > 4)$; and for V + A, $R^{\bar{v}}(E_H > 0) = 0.63 R^{\bar{v}}(E_H > 4)$. A comparison in Table 3B of the corrected experimental values of $\sigma_N^{\bar{v}}/\sigma_N^{v}$ with the expected values of this ratio for several admixtures of V and A is given. It is clear that V + A is ruled out. Furthermore, the experimental value for pure V or A is 3 standard deviations away from the value expected for either of those pure forms, while V − A is within 1 standard deviation of the expected value.

The best fit for the form of the weak neutral current is obtained by using the general forms of the $y \equiv E_H/E_v$ distribution expected for any V, A combination:

$$d\sigma_N^v/dy = a + b(1 - y)^2, \quad d\sigma_N^{\bar{v}}/dy = b + a(1 - y)^2.$$

For a V − A interaction, $a \approx 1$ and $b \approx 0$, while a V or A interaction has $a = b = \frac{1}{2}$. The best fit is obtained by varying a and b until the expected value of $\sigma_N^{\bar{v}}/\sigma_N^{v}$ agrees with the corrected experimental value. For the best fit, $\sigma_N^{\bar{v}}/\sigma_N^{v} = 0.48 \pm 0.20$, $a = 0.85$, and $b = 0.15$.

A summary of R^v, $R^{\bar{v}}$, and $\sigma^{NC}(\bar{v})/\sigma^{NC}(v)$ from all three experiments is given in Table 3C; only the HPWF and GGM cross-section ratios have been fully corrected.

Table 3C   Determination of R_{vN} and $R_{\bar{v}N}$ in three experiments

	GGM	HPWF	CITF
R_{vN}	0.25 ± 0.04	0.31 ± 0.06	0.24 ± 0.02
$R_{\bar{v}N}$	0.39 ± 0.06	0.39 ± 0.10	0.39 ± 0.06
$\dfrac{\sigma^{NC}(\bar{v}N)}{\sigma^{NC}(vn)}$	0.59 ± 0.14	0.61 ± 0.25	—
·Average $v, \bar{v}$ energy (GeV)	$\langle E \rangle \simeq 2$	$\langle E_v \rangle \simeq 53, \langle E_{\bar{v}} \rangle \simeq 41$	$\langle E \rangle \simeq 50$

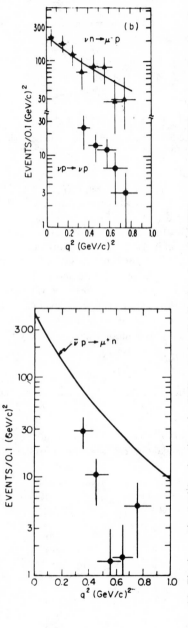

Figure 16   Elastic neutrino (*right*) and antineutrino (*left*) scattering distributions from the HPW experiment at BNL. A comparison is made with quasi-elastic scattering.

The CITF group has fit their data to helicity amplitudes: the results can be found in (32). All experiments are in agreement that $\sigma^{NC}(v) > \sigma^{NC}(\bar{v})$, which suggests that both V and A currents are involved in the weak neutral currents.

6.3   Elastic Scattering from Protons and Electrons

Two experiments (HPW and CIR groups) have reported evidence for the process (12, 33)

$$v_\mu + p \rightarrow v_\mu + p,$$

and the process

$$\bar{v}_\mu + p \rightarrow \bar{v}_\mu + p$$

has been observed by the HPW group. The CIR group used a Al spark-chamber detector, whereas the HPW group used a large, segmented liquid-scintillation calorimeter detector (see Figure 2B) Figure 16 shows the observed q^2 distribution for the HPW data on elastic neutrino and antineutrino scattering (12, 34). The ratio of elastic to quasi-elastic scattering cross sections for the q^2 interval of 0.3–0.9 GeV/c^2 are (33, 34)

neutrino : 0.17 ± 0.05 HPW 0.25 ± 0.07 (CIR)

antineutrino : 0.2 ± 0.1 HPW

and the ratio of antineutrino to neutrino elastic scattering in the same q^2 interval is 0.4 ± 0.12. If the neutral current were parity-conserving, this value would be expected to be 1.

By far the cleanest neutral-current reaction from the theoretical point of view is the scattering of v_μ, $\bar{v}_\mu$, v_e, v_e, and $\bar{v}_e$ from electrons. The GGM and Aachen-Padova groups have reported on $v_\mu e^-$ and $\bar{v}_\mu e^-$ scattering, and Reines and collaborators have reported $\bar{v}_e e^-$ (35–37). These data have been reviewed by Williams and we follow his report here (38). The cross sections for the various processes are recorded in Table 4. As can be noted from this table, there is good agreement between the GGM and A-P results if the latter make a cut on the electron energy of 0.4 GeV. In this case the $\bar{v}_\mu e^-$ and $\bar{v}_\mu e^-$ cross sections are compatible with equality.

Table 4   Elastic scattering from electrons

Process	Cross section (cm²/electron)	Comments	Group
$\bar{v}_\mu + e^- \rightarrow \bar{v}_\mu + e^-$	$1 {}^{+2.1}_{-0.9} \times 10^{-42} E_v$	bubble-chamber data	GGM
$\bar{v}_\mu + e^- \rightarrow \bar{v}_\mu + e^-$	$5.4 \pm 1.7 \times 10^{-42} E_v$	no E_e cut	A.P.
$\bar{v}_\mu + e^- \rightarrow \bar{v}_\mu + e^-$	$2.4 \pm 1.3 \times 10^{-42} E_v$	$E_e > 0.4$ GeV	A.P.
$v_\mu + e^- \rightarrow v_\mu + e^-$	$\leq 2.6 \times 10^{-42} E_v$	90% confidence	GGM
$v_\mu + e^- \rightarrow v_\mu + e^-$	$2.4 \pm 1.3 \times 10^{-42} E_v$	no E_e cut	A.P.
$v_\mu + e^- \rightarrow v_\mu + e^-$	$2.1 \pm 1.2 \times 10^{-42} E_v$	$E_e > 0.4$ GeV	A.P.

6.4   Limits on Flavor-Changing Neutral-Current Processes

The search for strangeness-changing neutral currents has been going on for more than 12 years without success. The discovery of one or more new hadronic quantum numbers brings to the fore the question of whether there are neutral-current processes in which a new quantum number is changed ($\Delta F \neq 0$, where F is any quantum number "beyond" strangeness) (see Section 7). There are a number of neutrino processes in which the $\Delta F \neq 0$ neutral-current process can be detected. These processes are shown schematically in Figure 17. Note that these signatures are not limited to the charm quantum number, but would serve equally well any new quantum number excited by neutrino or antineutrino beams. There are a number of possible background processes that can simulate, at one level, the existence of flavor-changing neutral-current processes. A schematic representation of some backgrounds is shown in Figure 18.

Perhaps the cleanest channel in which to search for charm-changing neutral currents is

$$\nu_\mu + N \to \nu_\mu + (\text{new hadron}) + X \to \begin{Bmatrix} \mu^+ + \nu_\mu + \dots \\ e^+ + \nu_e + \dots \end{Bmatrix}$$

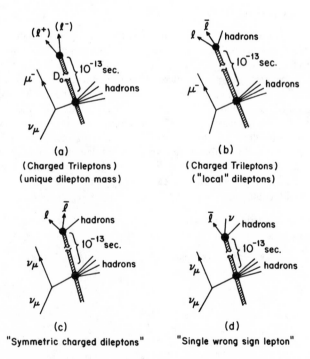

Figure 17   Signature for charm-changing neutral currents or for the neutral-current decay of particles with new flavors.

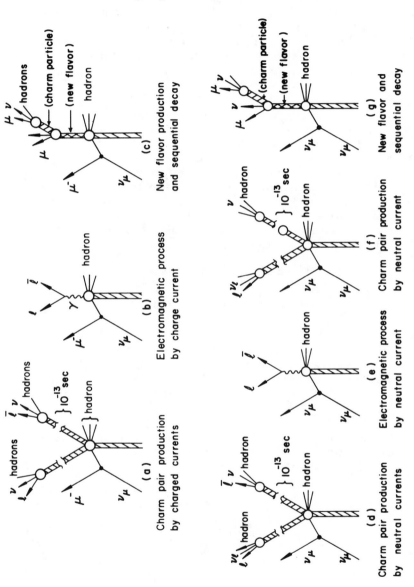

Figure 18   Mechanisms for backgrounds in the search for charm-changing neutral currents.

These events are characterized by the appearance of "wrong-sign" single-lepton production. A previous search for single μ^+ production by neutrino beams has been carried out by the CITF group (39). A negative search for single-e^+ production has been carried out by the GGM group. More recently, in the Fermilab neon bubble-chamber experiments, a preliminary search for e^+ events has been carried out. These events could be due to misidentified e^+ events but this seems unlikely. The most serious background comes from $\bar{\nu}_e$ events.

$$\bar{\nu}_e + N \rightarrow e^+ + X.$$

However, this rate is expected to be suppressed in a neutrino beam because the (π^-, K^-) parents are defocused. A test of the $\bar{\nu}_e$ origin of the events can be made by plotting the y distribution of the events, on the hypothesis that they originate from a $\bar{\nu}_e$ interaction. The distribution is shown in Figure 19.

The average energy of the events (on the $\bar{\nu}_e$ origin hypothesis) is $\langle E_{\bar{\nu}_e} \rangle$ 40 GeV. At this energy, the measured distribution for $\bar{\nu}_\mu$ events follows closely the distribution $(1-y)^2$. As shown in Figure 19, a comparison with the $(1-y)^2$ distribution indicates that the most likely explanation for the bulk of the events is the $\bar{\nu}_e$ origin.

We use these events as an upper limit to obtain a limit on the production of single charmed particles by neutral currents (40)

$$\frac{R(\nu_\mu \rightarrow e^+)}{R(\nu_\mu \rightarrow \nu_\mu)} \simeq \frac{\Gamma(D/F \rightarrow e\nu_e \ldots)}{\Gamma(D/F \rightarrow \text{all})} \cdot \frac{\sigma_{\text{NC}}(D/F)}{\sigma_{\text{NC}}(\text{all})} \lesssim 6 \times 10^{-3} \text{ (90\% confidence level)}$$

where D/F denotes charmed mesons. From this we estimate

$$\frac{\sigma_{\text{NC}}(D/F)}{\sigma_{\text{NC}}(\text{all})} \lesssim 6 \times 10^{-2}.$$

If, on the other hand, we assume the three high-y events in Figure 19 are real, the neutral-current rate

$$\frac{\sigma_{\text{NC}}(D/F)}{\sigma_{\text{NC}}(\text{all})} \sim 10^{-1},$$

which is consistent with the limit on the rate for charmed-particle production by charged currents.

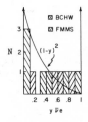

Figure 19 $y^{\bar{\nu}e}$ distribution for muonless e^+ candidates (BCHW, FIMS data) reported at a recent APS meeting at BNL.

Another potentially clean signature for charm-changing neutral currents comes from the neutral-current decay of a charmed particle produced either by charged currents

$$\nu_\mu + N \rightarrow \mu^- + (D/F) + X \rightarrow \mu^+ + \mu^- + \dots \quad \text{(trimuons)}$$

or by neutral currents

$$\nu_\mu + N \rightarrow \nu_\mu + (D/F) + X \rightarrow \mu^+ + \mu^- + \dots \quad \text{(symmetric dimuons)}$$

Trimuon events have been observed by the CITF and HPWF groups with the rate

$$\frac{R(\mu^- \mu^+ \mu^-)}{R(\mu^- \mu^+)} \sim (10^{-1} - 4 \times 10^{-2}).$$

Using this value, we can estimate a limit for the specific case of charm:

$$\frac{\Gamma(D/F \rightarrow \mu^+ \mu^- \dots)}{\Gamma(D/F \rightarrow \mu^+ \nu \dots)} \lesssim 2 \times 10^{-1}.$$

The low rate for single-e^+ production and for trimuon production by neutrinos or antineutrinos compared to the rate for dimuon production, which signals either charm or other heavy hadrons, indicates that flavor-changing neutral currents are suppressed relative to flavor-changing charged currents. It is still possible that the neutral currents are within 10% of the charged-current rates and sensitive searches for the processes depicted in Figure 17 are of great importance.

6.5 Isotopic-Spin Decomposition of the Weak-Neutral-Current Amplitude

There is considerable interest in the isotopic-spin decomposition of the weak-neutral-current amplitude. Three classes of experiments bear on this determination.

1. Single-pion production at low energy (the ratio of $\pi^+/\pi^\circ/\pi^-$ production from various targets) or $\Delta(3,3)$ production.
2. The $h^\pm$ spectrum from inclusive processes

$$\nu_\mu + (I = 0) \rightarrow \nu_\mu + h^\pm + X.$$

3. Comparison of reactions on neutrons and protons.

The data available so far on the πN mass distribution in the Δ region are still inconclusive as to the presence or absence of Δ production (33). One piece of information relevant to the isospin properties of the neutral currents has been reported (44). Recently, the GGM Collaboration has examined the pion-charge ratio in (44)

$$\nu + N \rightarrow \nu + \pi^\pm + N'.$$

If the current is pure isoscalar and if the target is also isoscalar, the ratios are expected to be

$$\sigma(\pi^+) : \sigma(\pi^\circ) : \sigma(\pi^-) = 1 : 1 : 1.$$

Table 5   Isovector and isoscalar amplitudes in neutral currents

		GGM experimental data on π°/π^- production			
Type of current	Number of events with π°	Number of events with π^-	π° Efficiency	π^- Efficiency	$\pi^\circ:\pi^-$ Corrected
Neutrino	142	60	$(86\pm7)\%$	$(49\pm6)\%$	1.4 ± 0.2
Antineutrino	152	42			2.1 ± 0.4

Predictions of models for π°/π^- production

$\pi^\circ/\pi^- = 0.9$	isoscalar
$\pi^\circ/\pi^- = 4$	isovector—no nuclear corrections
$\pi^\circ/\pi^- = 1.8$	dominant isovector—nuclear corrections included

The data of the GGM collaboration on the π°-to-π^- ratio is reported in Table 5. These numbers should be compared to the isoscalar-current prediction of 0.9 for freon. Table 5 also gives the expected ratio of π°/π^- rate for the Weinberg-Salam model, including various nuclear corrections (45).

In addition to single-pion production, pion semi-inclusive reactions

$$(\bar{\nu}) + N \rightarrow (\bar{\nu}) + \pi^{\pm,0} + X$$

can provide information on the isoscalar-isovector interference. To extract the relevant coupling constants, the details of parton-fragmentation models are used (46). We define $Z^\pm$ as

$$Z^\pm = E\pi^\pm/EH$$

If ν or $\bar{\nu}$ collide with an isoscalar target, the difference between the Z^+ and Z^- distributions will indicate the presence of interference between isoscalar and iso-

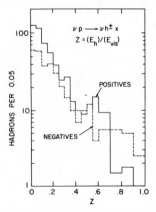

Figure 20   The $Z^\pm$ distributions for a neutral-current reaction on hydrogen.

vector amplitudes. The $Z^{\pm}$ distributions for neutrino scattering from hydrogen are shown in Figure 20. There is a large difference in these distributions, which results in more high-energy π^- than π^+. This may be due to background, the $T = \frac{1}{2}$ nature of the target, or isoscalar-isovector interference. Data on the $Z^{\pm}$, distribution for neon experiments will be very important in determining the exact amount of isoscalar and isovector interference.

There is some evidence for an appreciable isoscalar component from measurements on the reaction (38, 47)

$$v_\mu p \to v_\mu p \pi^+ \pi^-,$$

$$v_\mu n \to v_\mu n \pi^+ \pi^-.$$

Any real difference between scattering on neutrons and protons can only result from interference between isoscalar and isovector amplitudes. The preliminary experimental data from the Brookhaven National Laboratory (BNL) 7-ft bubble chamber gives

$$\frac{\sigma(vn \to vn\pi^+\pi^-)}{\sigma(vp \to vp\pi^+\pi^-)} = 0.49 \pm 0.2.$$

The data provides a hint that there is an appreciable isoscalar component as well as an isovector one.

6.6   Comparison of the Data with the Weinberg-Salam Model

In volume 24 of the *Annual Review of Nuclear Science*, an excellent review of the Weinberg-Salam gauge theory was presented (48)[2]. Our purpose here is to describe the present experimental comparison with the model. The model has two free parameters, the Weinberg angle ($x_w = \sin^2 \theta_w$) and an overall weak-coupling constant g. From observed rates for weak processes, g can be expressed in terms of G^2 and the mass of the intermediate vector bosons. Conversely, the mass of the intermediate vector boson can be predicted if $\sin^2 \theta_w$ and G^2 are known. The ultimate test of the model is to discover the intermediate vector boson and compare its mass with the value predicted from the measured $\sin^2 \theta_w$ values. Until this possibility occurs (with high-energy storage ring etc), consistency tests of the model are valuable. This involves "measuring" the value of $\sin^2 \theta_w$ from many different neutral-current reactions and comparing the values. Unfortunately, nearly all reactions are rather insensitive and thus the experimental precision must be greatly improved to make meaningful tests. Such a comparison has been made by Barnett for both the Weinberg-Salam model and another gauge-theory model that incorporated new right-handed currents and quarks (CHYM) (49, 50). A comparison of the models and the data is given in Figures 21A and B. At present, the atomic physics parity-violation experiments appear to be in disagreement with the W-S model, whereas all neutrino processes are consistent with either the W-S model or many other possible models such as the CHYM model (51).

It is perhaps significant that the neutral current appears to contain both isovector

[2] Beg, M. A. B., Sirlin, A. 1974, *Ann. Rev. Nucl. Sci.* 24:379–449.

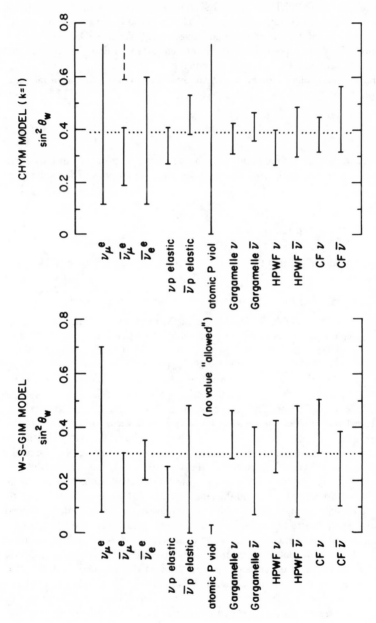

Figure 21 Comparison of existing neutral-current experimental data with two gauge-theory models.

as well as isoscalar components. The present atomic physics measurements study the interference between the photon and the isoscalar axial-vector weak neutral current. If the atomic physics results continue to be negative, the cause is perhaps due to a lack of axial-vector component in the electron vector of the weak neutral current and not due to the absence of isoscalar components. This would suggest that the processes

$$\nu_\mu e^- \to \nu_\mu e^-,$$

$$\bar{\nu}_\mu e^- \to \nu_\mu e^-$$

will have equal cross sections, and leaves open the possibility of a large isoscalar vector component of the weak neutral current in the neutrino sector. The latter is needed if supernovae are blown apart by neutrino pressure.

7 EXCITATION OF STATES WITH NEW QUANTUM NUMBERS

7.1 Multilepton Production by Neutrinos

The high neutrino energy and the new generation of neutrino detectors are both well suited to the observation of neutrino-induced events with more than one charged lepton in the final state (41). Counter experiments have so far obtained evidence for dimuon production

$$\nu_\mu + N \to \mu^- + \mu^+ + \nu_\mu + X, \qquad\qquad 14.$$

$$\bar{\nu}_\mu + N \to \mu^+ + \mu^- + \bar{\nu}_\mu + X, \qquad\qquad 15.$$

whereas bubble chambers have been used to observe events with electrons or positrons in the final state, i.e. (52, 53)

$$\nu_\mu + N \to \mu^- + e^+ + \nu_e + X, \qquad\qquad 16.$$

$$\nu_\mu + N \to \mu^+ + e^- + \bar{\nu}_e + X. \qquad\qquad 17.$$

The identification of muons requires high-energy muons to separate the particles from hadrons. This bias is well matched by the counter experiments with their much larger target mass. For example, the first two $\mu\mu$ events observed were at high neutrino energy and had relatively high-energy muons (4).

Since the early observation of dimuons, many groups have reported evidence for multilepton production. Table 6 gives a summary of the number of events reported so far (autumn 1976) (40). New experiments in progress at Fermilab or soon to start at the CERN-SPS will increase the number of events by a factor of 10–100.

The most important aspect of multilepton production is the relative rate compared to single-muon production. Figure 22 shows a compilation of the neutrino rate versus visible energy (40).

This figure shows the observed rate corrected (in some cases) for angular acceptance, but uncorrected for the experimental energy cutoff of the charged leptons. This correction cannot be made until the real spectrum is known. Note

Table 6   World observation of multilepton events (1976)

Final state	Number of events	Group	Ref.
$\nu_\mu N \rightarrow \mu^{\mp}\mu X$	84	HPWF	6
$\nu_\mu N \rightarrow \mu^- e X$	$3+14\pm8$?	GGM	7
$\nu_\mu N \rightarrow \mu^- e X$	11	BCHW	7
$\nu_\mu N \rightarrow \mu^{\mp}\mu X$	115[a]	CITF	29
$\nu_\mu N \rightarrow \mu^- e^+ X$	61	BNAL-Columbia	40
$\nu_\mu N \rightarrow \mu^{\mp}\mu X$	10??	ITEP-Serpukhov	40
$\bar{\nu}_\mu N \rightarrow \mu^+ e^- X$	1?	FIMS	16
$\nu_\mu N \rightarrow \mu e X$	4	CIR-BNL	33
$\nu_\mu N \rightarrow \mu^- \mu X$	1?	ANL-Purdue	40
Total	300 ± 20[b]		

[a] 21 with both muons identified.
[b] $\sim 10^{-39}$ cm^2/event.

that multilepton events have been observed down to very low energy. The real rate at these low energies is very uncertain because of the severe experimental cuts. We have attempted to guess the limits from the available data. It may be that the actual rate is compariable to that observed at much higher energies (~ 40 GeV) without an appreciable threshold effect.

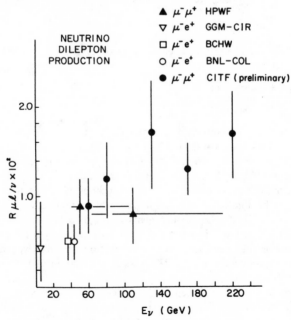

Figure 22   Multilepton rate for neutrinos (uncorrected for the $[\mu^+, e^+]$ energy bias).

The observation of copious dimuon production relative to trimuon production provides strong evidence that the dimuons have in addition a missing neutral lepton. A missing neutrino is a clear choice. Direct evidence in this regard is provided by a two-prong bubble-chamber event with a $\mu^- e^+$ and no additional charged lepton of energy greater than 5 MeV (Figure 23). A missing neutrino in nearly every event is strongly implied by the existing multilepton data. There is also strong evidence

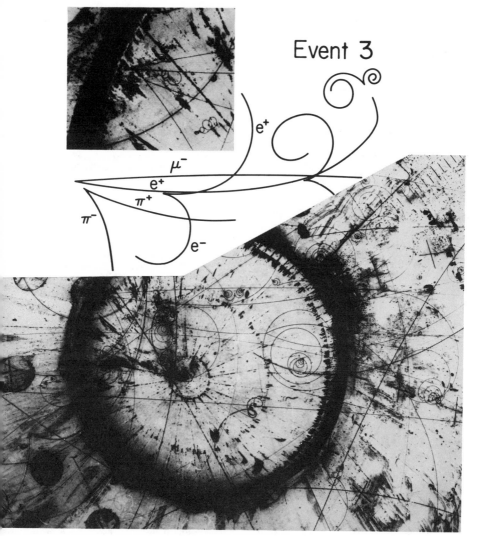

Figure 23   Unusual $\mu^- e^+$ event from light-neon bubble-chamber experiment at Fermilab.

for more strange-particle production in the μe events. The exact amount is still controversial.

Besides the rate and strange-particle association, another striking feature of multilepton production is the leading muon effect. In most high-energy events recorded so far, the μ^- carries a much higher energy than the μ^+ or e^+. Figure 24 shows data from the counter and bubble chamber that indicate the leading muon trend of the dimuon data. Another feature of the data is the predominance of μ^- e^+ events over $\mu^- e^-$ events (54).

The characteristics of the multilepton events lead to the unambiguous conclusion that some new-particle production is involved (55–58). In essence, this conclusion stems from two facts:

1. The rate is much too high to come from a 4-fermion process.
2. The μ asymmetry and neutrino in the final state indicate that a second weak interaction has occurred (the rate for higher-order weak processes is expected to be $\sim G^2$ down). The only known way to enhance the weak process is to produce a long-lived state that subsequently decays semileptonically.

The relatively low rate of $\mu^- \mu^-$ production compared to $\mu^- \mu^+$ production indicates the operation of a selection rule in the decay of the new particles, i.e.

$$\frac{\Gamma(Y \to \mu^- + \bar{v}_\mu + \ldots)}{\Gamma(Y \to \mu^+ + v_\mu + \ldots)} \approx \frac{R^v(\mu^- \mu^-)}{R^v(\mu^- \mu^+)} \lesssim 0.15,$$

$$\frac{\Gamma(\bar{Y} \to \mu^+ + v_\mu + \ldots)}{\Gamma(Y \to \mu^- + \bar{v}_\mu + \ldots)} = \frac{R^{\bar{v}}(\mu^+ \mu^+)}{R^v(\mu^+ \mu^-)} \lesssim 0.2.$$

At the same time, the reduced same-sign rate indicates that predominantly single

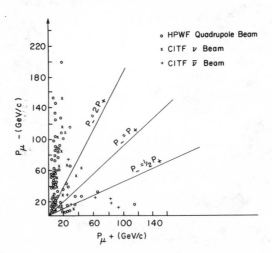

Figure 24   A scatter plot of P_{μ^+} vs P_{μ^-} for the existing dimuon events obtained in the HPWF and CITF experiments.

Table 7   Comparison of $\bar{v}_\mu$ and v_{μ^-} induced $\mu\mu$ events

	$\mu^-\mu^+v$ HPWF	$\mu^+\mu^-\bar{v}$ HPWF	$\mu^-\mu^+v$ CITF	$\mu^+\mu^-\bar{v}$ CITF
$\langle P_\mu +/-\rangle$	68.2[a]	39.0[a]	67.2	53.0
$\langle P_\mu -/+\rangle$	11.2[a]	13.0[a]	18.5	25.5
$\langle E_{\rm vis}\rangle$	115.6[a]	80.0[a]	~ 150.0	~ 130.0
$\langle y\rangle_{-/+} = \dfrac{\langle E_{\rm vis}\rangle - \langle P_{\mu^-}\rangle}{\langle E_{\rm vis}\rangle}$	0.4	0.51	0.5	0.63
$\langle v\rangle_{-/+}$	0.09	0.05		
$\langle x\rangle_{-/+}$	0.23	0.1	0.1–0.2	0.1–0.2
$\dfrac{\langle P_\mu +/-\rangle}{\langle P_\mu -/+\rangle}$	6.1	3.0	3.63	2.1
$\langle P_{t\mp}\rangle$	0.3[a]	0.60[a]		
$\dfrac{\langle x\rangle_{\mu^-}}{\langle x\rangle_{\mu^+}} = \dfrac{0.23}{0.1} \simeq 2.3$				
$\langle P_{t^-}\rangle/\langle P_{t^+}\rangle \simeq 0.3/0.15 \sim 0.5$				

[a] in GeV/c.

new hadrons are being produced by neutrinos and antineutrinos. There is evidence for three additional classes of multilepton events.

$$\left.\begin{cases} v_\mu + N \to \mu^-\mu^-\bar{v}_\mu + X \\ \bar{v}_\mu + N \to \mu^+\mu^+ v_\mu + X \\ v_\mu + N \to \mu^-\mu\mu\cdot + X \end{cases}\right\}.$$

Evidence for $\mu^-\mu^-$ production has been obtained by the HPWF and CITF experiments (59, 42). In the former experiment, 7 ($\mu^-\mu^-$) and 3 ($\mu^+\mu^+$) events were observed, and in the latter, 2 ($\mu^-\mu^-$) events were observed.

Tables 7 and 8 show some of the kinematic properties of $\mu^+\mu^-$ events and μ^-e^+ events obtained in the counter and bubble-chamber experiments (40, 60). The $(x, y)_{\rm visible}$ properties of the events obtained by the HPWF group and CITF group are shown in Figures 25A and B, respectively (40, 42). We return to these distributions later.

7.2   Observation of Trimuon Events

Two experiments have reported examples of trimuon production by neutrinos. The CITF group has observed two events and the FHPRW group has observed at least three such events (40, 42, 43). Figure 26 shows the trimuon-event signature superimposed on the neutrino detectors for these two experiments. The trimuon signal is enhanced over the expected background from $K_\pi \to \mu$ decays. The rate for trimuon production is very uncertain because of the experimental cut on the momentum of the three tracks in order to identify these as muons. The CITF group reports a rate of $\sim 3 \times 10^{-4}$ per neutrino interaction, whereas the FHPRW group gives

Table 8 Lepton and strange-particle kinematics for $v_\mu + N \cdot \mu^- + e^+ + \cdots$ (μ identified) (BCHW Group)[f]

Event number	E_{vis}[a]	E_{μ^-}	E_{e^+}	E_{K^0}	Q^2[b]	v	x[c]	y[d]	W^2	P_μ/P_{e^+}[e]
1	33	14	1.6	6.1	0.13	18.0	0.004	0.56	35	8.75
2	12	3.4	1.1	3.1	1.16	8.1	0.10	0.71	15	3.09
3	26	21.8	2.2	1.6	6.2	3.1	0.9	0.12	—	9.91
4	28	9.3	5.2	5.8	7.2	18.0	0.21	0.66	28	1.79
5	98	58.0	2.3	4.4	0.21	39.0	0.003	0.41	75	27.60
6	24	15.0	0.4	2.2	5.5	7.8	0.37	0.34	10	37.50
7	97	9.3	5.7	38.0	9.4	87.0	0.058	0.90	154	1.63
8	21	14.1	1.3	1.6 or Λ	1.2	5.7	0.11	0.29	10	10.85
9	30	26.2	1.4	(1.6 or Λ)	2.7	2.9	0.49	0.10	3.6	18.70
10	61	17.9	4.0	1.5 or Λ	0.70	42.0	0.009	0.70	78	4.50
11	42	33.6	0.80	(1.5 or Λ)	3.8	7.3	0.29	0.17	10	42.00
Average	43	20	2.4		3.47		0.23	0.45		

[a] $\langle E_{vis} \rangle = 43$ GeV.
[b] $\langle Q^2_{vis} \rangle = 3.5$ (GeV/c)².
[c] $\langle x_{vis} \rangle = 0.23$.
[d] $\langle y_{vis} \rangle = 0.45$.
[e] $\langle P_{\mu^-} F = 20$ GeV; $\langle P_{e^+} \rangle = 2.4$ GeV; and $\langle P_{\mu^-} \rangle / \langle P_{e^+} \rangle = 8.5$.
[f] Nine strange particles identified in eleven events.

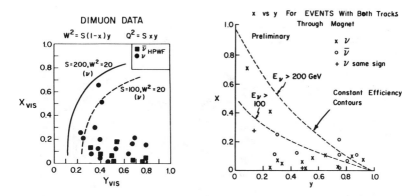

Figure 25 (A) X_{vis}, Y_{vis} scatter plot for HPWF (ν, $\bar{\nu}$) data. (B) X_{vis}, Y_{vis} scatter plot for CITF data.

an upper limit of $\sim 10^{-3}$ trimuons per single-μ production at very high energy. The physical origin of the trimuon events is not yet known, but may originate from charm pair production, heavy-lepton production, the production of states with another new quantum number, or some combination of these processes.

7.3   Direct Observation of a Charmed-Particle Event in a Nuclear Emulsion

The observation of a likely example of the decay of such a short-lived charged particle produced in a high-energy neutrino interaction in nuclear emulsion has been reported by a large group of European and American physicists (13). The experiment was performed in the wide-band neutrino beam at Fermilab, using a technique developed earlier, in which spark chambers are placed downstream of nuclear emulsion stacks. A neutrino interaction occurring in the emulsion can be located by predicting for the tracks of secondary particles observed in the spark chambers their point of origin in the emulsion. Stacks containing 18 liters of Ilford K5 emulsion, comprised of pellicles of dimensions 20 cm × 8 cm × 0.6 mm were placed in association with a double wide-gap spark chamber followed by a detector of electromagnetic showers and a rudimentary muon identifier. A veto counter upstream discriminated against interactions in the emulsion produced by charged particles. The experimental setup is shown schematically in Figure 2D.

The stacks were exposed to neutrinos produced by a total of about 7×10^{17} protons of energy 400 GeV on target. From the known neutrino flux it is estimated that some 150 neutrino interactions should have been produced in the emulsion. Approximately 250 candidates for these interactions have been seen in the spark-chamber pictures and a search has been made for about one third of them to date. The search involves scanning a volume of the emulsion averaging about 0.7 cm^3, around the position of the vertex predicted from the measurements of the spark-chamber film. So far, 16 interactions have been located in the emulsion and analyzed. Among these has been found an event in which one of the secondary particles presents the features expected of the decay of a short-lived particle. Figure

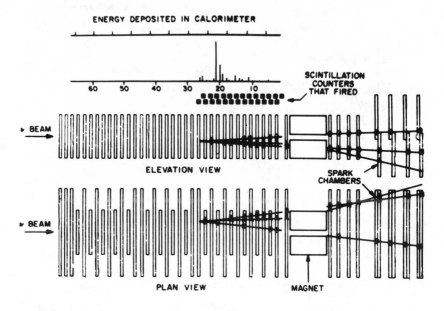

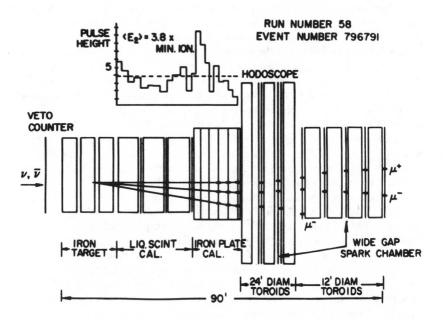

Figure 26   Trimuon events observed in the CITF detector and the rebuilt HPWF detector.

27A shows a photomicrograph of the event as seen in the emulsion. Figure 27B shows a schematic drawing based on the spark-chamber photographs associated with the event. At the neutrino interaction, vertex A in Figure 27, are seven tracks due to low-energy protons or nuclear fragments. There are in addition five tracks with a specific ionization, as measured in the nuclear emulsion, compatible with minimum ionization. One of these, track 4, is found to give rise to three further tracks of minimum ionization at vertex B after a distance of 182 μm. As there is no sign of a nuclear recoil track, the event has the characteristics expected of the decay of an unstable particle. The direction of the neutral V particle derived from the spark-chamber pictures makes it a good candidate for the missing neutral particle from the vertex B, removing at least part of the imbalanced transverse momentum. The muon identifier recorded the passage of at least one particle, correlated in time with the spark-chamber pictures, through the 1.35-m lead screen. The total pulse height recorded by the electron multipliers of the shower detector indicates the absence of any high-energy electron cascade.

The conclusion is that the event most probably represents a massive new short-lived particle decaying after an observed flight time of $\sim 6 \times 10^{-13}$ sec with characteristics compatible with those expected for the decay of a charmed particle. This is the first clean observation of a charmed-particle track.

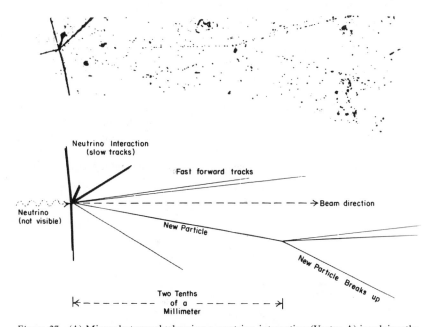

Figure 27 (A) Microphotograph showing a neutrino interaction (Vertex A) involving the emission of a short, charged particle (track 4) that undergoes a secondary process (nuclear interaction or decay) after 182 μm (Vertex B). (B) Schematic drawing based on the spark-chamber photographs associated with the event of the new-particle production.

7.4   *Violation of the* $\Delta S = \Delta Q$ *Rule and Strange-Particle Production*

Data on strange-particle production by neutrinos and antineutrinos has been obtained recently by the GGM group (61), and Argonne National Laboratory (ANL)-Purdue (62), a BNL (63), and a Fermilab-Michigan group (64). In the latter three cases, they were obtained from a H_2 bubble chamber and permit a clean study of strange-particle production without the influence of nuclear effects.

In the quark-parton model, single strange-particle production occurs by the processes

1. $\bar{v}_\mu + p \to \mu^+ + \lambda$,

2. $v_\mu + \bar{p} \to \mu^- + \bar{\lambda}$,

3. $v_\mu + \lambda \to \mu^- + p$,

where p and λ denote charge $+\frac{2}{3}$ and charge $-\frac{1}{3}$ quarks, respectively. Processes 2 and 3 would occur on partons or antipartons in the sea, whereas 1 would occur on valence partons as well. Present QPM models lead to the expectation that processes 2 and 3 would occur predominantly at low x. There are no firm predictions for associated production in the quark-parton model. The first evidence for single strange-particle production by neutrinos ($\Delta S = \Delta Q$) has been obtained by the ACMP group (two events of the type $vN \to \mu NK$) (62). These events, while at very low energy, may be examples of reactions 2 and 3.

Strange-particle events have been studied by the GGM group. Only K^+, K°, and Λ particles are reliably observed in this detector. Using these strange particles and neglecting Σ and K^- production, it is possible to obtain information on the fraction of the events that come from $\Delta S = 0(R_0)$ and $\Delta S = 1(R_1)$ transitions (61). In the neutrino case, the separation is not unique and leads to a correlation between the measured rates for the two processes. In Figure 28, the values of R_0 and R_1 for neutrinos and antineutrinos are recorded in the form of contour plots. For the neutrino case, R_1 represents an upper limit and R_0 a lower limit. The neutrino and antineutrino values of R_1 are in rough agreement with the value measured in H_2

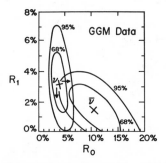

Figure 28   Determination of the relative fraction of $\Delta S = 1$ (R_1) and $\Delta S = 0$ (R_0) events in neutrino and antineutrino interactions. Since R_1 and R_0 are correlated, contour plots of likelihood probability are plotted. The data are from the GGM experiment.

at ANL. However, the errors on these measurements are such that the existence of inelastic $\Delta S = 1$ neutrino or antineutrino transitions cannot be firmly established because of the larger rate for associated production (61).

The best experimental signature for strange-particle production is obtained for processes with a V in the final state, either a Λ or K_s° decay. The Fermilab-Michigan group has reported the observation of events with one or more V's (64). There is no evidence from this sample of events for a violation of the $\Delta S = \Delta Q$ rule.

The cleanest test of the $\Delta S = \Delta Q$ rule would come from the observation of events of the kind

$$\nu_{\mu} + N \rightarrow \mu^- + \Lambda + X,$$

where X contains no strange particles. One such production event, interpreted as

$$\nu_{\mu} + p \rightarrow \mu^- + \Lambda + \pi^+ + \pi^+ + \pi^+ + \pi^-,$$

has been reported by the BNL group (63). An upper limit on the violation of the $\Delta S = \Delta Q$ rule can be obtained by taking the ratio of all events with Λ's compared to the total number of events without Λ's. In some cases additional information is available that allows the event to be classified as an example of associated production. A summary of the upper limit on number of single-Λ events and the limit on the ratio of the $\Delta S - \Delta Q$ rate to all neutrino interactions is given in (64). It appears that the violation of the $\Delta S = \Delta Q$ rule is likely no larger than a few $\times 10^{-2}$ even at the Fermilab neutrino energies. Detection of an unambiguous violation of the $\Delta S = \Delta Q$ rule requires numerous examples of the reactions without missing neutrals. At high energies, such processes would be expected to be a small fraction of all $\Delta S = \Delta Q$ events. The already stringent limits on the possible violation of $\Delta S = \Delta Q$ indicate the experimental difficulty in establishing a clear violation of this selection rule.

Although no additional $\Delta S = -\Delta Q$ events have so far been repeated and the credibility of one event is weak, it seems likely, in the face of the other indications for new quantum-number production by neutrinos (Sections 7.1 and 7.2) and the evidence from $e^+ e^-$ and photoproduction experiments, that the BNL event is real and an "effective" violation of the $\Delta S = \Delta Q$ rule occurs for high-energy neutrino collisions (63). From the BNL event, interpreted as the production of a new baryon, the charmed baryon mass is observed to be 2426 MeV. This value is in reasonable agreement with theoretical expectations and recent photoproduction results (8). Indication of new baryon production has been obtained in the HPWF experiment from structure on the W distribution, as shown in Figure 14 (24). It seems very likely that a new family of charmed baryons with mass in the vicinity of 2–3 GeV exists.

7.5   Quasi-Elastic Baryon Production and the Adler and GLS Sum Rules

A variety of sum rules have been derived that are applicable to neutrino and antineutrino scattering (65). Beyond the basic assumptions of current algebra, these sum rules depend on quantum numbers of the basic hadronic constituents. The most famous is the Adler sum rule, which in effect provoked the scaling hypothesis (65). This sum rule can be evaluated without the scaling hypothesis simply as a relation

$$\frac{\pi}{G^2}\left[\frac{d\sigma^{vn}}{dq^2} - \frac{d\sigma^{vp}}{dq^2}\right]_{E_v \to \infty} = 2,$$

or in the scaling region the sum rule is given by

$$\int_0^1 \frac{dx}{x}\left[F_2^{vn} - F_2^{vp}\right] = 2.$$

A test of the sum rule requires accurate H_2 and D_2 data and a separation of the structure function $F_2^{vp}(x)$ and $F_2^{vn}(x)$. If charge-symmetry invariance is assumed, antineutrino data can also be used to test the sum rule

$$\int_0^1 \frac{dx}{x}\left[F_2^{\bar{v}p} - F_2^{vp}\right] = 2$$

or

$$\frac{\pi}{G^2}\left[\frac{d\sigma^{\bar{v}p}}{dq^2} - \frac{d\sigma^{vp}}{dq^2}\right] = 2.$$

Since the validity of charge-symmetry invariance for high-energy neutrino reactions is questionable, the latter form of the sum rule is in doubt. The saturation of the Adler sum rule has been a question of great theoretical interest for some time, and serious doubts have been raised from time to time. The basic problem stems from the small size of the integral of $\int [F_2^{vp} + F_2^{vn}]\, dx \approx 1$ obtained from total cross-section measurements at moderate energies. In order for the sum rule to be saturated, $F_2^{vn}(x)$ must be greater than $F_2^{vp}(x)$ over an appreciable range of x and especially at low x. However, at small x, it has been suggested that $F_2^{vp} \to F_2^{vn}$ if diffraction scattering dominates. Phenomenological analyses using the available electro-production data have indicated that the sum rule will only be saturated at very high energies unless additional, nondiffractive contributions at low x make an appearance in the data. We note that one feature of the $F_2(x)$ obtained at high energy by the HPWF and bubble-chamber group (Figures 5 and 6) is an enhanced contribution over the electroproduction $F_2(x)$ at low x. This additional contribution could help effect the convergence of the sum rule. The data from $I = 0$ targets in conjunction with H_2 data can be used to evaluate the sum rule if it is rewritten as

$$\int_0^1 [F_2^{vp} + F_2^{vn}]\frac{dx}{x} - \int_0^1 F_2^{vp}\frac{dx}{x} = 2$$

in the scaling region. The first term in the left-hand side can be obtained from HPWF data. An evaluation of this integral from x to 1 for variable x indicates that the integral increases rapidly as x decreases and reaches 8 ± 1.5 for $x \to 0$, where the error is primarily determined by the present total cross-section measurements. While precision total cross-section measurement and F_2^{vp} will be needed to accurately evaluate the sum rule, it is encouraging that the $(I = 0)$ integral appears sufficiently large to allow the sum rule to be saturated at Fermilab energies. We note that if $\sigma^{vp}/\sigma^{vn} \approx \frac{1}{2}$ and if the x distribution obtained from the FNAL-Michigan group is taken for the shape of $F_2^{vp}(x)$, then the sum rule is approximately saturated, within the error in the present total cross-section measurement.

If charge symmetry is broken, a modified form of the Adler sum rule can be evaluated on $I = 0$ targets (66),

$$\frac{\pi}{G^2} \left[\frac{d\sigma^{\bar{v}}}{\overline{dq^2}} - \frac{d\sigma^{v}}{\overline{dq^2}} \right] = A.$$

For example, $\Delta S = 1$ processes give rise to a very small but nonzero value of A. The production of new particles with new hadronic quantum numbers can also contribute to a nonzero A. For example, the production of charmed baryons may give a nonzero value of A at low q^2.

The modified Adler sum rule has been evaluated by two groups, the GGM at low energy and the HPWF at high energy. Figure 29 shows both the GGM and HPWF results. The HPWF data (67, 15) indicate a nonzero value for A. One explanation of these latter results is that the sum rule has not yet converged in the kinematic region studied, and another explanation is that the value of A reflects some new hadronic quantum numbers that have been excited at higher energies; of course, experimental error is not ruled out either.

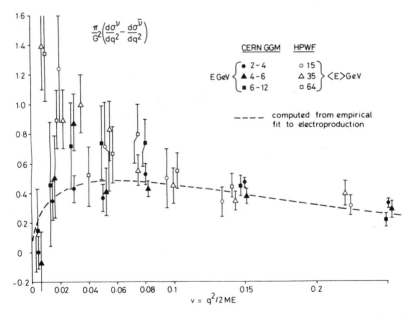

Figure 29 The difference between differential cross sections for neutrinos and antineutrinos, from CERN, GGM, and HPWF experiments. The Adler sum rule, in the limit of large E, predicts the intercepts at $v = 0$.

Another interesting sum rule that is directly measurable for $I = 0$ targets is the Gross-Llewellyn-Smith sum rule

$$\int [F_3^{\nu p} + F_3^{\nu n}] \, dx = B,$$

with $B = -6$ for the 4 quark-parton model. In some sense, B "counts" the number of quarks. If charge symmetry invariance is broken, the sum rule can be evaluated as (15)

$$\int [F_3^{\bar{\nu}} + F_3^{\nu}] \, dx = B'.$$

The existence of the y anomaly suggests that B' will deviate from the value of -6 at high energy. The GGM group has determined B in the low-energy region and find a value consistent with -6. Preliminary results from the HPWF group reported at the Palermo Conference suggest that the value of B' increases at high energy (15). This may be due to the production of new flavors.

7.6 Unified Explanation of the New Physics in Terms of Charmed-Particle Production

For the reasons given above, we conclude that the bulk of the dimuon and μe events arise from the production and decay of new particles. Several possible origins of dilepton events from new-particle production are shown in Figure 30. The intermediate vector-boson production is easily ruled out by the existing data. In order

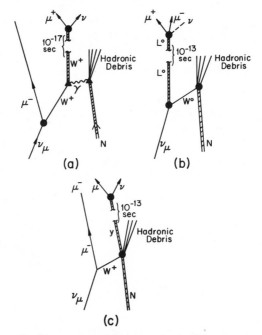

Figure 30   Three new-particle sources for the dimuon events.

to compare the heavy-lepton and new-hadron hypothesis, Pais & Treiman devised a test based on the ratio $\langle P_{\mu^-}\rangle/\langle P_{\mu^+}\rangle$ (56). Comparison of this test with available data is shown in Figure 31. It is concluded that the bulk of the multilepton events come from new-hadron production. There is now some evidence for the direct observation of these states from an emulsion exposure in the Fermilab neutrino beam. The multilepton events produced by neutrinos have the characteristics expected for charm production. In addition, an estimate of the mass of the charmed state following the model of Sehgel & Zerwas gives ~ 2 GeV, consistent with that directly measured for charm states (57). Furthermore, the μe events have associated the strange particle with an enhanced rate compared to non-μe events. These are the properties expected for the production of charmed particles. Until otherwise disproven, the only sensible conclusion from these is that a large fraction, if not all, the $\mu\mu$ and μe events produced by neutrinos come from one or more charmed particles produced, such as

$$\nu_\mu + N \rightarrow \mu^- + \left(\frac{D_{\text{°}}^{\,\ast}}{D}\right) + \text{hadrons}$$
$$\downarrow$$
$$\left\{\begin{matrix} \mu^+ + \nu_\mu \ldots \\ e^+ + \nu_e \ldots \end{matrix}\right\},$$

$$\nu_\mu + N \rightarrow \mu^- + F^+ + (\text{hadrons})$$
$$\downarrow$$
$$\left\{\begin{matrix} \mu^+ + \nu \ldots \\ e^+ + \nu_+ \ldots \end{matrix}\right\},$$

$$\nu_\mu + N \rightarrow \mu^- + \left(\begin{matrix} \text{charmed} \\ \text{baryon} \end{matrix}\right) + \text{hadrons}$$
$$\downarrow$$
$$\left\{\begin{matrix} \mu^+ + \nu_\mu + \ldots \\ e^+ + \nu_e + \ldots \end{matrix}\right\}.$$

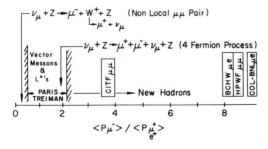

Figure 31   Pais-Treiman test with available $\mu\mu$ or μe data.

The production of these charmed states can occur in three different ways (68)

1. Production of valence quarks:

$$\nu_\mu + n \to \mu^- + p'.$$

2. Production of sea quarks:

$$\nu_\mu + \lambda \to \mu^- + p'.$$

3. Diffractive production of charmed vector mesons

$$\nu_\mu + N \to \mu + W^+ + \ldots$$
$$ \hookrightarrow F^{+*} \to F^+ + \gamma$$
$$ \hookrightarrow D^{+*} \to D + \gamma.$$

Process 1 should show the same characteristics as ordinary μ_ν events, whereas process 2 should be characteristic of the strange-particle sea. In this case, presumably the distribution will be peaked at very small x values. In both cases, the y distribution is expected to be flat at energies well above the threshold.

A theoretical lower limit on the production rate of charmed particles is obtained by considering only the valence contribution (68). Using the observed dimuon rate at high energy, far above the threshold (Figure 22), an upper limit on the semileptonic branching ratio can be obtained, i.e.

$$B_c \to \mu\nu\ldots\cdot\frac{\sigma^\nu_{\text{charm}}}{\sigma^\nu_{\text{all}}} \simeq 1.5 \times 10^{-2}, E > 100 \text{ GeV},$$

$$\frac{\sigma_{\text{charm}}}{\sigma_{\text{all}}} \geqq 0.05; B_c \to \mu\nu < 0.3.$$

A lower limit on the branching ratio is obtained from the limits on the deviation from Bjorken scaling in neutrino deep inelastic scattering, which gives

$$\frac{\sigma^\nu_{\text{charm}}}{\sigma^\nu_{\text{all}}} < 0.15,$$

leading to

$$B_c \to \mu\nu\ldots > 0.1.$$

Thus the semileptonic branching ratio to $\mu\nu\ldots$ lies between 10 and 30%. These results are consistent with that obtained from e^+e^- inclusive excitation of charm and subsequent semileptonic decay.

The characteristics of $\mu\mu$ and μe events from the three experiments are given in Tables 7 and 8. Both the $\mu^-\mu^+$ and μ^-e^+ events tend to have relatively large X_{vis}; y_{vis} is consistent with 0.5. This suggests that the bulk of the events come from production from valence quarks. An appreciable yield of events from diffraction production is not yet ruled out, however.

The mass of the charmed state can be estimated from the transverse-fit momentum distribution following Sehgel & Zerwas. The assumptions of this model (57) are as follows:

1. The deep inelastic production of the charmed particle via

$$\nu_\mu + N \rightarrow \mu^- + (D/F) + \text{hadrons}$$

follows the form

$$\frac{d^3\sigma(D/F)}{dx\,dy\,dz} = \frac{G^2 M E_\nu}{\pi} F(x,y)D(z),$$

where $F(x,y)$ is the inclusive distribution for the production reaction and $D(Z)$ is the fragmentation function of the variable. The form of $D(Z)$ is abstracted from the measurements of pion production by neutrinos.

2. For the decay of the D/F particle into

$$D/F \rightarrow \mu + \nu + \text{hadrons},$$

two approximations can be made: (a) two-body decay into $\mu + \nu$, (b) n-body decay into 3 or more particles ($\mu + \nu + \text{hadron}$). The transverse momentum of the μ^+ relative to the $\nu_\mu +$ plane is closely related to the decay kinematics so that the final distribution for the production of the (D/F) particles is assumed to follow

$$\frac{d\sigma(\mu^-\mu^+)}{dx\,dy\,dz\,dP_t} = \frac{G^2 M E}{\pi} F(x,y)D(Z)BH(P_t),$$

where B is the branching fraction

$$B = \frac{\Gamma_{D/F} \rightarrow \mu + \nu + \text{hadrons}}{\Gamma_{D/F} \rightarrow \text{all}},$$

and $H(P_t)$ is what describes the P_t distribution of the μ^+. $H(P_t)$ will in general depend on the mass of the decaying particle and the number of particles in the final state of the decay.

Figure 32 shows the χ^2 for the fit as a function of D/F-particle mass (69). The best fit is for a mass of 2 ± 0.4 GeV. This measured mass is in good agreement with that obtained directly from nonleptonic decays.

A characteristic prediction for charm is the $\Delta S = \Delta C$ rule for both D and F decays (68)

$$\frac{\Gamma(D/F \rightarrow \mu^- + \bar{\nu}_\mu \ldots)}{\Gamma(D/F \rightarrow \mu^+ + \nu_\mu \ldots)} \leq 0.3$$

to 90% confidence. Although this limit is not yet restrictive, it is consistent with the standard theory of charm production.

7.7 Multilepton Production by Antineutrinos

Charmed-particle production by antineutrinos is expected to be suppressed, since these processes presumably occur on the strange-quark sea, i.e.

$$\bar{\nu}_\mu + \begin{Bmatrix} \bar{\lambda} \\ \lambda \end{Bmatrix} \rightarrow \begin{Bmatrix} \bar{p} \\ \lambda \end{Bmatrix} + \mu^+.$$

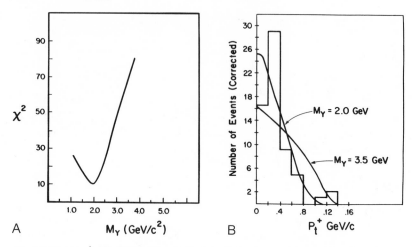

Figure 32 (A) χ^2 distribution for fit to Sehgal-Zerwas model as a function of charm-particle mass. (B) Comparison of theory and data on P_t distribution.

Furthermore, these events should show copious strange-particle production, which would provide a unique signature for the charm association. The fraction of strange quarks in the parton-antiparton sea is not known, but could be very small and thus, in contrast to the neutrino situation, it is not possible to predict even roughly the yield of charmed particles in antineutrino experiments. There already exists considerable evidence obtained from the single μ^+ channel that new phenomena (y anomalies) are produced by $\bar{\nu}_\mu$. We review here the evidence obtained from dimuon production and the possible relation of this evidence to the existence of quantum numbers beyond charm.

Three experiments have reported limits on or evidence for multilepton production by antineutrinos. The low-energy $\mu^+ e^-$ search has been carried out in bubble chambers, whereas $\mu^+ \mu^-$ events have been observed in the counter experiments at higher energies (40, 41, 70, 71).

Figure 33 summarizes the measured rates relative to single-μ^+ production (40). The rate appears to change with energy, but it is not possible to ascribe this uniquely to a clear threshold effect, since a similar rise in rate with energy may be observed for neutrinos, presumably due to scale breaking attributed to the charm particle mass. The overall rate for antineutrino dimuon production is surprisingly large at high energy.

We now turn to the (x, y) distributions for the antineutrino multilepton events. Figure 25 shows a (X_{vis}, Y_{vis}) scatter plot for the HPWF and CITF data as compared to the X_{vis}^v distribution. The X_{vis} distribution for the $\bar{\nu}_\mu$ that was obtained by the HPWF group appears rather sharp. The mean antineutrino energy for these data is ~ 80 GeV. There are at least two possible explanations for the sharp X_{vis} distributions.

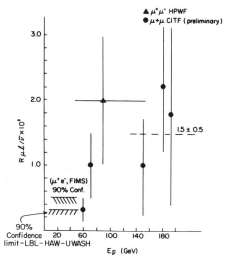

Figure 33 Compilation of the antineutrino-induced μ^+ ($\mu^- e^-$) event rate as a function of visible energy. The data have not been corrected for the (e^-, μ^-) energy bias.

1. The dimuons are produced on a sea of strange quarks, which have a sharp x distribution.
2. The dimuons are produced with a flat y distribution and a high mass threshold, which kinematically forces x to be small at low energy.

Thus the sharp (X_{vis}) distribution may come either from antiquark dynamics or production kinematics. In order to distinguish the two cases, higher-energy data is required. The Y_{vis} distribution for the $\bar{\nu}$ dimuons is consistent with flatness, as shown in Figure 25.

The (x, y) values for the CITF data are shown in a scatter plot in Figure 25 (42). These data, which are merely preliminary, appear to have similar characteristics to the HPWF data. Note that the Y_{vis} distribution for both ν and $\bar{\nu}$ data is consistent with flatness. Since the average energy of this experiment is considerably higher than the HPWF, this might indicate that the sharp X_{vis}^{ν} distribution is a kinematic effect. Note that the Y_{vis} distribution is not peaked at low y.

A further comparison of the properties of neutrino- and antineutrino-induced dimuons is given in Table 7 (40). It is notable that the average P_{t^-} for the $\bar{\nu}$ events is about twice that for the ν events. This could indicate that the new particles produced by antineutrinos are twice as heavy as those produced by neutrinos; for the b-quark model, this would give

$$M_b \gtrsim 4 \text{ GeV}.$$

7.8   *Present Search for Heavy Leptons Excited by Neutrinos*

Two kinds of heavy leptons have been sought in neutrino interactions involving the production and decay processes

$$\nu_\mu + N \to M^+ + X$$
$$\hookrightarrow \mu^+ + \dots$$
$$\hookrightarrow e^+,$$

$$\nu_\mu + N \to M^\circ + X$$
$$\hookrightarrow \mu^+ \mu^- \nu.$$

The first process has been sought by the CITF group and in two neon bubble-chamber exposures (39–41). No unambiguous examples of this reaction have been observed, but several candidates for anomalous e^+ production have been reported. These events could either arise from heavy-lepton production, flavor-changing weak neutral currents, or charm pair production (as discussed in Section 6.4). The present limit on anomalous e^+ production is $\sim 2 \times 10^{-3}$ per ν_μ interaction. A somewhat large limit exists for the μ^+ case. The mass limit for the M^+ lepton can be set safely above 6 GeV from these observations, even allowing for uncertainties in the branching ratio.

The search for M° production is far less conclusive, since the production of charmed particles and subsequent semileptonic decay leads to multilepton final states. Nevertheless, the observed characteristics of the dimuon events is compatible with only a relatively small "contamination" from M° production, as evidenced by the Pais-Treiman test or direct fits to the data with heavy-lepton–production models (Figure 31) (56). As an illustration of the possible existence of events that could arise from heavy-lepton decay, we show data from the HPWF experiment in Figure 34 (59). Dimuon events were selected within or outside the constraint

$$\tfrac{1}{2} \leqq P_{\mu^+}/P_{\mu^-} \leqq 2$$

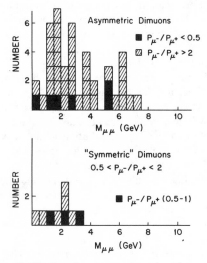

Figure 34 $\mu^+ \mu^-$ mass distributions for symmetric and anti-symmetric dimuon HPWF data.

and the dimuon mass distribution plotted in Figure 34 (40). There exist a sizable number of these "symmetric" dimuon events that could come from the production of an $M°$ at the level of 20% of the dimuon rate at high energy. Thus, heavy-lepton production is by no means ruled out in a fraction of the dimuon events.

8 SUMMARY AND SPECULATION

Multilepton production by neutrinos was the earliest observed process that followed the predictions of charm production. Apparently, charged-current neutrino interactions produce copious numbers of charmed particles (5–15% of all events). A semileptonic branching ratio of 10–30% is indicated by the data, which is surprisingly large compared to early predictions.

Multilepton production by antineutrinos promises to be a rich area of exploration because of the possible low rate of charm production. A measurement of the charm production at low energy could serve as a direct measurement of the characteristics of the strange-quark sea. There are indications that new phenomena, not simply related to charm, are occurring in $\bar{v}$ single-μ reactions. Dimuon production by $\bar{v}$ appears to have similar characteristics. If this trend continues, the next evidence

Table 9 Summary of new phenomena observed in neutrino interactions

Process	Rate	First observed	Present interpretation
$v_\mu + N \rightarrow \mu + K$	~ 1	~ 1950–1960	charged currents
$v_\mu + N \rightarrow v + X$	~ 0.2–4.0	1973	neutral currents
$\bar{v}_\mu + N \rightarrow \mu^+ + X^a$	~ 0.15–0.5	1974	new-particle production and hadronic decay(?)
$\sigma_c(\bar{v})/\sigma_c(v) \sim (0.5$–0.6)	~ 0.3–0.5	1976	new-particle production and hadronic decay—V + A currents(?)
$v_\mu + N \rightarrow \mu^- \mu^+ (v_\mu) X$	0.01–0.02	1974	new particle production and semileptonic decay (charmed particles)
$\bar{v}_\mu + N \rightarrow \mu^+ \mu^- (\bar{v}_\mu) X$	0.01–0.05	1975	new-particle production and semileptonic decay (b quark?)
$v_\mu + N \rightarrow \mu^- \mu^- (\bar{v}_\mu) X$	$\sim 10^{-3}$	1975	unknown
$\left\{ \begin{array}{l} v_\mu + N \rightarrow \mu^- e^+ (v_e) X \\ X \rightarrow \text{multistrange} \\ \text{particles} \end{array} \right\}$	0.01–0.02	1975	new-particle production and decay—new quantum number related to strangeness quantum number (charm)
$v_\mu + N \rightarrow \mu^- \mu^+ X^b$	2×10^{-3}	1976	possible neutral heavy-lepton production (Lo, No)
$v_\mu + N \rightarrow \mu\mu\mu X$	10^{-3}–10^{-4}	1976	unknown

a Excess at large y.
b Symmetric dimuons.

for new quarks may come from the study of $\bar{v}$ multilepton production. Some specific measurements that could unambiguously indicate the existence of another hadronic quantum number and possibly the existence of V + A currents are discussed below.

Limits on the W mass have been set and suggest that the W may not be observable with neutrino beams in the near future.

A summary of the new processes observed in neutrino interactions is given in Table 9. Note that one or more new phenomena have been observed for every decade decrease in the rate of the process. We now turn to speculations concerning the origin of additional new physics in light of the existence of charm.

8.1   The y Anomaly, the b Quark, and V + A Currents in Antineutrino Reactions

In charged-current, deep inelastic neutrino scattering ($vN \rightarrow \mu + X$), the ratio R_c of antineutrino to neutrino cross sections appears to become about double that expected at the highest energies. In addition, the antineutrino y dependence changes as a function of energy, (see Figures 9 and 10). The HPWF and CITF data both show these effects. It is difficult to understand these two phenomena (R_c and $\langle y \rangle$) without right-handed currents; even the increasing sea contributions due to asymptotic-freedom corrections appear to be inadequate (see section 8.3). It has been assumed for most models that there must be a right-handed coupling of the u quark to a heavy, $-\frac{1}{3}$ quark (50, 72).

Crucial to understanding the energy dependence of the $\bar{v}$ cross sections and of their anomalous y dependence is the manner in which scaling is resumed after passing quark mass thresholds (72, 73). The results perhaps correspond to a slow rescaling and differ drastically from previous work, which assumed a fast rescaling. With this slow rescaling, good agreement is obtained with all available data.

In the quark-parton model, it is assumed that the structure functions $F(z)$ are functions only of the scaling variable z. z is defined as the fraction of the target nucleon's momentum that is carried by the struck quark. It is further assumed that the quarks are quasi-free so that the produced quark is on the mass shell. If the exchanged W boson has momentum k, the struck quark has momentum zp, and the produced mass is neglected, then one finds

$$(k + zp)^2 \approx m_q^2 \approx 0 \rightarrow z \approx -k^2/2p \cdot k \equiv x.$$

The quantity x can be measured experimentally, so that in this case z is known.

However, for heavy, produced quarks ($m_q \gg 0.3$ GeV/c^2), it is not reasonable to neglect their mass, and then z cannot be directly measured (72, 73). Rather, when m_q is kept constant, it follows from Equation that

$$z \approx x \left(\frac{-k^2 + m_q^2}{-k^2} \right) = x \frac{m_q^2}{2ME_y},$$

where $y \equiv (E - E')/E$; $E(E')$ is the incoming or outgoing lepton energy, and M is the nucleon mass.

We now specifically address the b-quark model and compare the multilepton production rate with the theoretical calculations (74, 75). The major uncertainty is the semileptonic branching ratio (B_l) of charmed particles and the new particles (called y particles) that carry b quarks (75). The processes considered are, in the quark-parton notation,

$$v_\mu + n \to \mu^- + c \qquad\qquad (\text{V} - \text{A})$$
$$\hookrightarrow \mu^+ + v_\mu + \lambda \text{ and}$$

$$\bar{v}_\mu + p \to \mu^+ + b \qquad\qquad (\text{V} + \text{A})$$
$$\phantom{\bar{v}_\mu + p \to \mu^+}\hookrightarrow \mu^- + v_\mu + p,$$

and the specific processes are

$$v_\mu + N \to \mu^- + D + (\text{anything})$$
$$\hookrightarrow \mu^+ + v_\mu + K^* \text{ and}$$

$$\bar{v}_\mu + N \to \mu^+ + y + (\text{anything})$$
$$\phantom{\bar{v}_\mu + N \to \mu^+}\hookrightarrow \mu^- + v_\mu + \rho,$$

where N is an isoscalar nucleus and K^* is a strange vector meson and ρ is a non-strange vector meson. The y-particle mass is taken as 5 GeV.

The charmed-particle production is calculated for a combination of valence and sea contributions, whereas the b-quark production is only from the valence quarks. Slow rescaling is used, following the above discussion. The results of the calculation of Barget et al are shown in Figure 35 (75). In these calculations, the energy

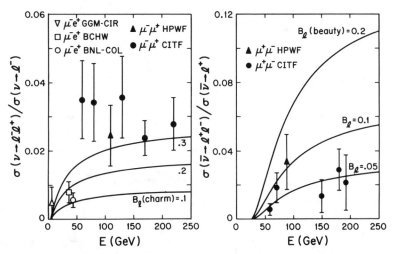

Figure 35 (A) Ratio of dilepton to single-lepton rates versus incident-beam energy for neutrino- and antineutrino-induced dileptons. The experimental rates have been corrected according to the estimated slow-lepton detection efficiencies. The curves are based on (A) GIM charm production with the indicated values B_l for the semileptonic inclusive branching ratio of D and (B) on b-quark production with an appropriate B_l value.

spectrum of the charmed or y particles is assumed to be the same as that of other hadrons produced in neutrino collisions. Since various experiments have different energy cuts on the leptons, a correction for these cuts has been applied to the data. For neutrino data below 50 GeV, there is good agreement with the charm model with a semileptonic branching ratio of $B_l \sim (0.1-0.2)$. At higher energy, there is a serious deviation from the model, the origin of which is not yet known. For the production of y particles by antineutrinos (Figure 35B), an assumption has been made about the relative fraction of y-particle production to the total inelastic contribution. The y anomaly and the deviation of $\sigma^{\bar{v}}/\sigma^v$ from the value of 0.38 were used as input. Some caution is required here, since the total cross-section measurements are not very precise. The semileptonic branching ratio for the y particle is observed to be ~ 0.05 (Figure 35B), but could be larger by a factor of 2 if the uncertainties in the calculations are included.

The calculations of y-particle production, using slow rescaling and $V + A$ currents, are in reasonable agreement with the data. A need remains, therefore, for a direct exprimental observation of these states that will open up a new chapter in the spectroscopy of hadrons and add a new element to the search for a theory of hadronic matter.

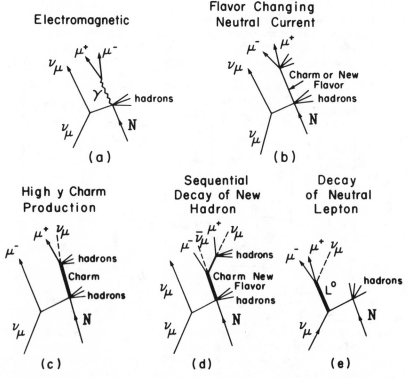

Figure 36   Processes that may contribute to the production of symmetric dimuon pairs.

8.2 *Separation of New Physics in Dimuon and Trimuon Final States*

There is new convincing evidence for the existence of trimuon events produced by neutrinos and of a class of "symmetric" dimuon events. The physical origin of these events is not yet known, nor are the detailed characteristics of the final states. Since there is already compelling evidence for the production of hadronic-state (charmed- and b-quark) new quantum numbers, it is important to separate possible new physical effects, such as additional new hadronic states carrying new quantum numbers and new leptons, from the charm background. It is very likely that the new hadronic states can be partially separated from new leptonic states by the behavior of the $P\mu^+$ and $P\mu^-$ spectrum of the events. New hadronic states are expected to result in events with a large $P\mu^-/P\mu^+$ asymmetry. New leptonic states should give symmetric, hard dimuon events. Figure 36 summarizes the processes that might give symmetric dimuon events, and Figure 37 shows where these various processes are concentrated on a $P\mu^+$-$P\mu^-$ plot (we take the case of dimuon production here so that a comparison with Figure 24 can be made). Basically, the new leptons and new hadronic effects are separated because of kinematics and the dynamics of leptonic production. Bubble-chamber measurements of the characteristics of had-

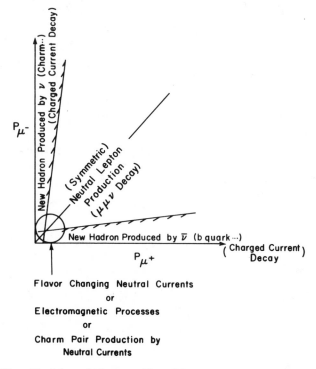

Figure 37   Schematic decomposition of the new physics on a P_μ^+-P_μ^- plot.

ronic production indicate a very rapid falloff of the production probability with the variable Z, $(Z = E_{hadron}/E_{all\ hadrons})$. In addition, when charmed particles or y particles decay semileptonically, the energy is divided. Finally, for the decay of psuedoscalar particles, there are matrix-element effects. All of these characteristics tend to concentrate the new hadronic effects in the bands shown in Figure 37. Other, more speculative processes, such as flavor-changing neutral currents, are expected to be concentrated at low values of $P\mu^+$ and $P\mu^-$ because of the Z effect mentioned above.

In contrast, neutral-lepton production (process e in Figure 36) is an efficient way to transfer energy from the neutrino beam to the multileptonic final states. For spin-$\frac{1}{2}$ leptons, the local nature of the decay ensures that there are bounds on the ratio of $\langle P\mu^-\rangle/\langle P\mu^+\rangle$. Thus the new leptons tend to produce hard, symmetric dimuons. As mentioned in Section 7.8, such events have been observed in the HPWF and CITF experiments.

Charged-trilepton production holds the promise of unique new physics. However,

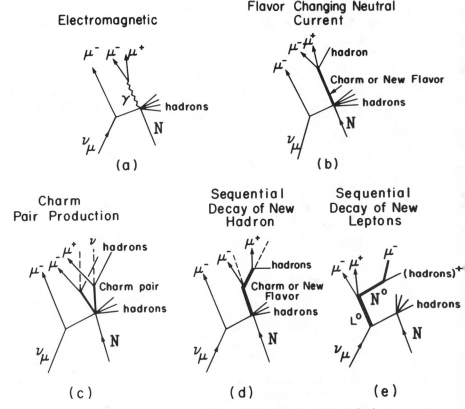

Figure 38   Processes that may contribute to the production of $\mu^-\mu^-\mu^+$ trimuon events.

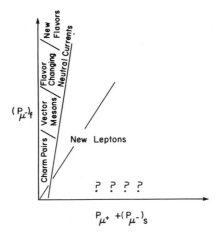

Figure 39   Schematic decomposition of the new physics in $\mu^- \mu^- \mu^+$ production by neutrino beams. $P_{\mu_f^-}$ and $P_{\mu_s^-}$ refer to the highest-energy and lowest-energy μ^- respectively.

the new physics must be separated from the old (charm) physics. The processes that can give rise to trimuon final states are shown in Figure 38. Some of these processes are expected to be very unlikely (charm-changing neutral currents or charm pair production), but nevertheless, their exclusion from the new physics has yet to be observed. It will be useful to plot the trimuon data in a manner analogous to the dimuon data; for example, for neutrino-produced trimuons,

$$\nu_\mu + N \rightarrow \mu_f^- + \mu_s^- + \mu^+ + (\text{anything})$$

we combine the momenta of the slow μ^- and the μ^+, and compare them with the fast-μ^- momentum. A scatter plot of these variables can be used to separate hadronic processes from leptonic processes, as shown in Figure 39. Such detailed characteristics for a large sample of trimuon events could reveal the production of heavy leptons. (Note that two heavy leptons are required.) Comparison with the symmetric dimuon yield (or other dilepton final states with $e^+ e^-$, for example) could make a convincing argument for the existence of such states. Note that the neutrino production of $\mu^- \mu^-$ events would also occur for some heavy-lepton processes.

The existence of symmetric dimuon and trimuon events promises observation of additional new physics in high-energy neutrino collisions. If new leptons are found to be produced in this way, a remarkable new chapter in neutrino physics will have been started.

8.3   Deviation from Scale Invariance in Neutrino and Charged-Lepton Scattering

The HPWF group first reported that the x distribution for neutrino scattering appeared to be sharper than the expected distribution based on the SLAC electro-

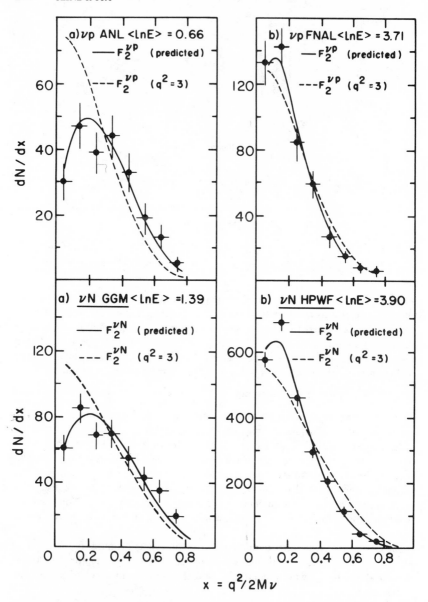

Figure 40 (A, B) Observed x distribution for neutrino scattering on hydrogen at two energies compared with predictions based on scale-invariance violation observed in the SLAC and Fermilab leptoproduction data. (C, D) Same as (A, B), except that the neutrinos scatter from an approximate isoscalar target. These points are taken from the work of Perkins, Schreiner & Scott.

production results (see Figure 15). Subsequently, bubble-chamber data for $v+p$ scattering showed a similar trend (76). The low-energy GGM data, on the other hand, compare well with the SLAC data. Clear evidence for a deviation from the expectations of scale invariance has been reported for muon interactions on both heavy targets and hydrogen (77). Similar effects were observed in electron scattering at SLAC (78). There is considerable evidence for a deviation from Bjorken scaling in both charged-lepton and neutrino scattering. It is significant to emphasize that this type of scaling deviation is considerably smaller than that reported for anti-neutrino interactions (the y anomaly). Nevertheless, if there is an additional scaling violation, the analysis of the y anomaly will be more complicated. It is important to learn whether the scaling violations in charged-lepton and neutrino scattering have a common origin or are related in any way. Progress in this regard has been made by an analysis of the available data by Perkins, Schreiner & Scott (79). Although several plausible assumptions are made in this work, it strongly suggests that there is a close relation of the scale violation observed in these lepton-induced processes. Using a parameterizer of the charged-lepton scaling violation, it is possible to predict the x distribution for several sets of data; neutrino scattering on hydrogen at low energy (ANL) and high energy (FNAL); and neutrino scattering on an isoscalar target at low energy (GGM) and high energy (HPWF). These predictions are shown in Figure 40. There is very good agreement between the model and the data.

It is unlikely that these scaling violations are completely related to new hadronic-particle production, since single new-particle production is possible with neutrinos, whereas pairs are produced by charged leptons. Nevertheless, we expect some small scaling violations (5–10%) in neutrino interactions due to the production of charmed baryons and charmed mesons. Small deviations are expected in electroproduction, due to the production of charmed pairs and y particles. It is likely that the bulk of the scaling violation comes from a fundamental deviation from Bjorken scaling due to the q^2 dependence of the quark-gluon coupling constant. Such a dependence is expected in asymptotically free field theories of inelastic scattering (80). In essence, at large q^2, the strong coupling constant in quark-quark scattering decreases and the strong interaction gets weaker. Thus at large q^2, the x distribution gets sharper, since a larger fraction of the proton's momentum is carried by the glue and less by the quarks.

There is one additional ramification of this result on neutrino scattering. In the search for nonlocality due to intermediate vector-boson exchange, it is necessary to know the momentum spectrum of the quarks at large q^2, in order to observe a propagator-cutoff effect. The HPWF analysis (Figure 7) will not be appreciably modified, since the observed x distribution in the region of 50–100 GeV was used as an input. Thus, there is still convincing evidence that the Λ values are greater than 20 GeV in light of the scaling violation. Nevertheless, in order to search for higher-mass propagators, it will be necessary to understand the deviations from scale invariance at a much better level than is presently known. It may be difficult to extend the search for the W through the propagator modification above 50 GeV with the present generation of proton accelerators.

ACKNOWLEDGMENTS

We wish to thank the very large number of people who have explained their data or theories to us. A partial list of these people includes C. Albright, J. D. Bjorken, C. Baltay, M. Barnett, V. Barger, B. Barish, A. Garfinkel, O. Facker, F. Halzen, G. Kane, B. Lee, R. Schrock, L. Stutte, H. H. Williams, and the members of the HPWF and BCHW groups.

One of us (D.C.) would like to thank his wife, Ludy, for her understanding through the long HPWF years at Fermilab.

Literature Cited

1. Danby, G. et al 1962. *Phys. Rev. Lett.* 9:36
2. Bienlein, J. K. et al 1964. *Phys. Lett.* 13:80
3. Benvenuti, A. et al 1973. *Phys. Rev. Lett.* 30:1084
4. Barish, B. C. et al 1973. *Phys. Rev. Lett.* 31:565
5. Hasert, F. J. et al, Benvenuti, A. et al, 1973. submitted to Int. Symp., 6th, Bonn; Hasert, F. J. et al 1973. *Phys. Lett. B* 46:138; Benvenuti, A. et al 1974. *Phys. Rev. Lett.* 32:800
6. Aubert, B. et al 1974. In *Proc. Int. Conf. High Energy Phys., 17th, London; Neutrinos—1974,* AIP. Conf. Proc. No. 22, ed. C. Baltay, p. 201. New York: Am. Inst. Phys.; Benvenuti, A. et al 1975. *Phys. Rev. Lett.* 34:419, 597; Barish, B. C. et al 1976. *Phys. Rev. Lett.* 36:939
7. Blietschau, J. et al 1976. *Phys. Lett. B* 60:207; von Krogh, J. et al 1976. *Phys. Rev. Lett.* 36:710
8. Cazzoli, E. G. et al 1975. *Phys. Rev. Lett.* 34:1125; Goldhaber, G. et al 1976. *Phys. Rev. Lett.* 37:255; Peruzzi, I. et al 1976. *Phys. Rev. Lett.* 37:569; Knapp, B. et al 1976. *Phys. Rev. Lett.* 37:882
9. Perkins, D. H. 1976. *Proc. 1975 Int. Symp. Lepton Photon Interactions High Energies,* ed. W. T. Kirk, p. 571. Stanford, Calif: Stanford Univ. Press; Bjorken, J. D. 1977. *Proc. 1976 SLAC Summer Inst.,* ed. G. S. Leith, F. Gilman. Stanford, Calif: Stanford Univ. Press; Lee, B. 1976. *Neutrino Conf. Aachen 1976.* To be published Wolfenstein, L. 1976. *Proc. 1975 Int. Symp. Lepton Photon Interactions High Energies,* ed. W. T. Kirk, p. 613. Stanford, Calif: Stanford Univ. Press
10. Aubert, B. et al 1975. Fermilab prepr. Conf. 75/31-Exp, 1975. *Paris Neutrino Conf. Proc.,* p. 385; Humphrey, T. 1975. PhD thesis. Calif. Inst. Tech.
11. Parker, S. I. 1975. *Bull. Am. Phys. Soc.* 20:593
12. Entenberg, A. et al 1976. *Phys. Rev. Lett.* 37:252
13. Burhop, E. H. S. et al 1977. *Phys. Lett.* To be published
14. Deden, H. et al 1975. *Nucl. Phys. B* 85:269
15. Benvenuti, A. et al 1974. *Phys. Rev. Lett.* 32:125; Cline, D. 1975. *Proc. Palermo Conf.*
16. Nezrick, F. 1976. *Proc. Aachen Conf., June;* Scott, W. G. 1976. Vanderbilt Conf., April
17. Roe, B. 1976. *Proc. Part. Fields 1976 Conf., Brookhaven Natl. Lab. Oct. 6–8, 1976*
18. HPWF group; unpublished data
19. Barish, B. 1975. *Proc. Hawaii Summer Sch.*
20. Barish, S. J. 1975. ANL Prepr. HEP-CP-75-39
21. Benvenuti, A. et al 1974. *Phys. Rev. Lett.* 33:984; Benvenuti, A. et al 1976. *Phys. Rev. Lett.* 36:1478
22. Eichten, T. et al 1973. *Phys. Lett. B* 46:274
23. Barish, B. C. et al 1975. *Phys. Rev. Lett.* 35:1316
24. Benvenuti, A. et al 1976. *Phys. Rev. Lett.* 37:1095
25. Benvenuti, A. et al 1976. *Phys. Rev. Lett.* 37:189
26. Barish, B. et al 1977. *Phys. Rev. Lett.* 38:314
27. Camerini, U., Cline, D., Fry, W., Powell, W. M. 1964. *Phys. Rev. Lett.* 13:318; Cline, D. 1965. PhD thesis. Univ. Wis., Madison
28. Aubert, B. et al 1974. *Phys. Rev. Lett.* 32:1454
29. Barish, B. C. et al 1975. *Phys. Rev. Lett.* 34:538
30. Benvenuti, A. et al 1976. *Phys. Rev. Lett.* 37:1039
31. Blietschau, J. et al 1977. CERN prepr.

Submitted to *Nucl. Phys. B*

32. Barish, B. II 1977. In *Neutrino Conf. Aachen 1976*. To be published
33. Lee, W. et al 1976. *Phys. Rev. Lett.* 37:186; Lee, W. 1976. *Neutrino Conf. Aachen 1976*. To be published
34. Williams, H. H. et al 1977. *Proc. 1976 SLAC Sumner Inst., Stanford, Calif*, ed. D. W. G. S. Leith, F. Gilman. Stanford, Calif: SLAC; Sulak, L. et al 1976. *Neutrino Conf. Aachen 1976*. To be published; Sulak, L. 1975. *Proc. Calorimeter Workshop*, ed. M. Atac, p. 155. Batavia, Ill: Fermi Natl. Accel. Lab.
35. Blietschau, J. et al 1976. CERN prepr. Submitted to *Nucl. Phys. B*
36. Bobisut, F. et al 1977. In *Neutrino Conf. Aachen 1976*. To be published
37. Reines, F., Gurr, H. S., Sobel, H. W. 1976. *Phys. Rev. Lett.* 37:315
38. Williams, H. H. 1976. Part. Fields 1976 Conf., Brookhaven Natl. Lab. Oct. 6–8, 1976. To be published; Univ. Penn. prepr.
39. Barish, B. C. et al 1974. *Phys. Rev. Lett.* 32:1387
40. Cline, D. 1976. Part. Fields 1976 Conf., Brookhaven Natl. Lab., Oct. 6–8, 1976. *Brookhaven Natl. Lab. Rep.* 50598
41. Sinclair, D. 1976. *Proc. Wis. Conf. New Quantum Numbers*
42. Facker, O. 1976. Part. Fields 1976 Conf., Brookhaven Natl. Lab., Oct. 6–8, 1976. *Brookhaven Natl. Lab. Rep.* 50598
43. HPWF experiment (Fermilab-Harvard-Penn.-Rutgers-Wisconsin collaboration), unpublished data
44. Bertrand Coremans, G. H. et al 1976. *Phys. Lett. B* 61:207
45. Adler, S. L. 1975. *Phys. Rev. Lett. D* 12:2644; Adler, S. L. 1975. *Proc. Hawaii Top. Conf. Part. Phys., 6th, Honolulu*, ed. P. N. Dobson Jr., p. 5. Honolulu: Univ. Press Hawaii
46. Albright, C., Cleymans, J. 1974. *Nucl. Phys. B* 76:48; Sehgal, L. M. 1974. *Nucl. Phys. B* 74:285
47. Murtagh, M. 1977. Private communication to H. H. Williams
48. Weinberg, S. 1967. *Phys. Rev. Lett.* 19:1264; Salam, A. 1968. In *Elementary Particle Physics*, ed. N. Svortholm, p. 367. Stockholm: Almquist & Wiksells
49. Barnett, R., private communication
50. Barnett, R. M. 1977. Harvard prepr. To be published in *Phys. Rev. D*; Barger, V., Nanopoulos, D. V. 1977. Univ. Wis. prepr.
51. Baird et al 1976. *Nature* 264:528
52. Bleitschau, J. et al 1976. *Phys. Lett. B* 60:207

53. von Krogh. J. et al 1976. *Phys. Rev. Lett.* 36:710
54. Baltay, C., private communication
55. Benvenuti, A. et al 1975. *Phys. Rev. Lett.* 35:1203
56. Pais, A., Treiman, S. B. 1975. *Phys. Rev. Lett.* 35:1206
57. Sehgal, L., Zerwas, P. 1976. *Phys. Rev. Lett.* 36:399; Aachen prepr. 1976; Barger, V., Phillips, R. J. N. 1976. Univ. Wis. prepr. COO-495; LoSecco, J. 1976. *Phys. Rev. Lett.* 36:336; Derman, E. 1976. Oxford prepr. 17/1/76, 16/2/76
58. Barger, V., Phillips, R. J. N. 1976. Univ. Wis. Madison rep. COO-495, 504; Chang, L. N., Derman, E., Ng, J. N. 1975. *Phys. Rev. Lett.* 35:1252; Lee, B. W. 1975. Fermilab-Conf-75/78 THY; Einhorn, M. B., Lee, B. W. 1975. Fermilab-Pub-75/66-THY; Pais, A., Treiman, S. B. 1975. *Phys. Rev. Lett.* 35:1206, 1556; Arbuzov, B. A. et al 1975. Serpukhov prepr. IHEP 75-11, 75-25; Cerny, V., Pisut, J. 1975. *Proc. Neutrino Conf., Balatonfured*, ed. A. Frenkel, G. Marx, 2:244
59. Benvenuti, A. et al 1975. *Phys. Rev. Lett.* 35:1199
60. Berkeley - CERN - Hawaii - Wisconsin Bubble Chamber collaboration, unpublished data
61. Deden, H. et al 1975. *Phys. Lett. B* 58:361
62. Barish, S. J. et al 1974. *Phys. Rev. Lett.* 33:1446
63. Cazzoli, E. G. et al 1975. *Phys. Rev. Lett.* 34:1125; Murtagh, M. J. 1976. *Proc. Int. Conf. Prod. Part. New Quantum Numbers*, ed. D. Cline, J. J. Kolonko, p. 370. Madison, Wis: Univ. Wis. Press
64. Berge, J. P. et al 1976. *Phys. Rev. Lett.* 36:127
65. Adler, S. L. 1966. *Phys. Rev.* 143:1144
66. DeRujula, A., Glashow, S. L. 1973. *Phys. Lett. B* 46:377
67. Deden, H. et al 1975. *Nucl. Phys. B* 85:269
68. Glashow, S. L., Iliopoulos, J., Maiani, L. 1975. *Phys. Rev. D* 2:1285: Gaillard, M. K., Lee, B. W., Rosner, J. L. 1975. *Rev. Mod. Phys.* 47:277; Altarelli, G., Cabbibo, N., Maiani, L., Pretronzio, R. 1974. *Phys. Rev. B* 48:435; DeRujula, A. et al 1974. *Rev. Mod. Phys.* 46:3919
69. HPWF, unpublished data
70. Benvenuti, A. et al 1975. *Phys. Rev. Lett.* 35:1206
71. Bingham, H. H., private communication for the Berkeley-Hawaii-Seattle group
72. Barnett, R. M., Georgi, H., Politzer, H. D.

1976. *Phys. Rev. Lett.* 37:3131 and references contained therein
73. Zee, A., Wilczek, F., Treiman, S. B. 1976. *Phys. Rev. D* 10:2881; Altarelli, G., Petronzio, R., Parisi, G. 1976. *Phys. Lett. B* 63:183
74. DeRujula, A., Georgi, H., Glashow, S. L., Quinn, H. R. 1974. *Rev. Mod. Phys.* 46:391; Barnett, R. M. 1976. *Phys. Rev. Lett.* 36:1163, *Phys. Rev. D* 14:70; Pakvasa, S., Pilachowski, L., Simmons, W. A., Tuan, S. F. 1976. *Nucl. Phys. B* 109:469; Barger, V., Weiler, T., Phillips, R. J. N. 1976. *Phys. Rev. D* 14:1276; Kaplan, J., Martin, F. 1976. *Nucl. Phys. B* 115:333; Albright, C. H., Schrock, R. E. 1976. Fermi Rep. 76/50-THY. Submitted to Int. Conf. High Energy Phys., 18th, Tbilisi, USSR

75. Barger, V., Gottshak, R., Phillips, R. J. N., private communication
76. Berge, J. P. et al 1976. *Phys. Rev. Lett.* 36:639
77. Fox, D. J. et al 1974. *Phys. Rev. Lett.* 33:1504; Watanabe, Y. et al 1975. *Phys. Rev. Lett.* 35:898; Anderson, H. L. et al 1976. *Phys. Rev. Lett.* 37:4
78. Riordan, E. M. et al 1975. SLAC-Pub-1634
79. Perkins, D., Schreiner, P., Scott, W. 1977. CERN prepr. EP/PHYG/77. Submitted to *Phys. Lett.*
80. Gross, D. J., Wilczek, F. 1973. *Phys. Rev. D* 8:3633; *Phys. Rev. D* 9:980; Politzer, H. D. 1974. *Phys. Rep. C* 14:129; Zee, A., Wilczek, F., Treiman, S. B. 1974. *Phys. Rev. D* 10:2881

Ann. Rev. Nucl. Sci. 1977. 27 : 279–332

K-SHELL IONIZATION IN HEAVY-ION COLLISIONS ✳5586

W. E. Meyerhof [1]
Department of Physics, Stanford University, Stanford, California 94305

Knud Taulbjerg
Institute of Physics, University of Aarhus, DK-8000 Aarhus C, Denmark

CONTENTS

[1] Work supported in part by the National Science Foundation.

1 INTRODUCTION

The study of electron excitation in heavy-ion collisions plays an important role in atomic as well as nuclear physics (1). Experimental investigations, especially of X-ray production by lighter projectiles, have expanded rapidly since Van de Graaff accelerators became commercially available in the 1950s. The theoretical understanding of the relevant inner-shell vacancy-production processes has grown correspondingly (2–4). For recent reviews see (5–8).

Impetus for collision studies with heavier ions was gained after the discovery by Armbruster (9, 10) and Specht (11) of regular fluctuations in K, L, and M X-ray production cross sections as a function of target atomic number, when a variety of target elements were bombarded with fission products. These variations showed enhanced X-ray production whenever the target and projectile energy levels matched. The effects are now understood (12–17) in terms of the transient formation of diatomic molecular orbitals (MO's) during the collision, as first suspected by Coates (18).

The formation of transient MO's was demonstrated directly in an experiment of Saris et al (19). An X-ray band was discovered that was assigned to the radiative filling of an inner-shell MO vacancy during the collision. Since then, many such quasimolecular X-ray bands have been found, due to vacancies in K, L, and M molecular orbitals.

This field has been stimulated by the prediction by Greiner and co-workers (20) and Popov and co-workers (21) that positrons as well as MO X rays might be emitted when a K MO vacancy is filled during a collision such as 1600-MeV U + U. The effect provides a new experimental test of quantum electrodynamics (22). This subject has been reviewed by Müller in the previous issue of this series (23). Obviously, an understanding of K-vacancy formation in heavy-ion collisions is crucial to the feasibility and interpretation of the positron experiment.

Finally we note that X-ray production in heavy-ion collisions can be used to identify compound nuclei in nuclear reactions (24) and in certain cases may be useful to determine lifetimes of compound nuclear states near 10^{-16} sec (25).

The present review is divided broadly into two parts, one dealing with vacancy formation in the atomic K shells of the collision partners leading to the emission of characteristic line radiation in the partners after they have separated, the other referring to studies of continuum radiations due to electronic transitions during ion-atom collisions. Background radiations are mentioned insofar as they are important in this context. In each case we present a theoretical basis, followed by experimental information and by interpretation in the light of the theory.

In this paper, collisions that may be treated by assuming the formation of transient molecular states are distinguished from those that can be described by assuming the projectile to be structureless and the target K electron to be essentially unperturbed by the projectile field. The two regions, which for brevity we call molecular and atomic, respectively, can be separated roughly by the following criteria. We denote the atomic number of the lower-Z and higher-Z collision partner by Z_L and Z_H, respectively, independent of which one is the projectile or the target. Then, if $Z_L \ll Z_H$, vacancy production in the higher-Z partner can be treated as the perturbative interaction of a charged point particle with the higher-Z electronic wave function (atomic regime). If $Z_L \simeq Z_H$ and if $v_1 \ll v_e$, where v_1 is the projectile velocity and v_e the Bohr velocity of the ejected electron, a molecular description becomes relevant. The velocity v_e is related to the binding energy U_e of the ejected electron by

$$v_e = (2U_e/m)^{1/2}, \qquad\qquad 1.1$$

where m is the rest mass of the electron. (For example, for 30-MeV Ne + Ne $v_1/v_K = 0.97$; for 30-MeV Br + Br $v_1/v_K = 0.12$.) The regime $Z_L \simeq Z_H$, $v_1 \gtrsim v_e$ is just starting to be explored (26).

It is worthwhile to contrast the study of atomic and nuclear collisions. In a nuclear collision, the reaction products have separated before γ emission occurs. Therefore, γ rays are sharp lines and cannot be used to explore the time development of a collision. In an atomic collision, due to the long-range nature of the Coulomb force, there is a small probability for X-ray emission during the collision. In principle, these X rays can therefore give information on the temporal evolution of the collision event.

2   SEPARATED-ATOM PHENOMENA

Typical collision times in heavy-ion collisions of interest here are of the order of a_K/v_1, where a_K is the K Bohr radius of the atom. For 30-MeV Br + Br, $a_K/v_1 \simeq 10^{-19}$ sec. On the other hand, typical decay times for K-shell vacancies are of the order of $10^{-10}/Z^4$ sec (27), which for Br is $\sim 10^{-16}$ sec. Since the outer electrons provide considerable shielding of the inner shells, one can separate the collision region from the post-collision region in a clear-cut way on the basis of time and strength of interaction (28). Sections 3–5 deal with the formation of K vacancies in the post-collision region and their decay.

3   MECHANISMS OF K-VACANCY FORMATION

The different regimes of K-vacancy formation in heavy-ion collisions described in the introduction lend themselves to different theoretical approaches. The interaction of an incident projectile with the tightly bound K electrons of the target atom is weak in the atomic regime, where $Z_L \ll Z_H$, so that a perturbation approach may be applied. Inner-shell electrons are, on the other hand, strongly perturbed in close collisions of heavy ions of comparable nuclear charge numbers. Particularly in the molecular regime, i.e. in slow collisions, the electrons may adjust almost adiabatically to the combined field of the collision partners. Thus, excitation cross sections become very small except for those transitions that may occur under near resonant conditions during the collision. At low collision velocities, such selective processes may be dealt with in terms of coupled-state calculations including only a few channels. As the projectile velocity increases, the adiabatic relaxation becomes less complete. The complexity of the calculations increases correspondingly, since more channels must be included. However, since the interaction time becomes short in very fast collisions, ultimately a perturbation approach may be applied. Thus, the bridging of the two theoretical approaches is very important, but remains an outstanding problem.

3.1   Perturbation Theories

The ionization of an inner-shell electron by an incident charged particle conventionally is referred to as Coulomb ionization. Several models for this process are in current use; the most distinctive feature is whether the nuclear motion is treated quantally or classically. The electronic system is generally described in the independent electron picture, i.e. all target electrons, except the ionized one, are considered to be spectators in unperturbed orbitals. In fact, except for a few calculations on particular collision systems (29, 30), hydrogenic wave functions have been employed in all extensive calculations (31, 32). Screening is accounted for in the hydrogenic model by choosing an effective nuclear charge $Z_K = Z_2 - 0.3$, where Z_2 is the atomic number of the target, and by adding a constant screening term V_{sc} to the effective potential seen by the ionized electron:

$$V(r) = -Z_K e^2/r + V_{sc}. \qquad 3.1$$

This expression takes into partial account the screening due to outer electrons, which mainly have the effect of lowering the K-shell ionization potential, as compared to the hydrogenic binding energy (2). Naturally, V_{sc} is chosen in order to reproduce the experimental binding energy:

$$V_{sc} = Z_K^2 e^2/2a_0 - U_K \qquad 3.2$$

$$\equiv (1 - \theta_K)Z_K^2 e^2/2a_0, \qquad 3.3$$

where a_0 is the hydrogenic Bohr radius and U_K is the binding energy of the target K electron. The screening parameter $\theta_K = U_K/(Z_K^2 e^2/2a_0)$ has been introduced for later convenience.

3.1.1 PLANE-WAVE BORN APPROXIMATION   In the plane-wave Born approximation (PWBA) the inelastic scattering amplitude for a transition from the initial state $\psi_0(r)$ to the final state $\psi_n(r)$ may be expressed as (8)

$$f_n(\Omega) = -(2Me^2/\hbar^2 q^2)Z_1\langle\psi_n|\exp(i\mathbf{q}\cdot\mathbf{r})|\psi_0\rangle, \qquad\qquad 3.4$$

where $\Omega(\theta, \phi)$ are the polar coordinates of the final relative nuclear momentum $\mathbf{K}_n$, M is the reduced nuclear mass, Z_1 is the projectile atomic number, and $\hbar\mathbf{q} = \mathbf{K}_0 - \mathbf{K}_n$ is the momentum transfer. It should be noted that although the scattering amplitude, Equation 3.4, formally depends on the projectile scattering angles, it does not furnish a realistic angular dependence of the inelastic cross section, since the scattering due to the nucleus-nucleus interaction is not incorporated in the first Born approximation. In fact, Equation 3.4 is independent of this interaction. To calculate the angular dependence in a quantal treatment, the theory must be evaluated in a higher Born approximation, or, more appropriately, in the Coulomb distorted-wave approximation (33). Hence only the integrated cross section is relevant in the PWBA.

Details of the evaluation of the total ionization cross section in the PWBA were discussed by Madison & Merzbacher (6a,b,c). It is readily shown that the cross section obeys the scaling law

$$\sigma_K = (Z_1^2/Z_K^4)F_K(\eta_K, \theta_K), \qquad\qquad 3.5$$

where the screening parameter θ_K is defined in Equation 3.3 and η_K is a dimensionless projectile velocity parameter:

$$\eta_K = (v_1/Z_K v_0)^2. \qquad\qquad 3.6$$

Here $v_0 = e^2/\hbar$ is the hydrogenic Bohr velocity. Extensive tables of F_K have been compiled by Khandelwahl et al (31). At low velocities, i.e. $\eta_K/\theta_K \gtrsim 1$, σ_K obeys approximately a one-parameter scaling law (2)

$$\sigma_K \simeq (Z_1^2/Z_K^4)\,\theta_K^{-1}G_K(\eta_K/\theta_K^2), \qquad\qquad 3.7$$

where G_K is a function given in (2).

In estimating the validity of the PWBA for K-shell ionization, it is important to realize that the first-order treatment of the projectile-electron interaction and the plane-wave treatment of the nuclear motion impose separate conditions that must be fulfilled simultaneously. In general, the electron-projectile interaction may be considered a small perturbation if

$$Z_1 \ll Z_K. \qquad\qquad 3.8$$

This condition may be relaxed only in the high-velocity limit. The restriction imposed by the plane-wave treatment of the nuclear motion may be based upon the Massey criterion (34) and becomes (8)

$$\eta_K^{3/2} \gg 1/(2\cdot 1836). \qquad\qquad 3.9$$

This condition is far less restrictive than the condition for the validity of the Born approximation for elastic scattering, which requires $\eta_K \gg 1$ (34).

3.1.2 BINARY ENCOUNTER APPROXIMATION   The binary encounter approximation (BEA) treats the encounter between the projectile and the target electron under consideration as a classical two-body collision. It has received considerable attention as an independent model for inner-shell ionization. Although formulated entirely on classical grounds, the BEA may also be thought of as a simplified version of the PWBA (35, 36). The exact form of the classical BEA ionization cross section can be obtained from the PWBA merely by replacing the final state of the ionized electron in the field of the target nucleus (i.e. a Coulomb wave function in a hydrogenic model) by that of a free electron of the same momentum (i.e. a plane wave). Although this replacement is of dubious validity, BEA and PWBA ionization cross sections differ by less than 15% if outer screening is incorporated properly into the BEA (37). Then the BEA and PWBA have identical scaling properties, as expressed by Equations 3.5 and 3.7. A different BEA scaling rule (5) has been used frequently, but is basically invalid, unless the effect of outer screening is negligible, in which case the scaling becomes identical to Equation 3.5.

3.1.3 IMPACT-PARAMETER METHOD: SEMICLASSICAL APPROXIMATION   Another model for calculating ionization cross sections is furnished by the impact-parameter method with a first-order perturbation (Born) treatment of the projectile-electron interaction (IPM-Born). In this approximation the transition amplitude to a final electron state n appears as a time integral of the interaction between the projectile and the target electron along a classical trajectory $\mathbf{R} = \mathbf{R}(\mathbf{b}, \mathbf{v}_1 ; t)$, where $\mathbf{b}$ is the classical impact parameter. The ionization probability $P_n(\mathbf{b})$ is obtained as the square of the transition amplitude, and the ionization cross section is obtained by integration over impact parameter

$$\sigma_n = \int \, \mathrm{d}\mathbf{b} P_n(\mathbf{b}).$$ 3.10

Explicit calculations within this formalism were performed by Hansteen and co-workers (3, 32, 38). In most cases straight-line trajectories $\mathbf{R} = \mathbf{b} + \mathbf{v}_1 t$ were used.

Then, the K-shell ionization probability obeys the scaling law (7)

$$P(b) = (Z_1^2/U_K) f_K(b/r_{\mathrm{ad}}, r_{\mathrm{ad}}/a_K),$$ 3.11

where f_K is a universal function; U_K and a_K are the binding energy and the orbital radius of a K-shell electron, respectively; and the so-called adiabatic distance $r_{\mathrm{ad}} = \hbar v_1/U_K$ is a convenient projectile velocity parameter since it is found (3) that $b P(b)$ peaks near $b \simeq r_{\mathrm{ad}}$. It may be noted that the total cross section obtained when Equation 3.11 is inserted into Equation 3.10 fulfills the scaling law of Equation 3.7.

The equivalence between IPM-Born and the PWBA for total cross sections is well established (39). Recently, a more general equivalence between the two methods has been obtained (40) by the application of a semiclassical approximation (SCA). This approximation is well known from potential scattering theory (41). It is based on a partial-wave expansion of the scattering wave functions and on the assumption that a large number of partial waves is required for convergence of the expansion in the interaction region. Then the l-values of the partial waves may be considered a quasicontinuous variable that is related to the classical impact parameter b by the relation

$$bMv_1 = [l(l+1)]^{1/2}\hbar \simeq l\hbar. \qquad 3.12$$

Several hundred partial waves are required at typical impact velocities in inner-shell studies. However, since the inelastic-collision problem generally does not possess azimuthal symmetry, in the development of the SCA for ionization one must use spherical harmonics Y_l^m rather than Legendre polynomials in order to expand the nuclear wave functions in the initial and final channels. Utilizing the PWBA expression, Equation 3.4, and the identification, Equation 3.12, one can then obtain the following expression for $P_n(b)$ (40)

$$P_n(b) = (Z_1^2 M^4 v_1^2 e^4 / \pi^2 \hbar^6) \cdot \sum_m \left| \int d\Omega \langle \psi_n | \exp(i\mathbf{q} \cdot \mathbf{r}) | \psi_0 \rangle J_m(q_\perp b) \exp(-im\phi)/q^2 \right|^2. \qquad 3.13$$

Here J_m is the Bessel function of order m and $\mathbf{q}_\perp$ is the component of $\mathbf{q}$ perpendicular to the incident direction. Since electrons are ejected in low angular-momentum states (42), only low values of $|m|$ are effective. Beloshitsky & Nikolaev (43) arrived at a result similar to Equation 3.13 with $m = 0$.

Comparing Equation 3.13 with the IPM-Born ionization probability [see e.g. (6a)] one can show (40) that, if the latter is averaged over the azimuthal angle of the impact parameter, it becomes identical to Equation 3.13. Thus, the PWBA and the IPM-Born are equivalent not only with respect to integrated ionization cross sections, but more generally, as long as the SCA is valid.

A formulation of the BEA in impact-parameter form has been attempted by Hansen (44) and McGuire (45). However, their formulation is defective. In order to express the target-electron velocity distribution, which is central in the BEA, as a function of the electron coordinate, a relation between spatial and momentum coordinates for the K-shell electron is required, in apparent conflict with the Heisenberg uncertainty principle. A less inhibited route to an IPM formulation of the BEA might be based on a substitution of plane waves for the ionized electron state in Equation 3.13, following the work of Bates & McDonough (35, 36).

3.1.4 EXTENSIONS OF PERTURBATION THEORIES   The simple theoretical models described here furnish a basic understanding of the Coulomb-ionization process. However, in several situations discrepancies have been found that illustrate the limitations of the models.

1. The nonrelativistic hydrogenic model becomes inadequate in the description of ionization of high-Z atoms, particularly at low impact velocities, where the high-momentum components of the initial electron wave function largely decide the transition amplitude. This implies that relativistic corrections should increase the ionization cross section, as confirmed by the relativistic calculations of Amundsen (46, 47). At GeV projectile energies, the current interaction with the electromagnetic field becomes important (48).

2. In low-velocity collisions, the inequality, shown in Equation 3.9 is violated.

Then, the internuclear repulsion becomes important (3). This effect reduces the cross section, particularly because of the slowing down of the projectile in the spatial region where the high-momentum components of the initial wave function are concentrated (49). Somewhat misleadingly, this effect is usually referred to as the Coulomb deflection effect.

3. Except at high velocities, a perturbation treatment of the ionizing interaction is invalid if the inequality shown in Equation 3.8 is violated. Although in most situations a perturbation treatment of the coupling to the continuum remains valid, the distortion of the initial wave function becomes significant. In the case of a K-shell electron, this effect reduces the cross section, since the relaxation of the electron wave function, to first order, is manifested by an increase in the binding of the electron.

Brandt, Basbas, and co-workers (50–52) have devised cross-section expressions that incorporate corrections to the Born approximation for Coulomb deflection (3) and the binding effect. In a more advanced treatment, the distortion of the target wave function must be incorporated. At low velocities, this is accomplished by accounting for the relaxation of the K-shell radius (53), whereas at higher energies the polarization of the wave function becomes important (54, 55). Anholt (56, 57) has attempted to incorporate binding, polarization, Coulomb deflection, and relativistic corrections in a semiempirical extension of the Basbas theory discussed in Section 4.3.1. It should be noted, though, that in some situations the relativistic correction, in particular, may compensate approximately for the correction terms due to binding and Coulomb deflection, so that it is not easy to separate these effects experimentally.

3.2 Molecular Theories

If the inequality, Equation 3.8, is no longer fulfilled, in addition to the just-mentioned distortion of the target-electron wave function by the projectile, the electron states on the projectile are accessible either as final states for electrons transferred from the target or as intermediate states in target-electron excitation processes. In fact, since the projectile may also become excited during the collision, there is no real distinction between the projectile ion and the target atom. This is the case particularly at low collision velocities, where the electronic system may relax more or less adiabatically during the collision in response to the combined field of the collision partners. This suggests that during the collision the electrons may be well represented as Born-Oppenheimer states (58) for the quasi-molecule composed of the two collision partners.

The near-adiabatic relaxation in slow collisions does not imply that inelastic processes are prohibited. The contrary is often brought about by the circumstance that molecular states, which in the (separated) atomic limit are spaced by a large energy gap, at finite nuclear separations may become near-degenerate, thus providing near-resonant conditions for excitation processes. A determination of the molecular energy levels of the combined system as a function of the internuclear distance, therefore, is crucial for the interpretation of excitation phenomena in slow ion-atom collisions and forms the basic starting point for explicit calculations.

3.2.1 MOLECULAR-ORBITAL ENERGY CALCULATIONS   As for Coulomb ionization, the independent-electron picture is also considered a valid approximation in the molecular description of inner-shell processes, because the nuclear Coulomb fields acting on an inner-shell electron are much stronger than the fields due to the other electrons. The basic electronic states thus may be expressed in terms of molecular orbitals (MO). The MO energies have been computed with various degrees of sophistication, the most elaborate ones being molecular Hartree-Fock calculations (59–63). More recently, Eichler & Wille (64, 65) have developed a method based on the Thomas-Fermi model, which is particularly well suited for heavier systems, for which Hartree-Fock calculations become expensive. For very heavy systems, relativistic effects become important. As a first step, Fricke et al (66) have included relativistic corrections in their nonrelativistic Hartree-Fock calculations. An analogous procedure has been adopted in the Eichler-Wille model (67). Recently, the first self-consistent, relativistic molecular calculation has been reported (68).

A drastically simplified approach is suggested by realizing that the effective field for tightly bound electrons of the quasimolecule closely resembles a two-center Coulomb field, i.e. the MO's for inner-shell electrons are qualitatively similar to those of a single electron in the field of two nuclei. This problem has been solved nonrelativistically (69–74) and relativistically (75, 76). The solutions exhibit a phenomenon called promotion, namely the increase in principal quantum number of certain MO's during the transition from the separated-atom (SA) system ($R = \infty$) to the united-atom (UA) ($R = 0$) system, where R is the internuclear distance. Rules for correlating the SA and UA quantum numbers have been given by Gershstein & Krivchenkov (77). Promoted orbitals can cross other MO's at finite values of R, allowing resonance conditions for electron transfer from one MO to another.

3.2.2 ELECTRON PROMOTION AND EXCITATION PROCESSES   Electron promotion is the basic ingredient in a model suggested by Fano & Lichten (12, 13, 65, 78a), which has been applied very successfully in a qualitative interpretation of various experimental findings in ion-atom collisions (1, 5, 15). The conceptual simplicity of the Fano-Lichten model is exemplified in Figure 1 (78b), which shows a selection of nonrelativistic one-electron MO energy curves for a nuclear charge ratio $Z_L/Z_H = 0.8$. (Z_L and Z_H are the atomic numbers of the lower-Z and higher-Z collision partners, respectively.) This type of diagram is often referred to as a correlation diagram. Some of the processes that may be of significance in producing SA K vacancies have been indicated. To produce K-shell vacancies in the higher-Z (H) or lower-Z (L) collision partners, electrons must be removed from the $1s\sigma$ or $2p\sigma$ MO, respectively, during the collision. A resonance condition for the latter process is seen to occur at small values of R (process A), where the $2p\sigma$ and $2p\pi$ MO may couple strongly by the *rotational* motion of the system. Thus, if $2p\pi$ vacancies are available, process A is expected to furnish an effective mechanism for $2p\sigma$ vacancy production, i.e. K-shell excitation in the lower-Z partner.

Figure 1 shows that the $1s\sigma$ MO remains isolated between $R = 0$ and $R = \infty$. Therefore, the $1s\sigma$ MO is expected to couple only weakly to the other MO, except for near-symmetric systems, where the relative *radial* motion of the system may

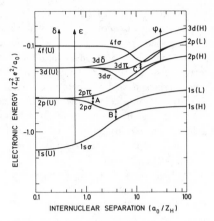

Figure 1   Molecular-orbital energy-level correlation diagram for an electron in the field of two nuclear charges Z_H and Z_L, for a charge ratio $Z_L/Z_H = 0.8$ (78b). Possible electronic transitions, discussed in the text, are shown. Note that $e^2/a_0 = 27.2$ eV, $a_0 = 5.3 \times 10^{-9}$ cm.

couple the $2p\sigma$ and $1s\sigma$ MO, as indicated by process B. Another process with the same characteristics has been labled as C on Figure 1. It corresponds to the radial coupling between the $3d\pi$ and $2p\pi$ MO and plays an important role in furnishing $2p\pi$ vacancies for the coupling process A in solid targets (see Section 4.2) (79–82). Since the location of the closest approach of the MO energy curves in processes B and C lies well outside the range of the near degeneracy of the $2p\pi$ and $2p\sigma$ MO, it is suggestive to consider the three processes A, B, and C as independent of each other (80). Then process C is operative early in the collision and governs the "initial" conditions for process A, whereas process B operates late in the collision and determines the sharing of the $2p\sigma$ vacancies between the two collision partners (83).

Some coupling processes involving large electronic energy defects have been indicated by Greek letters in Figure 1. Process ε (84, 85) can be related to the Coulomb ionization process, as discussed in Section 3.1.4 (57). The roles of processes δ (84) and ϕ (79, 86) are not yet clear (82).

Since calculations of the quasimolecular processes have been reviewed quite recently in detail by Briggs (17), we confine ourselves to a general discussion. The first calculations of $P(b)$ and σ (see Equation 3.10) for the $2p\sigma$-$2p\pi$ rotational excitation in one-electron collisions ($H^+ + H$) were made by Bates & Williams (87), later extended by others (88, 89). Using Hartree-Fock representations of Larkins (59), Briggs & Macek (90) solved the same problem for the many-electron collision of Ne + Ne. Due to the reflection symmetry of homonuclear collisions, certain simplifications occur, because transitions between gerade (even) and ungerade (odd) MO are prohibited. For example the $1s\sigma_g$ and $2s\sigma_g$ MO (the latter is almost degenerate with the $2p\pi$ MO shown on Figure 1), do not couple to the $2p\sigma_u$ and $2p\pi_u$. In principle, in a heteronuclear collision a more extended many-channel problem should be solved, but a detailed evaluation (63, 91, 92) of the $1s\sigma$-$2s\sigma$-$2p\sigma$-$2p\pi$ coupled-channel problem for Ne + O has shown that certain simplifications can be made. First, inclusion of the $2s\sigma$ channel affects the O and Ne K-excitation probabilities by at most a few percent. Second, the $2p\sigma$-$2p\pi$ coupling region is found to be separated from the $2p\sigma$-$1s\sigma$ coupling region, as had been surmised by Meyerhof (83) in a model calculation (see Section 3.2.4). Only at very high projectile velocities do the $2p\sigma$-$2p\pi$ and $2p\sigma$-$1s\sigma$ coupling regions overlap so much that a separation of the two processes becomes invalid (74).

The interaction with orbitals higher than the $2p\pi$ MO has usually been ignored in K-shell excitation calculations. Furthermore, the initial configuration is assumed to contain a single vacancy in the $2p\pi_x$ MO, where π_x refers to the π-component in the scattering plane, perpendicular to the internuclear vector. Therefore, the calculated values of $P(b)$ and of σ must be multiplied by the number N_π of such vacancies. In a static situation, N_π may be calculated from the simple overlap between atomic and molecular wave functions (93). Usually statistical arguments suffice; for example, for asymmetric systems $N_\pi = \frac{1}{3}$ for each vacancy in the $2p(H)$ state (see Figure 1), because this subshell correlates to the $2p\pi_x$, $2p\pi_y$, and $3d\sigma$ MO. Fastrup et al (86) suggested that the dynamic coupling of the $2p\pi_x$ MO to higher-lying MO (processes C or ϕ on Figure 1) may increase the effective number of $2p\pi_x$ vacancies and that the total number of $2p\pi_x$ vacancies should be given by

$$N_\pi(v_1) = N_\pi^{st} + N_\pi^{dy}(v_1),\qquad\qquad 3.14$$

where N_π^{st} is the static part, computed from the total number of initial $2p$ vacancies (93) and $N_\pi^{dy}(v_1)$ is the dynamic contribution, so far not computed ab initio.

Relativistic effects on the rotational-coupling cross section are twofold. First, the fine structure splitting ΔE_{2p} between the united atom (UA) $2p_{3/2}$ and $2p_{1/2}$ levels introduces an effective energy gap at $R = 0$ between the $2p\pi$ and the $2p\sigma$ MO, which correlate with the UA $2p_{3/2}$ and $2p_{1/2}$ levels, respectively. Second, the relativistic MO energies are quite different from the nonrelativistic energies (76), so that scaling relations based on nonrelativistic calculations (see Section 3.2.3) become invalid. The first effect is small as long as the fine-structure splitting is much smaller than the $2p\sigma$-$2p\pi$ energy difference at the effective range of the rotational-coupling process. The cross section depends upon the relative velocity v_1 and may be estimated by means of simple scaling relations developed by Taulbjerg et al (94). The rotational-coupling process (A) is found to be relatively unaffected by the fine-structure splitting ΔE_{2p} until the parameter (95)

$$\xi = (\Delta E_{2p}/I_0)[(Z_L-1)(Z_H-1)(Z_L+Z_H-2)^2(v_1/v_0)^2/5]^{-1/3}\qquad 3.15$$

exceeds unity. Here I_0 and v_0 are the ionization energy and Bohr velocity for H, respectively. For most collision systems with Z_L, $Z_H \gtrsim 30$, $\xi < 1$, but, e.g. for 15-MeV I + I, $\xi \simeq 2$ and σ is reduced by a factor of approximately 0.1 (95).

3.2.3   SCALING LAWS   Briggs & Macek (96) have discussed the scaling properties of the $2p\sigma$-$2p\pi$ rotational coupling for symmetric systems and introduced a common scaling constant Z_s, nearly equal to the common atomic number, to scale lengths and velocities. They showed that $P(bZ_s, v_1/Z_s)$ and $Z_s^2\sigma(v_1/Z_s)$ should be universal functions of the arguments. Similar scaling properties apply in general (74, 84) to excitation processes in one-electron systems as long as the nuclear charge ratio is kept fixed.

Utilizing the fact that the main rotational coupling between the $2p\sigma$ and $2p\pi$ MO's takes place at separations less than the K radius, where analytical expressions for the energy separation exist (77) and assuming that the nuclear trajectory is governed by an appropriately screened Coulomb potential, Taulbjerg et al (94) were able to show that in any kind of collision $P(b\lambda, v_1/\omega)$ and $\lambda^2\sigma(v_1/\omega)$ should be universal functions of the arguments, where the scaling parameters λ and ω are given by

$$\lambda^7 = (Z_L - 1)^2(Z_H - 1)^2[\tfrac{1}{2}(Z_L + Z_H - 2)]^4/(Z_L Z_H/A),$$ 3.16

$$\omega^7 = (Z_L - 1)(Z_H - 1)[\tfrac{1}{2}(Z_L + Z_H - 2)]^2(Z_L Z_H/A)^3.$$ 3.17

Here A is the reduced mass number of the nuclei. By comparison with available ab initio calculations and an exhaustive series of one-electron calculations, it was found that Equations 3.16 and 3.17 produce scaled values of $P(b, v_1)$ and $\sigma(v_1)$ accurate to within 10%, except for very low velocities (37). The resultant universal curves for $P(\lambda b, v_1/\omega)$ and $\lambda^2\sigma(v_1/\omega)$ have been tabulated (94). A selection of $P(b)$ curves is shown in Figure 2, as well as the σ curve. The narrow peak in $P(b)$ at small impact parameters at higher velocities is a trajectory effect that reflects the sudden change in orientation of the internuclear line, corresponding to center of mass scattering angles near 90°, and produces a perfect overlap of the $2p\pi_x$ and $2p\sigma$ wave functions (17, 90).

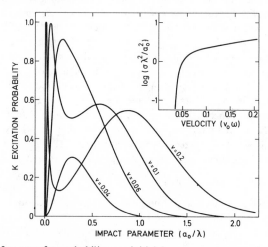

Figure 2 $2p\sigma$-$2p\pi$ transfer probability per initial $2p\pi_x$ vacancy versus impact parameter (94). The projectile velocity v is given in scaled atomic units $\omega v_0 = \omega e^2/\hbar$ ($e^2/\hbar = 2.19 \times 10^9$ cm sec^{-1}). The scaling parameters λ and ω are given by Equations 3.16 and 3.17. The inset shows the integrated cross section as a function of projectile velocity.

3.2.4 VACANCY SHARING   The separability of the $2p\sigma$-$2p\pi$ rotational-coupling region from the $2p\sigma$-$1s\sigma$ radial-coupling region, which is established through the ab initio calculations (63, 92), gives support to the concept of $2p\sigma$ vacancy sharing between the two collision partners (process B on Figure 1) (83). Then, a simple charge-transfer model of Demkov (97) may be applied. The simplicity of this model is due to the description of the appropriate MO as linear combinations of atomic orbitals, such that the coupling matrix element can be approximated by an exponential of the form $\exp(-\gamma R)$. The resulting coupled equations can be solved analytically at zero impact parameter and yield for the ratio of the K excitation probabilities (83)

$$r(v_1) \equiv P_H(b = 0, v_1)/P_L(b = 0, v_1) \qquad 3.18$$

$$= \exp\left[-\pi(U_H - U_L)/\hbar\gamma v_1\right], \qquad 3.19$$

where $U_H - U_L$ is the difference between the K-shell ionization energies. Briggs (98) has shown that the ratio, Equation 3.18, is approximately valid also for the integrated cross section ratio $\sigma_K(H)/\sigma_K(L)$, as originally assumed by Meyerhof (83). Replacing the range $1/\gamma$ of the exponential coupling by the reduced K-shell radius for the two collision partners (99), appropriately expressed in terms of ionization energies, the final expression for the sharing ratio can be written

$$r(v_1) = \exp(-2x), \qquad 3.20$$

$$x = \pi\left[U_H^{1/2} - U_L^{1/2}\right]/(2mv_1^2)^{1/2}. \qquad 3.21$$

For later purposes we also define a vacancy transfer probability

$$w = r/(1 + r). \qquad 3.22$$

It should be noted that the Demkov model appears to be a special case of a similar exponential model of Nikitin (100, 101). However, the parameters in Nikitin's model cannot be determined solely from the SA atomic energies, but depend on details of the interacting orbitals (102).

So far we have discussed only process B on Figure 1. Process C can also be considered a sharing of $2p$(H) or $2p$(L) incoming vacancies between the $2p\pi$ and $3d\pi$ MO's (80). From the point of view of the Demkov model, process C is just the inverse of process B and the same vacancy-sharing Equation 3.20 applies, except that in Equation 3.21, U_H and U_L refer to the respective SA $2p$ energies.

3.3   Charge Exchange

Since there is experimental evidence (26, 53, 103–108) that the transfer of electrons from the target atom to bound projectile states is an important process in inner-shell studies, a few comments on the theory of charge exchange are in order.

In the molecular approach, no distinction is made between target and projectile states. Therefore, charge-exchange phenomena are treated essentially on the same footing as direct-excitation processes between target electron states. In contrast to the molecular approach, the perturbation treatments discussed in Section 3.1 are based on a description in terms of target eigenfunctions. This is obviously inappropriate in accounting for electron-capture processes by the projectile. Rather, a consistent theory must be based on an appropriate expansion in terms of a complete set including the final projectile states of interest.

Brinkmann & Kramers (109) have used a perturbation treatment in the calculation of capture to a bare projectile nucleus from a one-electron target atom. Since the initial and final electron states are non-orthogonal, they are coupled not only by the electron-projectile interaction, but formally also by the internuclear potential. However, since the nuclear interaction does not occur in the exact expression for the electron-capture amplitude (110), the omission of this contribution in the Brinkmann-Kramers (BK) approximation can be justified. [An earlier, contrary assumption by Jackson & Schiff (111) is examined by Halpern & Law (112) and is shown to scale incorrectly with Z_1 and Z_2.]

Since the BK approximation is based on the first Born approximation, only the integrated cross section is relevant (see discussion in Section 3.1.1). To compute the differential cross section a distorted-wave approximation must be used or the cross section can be evaluated in semiclassical eikonal approximation (113).

To treat the target atom more realistically, Nikolaev (114) has introduced screening parameters into the BK theory in analogy with those introduced in the PWBA for ionization (Section 3.1.1). Nevertheless, the capture cross section is typically overestimated by an order of magnitude, even though it predicts correctly the ratios of capture to the various projectile shells (115a). Agreement with experiment is greatly improved by introducing a Coulomb deflection factor (115b).

The BK treatment has been questioned on theoretical grounds (110, 116, 117a). One finds, in particular, that even in the high-velocity limit, where a first-order treatment would be expected to be exact, the BK cross section does not converge to the charge-transfer cross section in a second Born approximation (116). It is clear that much more work is needed before this subject is fully understood.

Only little work has been done within the framework of the molecular model. The Demkov model (Section 3.2.4) has been used to estimate charge transfer in inner-shell excitation with some success (117b). A more complete calculation has also been attempted (118).

4   INTEGRAL CROSS-SECTION MEASUREMENTS

Vacancy production in inner atomic shells can be determined by measuring X rays or Auger electrons. For the K shell, the fluorescence yield ω_K is defined by

$$\omega_K \equiv \Gamma_R/\Gamma_K = N_x^K/N_K, \qquad\qquad 4.1$$

where Γ_R is the radiative width of the excited atomic state, Γ_K the total width, N_x^K the fractional number of K X rays emitted per atom, and N_K the fractional number of K vacancies per atom. The Auger width is given by $\Gamma_K - \Gamma_R$. Experimental X ray or Auger cross sections have to be divided by ω_K or $1 - \omega_K$, respectively, in order to convert them to vacancy cross sections that can be compared to theory (Section 3). For atoms with a single inner-shell vacancy, the tabulations of ω_K in (27) are useful.

The spectrometry and detection of X rays and of Auger electrons have been reviewed elsewhere in detail (119). Integral cross-section measurements are those in which one detects X rays or Auger electrons from target or projectile atoms, without detecting the inelastically scattered projectile in coincidence (1). In differential cross-section measurements, one determines the scattering angle and energy loss of the projectile (15) or the scattered projectile is detected in coincidence with X rays and Auger electrons (1, 120, 121). In the present section we discuss integral cross-section measurements; in Section 5, we discuss differential measurements.

4.1 Gas Targets

Differentially pumped gas targets have the advantage that a high degree of charge purity of the projectiles can be maintained throughout the entire target and con-

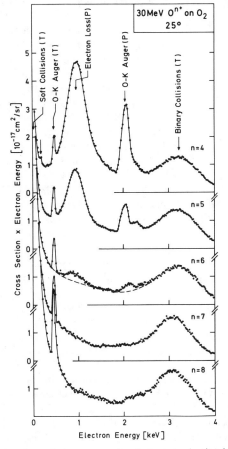

Figure 3   Cross section times electron energy for electron production in 30-MeV $O^{n+} + O_2$ collisions at an electron observation angle of 25° for different projectile charge states n (128).

ditions arranged so that only single-collision effects occur (79). Also, gas targets are needed for clean studies of Auger electron spectra (122–124). Unfortunately, the variety of monatomic elemental targets is limited, although in principle, vapors can be used (125, 126). In diatomic molecules, multiple-collision effects can occur, since the projectile can scatter from one atom in the molecule to the other one (127).

Indicative of the multitude of electronic processes that can occur during ion-atom collisions is Figure 3, which shows the cross section for electron production in 30-MeV $O^{n+} + O_2$ collisions (128). The continuum of ejected valence electrons from the target (T) dominates the overall spectrum. This continuum extends from the soft collision peak (T) to the binary collision peak (T), which corresponds to the maximum energy a nearly free electron can receive in the collision with a projectile. There is a similar continuum from the projectile [electron loss (P)], but it is kinematically shifted due to the projectile motion. The Auger electrons from the K shells of the target [O-K Auger (T)] and the projectile [O-K Auger (P)] are separated in energy for the same reason. For $n = 7$, 8, there are no L electrons left on the O projectile; therefore no Auger effect occurs. High-resolution studies ($\sim$1-eV resolution) of the Auger peaks reveal many lines due to the numerous possible Auger transitions and states of ionization (123, 126).

K X-ray cross-section measurements with gas targets have also been made, as summarized in (1, 79, 82). With low resolution, the spectra reveal the K_α and K_β lines with very little background, in contrast to the Auger-electron measurements. With high resolution, satellite peaks appear due to the energy shifts produced by various L and M vacancies (1). In low-resolution studies, these energy shifts give rise to a centroid energy shift of the K_α and K_β lines, which can be as large as a few hundred eV (129). The disadvantage of X-ray studies with low-Z targets is the small fluorescence yield [e.g. for Ne, $\omega_K = 0.018$ (27)], which is very sensitive to the state of ionization and the energy of the projectile (130–133).

Typical results for gas target cross sections are discussed in Section 4.3.

4.2 Solid Targets

The severe scattering of electrons in solids prevents accurate Auger electron measurements. K X-ray measurements with solid targets are very convenient, but must be interpreted with care (15, 82, 134).

When a projectile traverses a solid, it slows down and is scattered by elastic and inelastic collisions with target atoms. The projectile X-ray yield represents an integration over these collisions. For the X rays from target atoms, yield is also generated by recoiling target atoms traversing the target material (135, 136). In order to extract information on individual cross sections, the integration over collisions must be unfolded, not only with respect to the slowing down of the ion and the recoils, but also with respect to the various possible electronic states of the projectile and recoils prior to the violent, X-ray–producing collisions inside the solid. X-ray production cross sections may depend strongly on the charge states of the moving ion (projectile or recoiling target atom) for a variety of reasons that are discussed below.

As the projectile traverses a solid, its charge state is continuously changed in collisions with target atoms, resulting in a distribution of charge and excitation states (106, 137, 138). This distribution is often of great importance, since several excitation processes discussed in Section 3 depend strongly on the availability of vacancies in certain electron shells, prior to a violent collision. We call the resultant distribution of vacancies in the inner shells *steady-state*, although the word *residual* has also been used (105). As the thickness of a solid target is increased from very small values ($\leqq 1$ μg cm^{-2}), the steady-state vacancy distribution in the projectile tends to an equilibrium distribution, usually at a target thickness of ~ 50–500 μg cm^{-2} (107, 108, 139). Since for these target thicknesses the energy loss of the projectile may still be negligible, if one speaks of a "thin" target, one must clarify whether or not the steady-state projectile vacancies are equilibrated.

By "thick" target, one always means a target thick enough to stop the projectile. In this situation, the X-ray production in direct collisions may be adequately described by a simple cross section (135, 136) $\bar{\sigma}_x(E)$, which is an average over the equilibrium distribution of charge states at projectile energy E. Under certain circumstances, however, the production of X rays from target atoms may be entirely dominated (135, 136) by the X-ray generation due to energetic target recoils from primary collisions. In general, X-ray production may then be described by the cross section $\Sigma_x = \bar{\sigma}_x + \sigma_R$, where the recoil X-ray production cross section is given by

$$\sigma_R(E) = \int_0^{T_{max}} dT \kappa(E, T) Y_R(T). \qquad 4.2$$

Here $\kappa(E, T) dT$ is the differential cross section for energy transfer T to the target atom in a primary collision and $Y_R(T)$ is the X-ray yield that would be obtained if a target atom of energy T were incident on and stopped in the target material. The recoil effect is particulrly important in situations where promotion and other strong excitation mechanisms are inhibited, e.g. in collisions that are sufficiently asymmetric so that $2p\sigma$ vacancy-sharing (process B in Figure 1) does not contribute to the target X-ray yield, but sufficiently symmetric so that the target atom can receive an appreciable recoil energy. Typical examples are rather low-energy collisions of N and Ar on Al (136), O on Al (140), and Ar on Ni (82, 141). Approximate estimates of σ_R may be obtained from formulae given in (142). Preferentially, however, the recoil contribution, Equation 4.2, should be determined from a direct measurement of the yield function $Y_R(T)$. In special cases, the recoil effect is important even in differential measurements using thin, solid targets (143).

As mentioned above, even for the relatively simple processes of K-vacancy production (see Section 3), the steady-state vacancy distribution in a solid can affect the detected target and projectile yields in several different ways. First, the fluorescence yield can be influenced by outer-shell stripping or excitation, as the projectile traverses a solid (144). Auger widths, as well as radiative widths (Equation 4.1), of the different K satellites, each roughly corresponding to a different multiple L-vacancy configuration (1), can be altered (145, 146). Even the chemical environment of the target atom may influence the relative satellite intensities (147–149). Despite these complications, it is typical for $Z \gtrsim 18$ ($\omega_K \geqq 0.1$) to use single-vacancy fluorescence yields even in solid targets. To put this practice in perspective, we note that for Cl, the single-vacancy fluorescence yield is 0.096 (27), whereas for 45-MeV Cl traversing a solid Si target, the fluorescence yield is found to be 0.074 and, for 45-MeV Cl^{+12} traversing a SiH_4 gas target, the value is 0.144 (145). For higher values of Z, the differences are expected to become much smaller.

Even at relatively low projectile velocities, $(v_1 < v_K$; see Equation 1.1), the projectile can receive L vacancies in traversing a solid, that live long enough to be available in a subsequent collision, where $2p\sigma$-$2p\pi$ electron promotion may transfer the vacancies to the SA $1s$ levels (19, 150–152). Lennard & Mitchell (80) have found that the cross section for producing $2p\sigma$ vacancies by this multiple-collision process is

$$\sigma_{mc}(2p\sigma) = n(v_1)w(2p)\sigma_{rot}/3, \qquad 4.3$$

where $n(v_1)$ is the number of steady-state $2p$ vacancies per projectile, $w(2p)$ is a vacancy-sharing factor between the $2p(H)$ and $2p(L)$ states (process C, Figure 1) and σ_{rot} is the cross section for $2p\sigma$-$2p\pi$ electron promotion per $2p_x$ projectile vacancy, discussed in Section 3.2.2. The factor $1/3$ is a statistical factor relating the number of $2p$ vacancies to the number of $2p_x$ vacancies (93). The vacancy-sharing factor $w(2p)$ is very similar to Equation 3.22 (see end of Section 3.2.4). Equation 4.3 assumes that the projectile is the lower-Z collision partner; if it is the higher-Z partner, $w(2p)$ must be replaced by $1 - w(2p)$. For collisions close to symmetry, using solid targets, the process just described gives the major contribution to the observed projectile and target K X-ray yields (Section 4.3.2).

In the case of high-velocity projectiles $(v_1 \gtrsim v_K)$, there can be an additional effect in solids, because an appreciable number of steady-state projectile K vacancies can be created, in addition to L vacancies. In that case, capture of target electrons into the projectile K shell (Section 3.3) can influence K X-ray production in both collision partners (103, 106). Betz (103) has drawn attention to the fact that with very thin targets, in which the steady-state projectile K vacancies have not yet equilibrated, the dependence of the X-ray yield on target thickness allows an experimental determination of the projectile K-vacancy lifetime in the solid. Numerous studies of this type are under way at present, which give information not only on projectile lifetimes, but also on target-electron capture and other processes of projectile K-vacancy quenching (105, 107, 108, 117b, 146, 153).

4.3   Results and Interpretation

Figure 4 gives typical cross-section data for solid targets (134). Very similar trends are observed with gas targets. The projectile and target K-vacancy cross sections for a given projectile and bombarding energy are plotted against the target atomic number Z_2. Common trends can be recognized on these plots and are denoted by roman numerals (i) to (iv). They are discussed below.

Since, for most collision pairs the atomic numbers have comparable magnitudes and since $v_1/v_K < 1$ (see Equation 1.1), or, in other words, since the inequality shown in Equation 3.8 is not fulfilled, it seems appropriate to use the molecular model (Section 3.2), rather than atomic models (Section 3.1) for the interpretation of the cross-section trends. Referring to Figure 1, in this model the basic processes of K-shell excitation are (i) excitation of $1s\sigma$ electrons (process ε), (ii) excitation of $2p\sigma$ electrons (processes A, $C-A$, $\phi-A$ or δ), (iii) sharing of $2p\sigma$ vacancies (process B), and (iv) production of additional K vacancies in the lower-Z partner by $K-L$ level-matching effects, if the $1s(L)$ level approaches or passes the $2p(H)$ level. In terms of these processes, the experimental K-vacancy cross sections of the higher-Z and lower-Z partners, either of which could be the target or the projectile, are (57, 82, 134):

$$\sigma_K(H) = \sigma(1s\sigma) + w\sigma(2p\sigma), \qquad\qquad 4.4$$

$$\sigma_K(L) = (1 - w)\sigma(2p\sigma) + \sigma_{K-L}. \qquad\qquad 4.5$$

Here $\sigma(1s\sigma)$ is the cross section for $1s\sigma$ excitation, $\sigma(2p\sigma)$ is the cross section for $2p\sigma$ excitation, and σ_{K-L} is the cross section contribution due to $K-L$ level-

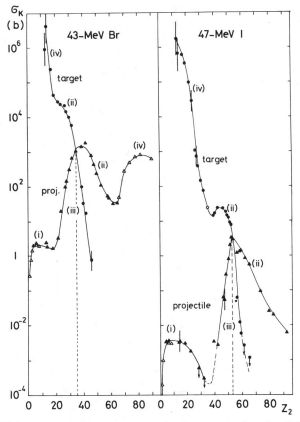

Figure 4   Projectile and target K-vacancy production cross sections for 43-MeV Br and 47-MeV I collisions as a function of target atomic number (134). Common features of the cross sections, discussed in the text, are denoted by Roman numerals (i) to (iv). Dashed vertical lines indicate symmetric collisions. Typical errors are $\pm 30\%$, except where shown.

matching effects. The vacancy-transfer probability w is given by Equation 3.22. Fortunately, each of the cross sections can be fairly well separated experimentally, because each of the terms in Equations 4.4 and 4.5 is dominant in a certain region of Z_2, for a given Z_1 and v_1.

4.3.1 REGIONS (i): $1s\sigma$ EXCITATION   The regions marked (i) in Figure 4 represent the K-vacancy cross sections of the higher-Z partner for collisions in which $Z_L \ll Z_H$. On each cross section plot there are two such possible regions, one, where $Z_L = Z_2$, and the other, where $Z_L = Z_1$. Since these regions fulfill the inequality 3.8, one would expect the cross sections to follow closely the prediction of perturbation theories (Section 3.1). Indeed, for $Z_L \gtrsim 7$, these theories give a good account of the v_1 dependence of the cross section for a given Z_1, Z_2 pair, as well as of the Z_2 dependence for a given Z_1, v_1 (4, 154, 155). Coulomb deflection of the projectile (3, 156a), binding energy effect of the projectile (50), polarization of the target-atomic wave function by the projectile (54, 55) as well as relativistic effects (46, 47) must be taken into account (Section 3.1.4). As an illustration, Figure 5 shows the experimental K X-ray cross section for Ni X rays produced in $C+Ni$ collisions, as a function of bombarding energy (156b). The curves give the theoretical vacancy cross sections, multiplied by the Ni K-fluorescence yield, for the PWBA (Section 3.1.1), as well as the effects of Coulomb deflection and binding energy perturbation (PWBABC) and of polarization (PWBABCP) on the PWBA.

It is also possible to incorporate the Coulomb deflection and binding energy

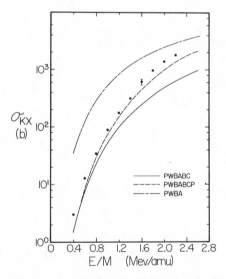

Figure 5 K X-ray production cross sections for ^{12}C ions on a thin Ni target (adapted from 156b). The PWBA, including corrections for binding and Coulomb deflection (PWBABC), as well as for polarization (PWBABCP) is shown. BEA and SCA cross sections are similar to the PWBA prediction.

perturbations into a universal form with which data from various projectiles and targets can be correlated (4, 51, 154). For $Z_L > 10$, increasingly greater deviations from this theory are found, especially as Z_H increases, which are partly due to an overestimate of the binding effect and partly due to a neglect of relativistic effects. Various empirical and semiempirical proposals have been made to remedy these defects (46, 47, 157, 158), the most successful of which seems to be that of Anholt (57). Anholt writes for the $1s\sigma$ cross section

$$\sigma(1s\sigma) = Z_L^2 \sigma_p [U_K(Z_H+1)/U_K(Z_H+Z_L)]^n CR, \qquad 4.6$$

where σ_p is the proton cross section for K-vacancy production in the atom Z_H at the projectile velocity v_1, $U_K(Z)$ is the K binding energy of atom Z, n is an empirically determined function that depends only on η_K (Equation 3.6), C is a Coulomb deflection factor similar to that used in (51), and R is a modified relativistic factor from (46, 47).

A recent development by Andersen et al (49, 159) may be somewhat more satisfactory, since it is based entirely upon theoretical grounds, the basic starting point being the scaling relation, Equation 3.11, from the impact-parameter formulation of Coulomb ionization. In addition to corrections for binding effects and Coulomb deflection, Andersen et al have incorporated appropriate corrections for relativistic wave functions and for the distortion of the initial K-shell electron wave function (53). Figure 6 illustrates the success of Equation 4.6 and of the method of Andersen et al in fitting the cross sections in regions (i).

Ab initio calculations for $1s\sigma$ excitation have so far been made with severe approximations for $H^+ + H$ (160, but see 161) and $U + U$ (85) collisions, but other calculations are under way (162).

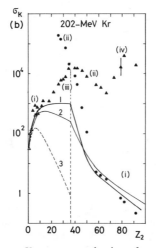

Figure 6   Projectile and target K-vacancy production of cross sections for 202-MeV Kr (compare Figure 4) as a function of target atomic number. The solid curves represent two proposals for computing the $1s\sigma$ cross section [regions (i)]: 1 is Equation 4.6 (57), 2 is the method of Andersen et al (49); and 3 is the theory of Basbas et al (51).

4.3.2 REGIONS (ii): $2p\sigma$ EXCITATION The regions marked (ii) in Figures 4 and 6 represent the K-vacancy cross sections for the lower-Z partner for not-too-asymmetric collisions. According to Figure 1 and Equation 4.5, in this region of Z_2, where K-L level-matching effects are small, the lower-Z partner receives vacancies from the $2p\sigma$ MO. This offers an experimental opportunity to study $\sigma(2p\sigma)$. More accurately, if one adds Equations 4.4 and 4.5 one finds in this region of Z_2

$$\sigma(2p\sigma) \simeq \sigma_K(\text{L}) + \sigma_K(\text{H}) \equiv \sigma_K^{\text{sum}}. \qquad 4.7$$

By comparing regions (i) in Figure 4 with regions (ii), one sees that $\sigma(1s\sigma) \ll \sigma(2p\sigma)$ (see also Figure 6), so that $\sigma(1s\sigma)$ could be omitted in Equation 4.7.

If $Z_H \leq 10$ and if gas targets are used so that only single-collision effects occur, one finds excellent agreement of the experimental summed cross section with the $2p\sigma$-$2p\pi$ electron-promotion calculations discussed in Section 3.2.2. Figure 7 gives a typical example for $\text{Ne}^+ + \text{Ne}^0$ collisions as a function of bombarding energy (132). The cross section follows the form

$$\sigma_K^{\text{sum}} = [N_\pi^{\text{st}} + N_\pi^{\text{dy}}(v_1)]\sigma_{\text{rot}}, \qquad 4.8$$

where the factor in brackets is taken from Equation 3.14 and σ_{rot} is the $2p\sigma$-$2p\pi$ electron-promotion cross section, which may be obtained from (94). As mentioned

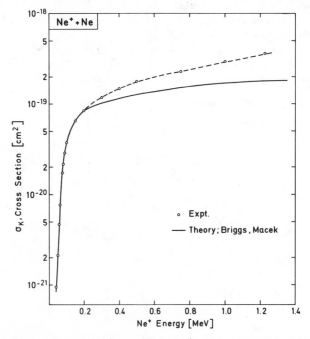

Figure 7 Total K-vacancy production section for $\text{Ne}^+ + \text{Ne}$ collisions (132). Solid line gives $2p\sigma$-$2p\pi$ rotational cross section (90); see inset on Figure 1. Deviation at higher energies is represented by Equation 4.8 (*dashed line*).

in Section 3.2.2, so far no ab initio calculations exist for $N_\pi^{dy}(v_1)$. Empirically one finds (79, 82, 132)

$$N_\pi^{dy} \simeq C \exp\left(-\alpha/v_1\right),\qquad\qquad 4.9$$

where $C \simeq 1$ and α varies with the collision pair.

For $Z_H > 10$, only very sparse gas-target data exist and no definite process has as yet been determined to explain the cross sections (79, 82). If solid targets are used, the multiple-collision effect discussed in connection with Equation 4.3 overwhelms the single-collision cross section. Figure 8 compares the experimental summed cross section (Equation 4.7) with Equation 4.3. The number $n(v_1)$ of $2p$ vacancies per projectile could be determined experimentally from the L X-ray yield and the appropriate thick-target formulation of Equation 4.3 was used (82). From Figure 8 it appears that near symmetry, multiple-collision effects dominate the cross section in solid targets and that the mechanism discussed in Section 4.2 is correct. The deviations in Figure 8 far from symmetry are due to K-L level-matching effects (Section 4.4.4).

4.4.3 REGIONS (iii): K-VACANCY SHARING According to the molecular model (Figure 1), on the outgoing part of the collision $2p\sigma$ vacancies are shared between the two collision partners. A model calculation gives Equation 3.20 for the sharing ratio $r(v_1)$ between the higher-Z and lower-Z partners. From Equations 3.22, 4.4, and 4.5, one obtains

$$r(v_1) = \left[\sigma_K(H) - \sigma(1s\sigma)\right]/\left[\sigma_K(L) - \sigma_{K-L}\right],\qquad\qquad 4.10$$

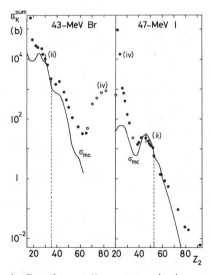

Figure 8 Summed projectile and target K-vacancy production cross sections for 43-MeV Br and 47-MeV I as a function of atomic number (compare Figure 4). Solid curves give computed multiple-collision contribution to the cross section (82).

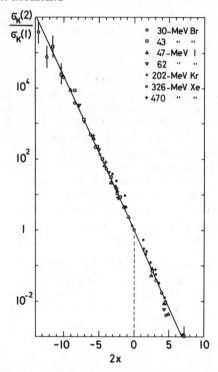

Figure 9 Ratio of K-vacancy production cross section in target to that in projectile versus parameter $2x$ given by Equation 3.21 (82). Predicted ratio 3.20 is shown by solid line. Vertical dashed line indicates symmetric collision.

which is equal to $\sigma_K(H)/\sigma_K(L)$ only in Z_1, Z_2 regions where $\sigma(1s\sigma) \ll \sigma_K(H)$ and $\sigma_{K-L} \ll \sigma_K(L)$. Figure 9 shows a variety of data in these restricted regions of Z_1, Z_2 in comparison with Equations 3.20 and 3.21. Agreement extends over nearly nine decades and is one of the most convincing pieces of evidence that the molecular model of Figure 1 is applicable to K-vacancy production in these collisions.

4.4.4 REGIONS (iv): K-L LEVEL MATCHING   In the regions marked (iv) on Figures 4 and 6 the $1s$ level of the lower-Z partner approaches or crosses the $2s, p$ levels of the higher-Z partner. A rapid rise in the $1s(L)$ vacancy cross section results (10, 11, 163), which has been ascribed to a sharing of $3d\sigma$ vacancies on the outgoing part of the collision between the $2p(H)$ and $1s(L)$ levels (13, 14, 158). It has been found empirically (164a) that the vacancy-sharing ratio can be represented by a generalization of Equation 3.20, first derived by Nikitin (100, 101). Although the Z_2 dependence of the steeply rising portions of $\sigma_K(L)$ in Figures 4 and 6 can be quantitatively reproduced by the generalized formula, the formula contains an adjustable parameter that so far has not been derived from basic considerations (164b).

In summary, the understanding of K-vacancy production in heavy-ion collisions in terms of the molecular model has advanced quite far. Nevertheless, only two of the many processes shown in Figure 1, namely electron promotion (A) and K-vacancy sharing (B) have been computed ab initio (Sections 3.2.2 and 3.2.4). It is imperative to extend such calculations to the remaining processes.

4.4   Double K Vacancies

The formation of double K vacancies in heavy-ion collisions is discussed by Richard (1). Such vacancies are normally filled by transitions of two uncorrelated electrons, the first of which experiences a slight energy shift due to decreased screening. However, in examining X-ray spectra from collisions with Fe and Ni projectiles and targets at ~ 40 MeV incident energies, Woelfli et al (165) found X-ray lines corresponding to correlated two-electron transitions to the $(1s)^{-2}$ state. The two-electron lines occur at an energy somewhat in excess of the summed energy of the single-electron $K\alpha$ lines. To satisfy parity change for an electric dipole transition, the jumping electrons must originate in different subshells, e.g. $2s$ and $2p$ (165–169). Correlated two-electron transitions had been predicted many years ago (170–172), but, because of their small intensity, had not been discovered previously. The intensity ratio of correlated two-electron to two uncorrelated one-electron transitions filling double K vacancies is of the order of 10^{-4} (173–175), in agreement with calculations (167, 176–180c).

5   DIFFERENTIAL CROSS-SECTION MEASUREMENTS

Studies of differential cross sections expose the finer details of theory to much more stringent tests than integral cross sections. Unfortunately, differential measurements are quite time-consuming. Two techniques, or a combination, are presently in use, namely the inelastic energy-loss (Q-value) method and the coincidence method, which we describe below. With either method the differential dependence of the cross section on (projectile or target) scattering angle can be obtained. The Q-value method even yields a doubly differential dependence on scattering angle and inelastic scattering energy.

5.1   Inelastic Energy-Loss Method

This method is usually based on electrostatic or magnetic analysis of the energy spectrum of scattered projectiles or recoiling target atoms [for a review see (15)]. A time-of-flight technique has also been employed, but so far only for outer-shell phenomena in very slow collisions (181).

Systematic investigations of K-shell excitation by Q-value measurements have been undertaken by Fastrup and co-workers (182–186). An example of an inelastic energy-loss spectrum is shown in Figure 10. The triple-peak structure is characteristic of these measurements. Each peak corresponds to a particular inner-shell process, broadened by an unresolved structure due to excitation and ionization of outer shells. The peak $\bar{Q}_\mathrm{I}$ may be identified as the "elastic" peak, i.e. no inner-shell excitation; $\bar{Q}_\mathrm{II}$ corresponds to the excitation of one K electron; and $\bar{Q}_\mathrm{III}$

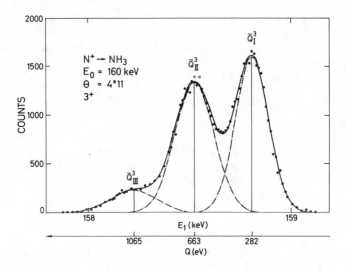

Figure 10 Energy spectrum of scattered 3+ nitrogen ions from 160-KeV $N^+ + NH_3$ collisions at an observation angle of 4.11° (186). Dashed lines show separated peaks: I—elastic, II—single K excitation, III—double K excitation. In this figure (and in Figure 11), the incident energy is denoted by E_0 and the scattered energy by E_1.

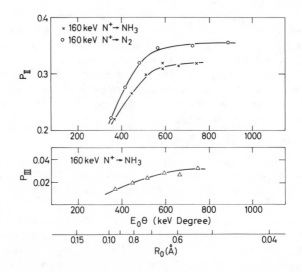

Figure 11 Probability of single (P_{II}) and double (P_{III}) K-vacancy production for the mean charge state in 160-KeV $N^+ + NH_3$ and $N^+ + N_2$ collisions (186). The data are plotted versus the reduced scattering angle $E_0\theta$ and distance of closest approach R_0 in the collision.

to double K-shell excitation (Section 4.4). Since each scattered charge state is resolved magnetically, the probabilities P_j^m corresponding to charge state m of the scattered ions for process j ($=$ I, II, or III) can be obtained from the energy-loss spectrum. Results for P_{II} and P_{III} averaged over the charge-state distribution in $N^+ + N$ collisions are shown in Figure 11 (186). Some dependence on the chemical composition of the target gas can be noticed. This is expected from the Fano-Lichten model (Section 3.2.2), since here the valence-shell population decides the effective number of $2p$ vacancies N_π^{st} entering into the $2p\sigma$-$2p\pi$ promotion process (Equations 3.14 and 4.8).

The inelastic energy-loss method has also been applied to very asymmetric collisions, but so far only with proton projectiles (187, 188). Due to straggling and multiple-collision processes in the target foil, it is not easy to resolve the inelastic part of the projectile energy spectrum caused by K-shell ionization from the elastic part. This difficulty has been overcome by recording the inelastic part in coincidence with target K X rays (189).

The inelastic spectrum is essentially a convolution of the elastic spectrum with the doubly differential cross section $d^2\sigma_K/dE_f \, d\theta$, where E_f is the energy of the ejected electron and θ is the projectile scattering angle. In principle, an unfolding of the inelastic spectrum will yield the doubly differential cross section, but in practice it is necessary to assume a parameterized form for the cross section such as (187)

$$d^2\sigma_K/dE_f \, d\theta \propto (1 + E_f/U_K)^{-s}, \qquad\qquad 5.1$$

where U_K is the target K-binding energy and s is empirically determined as a function of θ or impact parameter. For 2-MeV $p + $ Ag collisions, reasonable agreement with the SCA calculations of Bang & Hansteen (3) is found (187).

5.2   Coincidence Experiments

The detection of the emitted K X rays or Auger electrons in coincidence with the scattered projectile has been employed by several groups [for reviews see (120, 121)]. The technique is difficult, because the fractional number of K X rays produced per collision is very small and the random coincidence background can be appreciable. One can reduce the rate of random coincidences by discriminating against elastically scattered projectiles through magnetic energy analysis (189). Nevertheless, the experimental conditions must be carefully controlled, for example with respect to possible time structure in the beam current (159).

5.3   Results and Interpretation

For K excitation in the higher-Z partner perturbation theories (Section 3.1) with appropriate modifications for Coulomb deflection, binding energy, and relativistic corrections (Section 3.1.4) have been used. In the molecular model of Figure 1, this excitation would be called $1s\sigma$ excitation. For K excitation of either partner in symmetric or near-symmetric slow collisions, discussion in terms of the $2p\sigma$-$2p\pi$ electron-promotion process (Section 3.2.2) with K-vacancy sharing (Section 3.2.4) is appropriate. If energetic, highly stripped projectiles are used, charge exchange (Section 3.3) must be taken into account.

To facilitate a comparison between theory and experiment, the measured angular dependence of the cross section usually is converted into an impact-parameter dependence. According to Bohr (190), such a transformation requires the condition

$$2Z_1Z_2e^2/hv_1 \gg 1. \qquad 5.2$$

In inner-shell studies this condition is usually fulfilled. To determine the relationship between scattering angle θ and impact parameter b, one typically uses a procedure such as that of Everhart et al (191) based on a screened Coulomb field.

5.3.1   VERY ASYMMETRIC COLLISIONS   The first results of K-shell ionization studies in $H^+ + Ag$ collisions by the coincidence technique (189) compared quite favorably on an absolute scale with the SCA calculations of Hansteen & Mosebekk (38) (Section 3.1.3). Improved SCA calculations using relativistic hydrogenic wave functions give excellent agreement with experiment (46, 192). The significance of relativistic wave functions in collisions with very heavy elements is illustrated in Figure 12. The ionization probability $bP(b)$ (see Equation 3.10) is compared with nonrelativistic and relativistic calculations by Amundsen & Kocbach (193). The latter come within 10–20% of experiment, whereas the former differ by a factor of two or more.

In most experiments, the detected scattering angles are small enough so that a straight-line trajectory calculation is expected to be valid. However, as suggested by Ciochetti & Molinari (194), a determination of the large-scattering-angle behavior of the K-ionization cross section may become of interest. It may be used to study very short lifetimes of compound nuclei, by exploiting the fact that the retardation

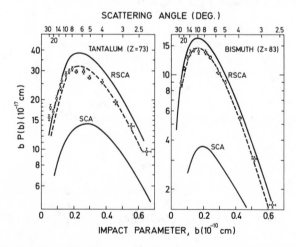

Figure 12   Probability for K-shell ionization (times impact parameter) for 4-MeV $H^+ + Ta$ and $H^+ + Bi$ collisions (adapted from 188). Solid curves show results of semiclassical approximation without (SCA) and with (RSCA) relativistic correction. Dashed curve shows RSCA calculation when normalized to the experimental integrated cross section.

of the projectile due to formation of a compound nucleus is signaled in the angular dependence of the interference pattern for the inner-shell ionization amplitude along the in- and outgoing parts of the trajectory. The predicted large-angle behavior of the ionization amplitude in H^+ + Cu scattering has been confirmed experimentally (159).

As the projectile atomic number is increased, the K ionization probability should follow the scaling law, Equation 3.11. Experimentally, very significant deviations from this simple scaling law have been found (53). For example, for 32-MeV O^{+8} + Cu, the data are found to fall by factors of $\frac{1}{3}$–$\frac{1}{2}$ below the curve obtained by scaling corresponding H^+ + Cu results. By modifying the scaling law (Equation 3.11) as mentioned in Section 3.1.4 Andersen et al (53) have been able to improve the SCA theory considerably and can account for the observed dependence on the atomic number and charge of the projectile.

5.3.2 SYMMETRIC AND NEAR-SYMMETRIC COLLISIONS   The coincidence and energy-loss methods have been employed successfully in slow collisions where the molecular theory (Section 3.2.2) is expected to apply. An interesting example of the former is given in Figure 13, which shows the oxygen K excitation probability for Ne^+ + O_2 collisions obtained from Auger-electron and from X-ray data (121, 195). Because of the small fluorescence yield for oxygen, the K Auger probability represents essentially the total K-vacancy formation probability and is in good agreement with the theoretical curve (91, 94) here and in other cases (133). More stringent selection rules cause the K X-ray emission probability to be more sensitive to outer-shell (i.e. L-shell) excitation processes, than the Auger emission probability (196). Therefore, the ratio of the two emission probabilities may be impact-parameter dependent, since it may be influenced by the impact-parameter dependence of outer-shell processes (197).

The bombarding energy dependence of $P(b)$ has been examined experimentally by Fastrup et al (186). Typical results are shown (37) in Figure 14. The solid curves on the figure represent the calculated probability

$$P(b) = N_\pi(v_1)P_{rot}(b),$$ 5.3

using the scaling theory of Taulbjerg et al (94) for the $2p\sigma$-$2p\pi$ rotational-coupling probability per $2p\pi_x$ vacancy $P_{rot}(b)$ and considering the number of $2p\pi_x$ vacancies $N_\pi(v_1)$ as a fitting parameter. Two important facts are found. First, since the data are fitted accurately by a function $N_\pi(v_1)$, which does not depend on impact parameter, it is implied that the dynamic part of the vacancy coupling into the $2p\pi$ MO (Equation 3.14) lies well outside the range of the $2p\sigma$-$2p\pi$ coupling. Second, for low projectile velocities, $N_\pi(v_1)$ is found to approach the static factor N_π^{st} predicted by Macek & Briggs (93), as expected (see Equation 4.9). It is also pleasing that $N_\pi(v_1)$ agrees with the value extracted from integral cross-section measurements (132, 198) by Equation 4.8 (37).

5.3.3 HIGHLY STRIPPED PROJECTILES   Cocke and co-workers (118, 199, 200) have used the coincidence method in an intermediate range of projectile velocities,

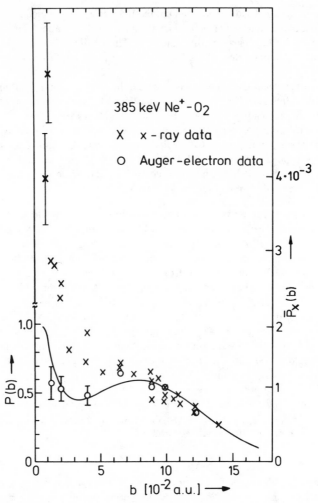

Figure 13   Probability for oxygen K X-ray and Auger-electron production for 385-keV $Ne^+ + O_2$ collisions (121, 195). Impact parameter is given in a.u. (1 a.u. = 5.3 × 10^{-9} cm). Curve is prediction for $2p\sigma$-$2p\pi$ rotational coupling (91, 94); see also Figure 2, $v = 0.1$.

where simple theoretical descriptions are expected to be invalid: the projectile velocity is of the order of relevant electron orbital velocities (Equation 1.1) and the projectile nuclear charge is not low enough to be considered a small perturbation on inner-shell electrons. Studies of the projectile charge-state dependence of $P(b)$ demonstrate that direct excitation and charge-transfer processes (Section 3.3) are highly competitive in such situations (118). This has also been demonstrated with

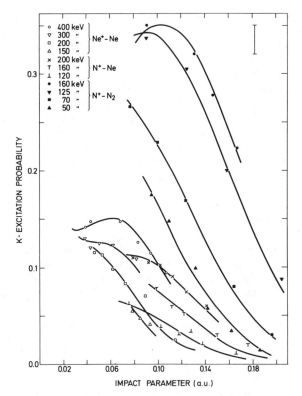

Figure 14 *K*-excitation probabilities from inelastic energy-loss measurements compared with Equation 5.3 (37, 94). Curves are least-squares fits to the data assuming an impact-parameter–independent value for $N_\pi(v_1)$ in each case. The experimental points are uncertain by ± 0.015.

total cross-section measurements (26, 104, 105, 115a, 117b, 201). Unfortunately, the available theory for charge transfer is not expected to be valid in these situations, and thus its role may not be fully ascertained at present (Section 3.3).

6 QUASIMOLECULAR PHENOMENA

In heavy-ion collisions, one finds not only emission of sharp, characteristic X-ray lines, as discussed in the preceding sections, but also of a variety of electronic and electromagnetic continua, sometimes called noncharacteristic radiations, even though some of them are quite characteristic of the collision system (19, 202). Often, these continua form unpleasant backgrounds in studies of inner-shell processes. For example, in Figure 3 electronic continua can be seen that inhibit the accurate study of Auger lines in heavy-ion collisions. In recent years one has

310   MEYERHOF & TAULBJERG

found, though, that the study of X-ray continua can furnish new insight into ion-atom collisions, because in some situations the collision is slow enough that radiative decay during the collision can trace out the time development during the collision. Similar developments are now taking place in the study of nuclear heavy-ion collisions [see e.g. (203)].

Since a confusing nomenclature has arisen in this field, we review briefly various processes that give rise to continuum X rays, even though our interest centers on those arising from electronic transitions filling K vacancies during a collision.

Even without the generation of inner-shell vacancies, bremsstrahlung can be emitted in the collision of two atoms due to the acceleration of nuclear and electronic charges. The former gives rise to nucleus-nucleus bremsstrahlung (NNB) (204–206), the latter to a collective radiation (207), which appears to be of minor importance experimentally (208). NNB depends strongly on the isotopic mass of the projectile and the target atoms.

If a target electron is emitted from an inner shell and radiates as it scatters in the field of the projectile (or vice versa), one speaks of radiative ionization (RI) (205, 209, 210). If target valence electrons radiate as they are scattered in the field of the projectile, one calls the process primary or inverse bremsstrahlung (211). If an emitted inner-shell electron from the target or the projectile scatters on target atoms other than those involved in the ionizing collision, the radiation is called secondary electron bremsstrahlung (205, 212, 213). If the initial ionization involves valence electrons that are rescattered on other target atoms, the radiation is called knock-on bremsstrahlung (214). Since K electrons have the largest momenta in the atom, they are ejected with the greatest kinetic energies. Therefore, electron bremsstrahlung in the spectral range of interest for K continua is usually dominated by the emission of K electrons.

As mentioned in Section 4.2, a projectile can have an appreciable number of steady-state inner-shell vacancies if it traverses a solid target. In fast collisions, there can then be radiative electron capture (REC) of target electrons into projectile vacancies (209, 215–217), in addition to nonradiative capture mentioned in Section 3.3. The predominant capture probability is to projectile K vacancies (218). In slow collisions, MO's are formed and the original projectile vacancies are filled by electronic transitions between the MO's. As pointed out by Briggs & Dettmann (219) RI, REC, and MO radiation can be treated on a unified theoretical basis (Section 7).

Lastly, we mention that on the low-energy side of $K\alpha$ lines produced in heavy ion collisions, various extended structures have been found (1, 220–223) that cannot be assigned to any of the above processes, most of which give rise to structure on the high-energy side of the characteristic K lines. The low-energy structure is believed to be due to radiation accompanying the Auger emission process (224, 225a), electron rearrangement (225b), and to a volume plasma excitation in a solid target (226, 227).

7 MECHANISMS OF RADIATIVE DECAY

Three of the radiative decay processes mentioned in Section 6, MO radiation, REC, and RI, involve the transition of an electron to or from an inner shell with the emission of a photon in a single collision. MO radiation and REC require the existence of an inner-shell vacancy in one of the collision partners, usually the projectile. Unless the projectile comes prepared with such a vacancy, one may have to incorporate the vacancy-formation process into the description of the radiative decay (28, 228, 229). Then one speaks of a one-collision process in which $1s\sigma$ vacancy formation (Section 4.3.1) and radiative decay of the $1s\sigma$ vacancy occur in a single collision. If a projectile K vacancy is produced in a primary collision and is transferred to the $1s\sigma$ MO, e.g. by vacancy sharing, in a secondary collision, where the $1s\sigma$ vacancy radiatively decays, one speaks of a two-collision process. Because of the spontaneous decay of the projectile K vacancy between collisions, one expects the two-collision process to operate only in solid targets, but not in monatomic gas targets in which the atoms are too far apart for the vacancy to survive.

The theory of radiative decay at present is a subject of ongoing research. In the case of MO radiation, the first basic calculations have just been done for the one-collision process (229). Experimentally the process seems to be established in solid targets (Section 8.2.1), but not in gas targets (Section 8.2.2). A controversy about the origin of an experimentally found, X-ray energy–dependent anisotropy of the MO spectrum with respect to the beam direction appears to be settled (Section 7.1.3). The proper incorporation of the natural line width, e.g. of the $K\alpha$ line (230a,b), has now been accomplished (219).

In atomic systems it is sufficient to use the dipole approximation for the radiative process. Then, within the impact-parameter formalism, the radiative transition amplitude is given as a time integral (28, 219, 228, 229)

$$a_f(\mathbf{b}, \mathbf{v}_1; \varepsilon, \omega) \equiv \varepsilon \cdot \mathbf{D}_f(\mathbf{b}, \mathbf{v}_1; \omega)$$

$$= (e/2\pi m)(\hbar/\omega)^{1/2} \int_0^\infty dt a_i(t) \exp(i\omega t)\langle \psi_f^-(t) | \varepsilon \cdot \nabla_r | \psi_i^+(t) \rangle. \qquad 7.1$$

Here $\mathbf{r}$ is the electron position vector (for simplicity we use a one-electron notation), ε is the polarization vector, and ω is the angular frequency of the emitted photon. Elliptic polarization is represented by a complex vector ε. In principle, ψ_i^+ and ψ_f^- are exact wave functions for the scattering problem, fulfilling appropriate boundary conditions at $t = 0$ and $t = \infty$, respectively. For example, in slow collisions, referring to Figure 1, $\psi_i^+(t = 0)$ may correspond to a state with an initial $1s\sigma$ [1s(H)] vacancy, whereas $\psi_f^+(t = \infty)$ may correspond to a final state with a $2p\sigma$ [1s(L)] vacancy or a state with a $2p\pi$ [2p(H)] vacancy.

All nonradiative processes are in principle assumed to be incorporated in the scattering wave functions ψ_i^+ and ψ_f^- in Equation 7.1, whereas the radiative decay of the scattering state ψ_i^+ is accounted for by the amplitude $a_i(t)$. This amplitude may be approximated by the form (28) $\exp(-\Gamma t/2)$, where Γ is the natural decay constant of the initial state. However, for transition frequencies in the continuum spectrum, away from characteristic atomic lines, the decay of ψ_i^+ may usually be ignored (231), i.e. one can set $a_i(t) = 1$.

The transition probability per unit frequency interval and per unit solid angle Ω

of photon emission is given by

$$d^2 P_f(\mathbf{b}, \mathbf{v}_1 ; \varepsilon, \omega)/d\omega \, d\Omega = |a_f|^2 \, d^3\mathbf{k}/d\omega \, d\Omega = |a_f|^2 \omega^2/c^3, \qquad 7.2$$

in terms of the amplitude, Equation 7.1. By integrating Equation 7.2 over impact parameter $\mathbf{b}$, the cross section $d^2\sigma_f/d\omega \, d\Omega$ is obtained.

7.1 Molecular-Orbital X Rays

Various radiative continuum processes are described by Equation 7.1. As discussed in Section 3.2, in slow ion-atom collisions with $Z_1 \simeq Z_2$, it is appropriate to represent the scattering states in terms of configurations of the quasimolecular system that is formed in close encounters:

$$\psi^{\pm}(t) = \sum_n c_n^{\pm}(t)\chi_n(t) \exp\left[-(i/\hbar) \int^t E_n(t') \, dt' \right] \qquad 7.3$$

where $\{\chi_n[R(t)]\}$ is a set of molecular eigenfunctions of the electronic Hamiltonian at fixed internuclear separation $\mathbf{R}$. The MO X-ray emission amplitude of Equation 7.1 may then be written

$$a_f = (e/2\pi m)(\hbar/\omega)^{1/2} \sum_{n,m} \int_0^\infty dt \, \exp(i\omega t) a_i (c_n^-)^* c_m^+$$

$$\cdot \langle \chi_n | \varepsilon \cdot \nabla_r | \chi_m \rangle \exp\left\{ -(i/\hbar) \int^t [E_m(t') - E_n(t')] \, dt' \right\}, \qquad 7.4$$

which is a coherent superposition of molecular transition dipoles.

The amplitude $c_{1s\sigma}^+(t)$, corresponding to a molecular state with a $1s\sigma$ vacancy, is of particular importance for K-MO continua. In a two-collision process, $c_{1s\sigma}^+(0) = 1$ corresponding to a situation where a $1s\sigma$ vacancy is brought into the collision, whereas $c_{1s\sigma}^+(0) = 0$ in a one-collision process. Obviously, the one- and the two-collision cross sections must be weighted by the probabilities for obtaining the corresponding initial conditions in a given collision situation. The time development of the set of amplitudes $\{c_m^+(t)\}$ is crucial (229). It may be obtained from coupled-state calculations (219, 231) subject to the above initial conditions, whereas the final-state amplitudes $\{c_n^-(t)\}$ must satisfy appropriate boundary conditions as t approaches infinity after the collision.

7.1.1 STATIONARY PHASE APPROXIMATION Although energy conservation is not implied in the interaction of the electron with the radiation field since energy may be freely exchanged between nuclear and electronic motion, the radiative transition amplitude (Equation 7.4) is significantly reduced unless the transition takes place subject to local energy conservation. This requires the existence of stationary phase points in Equation 7.4, i.e. times t_s (or internuclear separations R_s) at which

$$\omega_{nm}(t_s) \equiv [E_m(R_s) - E_n(R_s)]/\hbar = \omega, \qquad 7.5$$

where for K-MO radiation, E_m is the electronic energy of an MO configuration χ_m with a $1s\sigma$ vacancy and E_n is the energy of a configuration that may be connected to χ_m by a dipole transition. If the condition of Equation 7.5 can be fulfilled, Equation 7.4 may be evaluated approximately under the assumption that the population amplitudes $c_n^\pm$ and the MO dipoles are slowly varying functions of the time (219, 232)

$$a_f = (e/2\pi m)(\hbar/\omega)^{1/2} \sum_{n,m} \sum_{t_s} [(c_n^-)^* c_m^+ \langle \chi_n | \varepsilon \cdot \nabla_r | \chi_m \rangle]_{t_s}$$

$$\cdot (2\pi)^{1/2} | d\omega_{mn}/dt |_{t_s}^{-1/2} \exp\left[-(i/\hbar) \int^{t_s} (\omega_{nm} - \omega) \, dt \pm \pi/4\right], \qquad 7.6$$

where $\pm$ in the phase factor refers to the sign of $d\omega_{mn}/dt$. The summation Σ_{t_s} extends over all points of stationary phase. A particularly simple result for the emission cross section is obtained, if it is assumed that the various phases are random, so that cross terms in $|a_f|^2$ can be ignored. Explicit evaluations of Equation 7.6 have been made only in a two-state approximation. Then only a single dipole contributes and the stationary phase approximation, with the random phase assumption, may be written

$$\frac{d^2 \sigma_f(\mathbf{v}_1, \varepsilon)}{d\omega \, d\Omega} = \frac{e^2 \hbar \omega}{\pi m^2 c^3} \int d\mathbf{b} \left[\frac{|\langle \chi_n | \varepsilon \cdot \nabla_r | \chi_m \rangle|^2}{|v_R \, d\omega_{nm}/dR|}\right]_{R_s}, \qquad 7.7$$

where v_R is the radial internuclear velocity and the integration over $\mathbf{b}$ must be compatible with the stationary phase condition of Equation 7.5. After integration over emission angles and summation over polarization modes, one finds for the cross section (233, 234)

$$d\sigma(v_1)/d\omega = 2 \int_0^{b'} 2\pi b \, db \, | \tau_x(R) v_R \, d\omega_{nm}/dR |_{R_s}^{-1} \qquad 7.8$$

$$= 4\pi R_s^2 (1 - D/R_s)^{1/2} | \tau_x(R) v_1 \, d\omega_{nm}/dR |_{R_s}^{-1}. \qquad 7.9$$

In Equation 7.8, $b' = (R_s^2 - DR_s)^{1/2}$ is the largest impact parameter for which Equation 7.5 can still be fulfilled, D is the distance of closest approach in a head-on collision, and $\tau_x^{-1}(R)$ is the radiative transition rate at an internuclear separation R. Equation 7.8 is usually referred to as the static approximation. It may be obtained directly, if it is assumed that at each internuclear distance R a photon with frequency in the interval $\omega_{nm}(R)$ to $\omega_{nm}(R) + d\omega$ is emitted with a probability $\tau_x^{-1} \, dt = \tau_x^{-1} \, dR/v_R = d\omega (\tau_x v_R \, d\omega/dR)^{-1}$ at each of the two locations where the trajectory intersects the radial distance R_s corresponding to ω_{nm} (Equation 7.5). In this approximation the coherence of the X-ray emission along the trajectory and the uncertainty principle are neglected.

The transition energies and the MO dipole matrix elements are the decisive quantities in the static approximation, as well as in the dynamic theory (Section 7.1.2). In the case of K-MO X rays, the transition energies vary smoothly with R (see Figure 1), but the dipole matrix elements exhibit an interesting variation with R. The most important elements are shown in Figure 15. Although the $1s\sigma$-$2p\sigma$ element vanishes as $R \to \infty$, it is identical to the $1s\sigma$-$2p\pi$ element at $R = 0$ and even

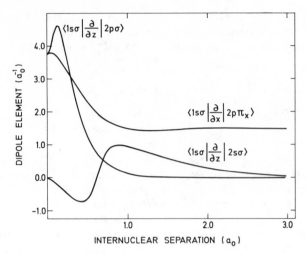

Figure 15 Dipole matrix elements for the Ne-O molecule based on the Hartree-Fock calculations of (63).

exceeds it at intermediate distances. Of course, as $R \to \infty$, the $1s\sigma$-$2p\pi$ element becomes the transition matrix element for the SA $K\alpha$ line. The $1s\sigma$-$2s\sigma$ matrix element vanishes identically for symmetric collisions, but is not negligible for asymmetric systems.

7.1.2   DYNAMIC CALCULATIONS   The first dynamic evaluations of Equation 7.4 were performed by Macek & Briggs (28). They accounted for the coherence of the amplitude and the Heisenberg broadening (16). The latter gives rise to a high-energy tail that exceeds the classical cutoff of the spectrum, but except for this tail the static and dynamic theories give closely similar spectra when integrated over the impact parameter (28).

If the impact parameter of the collision is specified, dynamic calculations reveal interference patterns in K-MO X-ray spectra, as first suggested by Lichten (16). Figure 16 shows a calculation of $dP_f/d\omega$ (Equation 7.2) by Müller (235), based on a model dependence for $\omega_{nm}(R)$ ($n \equiv 1s\sigma$, $m = 2p\pi$) and assuming a constant dipole matrix element. Whereas the quasistatic theory predicts a sharp cutoff at the impact parameter $b = b'$ (Equation 7.8), where $v_R = 0$, the dynamic theory gives an exponential falloff with a width Γ characteristic of the collision time ($\Gamma \propto v_1$). If the spectral distribution is integrated over impact parameter, the oscillatory features disappear and the high-energy tails of the spectra fall off exponentially with a width $\Gamma \propto v_1^{1/2}$, as first computed in a model calculation by Macek & Briggs (28) (see also 236).

Recently, multistate calculations have been made, taking into account the coherence effects due to dynamic couplings between the MO, which give rise to an explicit time dependence of the population amplitudes $c_n(t)$ (Equation 7.3) (219, 237).

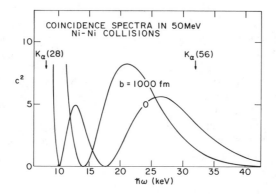

Figure 16 Differential probability for K-MO X-ray emission in 50-MeV Ni + Ni collisions (235). The quantity plotted is proportional to $|a_f|^2$ (see Equations 7.2 and 7.4). Curves are shown for two impact parameters. The SA and UA $K\alpha$ energies are indicated.

Also, Thorson & Choi (229) have examined the influence of a transient $1s\sigma$ vacancy population amplitude $c_{1s\sigma}^+(t)$ on the one-collision MO X-ray spectrum. Heinig et al (237, 238a) have shown that transitions to the $2p\sigma$ MO (Figure 1) contribute importantly to the MO spectral distribution just above the characteristic K lines (see Section 8.2.1). This has been verified by Anholt (238b).

7.1.3 ANGULAR DISTRIBUTION   As pointed out by Müller et al (239), the angular distribution of the continuum radiation with respect to the beam direction may serve as a sensitive tool in the identification of MO X rays. The original estimates of the anisotropy have to be modified considerably (219, 231, 240a,b, 241). The intensity I measured by a detector that is sensitive to radiation with polarization vector ε is proportional to the sum over final states of squared transition amplitudes (Equation 7.1):

$$I = C \sum_f |\varepsilon \cdot \mathbf{D}_f|^2 \qquad 7.10$$

Following Fano & Macek (242), it is possible to derive general expressions for the angular distribution of the MO radiation (231). In the simplest case of no polarization analysis and no detection of the scattered projectile, the total intensity becomes

$$I(\mathbf{v}_1; \omega, \theta) = I_0(\mathbf{v}_1, \omega)[1 - A_0 \tfrac{1}{2}(3 \cos^2 \theta - 1)], \qquad 7.11$$

where θ is the polar angle of the photon detector with respect to the beam. If only a single vector $\mathbf{D}_f$ is present in Equation 7.10, one finds (219)

$$A_0 = \int db(3|\mathbf{D}_{f\parallel}|^2 - |\mathbf{D}_f|^2)/2|\mathbf{D}_f|^2, \qquad 7.12$$

where $\mathbf{D}_{f\parallel}$ is the component of $\mathbf{D}_f$ parallel to $\mathbf{v}_1$. Explicit evaluations of A_0 show that it depends strongly on photon energy, increasing to values as high as $\sim 1/3$. near the UA limit (219, 231).

7.2   Radiative Electron Capture

In Section 7.1, radiative electronic transitions in collisions that are slow compared to typical orbital velocities of inner-shell electrons were considered. In most cases, such collisions are fast with respect to outer-shell target electrons. Hence these electrons respond impulsively to the electric field of the projectile, which makes a molecular approach inappropriate. Rather, as a first approximation, one might replace the exact scattering states in Equation 7.1 by their undistorted asymptotic forms. If capture occurs to a (vacant) inner projectile state, only projectile states need to be considered as final electron states. It is then convenient to adopt a frame of reference centered on the projectile nucleus:

$$\psi_f(t) \simeq \phi_f(\mathbf{r}) \exp\left[-(i/\hbar)E_f t\right], \tag{7.13}$$

$$\psi_i(t) \simeq \phi_i[\mathbf{r} + \mathbf{R}(t)] \exp\left[-(i/\hbar)m\mathbf{v}_1 \cdot \mathbf{r}\right] \cdot \exp\left[-(i/\hbar)(E_i + \tfrac{1}{2}mv_1^2)t\right], \tag{7.14}$$

where ϕ_f and ϕ_i are electron eigenstates of the projectile and target atom, respectively. For simplicity, a one-electron notation is used, $\mathbf{r} + \mathbf{R}$ being the electron coordinate with respect to the target nucleus and $\mathbf{R}$ being the internuclear vector $\mathbf{R} = \mathbf{b} + \mathbf{v}_1 t$. The additional phase factors, called translation factors, ensure that the initial electron wave function is properly Galileo-transformed from the target to the projectile frame (17, 243).

Upon expressing ϕ_i and ϕ_f in momentum representation, the dipole element, as well as the time integral in Equation 7.1, can be evaluated (219). After integration over impact parameter $\mathbf{b}$ and summation over polarization states, the result is

$$d^2\sigma_f/d\omega\,d\Omega = (2\pi e^2\hbar^2\omega/m^2c^3v_1)\int d^3\mathbf{q}\,|\chi_i(\mathbf{q})|^2\,|\chi_f(\mathbf{q} - m\mathbf{v}_1/\hbar)|^2$$
$$\cdot \sum_{\lambda=1,2} |\boldsymbol{\varepsilon}_\lambda \cdot (\mathbf{q} - m\mathbf{v}_1/\hbar)|^2 \delta[\hbar\omega - (E_i - E_f) - \tfrac{1}{2}mv_1^2 + \hbar\mathbf{q}\cdot\mathbf{v}_1], \tag{7.15}$$

where χ_i and χ_f are momentum representations of the initial and final wave functions. In the case of highly charged projectiles, the initial target-electron wave function is strongly distorted by the projectile field. This may be taken into account by replacing the plane wave factor $\exp\left[i\mathbf{q}\cdot(\mathbf{r} + \mathbf{R})t\right]$ in the momentum representation of $\phi_i(\mathbf{r} + \mathbf{R})$ (Equation 7.14) by projectile Coulomb waves (205, 219).

Equation 7.15 can be applied to radiative electron capture (REC) and to radiative ionization (RI). In REC, the target electron is captured into an inner electron shell (215). In this situation the momentum distribution $|\chi_i|^2$ limits the momentum variable $\mathbf{q}$ to small values: $q \ll mv_1/\hbar$, since the encounter is assumed to be fast with respect to the target electron orbital velocity. Equation 7.15 may then be written (244)

$$d^2\sigma_f/d\omega\,d\Omega = (2\pi e^2/\hbar c^3)\omega \sin^2\theta\,|\chi_f(-m\mathbf{v}_1/\hbar)|^2$$
$$\cdot F_i[(\hbar\omega + E_f - E_i - \tfrac{1}{2}mv_1^2)/\hbar v_1], \tag{7.16}$$

where F_i is the Compton profile

$$F_i(q_z) = \int\int dq_x\,dq_y\,|\chi_i(\mathbf{q})|^2. \tag{7.17}$$

Thus, as realized at the outset by Schnopper et al (215), the REC photon spectrum reflects the momentum distribution of the target electrons. The mean energy and the fluctuation around the mean, indicative of the width of the distribution, are (219)

$$\langle \hbar \omega \rangle = \tfrac{1}{2}mv_1^2 + E_i - E_f, \qquad\qquad 7.18$$

$$\hbar^2(\langle \omega^2 \rangle - \langle \omega \rangle^2) = (4/3)\tfrac{1}{2}mv_1^2\langle T \rangle, \qquad\qquad 7.19$$

where $\langle T \rangle$ is the average kinetic energy of the target electron. For N electrons in the target, $\langle T \rangle$ is replaced by $\langle T_N \rangle / N$, where $\langle T_N \rangle$ is the average kinetic energy of all N electrons. At high velocities, $\tfrac{1}{2}mv_1^2 \gg \langle T \rangle$, so that Equations 7.18 and 7.19 predict a REC photon distribution that is peaked at the energy of Equation 7.18. There is considerable uncertainty about the shape of the high-energy tail of the distribution (216, 245).

7.3 Radiative Ionization

Radiative electronic transitions from bound target states to the projectile continuum are called radiative ionization (RI) or direct electron bremsstrahlung. One can also think of this process as radiative capture to continuum states. The cross section may be obtained by substituting a projectile continuum electron wave function as the final state of the electron in Equation 7.14 or in the equivalent equation that takes into account the Coulomb distortion by the projectile (205). However, in this case the peaking approximation employed in Equation 7.16 is not valid and Equation 7.15 must be evaluated as it stands. To date there appear to be order-of-magnitude discrepancies in the evaluated cross sections (205, 210).

One can show that the cross section for RI is a convolution of the initial electron momentum-distribution function with the cross section for bremsstrahlung radiation of free electrons in the projectile field (205). The REC cross section is a corresponding convolution involving the cross section for radiative recombination of a free electron to a bound projectile state (219).

7.4 Other Radiative Processes

Of the other processes that emit continuum X rays and that can form backgrounds for MO X-ray spectra, we mention only secondary electron bremsstrahlung (SEB) and nucleus-nucleus bremsstrahlung (NNB).

SEB is a two-step process involving production of electrons and their subsequent slowing down in the target material (solid or gas). The cross section is given as the convolution integral of the cross section for the primary ionization process and for bremsstrahlung of free electrons in the target, appropriately weighted by slowing-down parameters. Explicit evaluations have been made, based on the BEA (213, 214, 246, 247) and the PWBA (205). Although the results compare favorably with experiment, Saris (202) has pointed to the need for evaluating the primary electron distribution using the molecular model (Section 3.2), if the collision is symmetric or near-symmetric (see also 212). This has been done only for $H^+ + H$ (160), but unfortunately, these calculations are not correct (161, 229). The SEB photon distribution falls rapidly with photon energy and is negligible in the region of K-MO spectra (247, 248).

NNB decreases much less rapidly with photon energy than SEB. In the dipole approximation the cross section is proportional to $(Z_1/A_1 - Z_2/A_2)^2$ (204). By proper choice of projectile and target isotopes, the NNB dipole term may be reduced to negligible proportions or eliminated. In such cases the dipole approximation is insufficient. If the dipole term is small but finite, an important contribution may occur through dipole-quadrupole interference (206).

8   MEASUREMENTS OF MOLECULAR-ORBITAL SPECTRA

The measurement of MO X-ray spectra does not differ in any essential way from the measurement of characteristic X-ray spectra (Section 4). It is very important, however, to avoid electronic pile-up effects (249), which is usually done by inserting suitable absorbers in front of the detector and by using anti–pile-up circuitry. If the resolution function of the detector is too broad, it may have to be unfolded from the experimental spectrum (234, 250, 251). Solid and gas target measurements in principle can lead to different mechanisms of MO X-ray production and are discussed separately (Sections 8.2.1 and 8.2.2).

As shown in Section 7.1, X-ray transitions between MO levels are continua because the transition energy $\hbar\omega_{nm}$ (Equation 7.5) varies with R_s and the transition can take place anywhere along the trajectory. Furthermore, there is a dynamic (Heisenberg) broadening due to the short collision time (Section 7.1.2). The first MO X-ray spectra were discovered by Saris et al (19) in collisions of Ar projectiles on Ar atoms implanted in various solid targets. The spectra were assigned to transitions in a two-collision process (Section 7), filling vacancies in the $2p\pi$ MO. Using a united-atom designation, such transitions are called L-MO X rays. M- and K-MO X rays have meanwhile also been found, first in I + Au (252) and C + C (253) collisions, respectively. We review only K-MO X rays; L- and M-MO X rays are reviewed in (254, 255).

In addition to the usual spectral measurements, in which the impact parameter of the collision is not defined, coincidence experiments (Section 5.2) have been made to fix the impact parameter and thereby test finer predictions of the theory. These experiments are discussed in Section 8.3. The measurement of the Doppler shift of continuum spectra is described in Section 8.4. It allows a model-independent determination of the velocity of the emitting system and thereby an identification of this system. Once the velocity of the emitting system is known, one can find the center-of-mass (c.m.) anisotropy of the radiation with respect to the beam direction and compare with theory (Section 7.1.3). This is described in Section 8.5.

8.1   Backgrounds

Since an understanding of continuum X-ray backgrounds is important in the assignment and analysis of MO X-ray spectra (256, 257) we review the present experimental status of the more important processes producing photon continua (Section 7).

8.1.1   SECONDARY ELECTRON BREMSSTRAHLUNG   The primary electron distribution shows a predicted knee at the free-collision energy $4mE_1/M_1$, and agrees with the BEA theory reasonably well above that energy, but deviates considerably from theory below that energy (212). Experimental tests of the BEA-based SEB theory (Section 7.4) indicate agreement within 20–30% of the predicted photon spectra (shape and intensity), especially if the angular anisotropy of the radiation is taken into account (213, 246, 258). In symmetric and near-symmetric collisions with solid targets, the SEB intensity is negligible in comparison with experimental continuum spectra (247). On the other hand, in a gas-target measurement with 40-MeV $Si^{+6} + SiH_4$, the high-energy portion of the continuum spectrum is attributed nearly entirely to SEB (260) but this is disputed by (261).

8.1.2 RADIATIVE ELECTRON CAPTURE   The peaking of REC spectra at the predicted energy (Equation 7.18) was observed in the earliest measurements with high-velocity, highly stripped projectiles on solid and gas targets (215, 262). The c.m. angular distribution follows closely the expected $\sin^2\theta$ distribution (Equation 7.16) (263; 264a,b) and the absolute magnitude of the cross section is in reasonable agreement with a free-electron calculation of Bethe & Salpeter (265), but not with the Born approximation (264a,b; 266a). The squared width of the photon distribution follows the expected form of Equation 7.19 (245), but the shape of the distribution is only in qualitative agreement with theory (216). The agreement improves, if, for the capture of target valence electrons, the *experimental* Compton profile is used, and if the projectile steady-state vacancy distribution is taken into account (266b).

8.1.3 NUCLEUS-NUCLEUS BREMSSTRAHLUNG   In situations in which the dipole part of this radiation is important, the absolute experimental intensity is in agreement with the theory of Alder et al (204, 234, 255). The theoretical angular anisotropy (206) and the important contribution that dipole-quadrupole interference can make to the intensity of the spectrum (206) have been verified experimentally (267). Therefore, one can compute the NNB background under MO X-ray spectra with considerable confidence. This is important, since NNB spectra fall off relatively slowly with energy and can interfere considerably with the measurement of MO X-ray spectra, especially in asymmetric collisions (234).

8.1.4 OTHER BACKGROUNDS   None of the other radiative continua mentioned in Section 7 appear to play a significant role as a background to MO X-ray spectra, although it should be pointed out that in the case of quite asymmetric collisions, the experimental continua lying above the characteristic K lines are not understood at present (268, 269).

Lastly, we mention that in the solid-state detectors usually employed for X-ray measurements, nuclear γ rays excited in heavy-ion collisions will produce approximately flat Compton distributions. By using radioactive sources emitting γ rays of comparable energies, it is possible to subtract such Compton backgrounds from heavy-ion photon spectra before analysis.

8.2   Results and Interpretations

As mentioned in Section 7.1, MO X rays are expected to be produced in one- and two-collision processes. Since the one-collision process is due to the direct creation of a $1s\sigma$ vacancy during a collision, it is convenient (229, 234) to normalize the spectral distribution $d^2\sigma_f/d\omega\, d\Omega$ (see sentence following Equation 7.2) to the cross section $\sigma(1s\sigma)$ discussed in Section 4.3.1. In fact, the bombarding energy dependence of the one-collision MO X-ray intensity is determined mainly by that of $\sigma(1s\sigma)$.

In the case of the two-collision process, the spectral distribution, which is given by Equation 7.9 in the static approximation, must be multiplied by the probability $P_{1s\sigma}$ of having available a $1s\sigma$ vacancy in the second collision. It is not difficult to show that independent of whether the projectile is the higher-Z or lower-Z partner, this probability is given by

$$P_{1s\sigma} = w(1-w)n\sigma(2p\sigma)v_1\tau_K, \qquad\qquad 8.1$$

where w is the K-vacancy sharing fraction given by Equation 3.22 and discussed in Section 3.2.4; n is the atomic target density; $\sigma(2p\sigma)$ is the $2p\sigma$ vacancy cross section discussed in Section 4.3.2; and τ_K is the lifetime of the projectile K vacancy in the target material. Equation 8.1 assumes that the projectile K-vacancy formation in the first collision is due only to $2p\sigma$ vacancy sharing. If that is not so, the more complete Equations 4.4 or 4.5 would have to be used.

The vacancy factor $w(1-w)$ in Equation 8.1 is a steep function of the asymmetry $|Z_1 - Z_2|$. This can be seen from Figure 9 by noting that $w(1-w) = r/(1+r)^2$, where r is the ratio plotted on the ordinate of Figure 9, and by realizing that the abscissa $2x$ (Equation 3.21) is approximately proportional to $(Z_1 - Z_2)$. Hence the two-collision process is expected to be important only for symmetric or near-symmetric collisions. But even in these cases, at sufficiently high bombarding energies, the one-collision process is expected to become more important than the two-collision process, because of the steeper bombarding energy dependence of $\sigma(1s\sigma)$ compared to $\sigma(2p\sigma)$.

8.2.1   SOLID TARGETS   Upon unfolding absorber and detection efficiency effects, spectra such as those shown in Figure 17 are obtained (234, 259, 270). In Figures 17a and b, the lower-energy portions of the spectra are suppressed, but in Figure 17c it can be seen that above the highest characteristic K line there are two quite distinct continua (271). The lower-energy continuum has been assigned to MO transitions to the $2p\sigma$ MO (237, 238a,b) and the upper one to transitions to the $1s\sigma$ MO.

In Figure 17a the lines T, TQ, and O give the estimated two-collision static (234), dynamic (235), and one-collision spectra (234), respectively, all calculated assuming a single $2p\pi$-$1s\sigma$ dipole transition. The line B gives the computed NNB spectrum. In this case, there is little difference between the static and dynamic calculations (Sections 7.1.1 and 7.1.2). In Figure 17b, spectra from various Ni + Ni collisions (270) are compared with dynamic calculations for the two-collision process (22). Here, the static spectra would cut off at the UA limits shown by the vertical arrows, whereas the dynamic calculations give a very satisfactory agreement with experiment and demonstrate the importance of the dynamic broadening of spectrum beyond the UA limit. Betz (236) and Müller (235) have shown that the $1/e$ falloff energy of the dynamic tail varies proportionally with $v_1^{1/2}$, as expected (Section 7.1.2).

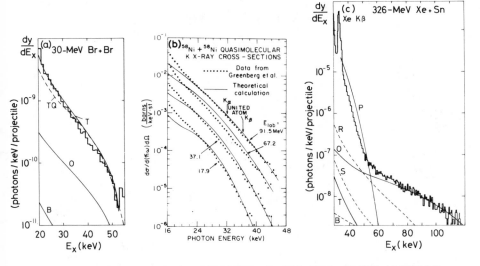

Figure 17   Continuum spectra for (a) 30-MeV Br + Br (234), (b) Ni + Ni (270), (c) 326-MeV Sn + Xe (259) collisions. In (a) and (b) the SA X-ray lines are not shown. See text for explanation of curves.

Figure 17c shows the spectral distribution for 326-MeV Xe + Sn collisions (259). Calculated SEB(S), NNB(B), and RI(R) backgrounds are given in dotted lines. In this case, the two-collision spectrum (T) cannot account for the remainder of the spectrum. The high-energy continuum follows a parameterized form for the one-collision process (0) (272). The low-energy continuum agrees well with the proposal (238a) that it is due to transitions to the $2p\sigma$ MO(P) (238b).

Following a suggestion of J. T. Aten, Meyerhof et al (234) have shown that it is possible to determine, from the target-density (n) dependence of the MO X-ray cross section, whether in a given collision the one- or two-collision process dominates. According to Equation 8.1, the two-collision cross section is proportional to n. The one-collision cross section is independent of n. (Thick-target yields would be proportional to n^2 and n, respectively.) Experiment confirms that for 30-MeV Br + Br collisions (Figure 17a), the two-collision process dominates (234).

8.2.2   GAS TARGETS   In principle, monatomic gas targets are well suited to study the one-collision process of K-MO X-ray production. Because of the low target density, the two-collision process is negligible (Equation 8.1). In polyatomic gases, there is a possibility for the projectile to obtain a K vacancy in a collision with one target atom and to carry it to a second collision within the same molecule (127). Indeed, continuum spectra found in 50–400-keV $N^+ + N_2$ collisions are consistent with the two-collision process (202, 273).

Unfortunately, measurements with gas targets have not to date yielded any definite information on the one-collision process. To contrast gas- and solid-target

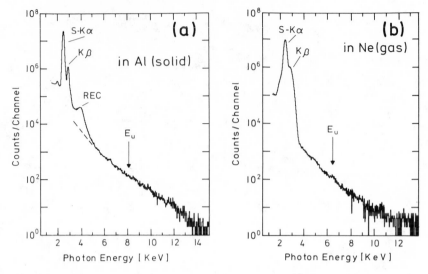

Figure 18   X-ray spectra for (*a*) 55-MeV S + Al, (*b*) 48-MeV S^{+7} + Ne collisions (274). The UA $K\alpha$ energies are indicated. In (*a*) the REC peak is clearly apparent.

spectra, we show in Figure 18a the spectrum from 55-MeV S bombarding a thin Al foil (100 μg cm^{-2}) and in Figure 18b the spectrum from 48-MeV S^{+7} bombarding a Ne gas target (0.9 torr) (274). In the solid, the REC peak appears at the energy predicted by Equation 7.18. This peak is absent in the gas target, because only in the solid is the number of steady-state projectile K vacancies sufficient to allow appreciable capture of target electrons. (Sections 4.2 and 7.2). The intensity of the S + Ne continuum is of the order of magnitude expected for the one-collision process (275); this is confirmed by other gas-target measurements (261). But there are also claims that with gas targets the high-energy continuum can be explained by background effects (260, 271).

8.3   *Coincidence Spectra*

According to the dynamic theory (Section 7.1.2), two-collision MO X-ray spectra should show interesting interference effects if the impact parameter of the collision is fixed (Figure 16). Unfortunately, because of the very low intensity of the X-ray continua, coincidence experiments have only been made to date at such high bombarding energies that the one-collision process becomes of major importance. But for the one-collision process, no dynamic calculations of $dP_f/d\omega$ (Equation 7.2) have yet been published (229), and the static approximation (234) is expected to be completely invalid at high bombarding energies.

Measurements of coincidence spectra of 32-MeV S + Cl and 35-MeV Cl + Cl collisions using solid targets show essentially featureless spectral distributions quite similar to the impact-parameter–integrated spectra (cf Figure 18a) (276, 277). This is in complete contrast to the predicted two-collision spectral distribution (Figure

16). As mentioned, no comparison with dynamic one-collision predictions could be made to date, although marked disagreement with the static one-collision predictions has been demonstrated (275, 276).

8.4 Doppler Shift of MO Spectra

In view of some of the uncertainties associated with the production of MO spectra, especially in the regime in which neither the two-collision process nor background continua can account for an experimental spectrum, it is of interest to show that the X-ray continuum is indeed radiated by the projectile-target quasimolecular system and not by the projectile or target alone. If the X-ray spectrum consists of monoenergetic lines, as is the case for the characteristic (SA) X rays, the Doppler shift of the X-ray energy provides an immediate clue to the velocity of the system from which the radiation is emitted (278). For a continuum spectrum, a Doppler shift measurement has to use the fact that in the c.m. system the radiation cannot show any fore-aft asymmetry, if the system is not polarized (Equation 7.11) (75, 219, 231, 240, 279). In the laboratory system, the Doppler shift affects the energy, angle, and solid angle of the radiation. Since the angular distribution in the c.m. system is not known, except for that fact that it is symmetric about 90° c.m., an iterative procedure must be used to determine the Doppler velocity (280). If the c.m. angular distribution is of the form of Equation 7.11, the laboratory angular distribution must be determined for at least three angles.

The Doppler velocity of continuum X rays lying above the characteristic K X rays has been determined for 200-MeV Kr + Zr (280). Figure 19a shows the X-ray spectrum at $\theta(\text{lab}) = 90°$. In this system nucleus-nucleus bremsstrahlung is negligibly small. The experimental spectrum is approximately an order of magnitude more intense than expected for the two-collision process, so it is presumably due to the one-collision process. The one-collision static prescription of (234) is shown by the solid line. The dashed line is the normalized line-broadening tail predicted by (236). The inset sketches the experimental arrangement. Corrections for the absorber (2.8-mm Al) and the efficiency of the detector have been applied to the spectrum shown in Figure 19a.

Figures 19b,c show two laboratory anisotropy ratios R as a function of the laboratory X-ray energy. As can be seen in Figure 19c, the spectral ratio at the 45° and 135° laboratory angles differ markedly from unity, whereas according to Equation 7.11, the ratio at the 45° and 135° c.m. angles should be equal to unity. In Figure 19d, the extracted Doppler velocity v_D is given in units of the c.m. velocity v_{ci} of the incident Kr + Zr quasimolecule. Because of the slowing down of the projectile, it is expected that $v_D/v_{ci} = 0.90$, compared to an experimental value of 0.92 ± 0.03 (280). If the target were the emitter of the radiation, one would expect $v_D \simeq 0$; if the projectile were the emitter, one would expect $v_D/v_{ci} \simeq 1.9$. Hence, the radiation definitely originates from the Kr + Zr quasimolecule. Since the spectral intensity cannot be accounted for as NNB or two-collision MO X rays, this case can be taken to confirm the existence of a one-collision MO process.

It has been shown that the Doppler velocity of the low-energy continuum (compare Figure 17c) in 67-MeV Nb + Nb collisions is also consistent with the velocity

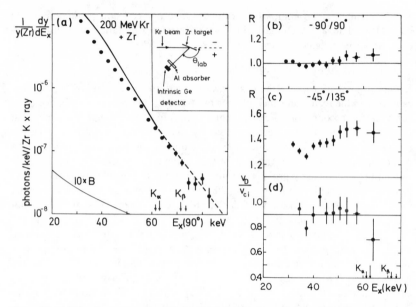

Figure 19 (*a*) Continuum spectrum for 200-MeV Kr + Zr collisions. Solid line gives one-collision prescription (234), dashed line shows predicted tail (236). UA $K\alpha$ and $K\beta$ energies are indicated. Inset gives schematic experimental arrangement. $10 \times B$ is computed NNB multiplied by 10. (*b*) Spectral ratio at 90° on either side of beam. No systematic asymmetry is apparent. (*c*) Spectral ratio at 45°–135° in lab. (*d*) Extracted Doppler velocity v_D in units of incident projectile-target c.m. velocity v_{ci}. Horizontal line gives expected velocity if continuum radiation is emitted by projectile-target quasimolecule (280).

of the projectile + target quasimolecule, but not with the velocity of either collision partner alone (271). This lends support to the interpretation of the low-energy continuum as MO X radiation leading to the $2p\sigma$ MO (237; 238a,b).

8.5 Anisotropy of MO Spectra

Except in cases where the projectile is stripped down to the K shell, the anisotropy of the characteristic K X rays with respect to the beam direction is small, typically less than 10% (104, 280). Much larger, X-ray energy–dependent anisotropies have been found for K-MO X rays (174, 271, 280–285). These measurements were inspired by calculations of Müller et al (75, 239) which predicted large, X-ray energy–dependent anisotropies for MO X-ray spectra. Meanwhile, these calculations have been shown to be quite incomplete (Section 7.1.2) and the qualitative agreement originally obtained between theory and experiment (75, 281) must be considered fortuitous.

Writing the expected c.m. angular distribution 7.11 in the form

$$1 + \eta \sin^2\theta, \qquad\qquad 8.2$$

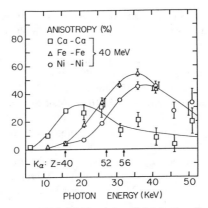

Figure 20   X-ray anisotropy $I(90°)/I(30°) - 1$ for 40-MeV Ca + Ca, Fe + Fe, and Ni + Ni collisions (282). Curves are drawn to guide the eye. UA $K\alpha$ energies are indicated.

Figure 20 shows a typical dependence of η on X-ray energy for various collisions (282). Wölfli et al (165, 174, 282, 286) have shown that the X-ray energy at the peak of η is closely related to the UA $K\alpha$ X-ray energy.

Wölfli et al (165, 174, 282) have also found a periodic fine structure in the X-ray energy dependence of η, which is as yet not firmly explained. Proposals have been made that the fine structure represents an interference effect between the incoming and outgoing parts of the amplitude a_f (Equation 7.4) (287) or that it represents a splitting of MO levels during the collision (174).

9 POSITRON CREATION IN SUPERHEAVY QUASIMOLECULES

The emission of positrons, rather than of X rays during the filling of a $1s\sigma$ vacancy in an atomic collision in which the binding energy of the $1s\sigma$ MO exceeds 1 MeV (20, 21, 288) has been reviewed recently by Müller (23). Hence we discuss only the question of $1s\sigma$ vacancy formation, which obviously influences the cross section of positron emission.

Similar to the MO X-ray transition amplitude (Equation 7.4), the positron emission amplitude appears as a time integral of a transition matrix element weighted by a $1s\sigma$ vacancy population amplitude $c^+_{1s\sigma}(t)$ (23). As in Section 7, both one-collision and two-collision processes may be considered. It is readily shown, however, that the probability of bringing a $1s\sigma$ vacancy into a second collision has a value of the order of 10^{-10} for 1600-MeV U + U (255), so that positron emission would be undetectable in the presence of the expected extraneous positron backgrounds (23).

For the one-collision process, the correct time dependence of $c^+_{1s\sigma}(t)$ must be evaluated and incorporated in the positron-emission amplitude. Such a program is presently in progress (23, 85, 162). Meanwhile, one can estimate the order of magnitude of $c_{1s\sigma}^+(t)$ at small internuclear separations by relating it to the K-vacancy production cross section in the same collision. Employing the semi-empirical Equation 4.6, one finds that $|c_{1s\sigma}^+(\infty)|^2$ is of the order of unity in 1600-MeV $U+U$ collisions (57), in agreement with the calculations in (85, 162). Earlier estimates, which did not take into account the relativistic effects on the $1s\sigma$ wavefunction, were 3–4 orders of magnitude smaller (158, 289). Recent experiments (290, 291) support the new calculations (85, 162). Positrons have been detected in ~ 4–6-MeV/N Pb+Pb, U+Pb, and U+U collisions (292). In principle, the positrons could be due to internal-pair conversion of gamma rays produced in nuclear Coulomb excitation (293, 294) or other nuclear reactions (295), or the positron could originate in other processes (23). The latter include direct positron production by the transient nuclear electric field in the collision (3, 296) and positron production from the filling of $1s\sigma$ and other MO vacancies by electrons from the negative-energy Dirac sea (23). So far, it has been established (292) that in 4–6-MeV/N Pb+Pb and U+Pb collisions positron production exceeds nuclear internal-pair production by nearly an order of magnitude and has the rough magnitude expected for direct positron production (296). For U+U collisions, the internal-pair conversion background is larger, but in a coincidence experiment with scattered (U) projectiles it has been demonstrated that the majority of the detected positrons in an energy range of ~ 500–900 keV are not due to nuclear processes (297). Hence, a detailed study of the originally proposed production of positrons in heavy-ion collisions (20, 21, 288) looks very hopeful and important tests of the underlying quantum-electrodynamic theory may be expected (23, 162).

ACKNOWLEDGMENTS

We would like to thank the many colleagues who have communicated to us their results prior to publication.

Literature Cited

1. Richard, P. 1975. In *Atomic-Inner Shell Processes*, ed. B. Crasemann, I: 73. New York: Academic
2. Merzbacher, E., Lewis, H. W. 1958. In *Encyclopedia of Physics*, ed. S. Flügge, 34:166. Berlin: Springer
3. Bang, J., Hansteen, J. M. 1959. *K. Dan. Vidensk. Selsk. Mat. Fys. Medd.* 31(13):1
4. Brandt, W. 1973. In *Proc. Int. Conf. Inner-Shell Ioniz. Phenom. and Fut. Appl.*, ed. R. W. Fink et al, p. 948. Oak Ridge, Tenn.: ERDA
5. Garcia, J. D., Fortner, R. J., Kavanagh, T. M. 1973. *Rev. Mod. Phys.* 45:111
6a. Madison, D. H., Merzbacher, E. 1975. See Ref. 1, p. 1
6b. Madison, D. H. 1976. *Proc. Int. Conf. Inner-Shell Ioniz. Phenom.* 2nd, ed. W. Mehlhorn, R. Brenn, p. 321. Freiburg, Germ.: Univ. Freiburg
6c. Merzbacher, E. 1976. See Ref. 6b, p. 1

7. Hansteen, J. M. 1975. *Advances in Atomic and Molecular Physics*, ed. D. R. Bates, 11:299. New York: Academic
8. Briggs, J. S., Taulbjerg, K. 1977. In *Topics in Modern Physics*, ed. I. Sellin. Berlin: Springer. To be published
9. Armbruster, P. 1962. *Z. Phys.* 166:341
10. Armbruster, P. et al 1964. *Z. Naturforsch.* 19:1301
11. Specht, H. J. 1965. *Z. Phys.* 185:301
12. Fano, U., Lichten, W. 1965. *Phys. Rev. Lett.* 14:627
13. Barat, M., Lichten, W. 1972. *Phys. Rev. A* 6:211
14. Armbruster, P., Mokler, P. H., Stein, H. J. 1973. See Ref. 4, p. 396
15. Kessel, Q. C., Fastrup, B. 1973. *Case Stud. At. Phys.* 3:137
16. Lichten, W. 1974. *Phys. Rev. A* 9:1458
17. Briggs, J. S. 1976. *Rep. Prog. Phys.* 39:217

18. Coates, W. M. 1934. *Phys. Rev.* 46:542
19. Saris, F. W. et al 1972. *Phys. Rev. Lett.* 28:717
20. Peitz, H., Müller, B., Rafelski, J., Greiner, W. 1973. *Lett. Nuovo Cimento* 8:37
21. Gershtein, S. S., Popov, V. S. 1973. *Lett. Nuovo Cimento* 6:593
22. Betz, W. et al 1975. *The Physics of Electronic and Atomic Collisions*, ed. J. S. Risley, R. Geballe, p. 531. Seattle: Univ. Wash. Press
23. Müller, B. 1976. *Ann. Rev. Nucl. Sci.* 26: 351
24. Bemis, C. E. 1974. In *Reactions Between Complex Nuclei*, ed. R. L. Robinson et al, 2:529. Amsterdam: North-Holland
25. Hardy, J. C. et al 1976. *Phys. Rev. Lett.* 37:133; Chemin, J. 1978. PhD Thesis. Univ. Bordeaux, France
26. Macdonald, J. R. et al 1976. *Phys. Rev. A* 14:1977; Rule, D. W. 1977. *Phys. Rev. A* 16:19
27. Bambynek, W. et al 1972. *Rev. Mod. Phys.* 44:716; Corrigendum 1974. 46: 853
28. Macek, J. H., Briggs, J. S. 1974. *J. Phys. B* 7:1312; Corrigendum 1975. 8:156
29. Choi, B.-H. 1975. *Phys. Rev. A* 11:2004
30. Aashamar, O., Kocbach, L. 1977. *J. Phys. B* 10:849
31. Khandelwahl, G. S., Choi, B. H., Merzbacher, E. 1969. *At. Data* 1:103; Choi, B. H., Merzbacher, E., Khandelwahl, G. S. 1973. *At. Data* 5:291
32. Hansteen, J. M., Johnson, O. M., Kocbach, L. 1975. *At. Data Nucl. Data Tables* 15:305
33. McDowell, M. R. C., Coleman, J. P. 1970. *Introduction to the Theory of Ion-Atom Collisions*, Chap. 6. Amsterdam: North-Holland
34. Mott, N. F., Massey, H. S. W. 1965. *The Theory of Atomic Collisions*, Chap. 19, paragraph 2. Oxford: Clarendon
35. Bates, D. R., McDonough, W. R. 1970. *J. Phys. B* 3:L83
36. Bates, D. R., McDonough, W. R. 1972. *J. Phys. B* 5:L107
37. Taulbjerg, K. 1976. See Ref. 6b, p. 130
38. Hansteen, J. M., Mosebekk, O. P. 1970. *Z. Phys.* 234:281; 1973. *Nucl. Phys. A* 201:541
39. Bethe, H. A., Jackiw, R. W. 1968. *Intermediate Quantum Mechanics*, p. 326. New York: Benjamin
40. Taulbjerg, K. 1977. *J. Phys. B* 10:L341
41. McDowell, M. R. C., Coleman, J. P. 1970. See Ref. 33, Chap. 2
42. Manson, S. T. 1972. *Phys. Rev. A* 6: 1013
43. Beloshitsky, V. K., Nikolaev, V. S. 1975.

Phys. Lett. A 51:97
44. Hansen, J. S. 1973. *Phys. Rev. A* 8:822
45. McGuire, J. H. 1974. *Phys. Rev. A* 9: 286
46. Amundsen, P. A. 1976. *J. Phys. B* 9: 971; 1977. *J. Phys. B* 10:1097
47. Amundsen, P. A., Kocbach, L., Hansteen, J. M. 1976. *J. Phys. B* 9:L203
48. Anholt, R. et al 1976. *Phys. Rev. A* 14: 2103
49. Andersen, J. U., Laegsgaard, E., Lund, M. 1976. Unpublished
50. Brandt, W., Laubert, R., Sellin, I. 1966. *Phys. Lett.* 21:518
51. Basbas, G., Brandt, W., Laubert, R. 1973. *Phys. Rev. A* 7:983
52. Basbas, G., Brandt, W., Ritchie, R. H. 1973. *Phys. Rev. A* 7:1971
53. Andersen, J. U. et al 1976. *Nucl. Instrum. Methods* 132:507
54. Basbas, G., Brandt, W., Laubert, R. 1971. *Phys. Lett. A* 34:277
55. Basbas, G. et al 1971. *Phys. Rev. Lett.* 27:171
56. Anholt, R. 1976. In *Abstr. Int. Conf. At. Phys. 5th*, ed. R. Marrus, M. H. Prior, H. A. Shugart, p. 172. Berkeley: Univ. Calif. Press
57. Anholt, R., Meyerhof, W. E. 1977. *Phys. Rev. A* 16:190
58. Born, M., Oppenheimer, J. R. 1932. *Ann. Phys.* 84:457
59. Larkins, F. P. 1972. *J. Phys. B* 5:571
60. Thulstrup, E. W., Johansen, H. 1972. *Phys. Rev. A* 6:206
61. Briggs, J. S., Hayns, M. R. 1973. *J. Phys. B* 6:514
62. Sidis, V., Barat, M., Dhuicq, D. 1975. *J. Phys. B* 8:474
63. Taulbjerg, K., Briggs, J. S. 1975. *J. Phys. B* 8:1895
64. Eichler, J., Wille, U. 1975. *Phys. Rev. A* 11:1973
65. Eichler, J. et al 1976. *Phys. Rev. A* 14: 707
66. Fricke, B. et al 1975. *Phys. Rev. Lett.* 34:273
67. Kaufmann, P., Wille, U. 1976. *Z. Phys. A* 279:259
68. Fricke, B. et al 1976. *Phys. Lett. A* 59: 375
69. Bates, D. R., Ledsham, K., Stewart, A. L. 1953. *Phil. Trans. R. Soc. (London) Ser. A* 246:215
70. Helfrich, J., Hartmann, H. 1970. *Theor. Chim. Acta* 16:263
71. Teller, E., Sahlin, H. L. 1970. In *Physical Chemistry*, ed. H. Eyring, D. Henderson, W. Jost 5:35. New York: Academic
72. Piacentini, R. D., Salin, A. 1974. *J. Phys. B* 7:L311

73. Piacentini, R. D., Salin, A. 1974. *J. Phys. B* 7:1666
74. Taulbjerg, K., Vaaben, J., Fastrup, B. 1975. *Phys. Rev. A* 12:2325
75. Müller, B., Greiner, W. 1974. *Phys. Rev. Lett.* 33:469
76. Müller, B., Greiner, W. 1976. *Z. Naturforsch.* 31a:1
77. Gershstein, S. S., Krivchenkov, V. D. 1961. *Sov. Phys. JETP* 13:1044
78a. Lichten, W. 1967. *Phys. Rev.* 164:131
78b. Vaaben, J., unpublished data
79. Fastrup, B. 1975. See Ref. 22, p. 361
80. Lennard, W. N., Mitchell, I. V. 1976. *Nucl. Instrum. Methods* 132:39; *J. Phys. B* 9:L317; Forester, J. S., Phillips, D. 1977. *J. Phys. B* 10:2199
81. Meyerhof, W. E., Anholt, R. 1976. Conf. Sci. Ind. Appl. Small Accel., 4th, ed. J. L. Duggan, I. L. Morgan. *IEEE Publ.* No. 76CH1175-9NPS, p. 60. New York: Inst. Electr. Electron. Eng.
82. Meyerhof, W. E., Anholt, R., Saylor, T. K. 1977. *Phys. Rev. A* 16:169
83. Meyerhof, W. E. 1973. *Phys. Rev. Lett.* 31:1341
84. Thorson, W. R. 1975. *Phys. Rev. A* 12:1365
85. Betz, W. et al 1976. *Phys. Rev. Lett.* 37:1046
86. Fastrup, B. et al 1974. *J. Phys. B* 7:L206
87. Bates, D. R., Williams, D. A. 1964. *Proc. Phys. Soc.* 83:425
88. Knudson, S. K., Thorson, W. R. 1970. *Can. J. Phys.* 48:313
89. Rosenthal, H. 1971. *Phys. Rev. Lett.* 27:835
90. Briggs, J. S., Macek, J. 1972. *J. Phys. B* 5:579
91. Briggs, J. S., Taulbjerg, K. 1975. *J. Phys. B* 8:1909
92. Briggs, J. S., Taulbjerg, K. 1976. *J. Phys. B* 9:1641
93. Macek, J. H., Briggs, J. S. 1973. *J. Phys. B* 6:841
94. Taulbjerg, K., Briggs, J. S., Vaaben, J. 1976. *J. Phys. B* 9:1351
95. Anholt, R., Meyerhof, W. E., Salin, A. 1977. *Phys. Rev. A* 16:951
96. Briggs, J. S., Macek, J. 1973. *J. Phys. B* 6:982
97. Demkov, Y. N. 1964. *Sov. Phys. JETP* 18:138
98. Briggs, J. S. 1974. Rep. No. T.P. 594. Harwell, Engl.: At. Energy Res. Establ.
99. Olson, R. E. 1972. *Phys. Rev. A* 6:1822
100. Nikitin, E. E. 1962. *Opt. Spectra. (USSR)* 13:761
101. Nikitin, E. E. 1970. In *Advances in Quantum Chemistry*, ed. P. O. Löwdin, 5:135. New York: Academic
102. Bøving, E. 1977. *J. Phys. B* 10:L63
103. Betz, H. D. et al 1974. *Phys. Rev. Lett.* 33:807
104. Pederson, E. H. et al 1975. *Phys. Rev. A* 11:1267
105. Hopkins, F. 1975. *Phys. Rev. Lett.* 35:270
106. Betz, H. D. 1976. *Nucl. Instrum. Methods* 132:19
107. Gray, T. J. et al 1976. *Phys. Rev. A* 14:1333
108. Gardner, R. K., Gray, T. J., Richard, P., Schmiedekamp, C., Jamison, K. A. 1977. *Phys. Rev. A* 15:2202
109. Brinkmann, H. C., Kramers, H. A. 1930. *Proc. Acad. Sci. Amsterdam* 33:973
110. Dettmann, K. 1971. *Springer Tracts in Modern Physics*, ed. G. Hoehler 58:119. Heidelberg, Germ.: Springer
111. Jackson, J. D., Schiff, H. 1953. *Phys. Rev.* 89:359
112. Halpern, A. M., Law, J. 1975. *Phys. Rev. A* 12:1776
113. Belkić, D., Salin, A. 1976. *J. Phys. B* 9:L397
114. Nikolaev, V. S. 1967. *Sov. Phys. JETP* 24:847
115a. Brown, M. D. et al 1974. *Phys. Rev. A* 10:1255
115b. Lapicki, G., Losonsky, W. 1977. *Phys. Rev. A* 15:896
116. Drisko, R M. 1955. PhD thesis. Carnegie Inst. Tech., Pittsburgh
117a. McDowell, M. R. C., Coleman, J. P. 1970. See Ref. 33, Chap. 8
117b. Cocke, C. L., Varghese, S. L., Curnutte, B. 1977. *Phys. Rev. A* 15:874
118. Randall, R. R. et al 1976. *Phys. Rev. A* 13:204
119. Crasemann, B., ed. 1975. *Atomic Inner-Shell Processes*. Vol. II, *Experimental Approaches and Applications*. New York: Academic
120. Lutz, H. O. 1975. See Ref. 22, p. 432
121. Lutz, H. O. 1976. See Ref. 6b, p. 104
122. Stolterfoht, N. 1973. In *The Physics of Electronic and Atomic Collisions*, ed. B. C. Cobič, M. V. Kurepa, p. 117. Belgrade, Yugoslavia: Inst. Phys.
123. Moore, R. C. 1975. See Ref. 22, p. 447
124. Krause, M. O. 1976. See Ref. 6b, p. 184
125. Stolterfoht, N. 1971. *Z. Phys.* 248:81
126. Stolterfoht, N. 1976. See Ref. 6b, p. 42
127. Saris, F. et al 1974. *J. Phys. B* 7:1494
128. Stolterfoht, N. et al 1974. *Phys. Rev. Lett.* 33:59
129. Betz, H. D. et al 1973. See Ref. 4, p. 1374
130. Stolterfoht, N. et al 1974. *Phys. Rev. Lett.* 33:1418

131. Burch, D. et al 1974. *Phys. Rev. Lett.* 32:1151
132. Stolterfoht, N. et al 1975. *Phys. Rev. A* 12:1313
133. Stolterfoht, N. 1977. See Ref. 8
134. Meyerhof, W. E. et al 1976. *Phys. Rev. A* 14:1653
135. Taulbjerg, K., Sigmund, P. 1972. *Phys. Rev. A* 5:1285
136. Taulbjerg, K., Fastrup, B., Laegsgaard, E. 1973. *Phys. Rev. A* 8:1814
137. Betz, H. D. 1972. *Rev. Mod. Phys.* 44:465
138. Bell, F., Betz, H. D. 1977. *J. Phys. B* 10:483
139. McDaniel, F. D. 1976. See Ref. 81, p. 52
140. Laubert, R., Losonsky, W. 1976. *Phys. Rev. A* 14:2043
141. Jones, K. W. et al 1975. *Phys. Rev. A* 12:1232
142. Brandt, W., Laubert, R. 1975. *Phys. Rev. A* 11:1233
143. Burch, D., Taulbjerg, K. 1975. *Phys. Rev. A* 12:508
144. Datz, S. 1976. *Nucl. Instrum. Methods* 132:7
145. Kauffman, R. L. et al 1976. *Phys. Rev. Lett.* 36:1074
146. Hopkins, F. 1976. See Ref. 81, p. 37
147. Watson, R. L., Chiao, T., Jensen, F. E. 1975. *Phys. Rev. Lett.* 35:254; Watson, R. L. et al 1977. *Phys. Rev. A* 15:914
148. Hopkins, F., Sokolov, J., Little, A. 1976. *Phys. Rev. A* 14:1907
149. Hopkins, F. et al 1976. *Phys. Rev. Lett.* 37:1100
150. Macek, J., Cairns, J. A., Briggs, J. S. 1972. *Phys. Rev. Lett.* 28:1298
151. Saris, F. 1973. In *Atomic Collisions in Solids*, ed. S. Datz, B. R. Appleton, C. D. Moak, p. 343. New York: Plenum
152. Saris, F. et al 1973. See Ref. 4, p. 1255
153. Feldman, L. C., Silverman, P. J., Fortner, R. J. 1976. *Nucl. Instrum. Methods* 132:29
154. Brandt, W. 1973. In *Atomic Physics 3*, ed. S. J. Smith, G. K. Walters, p. 155. New York: Plenum
155. McDaniel, F. D., Duggan, J. L. 1976. In *Beam-Foil Spectroscopy*, ed. I. A. Sellin, D. J. Pegg, 2:519. New York: Plenum
156a.Kocbach, L. 1976. *Phys. Norv.* 8:187
156b.Gray, T. J. et al 1976. *Phys. Rev. A* 13:1344
157. Meyerhof, W. E., Anholt, R. 1976. In *Abstr. Int. Conf. Inner-Shell Ionization Phenom. 2nd*, ed. W. Mehlhorn, p. 56. Freiburg, Germ.: Univ. Freiburg
158. Foster, C. et al 1976. *J. Phys. B* 9:1943
159. Andersen, J. U. et al 1976. *J. Phys. B* 9:3247
160. Sethuraman, V., Thorson, W. R., Lebeda, C. F. 1973. *Phys. Rev. A* 8:1316
161. Anholt, R. 1975. PhD thesis. Univ. Calif., Berkeley. Published as Lawrence Rad. Lab. Rep. LBL-4312
162. Betz, W., Heiligenthal, G., Müller, B., Oberacker, V., Reinhardt, J., Schäfer W., Soff, G., Greiner, W. 1976. *Int. Summer Sch. Nucl. Phys.*, Predal, Romania. Unpublished
163. Kavanagh, T. M. et al 1970. *Phys. Rev. Lett.* 25:1473
164a.Meyerhof, W. E. 1976. See Ref. 56, p. 64
164b.Meyerhof, W. E., Anholt, R., Eichler, J., Salop, A. 1977. *Phys. Rev. A.* To be published
165. Woelfli, W. et al 1975. *Phys. Rev. Lett.* 35:656
166. Hoogkamer, Th. P. et al 1976. *J. Phys. B* 9:L145
167. Åberg, T., Jamison, K. A., Richard, P. 1976. *Phys. Rev. Lett.* 37:63
168. Briand, J. P. 1976. *Phys. Rev. Lett.* 37:59
169. Knudson, A. R. et al 1976. *Phys. Rev. Lett.* 37:679
170. Heisenberg, W. 1925. *Z. Phys.* 32:841
171. Condon, E. U. 1930. *Phys. Rev.* 36:1121
172. Goudsmit, S., Gropper, L. 1931. *Phys. Rev.* 38:225
173. Woelfli, W., Betz, H. D. 1976. *Phys. Rev. Lett.* 37:61
174. Woelfli, W. et al 1976. See Ref. 6b, p. 272
175. Stoller, Ch. et al 1976. *Phys. Lett. A* 58:18
176. Vinti, J. P. 1932. *Phys. Rev.* 42:632
177. Kelly, H. P. 1976. *Phys. Rev. Lett.* 37:386
178. Khristenko, S. V. 1976. *Phys. Lett. A* 59:202
179. Kawatsura, K. et al 1976. *Phys. Lett. A* 58:446
180a.Moiseyev, N., Katriel, J. 1976. *Phys. Lett. A* 58:303
180b.Greenberg, J. S., Vincent, P., Lichten, W. 1977. *Phys. Rev. A* 16:964
180c.Mitchell, I. V., Lennard, W. N., Phillips, D. 1977. To be published
181. Brenot, J. C. et al 1975. *Phys. Rev. A* 11:1245, 1933
182. Fastrup, B., Hermann, G., Smith, K. J. 1971. *Phys. Rev. A* 3:1591
183. Fastrup, B., Hermann, G. 1971. *Phys. Rev. A* 3:1955
184. Fastrup, B., Hermann, G., Kessel, Q. C. 1971. *Phys. Rev. Lett.* 27:771
185. Fastrup, B., Crone, A. 1972. *Phys. Rev. Lett.* 29:825

186. Fastrup, B. et al 1974. *Phys. Rev. A* 9:2518
187. Laegsgaard, E., Andersen, J. U., Lund, M. 1974. *Phys. Fenn.* 9:Suppl. 1, p. 49
188. Clark, D. L. et al 1975. *J. Phys. B* 8:L378
189. Laegsgaard, E., Andersen, J. U., Feldman, L. C. 1972. *Phys. Rev. Lett.* 29:1206
190. Bohr, N. 1948. *Mat. Fys. Medd. Dan. Vid. Selsk.* 18, No. 8
191. Everhart, E., Stone, G., Carbone, R. J. 1955. *Phys. Rev.* 99:1218
192. Vader, R. J. et al 1976. *Phys. Rev. A* 14:62
193. Amundsen, P. A., Kocbach, L. 1975. Unpublished
194. Ciochetti, G., Molinari, A. 1965. *Nuovo. Cimento B* 40:69
195. Luz, N., Sackmann, S., Lutz, H. O. 1976. *Verh. Deutsch. Phys. Ges.* 11:120
196. Garcia, J. D. 1977. Private communication
197. Schmid, G. B., Garcia, J. D. 1976. *J. Phys. B.* 9:L219
198. Bøving, E. 1976. Unpublished
199. Cocke, C. L., Randall, R. 1973. *Phys. Rev. Lett.* 30:1016
200. Cocke, C. L. et al 1976. *Phys. Rev. A* 14:2026
201. Hopkins, F. et al 1976. *Phys. Rev. A* 13:74
202. Saris, F., Hoogkamer, Th. P. 1977. In *Atomic Physics 5*, ed. R. Marrus, M. H. Prior. H. A. Shugart, p. 509. New York: Plenum
203. Yamaji, S. et al 1976. *Z. Phys. A* 278:69; *J. Phys. C* 2:L189
204. Alder, K. et al 1956. *Rev. Mod. Phys.* 28:432
205. Jakubassa, D. H., Kleber, M. 1975. *Z. Phys. A* 273:29
206. Reinhardt, J., Soff, G., Greiner, W. 1976. *Z. Phys. A* 276:285
207. Chen, J. C. Y., Ishihara, T., Watson, K. M. 1975. *Phys. Rev. Lett.* 35:1574
208. Anholt, R., Salin, A. 1977. *Phys. Rev. A* 16:799
209. Kienle, P. et al 1973. *Phys. Rev. Lett.* 31:1099
210. Anholt, R., Saylor, T. K. 1976. *Phys. Lett. A* 56:455
211. Schnopper, H. W. et al 1974. *Phys. Lett. A* 47:61
212. Folkmann, F. et al 1975. *Z. Phys. A* 275:229
213. Ishii, K., Morita, S., Tawara, H. 1976. *Phys. Rev. A* 13:131
214. Sohval, A. R. et al 1975. *J. Phys. B* 8:L426
215. Schnopper, H. W. et al 1972. *Phys. Rev.*

216. Betz, H. D. et al 1975. See Ref. 22, p. 520
217. Kleber, M., Jakubassa, D. H. 1975. *Nucl. Phys. A* 252:152
218. Sohval, A. R. et al 1976. *J. Phys. B* 9:L47
219. Briggs, J. S., Dettmann, K. 1977. *J. Phys. B* 10:1113
220. Richard, P., Moore, C. F., Olsen, D. K. 1973. *Phys. Lett. A* 43:519
221. Åberg, T., Utriainen, J. 1969. *Phys. Rev. Lett.* 22:1346
222. Åberg, T. 1975. See Ref. 1, p. 353
223. Presser, G. 1976. *Phys. Lett. A* 56:273
224. Åberg, T. 1971. *Phys. Rev. A* 4:1735
225a. Scofield, J. H. 1974. *Phys. Rev. A* 9:1041
225b. Jamison, K. A. et al 1976. *Phys. Rev. A* 14:937
226. Oona, H., Garcia, J. D. 1973. In *Abstr. Int. Conf. Phys. Electron. At. Collisions, 8th,* ed. B. B. Čobič, M. V. Kurepa, p. 716. Belgrade: Inst. Phys.
227. McWherter, J. et al 1973. *Phys. Lett. A* 45:57
228. Smith, R. K., Müller, B., Greiner, W. 1975. *J. Phys. B* 8:75
229. Thorson, W. R., Choi, J. H. 1977. *Phys. Rev. A* 15:550
230a. Anholt, R. 1976. *J. Phys. B* 9:L249
230b. Heinig, K. H., Jäger, H. U., Münchow, L., Richter, H., Woittennek, H. 1977. *Phys. Lett.*
231. Briggs, J. S., Macek, J. H., Taulbjerg, K. 1977. *Abstr. Int. Conf. Phys. Electron. At. Collisions, 10th,* ed. M. Barat, J. Reinhardt, p. 908. Paris: Commisariat à l'Energie Atomique
232. Smith, R. K., Greiner, W. 1975. Unpublished
233. Briggs, J. S. 1974. *J. Phys. B* 7:47
234. Meyerhof, W. E. et al 1974. *Phys. Rev. Lett.* 32:1279
235. Müller, B. 1975. See Ref. 22, p. 481
236. Betz, H. D. et al 1975. *Phys. Rev. Lett.* 34:1256
237. Heinig, K. H., Jäger, H. U., Richter, H., Woittennek, H., Frank, W., Gippner, P., Kaun, K. H., Manfrass, P. 1977. *J. Phys. B* 7:1321
238a. Heinig, K. H. et al 1976. *Phys. Lett. B* 60:249
238b. Anholt, R., Meyerhof, W. E. 1977. *Phys. Rev. A* 16:913
239. Müller, B., Smith, R. K., Greiner, W. 1974. *Phys. Lett. B* 49:219
240a. Gros, M., Müller, B., Greiner, W. 1976. *J. Phys. B* 9:1849
240b. Gros, M., Greenland, P. T., Greiner, W. 1977. *Z. Phys. A* 280:31

241. Dettmann, K. 1976. See Ref. 6b, p. 57
242. Fano, U., Macek, J. H. 1973. *Rev. Mod. Phys.* 45:553
243. Bates, D. R., McCarroll, R. 1958. *Proc. R. Soc. London Ser. A* 245:175
244. Briggs, J. S., Dettmann, K. 1974. *Phys. Rev. Lett.* 33:1123
245. Sohval, A. R. et al 1976. *J. Phys. B* 9: L25
246. Folkmann, F. et al 1974. *Nucl. Instrum. Methods* 116:487
247. Gippner, P. 1975. Joint Inst. Nucl. Res., Dubna, USSR, Rep. E7-8843
248. Anholt, R. 1975. Unpublished
249. Meyerhof, W. E. et al 1973. *Phys. Rev. Lett.* 30:1279; Corrigendum 1974. 32:504
250. Mollenauer, J. F. 1962. *Phys. Rev.* 127:867
251. Uhlig, R. P. 1964. *Natl. Bur. Stand. J. Res. A* 68:401
252. Mokler, P. H., Stein, H. J., Armbruster, P. 1972. *Phys. Rev. Lett.* 29:827
253. Macdonald, J. R., Brown, M. D., Chiao, T. 1973. *Phys. Rev. Lett.* 30:471
254. Mokler, P. H. et al 1975. In *At. Phys. 4*, ed. G. zu Putlitz, E. W. Weber, A. Winnacker, p. 301. New York: Plenum
255. Meyerhof, W. E. 1976. *Science* 193:839
256. Saylor, T. K. 1973. Unpublished
257. Davis, C. K., Greenberg, J. S. 1974. *Phys. Rev. Lett.* 32:1215
258. Tawara, H., Ishii, K., Morita, S. 1976. *Nucl. Instrum. Methods* 132:503
259. Anholt, R., Meyerhof, W. E. 1976. See Ref. 56, p. 60
260. Laubert, R. et al 1976. *Phys. Rev. Lett.* 36:1574
261. Schmidt-Böcking, H., Bethge, K., Lichtenberg, W., Nolte, G., Schuch, R., Schulé, R., Specht, H. J., Tserruya, I. 1977. See Ref. 231, p. 922
262. Schnopper, H. W., Delvaille, J. P. 1973. See Ref. 151, p. 481
263. Betz, H.-D. et al 1974. *Abstr. Int. Conf. At. Phys. 4th*, ed. G. zu Putlitz, E. W. Weber, A. Winnacker, p. 670. Heidelberg, Germ.: Univ. Heidelberg
264a. Schulé, R. et al 1976. *Abstr. Int. Conf. Phys. X-Ray Spectra*, ed. R. D. Deslattes, p. 289. Gaithersburg, Md: Natl. Bur. Stand.
264b. Schulé, R., Schmidt-Böcking, H., Tserruya, I. 1977. *J. Phys. B* 10:889
265. Bethe, H. A., Salpeter, E. E. 1957. *Quantum Mechanics of One- and Two-Electron Atoms*, p. 408. New York: Academic
266a. Tanis, J. A. et al 1976. See Ref. 264a, p. 186
266b. Betz, H. D., Spindler, E., Bell, F. 1977.

See Ref. 231, p. 928
267. Trautvetter, H. P., Greenberg, J. S., Vincent, P. 1976. *Phys. Rev. Lett.* 37:202
268. Stott, W. R., Waddington, J. C. 1976. *Phys. Lett. A* 56:258
269. Anholt, R., Saylor, T. K. 1976. *Phys. Lett. A* 56:455
270. Greenberg, J. S. 1975. Unpublished
271. Kaun, K. H., Frank, W., Manfrass, P. 1976. See Ref. 6b, p. 68
272. Anholt, R. 1976. Unpublished
273. Meyerhof, W. E., Hoogkamer, Th. P., Saris, F. 1976. See Ref. 56, p. 56; Hoogkamer, Th. P. 1977. PhD thesis. FOM Inst., Amsterdam, Neth.
274. Bell, F. et al 1975. *Phys. Rev. Lett.* 35:841
275. Meyerhof, W. E. 1975. See Ref. 22, p. 470
276. Tserruya, I. et al 1976. *Phys. Rev. Lett.* 36:1451
277. Schmidt-Böcking, H. et al 1976. *Nucl. Instrum. Methods* 132:489
278. Robertson, H. P. 1949. *Rev. Mod. Phys.* 21:374
279. Yang, C. N. 1949. *Phys. Rev.* 74:764
280. Meyerhof, W. E. et al 1975. *Phys. Rev. A* 12:2641
281. Greenberg, J. S., Davis, C. K., Vincent, P. 1974. *Phys. Rev. Lett.* 33:473
282. Wölfli, W. et al 1976. *Phys. Rev. Lett.* 36:309
283. Frank, W. et al 1976. *Z. Phys. A* 277:333
284. Thoe, R. S. et al 1975. *Phys. Rev. Lett.* 34:64
285. Thoe, R. S., Sellin, J. A., Peterson, R. S., Liao, K. M., Pegg, D. J., Forrester, J. P., Griffin, P. M. 1976. See Ref. 155, p. 477
286. Stoller, Ch., Wölfli, W., Bonani, G., Stöckli, M., Suter, M. 1977. See Ref. 231, p. 912
287. Betz, W. et al 1976. See Ref. 6b, p. 79
288. Smith, K. et al 1974. *Phys. Rev. Lett.* 32:554
289. Meyerhof, W. E. et al 1975. *Phys. Rev. A* 11:1083
290. Armbruster, P. 1976. See Ref. 6b, p. 21
291. Behnke, H. H., Armbruster, P., Folkmann, F., Macdonald, J. R., Mokler, P. H. 1977. See Ref. 231, p. 156
292. Backe, H., Berdermann, E., Bokemeyer, H., Greenberg, J. S., Kaŕkeléit, E., Kienle, P., Kozhuharov, Ch., Handschug, L., Nakayama, Y., Richter, L., Stettmeier, H., Weik, F., Willwater, R. 1977. See Ref. 231, p. 162
293. Backe, H. et al 1975. Rep. No. 67, Lab. Nucl. Phys., Tech. Hochsch., Darmstadt, Germ.

294. Oberacker, V., Soff, G., Greiner, W. 1976. *Phys. Rev. Lett.* 36:1024
295. Meyerhof, W., Anholt, R., El Masri, Y., Cline, D., Stephens, F. S., Diamond, R. 1977. *Phys. Lett. B* 69:41
296. Soff, G. et al 1977. *Phys. Rev. Lett.* 38:592

297. Kozhuharov, Ch., Kienle, P., Berdermann, E., Bokemeyer, H., Greenberg, J. S., Nakayama, Y., Vincent, P., Backe, H., Handschug, L., Kankeleit, E. 1977. Submitted to *Int. Conf. Nucl. Struct.*, Tokyo

Ann. Rev. Nucl. Sci. 1977. 27: 333–51
Copyright © 1977 by Annual Reviews Inc. All rights reserved

DELAYED PROTON RADIOACTIVITIES

Joseph Cerny[1]
Department of Chemistry and Lawrence Berkeley Laboratory, Berkeley, California 94720

J. C. Hardy[2]
Chalk River Nuclear Laboratories, Chalk River, Ontario K0J 1J0, Canada and CERN, Geneva, Switzerland

CONTENTS

1 INTRODUCTION

With continuing advances in the development of accelerators and experimental techniques, the study of nuclear properties has steadily expanded from where it began, at the naturally occurring stable isotopes, outwards towards nuclei with radically different relative numbers of constituent neutrons and protons. Although many such exotic nuclei are now accessible, it is only among those that are proton-rich where we reach to the limits of nucleon stability in any but the lightest elements. As we approach this limit, successively more neutron-deficient isotopes of a given element show rapidly increasing isobaric mass differences and marked decreases in their proton-binding energies. This leads to the appearance of radioactive decay modes, unobservable nearer stability, that involve the emission of protons.

[1] Work supported in part by the United States Energy Research and Development Administration.
[2] Permanent address: AECL, Chalk River.

333

Three types of proton decay have been observed or are predicted to arise: β-delayed proton decay, proton radioactivity, and two-proton radioactivity. The first of these is a two-step process in which a nucleus (the precursor) β decays to states in its daughter (the emitter) that are unbound to prompt proton emission; consequently, the decay process is characterized by the appearance of energetic protons possessing the half-life of the initial β decay. The latter two decay modes (with two-proton radioactivity still unobserved) involve the direct emission of energetically unbound protons or proton pairs from a nucleus, with a decay time determined by the Coulomb and angular momentum barriers; for these processes, an experimentally observable lifetime will usually be associated with low-energy protons.

The subject of delayed proton radioactivities was comprehensively treated a little more than a decade ago by Goldanskii (1); the purpose of the present brief review is to provide a summary and guide to the developments in this field since that time, as well as to note a few of its more recent highlights. Additional major reviews of this rapidly growing research area have also appeared by Hardy (2, 3) and by Karnaukhov (4, 5). General studies of the properties of nuclei far from stability are covered in three conferences (6–8) and in a book by Baz et al (9).

Very substantial growth has occurred in our knowledge of β-delayed proton precursors since Goldanskii's review (1). The number of known precursors has grown from 10 to 42, and the phenomenon itself has been shown to be not just a curiosity, but a rich and diverse source of nuclear information. In what follows we divide our discussion into two sections, separating the lighter precursors, for which $Z > N$, from the heavier ones with $Z < N$. This is a natural separation, since the corresponding proton spectra and their analyses are qualitatively different. Among light nuclei, where states in the emitter are well separated, individual transitions can be clearly resolved and the intensity of each peak in the proton spectrum can be directly related to the intensity of the preceding β transition. Studies of such nuclei focus on the spectroscopy of these β transitions, which include the superallowed transition to the isobaric analog state. In several cases, this has even led to a determination of the isospin purity of excited states in the emitter.

For the heavier precursors with $Z < N$, the high level-densities in the emitter, together with the absence of a superallowed branch, lead to proton energy spectra that form a bell-shaped continuum upon which may be seen peaks created by Porter-Thomas fluctuations in the β decay transition probabilities. Analysis of these spectra follows a statistical approach that yields information on β-decay strength functions, and the average properties of excited states in the emitters.

With regard to the other two radioactive decay modes leading to proton emission, the first and only unambiguous example of proton radioactivity was observed in 1970 (10, 11) in the decay of 247-msec ^{53m}Co, a $J^{\pi} = 19/2^-$ many-particle isomer. The experimental results are discussed below, together with recent theoretical predictions of other possible many-particle isomeric states that may provide additional examples of proton radioactivity (12) or of the still-sought two-proton radioactivity (13).

2 BETA-DELAYED PROTON PRECURSORS WITH $Z > N$

2.1 Nuclei in the $4n+1$ Mass Series from ^9C to ^{61}Ge

The series of $A = 4n+1$, $T_z = -\frac{3}{2}$ β-delayed proton precursors, first discovered in 1963 by Barton et al (14), is particularly favored by strong proton branches that range from 12 to 100% per disintegration. The employment of proton and ^{3}He beams in $(p, 2-3n)$, and $(^3He, 2n)$ reactions permitted the observation of the ^{9}C through ^{41}Ti precursors by the time of Goldanskii's review (1). Recent research has focused on (a) the reinvestigation of these nuclides under high-resolution, low-background conditions using gas-sweeping and helium-jet techniques (15), and (b) extension of the series through ^{61}Ge (16) by the utilization of heavy-ion beams in reactions such as ^{40}Ca$(^{24}Mg, 3n)^{61}$Ge.

Table 1 (17–28) summarizes some of the properties of the known β-delayed proton precursors with $T_z = -\frac{3}{2}$, including ^{23}Al, which is not discussed until section 2.2. (In the table, $Q_\beta - B_p$ gives a measure of the energy available for proton emission, Q_β being the total β-decay energy of the precursor and B_p the proton separation

Table 1 Observed β^+-delayed proton precursors with $T_z = -\frac{3}{2}$

Precursor	Production reaction[a]	$t_{1/2}$ (msec)	$Q_\beta - B_p$ (MeV)[b]	Proton branching ratio	Reference[c]
N odd					
$^9_6C_3$	^{10}B$(p, 2n)$	127 ± 1	16.68	~ 1.0	17
$^{13}_8O_5$	^{14}N$(p, 2n)$	8.9 ± 0.2	15.82	0.12	18
$^{17}_{10}Ne_7$	^{16}O$(^3He, 2n)$	109 ± 1	13.93	0.99	19
$^{21}_{12}Mg_9$	^{20}Ne$(^3He, 2n)$	123 ± 3	10.67	0.33	20
$^{25}_{14}Si_{11}$	^{24}Mg$(^3He, 2n)$	221 ± 3	10.47	0.32	21
$^{29}_{16}S_{13}$	^{28}Si$(^3He, 2n)$	188 ± 4	11.04	0.47	22
$^{33}_{18}Ar_{15}$	^{32}S$(^3He, 2n)$	174 ± 2	9.34	0.34	19
$^{37}_{20}Ca_{17}$	^{36}Ar$(^3He, 2n)$	175 ± 3	9.78	0.76	23
$^{41}_{22}Ti_{19}$	^{40}Ca$(^3He, 2n)$	80 ± 2	11.77	1.0	23
$^{45}_{24}Cr_{21}$	^{32}S$(^{16}O, 3n)$	50 ± 6	10.80	0.25	24
$^{49}_{26}Fe_{23}$	^{40}Ca$(^{12}C, 3n)$	75 ± 10	11.00	0.60	25
$^{53}_{28}Ni_{25}$	^{40}Ca$(^{16}O, 3n)$	45 ± 15	11.63	0.45	26
$^{57}_{30}Zn_{27}$	^{40}Ca$(^{20}Ne, 3n)$	40 ± 10	13.99	0.65	26
$^{61}_{32}Ge_{29}$	^{40}Ca$(^{24}Mg, 3n)$	~ 40	~ 12.2	~ 0.50	16
N even					
$^{23}_{13}Al_{10}$	^{24}Mg$(p, 2n)$	470 ± 30	4.66	—	27

[a] The most recently used production mode is given. A detailed list of other options appears in (3).

[b] With one exception, the values quoted are either direct experimental measurements or the results of applying the isobaric multiplet-mass equation where three members of an isospin quartet are known. The value for ^{61}Ge was derived from Coulomb energy systematics.

[c] Where more than one reference exists, the one listed has been chosen to best illustrate the proton energy spectrum. See (3) and (28) for a complete list.

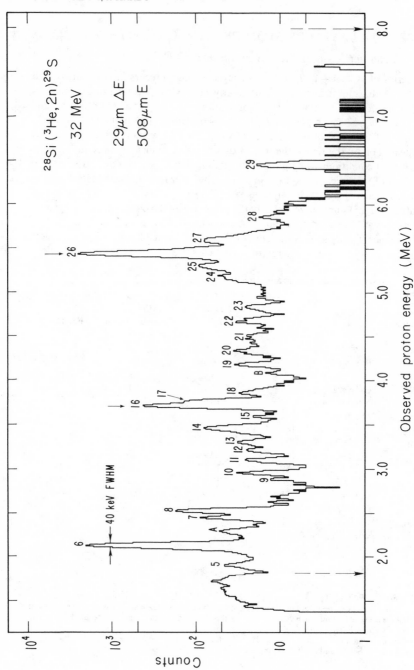

Figure 1   The proton spectrum observed following the decay of ^{29}S. Solid vertical arrows denote proton groups arising from the decay of the isobaric analog state in ^{29}P, while the dashed vertical arrows indicate the energy region over which protons could be reliably observed. (Groups A and B arise from contaminants.)

energy of the emitter.) As an example of the proton spectrum observed in the decay of nuclides in this series, Figure 1 presents results from a recent, high-resolution study of 188-msec ^{29}S, produced via the ^{28}Si $(^3$He$,2n)$ reaction at 32 MeV, and studied using helium-jet techniques and ΔE–E detector telescope identification of the protons (22). The superallowed β-decay transition from the $(T = \frac{3}{2}, T_z = -\frac{3}{2})$ precursor populates the isobaric analog state $(T = \frac{3}{2})$ in the $T_z = -\frac{1}{2}$ emitter, which promptly decays by isospin-forbidden proton emission; the corresponding proton groups are marked by arrows in the figure. All other proton groups arise from isospin-allowed proton emission from $T = \frac{1}{2}$ states in the emitter fed by allowed (Gamow-Teller) β decay. Figure 2 illustrates the decay scheme of ^{29}S. Analysis of these results, and others like them, on energy levels and transition rates leads to useful tests of nuclear-model wave functions and to investigations of isospin mixing in excited states of the emitter. We discuss both topics, using ^{29}S as an example.

Recent studies of these β-delayed proton precursors (20, 22) have been used to test the β-decay transition rates predicted by Wildenthal and his collaborators, using wave functions determined from large-basis shell-model calculations. A comparison of experimental and theoretical excitation energies, J^π values, and log ft values for states in ^{29}P that are fed in the β decay of ^{29}S is also shown in Figure 2. To make the calculations tractable, it was necessary to truncate the complete sd-shell basis space by requiring six or more particles in the $d_{\frac{5}{2}}$ subshell and, at least partly for this reason, the theoretical levels lie too high in energy by 0.5–1.0 MeV. Apart from that, though, there is generally excellent agreement between experiment and theory up to an excitation energy of 5.9 MeV; rectifying the discrepancies in the log ft values for the highly hindered transitions to the states at 3.1 and 4.1 MeV would require only minor changes in the wave functions. Furthermore, of the 21 $J^\pi = \frac{3}{2}^+, \frac{5}{2}^+, \frac{7}{2}^+, T = \frac{1}{2}$ levels predicted to lie above the proton-separation energy in ^{29}P and to be fed with at least a 0.1% branch by allowed Gamow-Teller β decay, 18 levels with appropriate log ft values were observed (and a few proton groups could not be assigned).

Similar comparisons of transition rates to a large number of levels are not so far possible for the heavier precursors in the fp shell (^{45}Cr–^{61}Ge). This is a consequence of the drastic reduction in yield found in the (heavy-ion, 3n) reactions: while the cross section for the reaction ^{20}Ne $(^3$He$,2n)^{21}$Mg is ~ 700 μb, that for ^{40}Ca$(^{24}$Mg$,3n)^{61}$Ge is ~ 50 nb (16). At the same time, the superallowed transition to the analog state becomes more predominant (up to $\sim 50\%$ of the total decay strength) with increasing precursor mass, so it is primarily proton groups from the decay of the analog state that are observed. However, the knowledge this yields of the excitation energy of the isobaric analog state often permits an excellent prediction of the mass of the precursor from the very reliable isobaric multiplet mass equation (29, 30). This mass can then, in turn, be used to evaluate other, more general approaches towards mass predictions of proton-rich nuclei far from stability.

The superallowed transition from the precursor to the isobaric analog state in the emitter has been observed in all the nuclei in this series from ^{17}Ne on. In

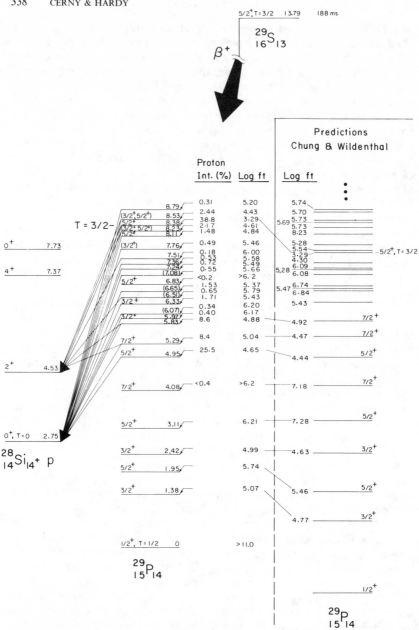

Figure 2   The observed decay scheme of ^{29}S. Shell-model calculations by W. Chung and H. Wildenthal (private communication) are shown for comparison. The log *ft* values for states below 3.5 MeV were taken from β^- decay of ^{29}Al to mirror levels in ^{29}Si.

several cases, where β decay to proton bound states is weak, a very accurate measurement of the absolute transition intensity, or log ft value, for the super-allowed branch has been obtained. A comparison of this absolute transition rate with the theoretically expected rate (2) for a transition between perfect $T = \frac{3}{2}$ analog states permits one to extract the isospin "purity" of the lowest analog state in the emitter; in this way the extent of mixing between this high-lying $T = \frac{3}{2}$ state and the large number of nearby $T = \frac{1}{2}$ states can be ascertained.

The isospin purities of the lowest $T = \frac{3}{2}$ states in ^{17}F, ^{33}Cl, and ^{41}Sc have been determined to be $>95\%$ (19), $81\pm9\%$ (19), and $91\pm4\%$ (23), respectively. The substantial isospin mixing of the $T = \frac{3}{2}$ state in ^{33}Cl with nearby levels of the same J^{π}, but lower isospin, may also reflect itself in enhanced transition rates to these $T = \frac{1}{2}$ levels. Such an effect can be seen in Figure 2 for the decay of ^{29}S. The isospin purity of the 8.38 MeV $T = \frac{3}{2}$ state in ^{29}P can be estimated to be $95\pm^{5}_{8}\%$ (22). If that were to indicate significant mixing, it could imply that the rather low log ft values observed to states at 8.11, 8.23 and 8.53 MeV (all, like the analog state, with known or possible $J^{\pi} = \frac{5}{2}^{+}$) are the result of analog-state admixtures. At best, the evidence for this effect is so far only circumstantial, and it is complicated by the fact that Gamow-Teller β transitions too are expected to be enhanced in the region of the analog state (31).

Finally, it should be noted that it is through isospin impurities in these $T = \frac{3}{2}$ states that they have also been produced by proton scattering. Such experiments, guided by earlier delayed-proton results, have produced precise values for their excitation energies and proton partial widths (29, 32, 33).

2.2   Other Light-Mass Precursors

A few additional, weak β-delayed proton precursors have been observed in the light elements among nuclei with $T_z = -1$ and $-\frac{3}{2}$ ($A = 4n+3$), and strong precursors are expected in the still undiscovered $T_z = -2$ series. Hardy (2, 3) reviewed in detail the data on those odd-odd $T_z = -1$ nuclei between ^{8}B and ^{44}V that are known to be β-delayed proton or α precursors. So far, only ^{32}Cl (34) and ^{40}Sc (35) have been observed to emit protons, both relatively weakly (0.5 $\times$ 10^{-3} and 5.0 $\times$ 10^{-3} protons per disintegration, respectively).

After mass measurements via the ^{24}Mg$(p, ^{6}$He$)^{19}$Na and ^{28}Si$(p, ^{6}$He$)^{23}$Al reactions (36) established that ^{23}Al was the lightest nucleon-stable member of the $A = 4n+3$, $T_z = -\frac{3}{2}$ series, this nuclide was observed via its weak β-delayed proton decay (see Table 1). Only a single proton group at low energy was detected. This is because the proton-binding energy in the emitters in the $4n+3$ mass series is much larger than in the $4n+1$ series already discussed, and few states at quite high excitation are strongly fed by β decay. Three more members of this $4n+3$ mass series (^{27}P, ^{31}Cl, and ^{35}K) are known to be nucleon-stable, but have not been otherwise characterized (30); ultimately, they can perhaps also be detected through their weak emission of low-energy, β-delayed protons. Unfortunately, though, the high proton separation energy precludes any observation of proton emission following the superallowed β-decay branch from most members of this series (2).

Although nuclei in the series with $T_z = -2$, $A = 4n$, beginning with ^{20}Mg and extending beyond ^{52}Ni, are expected to be nucleon-stable (30, 37) and to be strong β-delayed proton precursors—in fact, quite similar to those with $T_z = -\frac{3}{2}$ and $A = 4n+1$—numerous attempts to observe them have so far been unsuccessful. The most recent search, for ^{24}Si, by Robertson et al (38) via the ^{24}Mg(^{3}He, $3n$)^{24}Si reaction at ~ 60 MeV indicates that the cross section for producing ^{24}Si at this bombarding energy is probably less than 2% of the ^{24}Mg(^{3}He, $2n$)^{25}Si cross section. In their experiment, they employed a recoil-fragment time-of-flight detector to determine the mass of the nucleus recoiling after proton emission; their results clearly show that in any future searches such mass identification of the recoiling nucleus or of the precursor itself (by an on-line mass analyzer) is necessary in order to detect decays from the $T_z = -2$ nuclide of interest in a very high proton background from the $T_z = -\frac{3}{2}$ precursors closer to stability.

3   BETA-DELAYED PROTON PRECURSORS WITH $Z < N$

3.1   General Features

Even for a precursor as light as ^{29}S, it is evident from Figure 2 that many β transitions contribute to the experimental proton spectrum. Where detailed model calculations are possible, a comparison with such experimental data can provide useful theoretical constraints, but where calculations must be severely truncated or are not possible at all, it is more natural to view the experimental data in terms of average nuclear properties. This means, for example, that one abandons the idea that each β-transition matrix element must be measured and understood separately; and replaces it with a broader view, focusing instead on the average behavior of the β matrix element (squared) per unit energy interval [i.e. the strength function (39)], regardless of the number of final states involved. Among light precursors, this approach has been followed with some success (31), but its application remains a matter of philosophical preference. For the heavier precursors, though, one has no other choice, since the proton spectra are essentially continuous: the density of states is so high that individual transitions cannot be resolved from one another. Because the experiment is then sensitive only to average properties, its analysis must reflect that sensitivity.

Conceptually, this approach is uncomplicated. For an individual proton transition between a state i in the emitter and a state f in the daughter, the intensity I_p^{if} is determined by two factors: (a) the beta-decay (including electron capture) branching ratio, I_β^i, from the precursor for populating state i; and (b) the branching ratio for subsequent proton emission from that level to state f. Specifically,

$$I_p^{if} = I_\beta^i \frac{\Gamma_p^{if}}{\Gamma_p^i + \Gamma_\gamma^i},$$

with

$$\Gamma_p^i = \sum_f \Gamma_p^{if}, \qquad\qquad\qquad 1.$$

where Γ_p^{if} is the partial width for proton emission between states i and f, Γ_p^i is the total proton decay width of state i, and Γ_γ^i is its γ-decay width. Where individual transitions cannot be resolved, a statistical average of the individual I_p^{if} is observed, i.e.

$$I_p(E_p) = \sum_{if} \langle I_p^{if} \rangle_{E_p}.$$ 2.

Here, $\langle \ \rangle$ denotes the statistical mean, with the sum being extended over all pairs of states i and f between which protons of energy E_p can be emitted.

To understand the behavior of $I_p(E_p)$, one must establish, in addition to the functional form of I_p^{if}, the statistical distributions governing the fluctuations of its component parts. Both I_β^i and Γ_p^{if} are proportional to the square of a nuclear matrix element; they are thus expected to scatter with a Porter-Thomas distribution (40). On the other hand, the γ-decay width Γ_γ^i involves decay to many states, so its fluctuations are small and can be neglected. The effects of these distributions on the average in Equation 2 have been considered in detail elsewhere (28, 41); the major conclusion for the present discussion is that one is led to expect significant fluctuations in the proton spectrum $I_p(E_p)$, giving rise to what may appear to be peaks above the continuum, simply as a result of the Porter-Thomas distribution of matrix elements and quite independently of any specific features of the nuclear structure. Therefore, a spectral analysis must begin, at least, by actually ignoring any apparent peak structure in the spectrum and treating only the average behavior.

The procedure has then been to use a parameterized functional form for I_β, Γ_p, and Γ_γ that has been determined independently, to calculate the proton spectrum using Equations 1 and 2, and finally, by adjusting a few of the parameters, to bring the calculated spectrum into agreement with experiment. This method, which is illustrated by an example in Section 3.2.1, leads to remarkably good agreement after a minimum of adjustment, but it does suffer from the difficulty that it only tests a combination of several components, which individually may not necessarily be well determined. It is now evident that for certain cases this difficulty can be effectively reduced by the addition of a direct experimental determination of Γ_p, made with a new technique (42) that involves the measurement of coincidences between delayed protons and X rays. This is described in Section 3.2.2.

3.2   Nuclei in the $4n+1$ Mass Series from ^{65}Ge to ^{81}Zr

3.2.1   DECAY PROPERTIES   Until recently, β-delayed proton precursors studied among the heavier nuclei have not fitted so neatly into a regular pattern as have the light precursors. With the observation (31, 43) of five members of a new series of precursors having $A = 4n+1$ and $T_z = +\frac{1}{2}$, the possibility of systematic studies above $A = 65$ now exists as well. The properties so far determined for these precursors are listed at the top of Table 2 (43–53), and the spectrum observed for ^{69}Se is shown in Figure 3a. All but ^{65}Ge were produced in heavy-ion–induced reactions; following bombardment, each target was mechanically transferred to a counting position and viewed by a counter telescope (for protons), as well as X-ray and γ-ray detectors, all in very close geometry.

Qualitatively, the proton spectrum of Figure 3a may easily be understood from Equations 1 and 2. The low-energy part of the spectrum reflects the increasing magnitude of Γ_p relative to Γ_γ, while at higher energies, where $\Gamma_p \gg \Gamma_\gamma$, it is the β-decay properties that predominate. A detailed calculation of the proton-decay properties has also been made (41–43). The beta intensity I_β was derived assuming allowed decay and a Gaussian strength function; this prescription for the strength function is based on sound, though general, theoretical principles (54) and its parameters established by fitting experimental β-decay lifetimes throughout the entire periodic table (55). The partial γ-decay widths Γ_γ were calculated assuming

Table 2 Observed β^+-delayed proton precursors with $Z < N$

Precursor	Production reaction[a]	$t_{1/2}$ (sec)	$Q_\beta - B_p$ (MeV)	Proton branching ratio	Reference[b]
$T_z = +\frac{1}{2}$ series					
$^{65}_{32}\text{Ge}_{33}$	$^{64}\text{Zn}(^3\text{He},2n)$	31 ± 1	2.30	1.3×10^{-4}	43
$^{69}_{34}\text{Se}_{35}$	$^{40}\text{Ca}(^{32}\text{S},2pn)$	27.4 ± 0.2	3.39	6×10^{-4}	43
$^{73}_{36}\text{Kr}_{37}$	$^{60}\text{Ni}(^{16}\text{O},3n)$	29 ± 1	3.60	7×10^{-3}	43
$^{77}_{38}\text{Sr}_{39}$	$^{40}\text{Ca}(^{40}\text{Ca},2pn)$	9.0 ± 1.0	3.85	$<2.5 \times 10^{-3}$	43
$^{81}_{40}\text{Zr}_{41}$	$^{52}\text{Cr}(^{32}\text{S},3n)$	5.9 ± 0.6	~ 5.0	—	d
Others, listed by element					
$^{99}_{48}\text{Cd}_{51}$	$\text{Sn}(p,3pXn)$	12.0 ± 3.0	4.27^c	—	e
$^{109}_{52}\text{Te}_{57}$	$^{96}\text{Ru}(^{16}\text{O},3n)$	4.4 ± 0.4	7.14	$\sim 3.0 \times 10^{-2\,c}$	44
$^{111}_{52}\text{Te}_{59}$	$^{102}\text{Pd}(^{12}\text{C},3n)$	19.3 ± 0.4	5.07	$\sim 1.0 \times 10^{-3\,c}$	45
$^{113}_{54}\text{Xe}_{59}$	$\text{Ce}(p,5pXn)$	2.8 ± 0.2	7.09^c	$\sim 4.0 \times 10^{-2\,c}$	46
$^{115}_{54}\text{Xe}_{61}$	$\text{Ce}(p,5pXn)$	18.0 ± 3.0	6.20	3.4×10^{-3}	47
$^{117}_{54}\text{Xe}_{63}$	$\text{Ce}(p,5pXn)$	65.0 ± 6.0	4.10	2.9×10^{-5}	48
$^{114}_{55}\text{Cs}_{59}$	$\text{La}(p,3pXn)$	0.7 ± 0.2	8.20^c	—	e
$^{116}_{55}\text{Cs}_{61}$	$\text{La}(p,3pXn)$	3.6 ± 0.2	6.43	2.7×10^{-3}	49
$^{118}_{55}\text{Cs}_{63}$	$\text{La}(p,3pXn)$	16.4 ± 1.2	4.70	4.2×10^{-4}	50
$^{120}_{55}\text{Cs}_{65}$	$\text{La}(p,3pXn)$	58.3 ± 1.8	2.65^c	7.0×10^{-8}	51
$^{117}_{56}\text{Ba}_{61}$	$^{92}\text{Mo}(^{32}\text{S},2p5n)$	1.9 ± 0.2	8.08^c	—	52
$^{119}_{56}\text{Ba}_{63}$	$^{92}\text{Mo}(^{32}\text{S},2p3n)$	5.3 ± 0.3	6.35	$9.0 \times 10^{-3\,c}$	49
$^{121}_{56}\text{Ba}_{65}$	$^{92}\text{Mo}(^{32}\text{S},2pn)$	29.7 ± 1.5	4.70	2.0×10^{-4}	49
$^{129}_{60}\text{Nd}_{69}$	$^{102}\text{Pd}(^{32}\text{S},2p3n)$	5.9 ± 0.6	5.95^c	—	52
$^{131}_{60}\text{Nd}_{71}$	$^{102}\text{Pd}(^{32}\text{S},2pn)$	24.0 ± 3.0	4.36^c	—	52
$^{133}_{62}\text{Sm}_{71}$	$^{106}\text{Cd}(^{32}\text{S},2p3n)$	32.0 ± 0.4	6.90^c	—	52
$^{135}_{62}\text{Sm}_{73}$	$^{106}\text{Cd}(^{32}\text{S},2pn)$	10.0 ± 2.0	5.30^c	—	52
$^{179}_{80}\text{Hg}_{99}$	$\text{Pb}(p,3pXn)$	1.09 ± 0.04	9.32^c	$\sim 2.8 \times 10^{-3}$	48
$^{181}_{80}\text{Hg}_{101}$	$\text{Pb}(p,3pXn)$	3.6 ± 0.3	~ 6.2	1.8×10^{-4}	48
$^{183}_{80}\text{Hg}_{103}$	$\text{Pb}(p,3pXn)$	8.8 ± 0.5	~ 5.0	3.1×10^{-6}	48

[a] The most recently used (or most favorable) production mode is given. A detailed list of other options appears in (3).

[b] Where more than one reference exists, the one listed has been chosen to best illustrate the proton energy spectrum. See (3) and (28) for a complete list.

[c] In the absence of any experimental measurement, we have shown a predicted value [(53) for $Q_\beta - B_p$, and (5) for proton-branching ratios].

[d] J. C. Hardy, unpublished.

[e] ISOLDE collaboration, unpublished.

$E1$ radiation with a Lorentzian strength function (56), while the proton widths were derived from the formula

$$\Gamma_p = T(2\pi\rho)^{-1}. \hspace{4cm} 3.$$

Here T is the total optical-model transmission coefficient for protons, which was computed with parameters derived from low-energy scattering data on nuclei in the same mass region; ρ is the density of relevant excited states in the emitter, calculated with the formulas of Gilbert & Cameron (57).

With ^{69}Se as an example, one can now see the application of such a calculation. Decay properties, such as the proton spectrum shape, proton branching ratio, and relative population of final states in ^{68}Ge, were calculated with various J^π values assumed for ^{69}Se. The known spins of neighboring odd-mass nuclei indicate $\frac{1}{2}^-$, $\frac{3}{2}^-$, or $\frac{5}{2}^-$ to be the most probable values, and indeed, the calculations for these spins yielded reasonable agreement with experiment, while higher spins produced order-of-magnitude discrepancies. The spectrum shape produced with the initial calculations (for $J^\pi = \frac{3}{2}^-$) is shown as the dashed line in Figure 3a. The calculation was then repeated, and the relative magnitude of Γ_p/Γ_γ varied to optimize agreement with the low energy portion of the proton spectrum. The optimized result is shown as the solid line in the same figure.

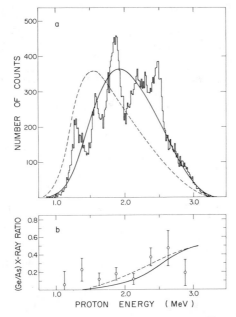

Figure 3   (a) Spectrum of protons observed following the decay of ^{69}Se; in the close geometry used, target-thickness effects led to an experimental energy resolution (FWHM) of $\sim$90 keV. (b) The ratio of Ge X rays relative to those from As (both measured in coincidence with protons), plotted as a function of coincident proton energy. The smooth curves in (a) and (b) are the results of calculations described in the text.

3.2.2.   LEVEL LIFETIMES   The experimental data, as they have already been described, yield a plausible method for determining the relative magnitudes of Γ_p and Γ_γ, but they are quite insensitive to the absolute values of either. However, the magnitude of Γ_p can be directly measured in a related experiment that compares the nuclear level lifetimes with the lifetime of an electron vacancy in the atomic K shell.

A heavy delayed-proton precursor decays to proton-unbound states in the emitter predominantly by electron capture. Any nucleus (with atomic number Z) that decays by electron capture produces simultaneously a vacancy in an atomic shell. If the excited states populated in the daughter $(Z-1)$ nucleus are unstable to proton emission, then the energy of the X ray emitted with the filling of the atomic vacancy will depend upon whether the proton has already been emitted (in which case the X ray would be characteristic of a $Z-2$ element) or not (a $Z-1$ element). If the nuclear and atomic lifetimes are comparable, then the K_α X rays observed in coincidence with protons will lie in two peaks whose relative intensities uniquely relate one lifetime with another. The measured result for the 0.06% proton branch from ^{69}Se is shown as the histogram in Figure 4.

To interpret these data, X-ray "standard" peak shapes were established for each relevant element using coincidences with specific known and prolifically produced γ rays, recorded at the same time as, but with much better statistics than, the p-X-ray coincidences. These are plotted, renormalized, as smooth curves in the figure. The X-ray peak intensity ratio could thus be determined as a function of the coincident proton energy and this result appears in Figure 3b. The observed ratios, when combined with the known (58) K-vacancy lifetime in arsenic, indicate nuclear lifetimes between 10^{-15} and 10^{-16} sec.

The calculations described in Section 3.2.1 were extended (41, 42) in order to

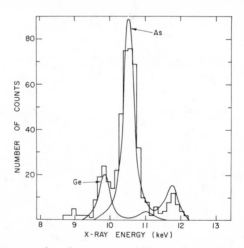

Figure 4   The histogram gives the spectrum of X rays observed in coincidence with all delayed protons from ^{69}Se. The smooth curves are X rays, measured simultaneously (with the same detector), in coincidence with specific known γ rays; they are normalized only in height to fit the histogram.

obtain the X-ray ratio as well. The dashed curve in Figure 3b corresponds to the original unoptimized parameters, while the solid line represents, as in Figure 3a, the final optimized result in which Γ_p/Γ_γ was determined from the singles data, and Γ_p itself from the ratio measurement. Actually, Γ_p was adjusted in the calculation by varying the level density (see Equation 3); the value derived for the density parameter, though somewhat different from the global prescription (57) originally employed, agreed well (41) with independent results on other odd arsenic isotopes. The optimized calculation also reproduced the experimental proton branching ratio and final state population with remarkable fidelity.

3.2.3 DECAY ENERGIES For high enough proton energies, competing γ decay can be neglected. Thus the upper part of the proton spectrum depends only upon β-decay properties and the "endpoint" energy should consequently give an accurate measure of $(Q_\beta - B_p)$. The validity of this conclusion can be seen from Figure 3a, where the endpoint energy used in the calculation was taken from an average of two other independent measurements involving quite different techniques (43). The agreement with the experimental spectrum is excellent, which indicates that in cases where other methods are impracticable, the value of $(Q_\beta - B_p)$ can, with care, be determined from the proton singles spectrum alone. Unfortunately, though, unexpected variations in the β-decay strength function could distort the spectrum shape (59), so it is necessary to examine carefully the spectrum in an extended region near the endpoint. The method is practical only in cases where the proton spectrum covers less than ~ 5 MeV.

3.2.4 CURRENT STATUS The conformity between experiment and theory for ^{69}Se has been demonstrated. Spectrum calculations for other precursors in the $T_z = \frac{1}{2}$ series appear equally successful (43), although full details are not yet available. When they are, one has reason to believe that a systematic case will have been made for using the techniques described here to determine not only average level widths, but also level densities, decay energies, and possibly even spins of nuclei that might otherwise be nearly inaccessible.

3.3 Other Heavy-Mass Precursors

The general method of analysis described in the preceding two sections did not originate with the $T_z = \frac{1}{2}$ series of precursor, but was developed much earlier (59, 60) to explain the decay of ^{111}Te and other isotopes of Xe and Hg. While it seemed successful, the precursors upon which it could be tested were scattered over such a wide region of masses that a systematic evaluation was effectively precluded. Even apart from the $T_z = \frac{1}{2}$ series, that situation is now changing. All known precursors with $A \geq 65$ are listed with a few of their properties in Table 2; evidently, quite recent work has added many new precursors in the region $48 \leq Z \leq 62$, so a broad systematic analysis of many neighboring decays will soon be possible.

Most of the precursors listed in the second part of Table 2 have been studied with the help of an on-line isotope separator, either the ISOLDE facility at CERN (46–51) or BEMS-2 at Dubna (52). They have been produced from spallation reactions initiated by 600-MeV protons or from heavy-ion reactions. All proton spectra exhibit the bell-shaped appearance already illustrated for ^{69}Se, but, characteristically, when moving to heavier nuclei—and higher-level densities—the occurrence of peaks in the spectra becomes reduced.

It is in the interpretation of the observed peak structure that the most interesting recent developments have occurred. If, as was noted in Section 3.2.1, the observed fluctuations in the proton spectrum are due entirely to the Porter-Thomas distribution of transition probabilities, then the variance in the proton intensity throughout the spectrum can be shown, for uncomplicated decay schemes, to depend only upon the experimental energy resolution and the density of levels in the emitter. Several techniques have been devised for determining the variance of an experimental spectrum, with the least ambiguous being one based upon autocorrelation functions (28). The experimental variance so obtained has been used to determine the level densities of ^{111}Sb and ^{115}I (from the decays of ^{111}Te and ^{115}Xe), with the results (28, 61) showing level densities that are somewhat higher than expected (57).

As these techniques mature, one can visualize the systematic investigation of level densities, which when combined with the X-ray lifetime measurements of Γ_p will also yield information on the average transition probability for proton emission over a wide range of excitation energies. The proton singles spectrum could then be used specifically for investigating the gross properties of nuclear β decay, a subject that not only has implications within nuclear physics, but also extends beyond it into astrophysics (See, e.g. Ref. 62).

4 PROTON RADIOACTIVITY

Proton radioactivity is the single-step emission of protons from a nuclear state (ground or isomeric) in which the proton has negative binding energy. It is directly analogous to α-particle radioactivity. Goldanskii's review (1) describes the early searches for nuclei decaying via this mode. Utilizing predicted ground-state masses, Goldanskii (see also Ref. 2) outlines the general region of nuclides in which penetration of a proton through the Coulomb (and centrifugal) barrier should lead to lifetimes longer than 10^{-12} sec, a possible lower limit for the process to be called radioactivity. His review also touches upon questions of relative lifetimes among the competing decay modes in nuclei so far from stability.

No nuclide whose ground state is proton-radioactive has yet been positively identified, though Karnaukhov and his collaborators (4, 63), in a search among very neutron-deficient rare-earth nuclei, advance the possibility that a light praseodymium (or lanthanum) isotope decays by proton emission from its ground state. In a series of difficult experiments, these authors obtained evidence that an emitter of 0.83-MeV protons with a ~ 1-sec half-life was produced in bombardments of ^{96}Ru with ^{32}S; though these results are best attributed to the decay of ^{121}Pr, further confirmatory experiments are required.

As noted in Section 1, however, this radioactive decay mode has been definitely observed in the decay of 247-msec ^{53m}Co (10, 11), a so-called spin-gap isomer (64), which arises from a $(f_{\frac{7}{2}})^{-3}$ configuration coupled to $J^\pi = \frac{19}{2}^-$. A proton energy spectrum (11) observed from the decay of this isomer following its production by the ^{54}Fe$(p, 2n)$ reaction is shown in Figure 5a; protons from recoils stopped in the

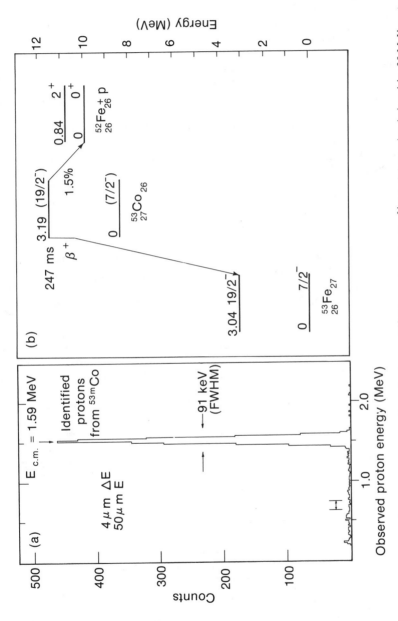

Figure 5 (*a*) An identified proton energy spectrum from the decay of ^{53m}Co produced by the ^{54}Fe(p, $2n$) reaction induced by 35-MeV protons. The horizontal arrow indicates the location of any possible transitions to the ^{52}Fe* (0.84-MeV) state. (*b*) The decay scheme of ^{53m}Co.

target were identified by a ΔE-E counter telescope. Only a single peak at a center-of-mass energy of 1.59 MeV was observed, in substantial contrast to the results from a β-delayed proton precursor, such as ^{29}S shown in Figure 1. This single-proton group, which was shown not to be in coincidence with positrons (10), can be attributed to a direct decay from ^{53m}Co to the ground state of ^{52}Fe, as shown in Figure 5b. The dominant decay mode of ^{53m}Co appears to be superallowed positron emission to its mirror ^{53m}Fe; the observed proton decay branch is estimated (11) to account for 1.5% of the decays, resulting in a partial half-life for proton radioactivity of ~ 17 sec.

This is a surprisingly long half-life for direct emission of a 1.59-MeV proton: transmission through the Coulomb and $l = 9$ centrifugal barriers leads to an expected half-life of ~ 60 nsec, so the ~ 17 sec partial half-life implies a reduced width of ~ 4 meV. Theoretical rates (Ref. 12 and G. Bertsch and G. Hamilton, private communication) have been calculated for this very hindered proton emission, assuming states with quite simple nuclear wave functions, to test one's knowledge of the high-momentum transfer and noncentral parts of the interaction. Bertsch and Hamilton predict the correct order of magnitude for this rate, finding their calculations to be insensitive to the single-particle potentials used to generate the wave functions, but highly sensitive to the residual nuclear interaction. A Yukawa or a central-plus-tensor G matrix, based on a realistic two-body potential, gave best agreement.

Although proton radioactivity from nuclear ground states should be a widespread phenomenon among very neutron-deficient nuclides, its experimental detection will be difficult. Proton-emitting isomeric states in nuclei that lie closer to stability should be much easier to produce and hence they may provide the primary source of information on this decay mode for some time to come. The general existence of proton-radioactive isomers arising from one- or two-particle configurations (1) or from many-particle configurations (65) has already been discussed. In addition, Peker et al (12) explicitly predict a number of three- and four-particle isomeric states with spin $\geq \frac{19}{2}$ in nuclei with $A < 100$ that might be proton-radioactive. Searches for such proton-unbound isomers will also be arduous; the reduction in proton decay rate due to the high angular-momentum barrier and the small reduced width must inhibit the total half-life enough to make proton emission experimentally observable, yet not so much that a competing decay mode overwhelmingly dominates the isomer's decay.

5 TWO-PROTON RADIOACTIVITY

The concept of two-proton radioactivity as a new mode of nuclear transformation was originally proposed by Goldanskii (66) in 1960. No experimental observations of this radioactive decay mode have as yet been reported. Since, in addition, very little theoretical work has appeared subsequent to Goldanskii's comprehensive review (67) of the approaches for detection and study of two-proton radioactivity, only a very brief discussion of this subject follows.

As a consequence of the increased binding energy gained by a nucleus that pairs two protons, it is probable that some very proton-rich nuclides will have a positive binding energy with respect to emitting a single proton, but will be energetically unstable to the simultaneous escape of a proton pair. Some will have half-lives on a radioactivity time scale; others will simply fail to "exist" by being unstable to prompt two-proton emission. Possible examples of the latter are ^{12}O, ^{16}Ne, and ^{19}Mg (67, 68). However, recent mass measurements of ^{16}Ne (69) indicate that its nonexistence, as in the case of ^{6}Be (see Ref. 1), can primarily be attributed to the consecutive sequential emission of two protons.

$$^{16}\text{Ne} \rightarrow p + {}^{15}\text{F} \rightarrow p + p + {}^{14}\text{O}.$$

Considering nuclei that "exist" and decay by this unique radioactivity, involving the simultaneous emission of two protons through the Coulomb and angular-momentum barriers surrounding the nucleus, many interesting theoretical questions arise regarding the rate for this decay mode as well as the energy and angular correlation of the emitted pair of protons. Particular interest lies in analyzing how these quantities are affected by the "final-state interaction" between the two protons and by the presence of a centrifugal barrier. Goldanskii's review (67) discusses the probable character of these phenomena associated with two-proton radioactivity for various general situations and also lists some specific possible examples of light two-proton emitters. For a more recent estimate of those light nuclides that are good candidates for decay via this mode, the predictions via the Kelson-Garvey approach (37) by Jelley and collaborators (70) of masses of nuclei with $A \leq 40$ and with $T_z = -2$, $-\frac{5}{2}$, and -3 can be used; through quite difficult to produce due to their extremely neutron-deficient character, searches for ^{22}Si and ^{31}Ar would appear to be particularly promising, since they are well-bound relative to single-proton emission and are both expected to have a low ($\lesssim 200$ keV) two-proton decay energy.

Finally, as noted earlier for the case of proton radioactivity, the two-proton decay of many-particle isomeric states, arising in nuclei substantially closer to stability than those capable of two-proton emission from the ground state, may provide the most experimentally accessible sources of this new decay mode. Goldanskii & Peker (13) suggest a number of possible three- and four-particle isomers in nuclides ranging from ^{47}Fe to ^{108}Te that might be particularly promising choices for these initial searches.

Literature Cited

1. Goldanskii, V. I. 1966. *Ann. Rev. Nucl. Sci.* 16:1–30
2. Hardy, J. C. 1974. In *Nuclear Spectroscopy and Reactions*, ed. J. Cerny, Pt. C, pp. 417–66. New York: Academic. 590 pp.
3. Hardy, J. C. 1972. *Nucl. Data Tables* 11:327–49
4. Karnaukhov, V. A. 1973. *Fiz. Elem. Chastits At. Yadra* 4:1018–76; 1974. *Sov. J. Part. Nucl.* 4:416–39

5. Karnaukhov, V. A. 1975. *Nucleonika* 19: 425–53
6. Forsling, W., Herrlander, C. J., Ryde, H., eds. 1966. *Nuclides Far Off the Stability Line*. Stockholm: Almqvist & Wiksell. 686 pp; 1967. *Ark. Fys.* 36:1–686
7. *Proc. Int. Conf. Prop. Nucl. Far Reg. Beta-Stability, Leysin, Switzerland*, Vols 1, 2. 1970. CERN Rep. 70-30. 602 pp., 549 pp.
8. *Proc. Int. Conf. Nucl. Far Stability, 3rd,*

Cargese, Corsica. 1976. CERN Rep. 76-13. 608 pp.

9. Baz, A. I., Goldanskii, V. I., Goldberg, V. Z., Zeldovich, Ya. B. 1972. *Light and Intermediate Nuclei near the Limits of Nucleon Stability.* Moscow: Izdatelstvo Nauka. 171 pp. (In Russian)

10. Jackson, K. P., Cardinal, C. U., Evans, H. C., Jelley, N. A., Cerny, J. 1970. *Phys. Lett. B* 33:281-83; Cerny, J., Esterl, J. E., Gough, R. A., Sextro, R. G. 1970. *Phys. Lett. B* 33:284-86

11. Cerny, J., Gough, R. A., Sextro, R. G., Esterl, J. E. 1972. *Nucl. Phys. A* 188:666-72

12. Peker, L. K., Volmyansky, E. I., Bunakov, V. E., Ogloblin, S. G. 1971. *Phys. Lett. B* 36:547-49

13. Goldanskii, V. I., Peker, L. K. 1971. *Zh. Eksp. Teor. Fiz. Pisma Red.* 13:577-78; 1971. *Sov. Phys.—JETP Lett.* 13:412-14

14. Barton, R., McPherson, R., Bell, R. E., Frisken, W. R., Link, W. T., Moore, R. B. 1963. *Can. J. Phys.* 41:2007-25

15. Macfarlane, R. D., McHarris, Wm. C. 1974. In *Nuclear Spectroscopy and Reactions,* ed. J. Cerny, Pt. A, pp. 243-86. New York: Academic. 518 pp.

16. Cerny, J. 1976. See Ref. 8, pp. 225-34

17. Esterl, J. E., Allred, D., Hardy, J. C., Sextro, R. G., Cerny, J. 1972. *Phys. Rev. C* 6:373-75

18. Esterl, J. E., Hardy, J. C., Sextro, R. G., Cerny, J. 1970. *Phys. Lett. B* 33:287-90

19. Hardy, J. C., Esterl, J. E., Sextro, R. G., Cerny, J. 1971. *Phys. Rev. C* 3:700-18

20. Sextro, R. G., Gough, R. A., Cerny, J. 1973. *Phys. Rev. C* 8:258-68

21. Reeder, P. L., Poskanzer, A. M., Esterlund, R. A., McPherson, R. 1966. *Phys. Rev.* 147:781-88

22. Vieira, D. J., Gough, R. A., Cerny, J. 1978. *Phys. Rev.* In press

23. Sextro, R. G., Gough, R. A., Cerny, J. 1974. *Nucl. Phys. A* 234:130-56

24. Jackson, K. P., Hardy, J. C., Schmeing, H., Graham, R. L., Geiger, J. S., Allen, K. W. 1974. *Phys. Lett. B* 49:341-44

25. Cerny, J., Cardinal, C. U., Evans, H. C., Jackson, K. P., Jelley, N. A. 1970. *Phys. Rev. Lett.* 24:1128-30

26. Vieira, D. J., Sherman, D. F., Zisman, M. S., Gough, R. A., Cerny, J. 1976. *Phys. Lett. B* 60:261-64

27. Gough, R. A., Sextro, R. G., Cerny, J. 1972. *Phys. Rev. Lett.* 28:510-12

28. Jonson, B., Hagberg, E., Hansen, P. G., Hornshøj, P., Tidemand-Petersson, P. 1976. See Ref. 8, pp. 277-98

29. Cerny, J. 1968. *Ann. Rev. Nucl. Sci.* 18:27-52

30. Benenson, W., Kashy, E., Mueller, D., Nann, H. 1976. See Ref. 8, pp. 235-45

31. Hardy, J. C., 1976. See Ref. 8, pp. 267-76

32. Temmer, G. M. 1974. In *Nuclear Spectroscopy and Reactions,* ed. J. Cerny, Pt. B, pp. 61-87. New York: Academic. 711 pp.

33. Ikossi, P. G., Clegg, T. B., Jacobs, W. W., Ludwig, E. J., Thompson, W. J. 1976. *Nucl. Phys. A* 274:1-27; McDonald, A. B. et al 1976. *Nucl. Phys. A* 273:451-92

34. Steigerwalt, J. E., Sunier, J. W., Richardson, J. R. 1969. *Nucl. Phys. A* 137:585-92

35. Verrall, R. I., Bell, R. E. 1969. *Nucl. Phys. A* 127:635-40

36. Cerny, J., Mendelson, R. A., Wozniak, G. J., Esterl, J. E., Hardy, J. C. 1969. *Phys. Rev. Lett.* 22:612-15

37. Kelson, I., Garvey, G. T. 1966. *Phys. Lett.* 23:689-92

38. Robertson, R. G. H., Bowles, T., Freedman, S. J. 1976. See Ref. 8, pp. 254-57

39. Hansen, P. G. 1973. In *Advances in Nuclear Physics,* ed. M. Baranger, E. Vogt, 7:159-227. New York: Plenum. 329 pp.

40. Porter, C. E., Thomas, R. G. 1956. *Phys. Rev.* 104:483-91

41. Macdonald, J. A., Hardy, J. C., Schmeing, H., Faestermann, T., Andrews, H. R., Geiger, J. S., Graham, R. L., Jackson, K. P. 1977. *Nucl. Phys.* In press

42. Hardy, J. C., Macdonald, J. A., Schmeing, H., Andrews, H. R., Geiger, J. S., Graham, R. L., Faestermann, T., Clifford, E. T. H., Jackson, K. P. 1976. *Phys. Rev. Lett.* 37:133-36

43. Hardy, J. C., Macdonald, J. A., Schmeing, H., Faestermann, T., Andrews, H. R., Geiger, J. S., Graham, R. L., Jackson, K. P. 1976. *Phys. Lett. B* 63:27-30

44. Bogdanov, D. D., Karnaukhov, V. A., Petrov, L. A. 1973. *Yad. Fiz.* 17:457-62; 1973. *Sov. J. Nucl. Phys.* 17:233-35

45. Karnaukhov, V. A., Ter-Akopyan, G. M. 1966. See Ref. 6, pp. 419-29

46. Hagberg, E., Hansen, P. G., Jonson, B., Jorgensen, B. G. G., Kugler, E., Mowinckel, T. 1973. *Nucl. Phys. A* 208:309-16

47. Hornshøj, P., Wilsky, K., Hansen, P. G., Jonson, B., Nielsen, O. B. 1971. *Proc. Int. Conf. Heavy-Ion Phys.,* Dubna Rep. D7-5769, p. 249

48. Hornshøj, P., Wilsky, K., Hansen, P. G., Jonson, B., Alpsten, M., Andersson, G., Appelqvist, Å., Bengtsson, B., Nielsen, O. B. 1971. *Phys. Lett. B* 34:591-93

49. Bogdanov, D. D., Dem'yanov, A. V., Karnaukhov, V. A., Petrov, L. A. 1975. *Yad. Fiz.* 21:233-238; 1975. *Sov. J. Nucl.*

Phys. 21:123–25

50. Jonson, B., Hansen, P. G., Hornshøj, P., Nielsen, O. B. 1973. *Proc. Int. Conf. Nucl. Phys., Münich,* ed. J. de Boer, H. J. Mang, 1:690. Amsterdam: North-Holland. 739 pp.

51. Hornshøj, P., Tidemand-Petersson, P., Bethoux, R., Caretto, A. A., Grüter, J. W., Hansen, P. G., Jonson, B., Hagberg, E., Mattsson, S. 1975. *Phys. Lett. B* 57:147–49

52. Bogdanov, D. D., Dem'yanov, A. V., Karnaukhov, V. A., Petrov, L. A., Plohocki, A., Subbotin, V. G., Voboril, J. 1976. See Ref. 8, pp. 299–303

53. Comay, E., Kelson, I. 1976. *At. Data Nucl. Data Tables* 17:463–66; 477–608

54. Koyama, S. I., Takahashi, K., Yamada, M. 1970. *Prog. Theor. Phys.* 44:663–88

55. Takahashi, K., Yamada, M., Kondoh, T. 1973. *At. Data Nucl. Data Tables* 12:101–42

56. Bartholomew, G. A., Earle, E. D., Ferguson, A. J., Knowles, J. W., Lone, M. A. 1973. See Ref. 39, pp. 229–324

57. Gilbert, A., Cameron, A. G. W. 1965. *Can. J. Phys.* 43:1446–96; Truran, J. W., Cameron, A. G. W., Hilf, E. 1970. See Ref. 7, 1:275–306

58. Sevier, K. D. 1972. *Low Energy Electron Spectrometry,* pp. 220–41. New York: Wiley Interscience. 397 pp.

59. Hornshøj, P., Wilsky, K., Hansen, P. G., Jonson, B., Nielson, O. B. 1972. *Nucl. Phys. A* 187:599–608, 609–23

60. Bogdanov, D. D., Darotsi, Sh., Karnaukhov, V. A., Petrov, L. A., Ter-Akopyan, G. M. 1967. *Yad. Fiz.* 5:893–900; 1968. *Sov. J. Nucl. Phys.* 6:650–55

61. Karnaukhov, V. A., Bogdanov, D. D., Petrov, L. A. 1973. *Nucl. Phys. A* 206:583–92

62. Hillebrandt, W., Takahashi, K. 1976. See Ref. 8, pp. 580–83

63. Bogdanov, D. D., Bochin, V. P., Karnaukhov, V. A., Petrov, L. A. 1972. *Yad. Fiz.* 16:890–900; 1973. *Sov. J. Nucl. Phys.* 16:491–96; Karnaukhov, V. A., Bogdanov, D. D., Petrov, L. A. 1970. See Ref. 7, 1:457–86

64. Auerbach, N., Talmi, I. 1964. *Phys. Lett.* 9:153–55; 10:297–99; Vervier, J. 1967. *Nucl. Phys. A* 103:222–24

65. Flerov, G. N., Karnaukhov, V. A., Ter-Akopyan, G. M., Petrov, L. A., Subbotin, V. G. 1964. *Zh. Eksp. Teor. Fiz.* 47:419–32; 1965. *Sov. Phys. JETP* 20:278–86

66. Goldanskii, V. I. 1960. *Zh. Eksp. Teor. Fiz.* 39:497–501; 1961. *Sov. Phys. JETP* 12:348–51. Goldanskii, V. I. 1960. *Nucl. Phys.* 19:482–95

67. Goldanskii, V. I. 1965. *Usp. Fiz. Nauk* 87:255–72; 1966. *Sov. Phys. Usp.* 8:770–79

68. Zeldovich, Ya. B. 1960. *Zh. Eksp. Teor. Fiz.* 38:1123–31; 1960. *Sov. Phys. JETP* 11:812–18

69. KeKelis, G. J., Zisman, M. S., Scott, D. K., Jahn, R., Vieira. D. J., Cerny, J., Ajzenberg-Selove, F. 1978. *Phys. Rev. C.* In press

70. Jelley, N. A., Cerny, J., Stahel, D. P., Wilcox, K. H. 1974. Lawrence Berkeley Lab. Rep. LBL-3414

Ann. Rev. Nucl. Sci. 1977. 27:353–92

HEAVY-ION ACCELERATORS ✗5588

Hermann A. Grunder and Frank B. Selph[1]

Lawrence Berkeley Laboratory, University of California, Berkeley, California 94720

CONTENTS

1 INTRODUCTION

Over the last two decades, heavy-ion beams have become increasingly important tools for scientific investigations, ranging from the synthesis of heavy "man-made" elements to the fields of nuclear physics and biology and medicine.

In the context of this paper, a heavy ion is the nucleus of any element with an atomic mass larger than 4 amu (atomic mass units), deliberately excluding hydrogen and helium. Heavy-ion accelerators are employed to accelerate and form the particles into pure and intense beams of the desired energy. Unstable isotopes are occasionally used if their half-life is long compared to the acceleration process.

The invention and development of the cyclotron in the 1930s by E. O. Lawrence, S. Livingston, and other collaborators started particle physics, nuclear physics.

[1] This work was done with support from the US Energy Research and Development Administration.

and nuclear chemistry on the road of expansion to the major scientific enterprise we know today. The development of beams of protons, deuterons, alphas, and lithium ions from 60-in cyclotrons throughout the world served a highly productive area of nuclear physics and chemistry. The Berkeley 60-in cyclotron ran very intense beams until 1961, when the Lawrence Berkeley Laboratory (LBL) 88-in isochronous cyclotron and the Oak Ridge Isochronous Cyclotron (ORIC) took its place with strong emphasis on nuclear physics. Livingston & Blewett (1) give a detailed history of accelerator development prior to 1962.

Nuclear chemistry research with heavy-ion linacs dates from 1957, when the Lawrence Radiation Laboratory, together with Yale University, developed and built the first two productive heavy-ion linear accelerators, delivering beams of 10 MeV/amu with masses up to 40 amu (2). A similar development took place in Dubna, USSR, in the Joint Institute for Nuclear Research (JINR).

Parallel developments with electrostatic accelerators, particularly the tandem Van de Graaffs (3), led to the attainment of increasingly higher terminal voltages, up to the presently envisioned 20 and 30 MV. With the refinement of negative heavy-ion sources (4), the tandem Van de Graaff accelerators entered the field of heavy-ion nuclear physics, and now complement the linacs and cyclotrons. The excellent phase-space characteristics of these beams, including energy resolutions of up to 10^{-4}, make them ideal tools for precision nuclear physics experiments. The Yale group, with its MP-tandem, led the way in the use of Van de Graaffs for heavy-ion research.

Accelerator experiments with heavy relativistic ions were started in 1970, with the conversion of the proton synchrotrons at Princeton (5) (Princeton-Penn accelerator, which discontinued operations in 1975) and Berkeley (6) (Bevatron), to the acceleration of heavy ions. Initially, beams were limited in intensity and ion species by injectors, which had been designed for protons. At Berkeley, the limits were extended by injecting the beam of a high-intensity, low-energy, heavy-ion accelerator (SuperHILAC) into the Bevatron. This combination (the Bevalac) yields particle currents of about 10^9/sec for ions up to neon, sufficient for counter and coincidence experiments with high-energy and particle-mass resolution, and beams of reduced intensity up to iron. Available energies range from 100 MeV to 2.4 GeV per nucleon.

The applications of heavy-ion beams in basic and applied research are numerous. At energies below the Coulomb barrier, very precise nuclear quantities (e.g. energy levels, matrix elements) can be extracted. With beams of ^{136}Xe, rotational states with $I = 24$ have been excited in ^{238}U, and heavier projectiles should excite even higher states. Above the Coulomb barrier, aspects of nuclear interactions, e.g. the ion-ion potential or the effects of deformations on scattering and reaction processes, are studied. The investigation of transfer reactions and compound-nucleus formation, as well as deep inelastic processes, advances our understanding of reaction mechanisms. The search for superheavy elements $(Z = 114\text{–}126)$ constitutes another branch of ongoing research at modest energies. In atomic physics, improved yields for the production of molecular-orbital X rays or the observation of the predicted auto-ionization of positrons are expected from collisions between very heavy ions.

Above 20 MeV/amu, a qualitatively new feature appears, in that now the estimated velocity of sound in nuclear matter is exceeded, opening up the possibility of studying nuclear matter at varying densities.

In the realm of nuclear physics at high energies, researchers are eagerly searching out novel features of interactions where a high number of baryons are simultaneously involved (7). Nucleons at high kinetic energies may be excited to the baryonic resonances, produce pions and kaons, or reabsorb them. In the process of a central collision, the initial ground-state configuration of nuclei will be immediately transformed into a very hot piece of nuclear matter, where the binding energy per nucleon is readily overcome.

High-energy (above several hundred MeV per amu), heavy-ion fragmentation cross sections are of particular interest to astrophysicists. They rely on these data to formulate models describing the propagation of cosmic rays in space and to approach an understanding of how the universe was created. Furthermore, very precise beams of heavy ions are required to calibrate the sensitive instruments flown in space studies.

One of the most exciting prospects is the generation of the predicted abnormal states of nuclear matter that can be tested when relativistic beams as heavy as Pb or U become available.

The application of heavy ions for radiation biology, radiation therapy, and diagnostic radiology has fostered an entire new field of heavy-ion work. The potential attractiveness of these beams for radiotherapy lies both in their radiobiological effects and the fact that a precise radiation dose can be delivered to a specific tumor volume with minimal damage to normal tissue, compared with conventional X-radiation therapy. At LBL's Bevalac facility, an active experimental program is investigating these fields (8). Also, the first steps in instituting an active patient-treatment program are being taken.

Low-charge-state heavy ions have been used in industry for some time. The major application is ion implantation—the doping of semiconductor materials by ion beams—to achieve far better uniformity and dose control, and in some cases to produce devices that were theretofore impossible to fabricate (9). Other potential uses of heavy ions now under active investigation are changing the optical properties of materials, changing magnetic properties of materials used for bubble memory, enhancing surface hardness and corrosion resistance, and altering properties of thin films (10). Ion implantation has also been used in the production of superconductors and modification of the coefficients of friction. At Oak Ridge National Laboratory (ORNL), heavy-ion beams are used to simulate radiation damage from reactor neutrons in the absence of sufficiently large neutron fluxes (11).

Among the many types of potential and operational heavy-ion accelerators, the choice of the proper machine depends upon the intended use. The precise beams of tandem electrostatic accelerators and cyclotrons are most suitable for experiments with high-resolution spectrometers. Linacs, with their relatively high intensity and velocities independent of mass, are excellent tools for chemistry-type experiments, overlapping with cyclotrons and electrostatic machines. Synchrotrons can produce high-energy beams of excellent quality at moderate cost. A survey of other component considerations and basic accelerator types is presented in this paper.

2 SOME FUNDAMENTAL CONSIDERATIONS OF PARTICLE ACCELERATORS FOR HEAVY IONS

The major characteristics of a particle beam are particle species, charge state (Sections 2.1 and 4.1), kinetic energy, energy spread, and time structure. Also of great importance are beam intensity, emittance, and the related concept of brightness (Section 2.2). The time structure can be both macroscopic (duty factor), which is the proportion of time the beam is on and microscopic, which is the intensity modulation due to the accelerating rf.

A simple accelerator is an evacuated tube with a different electrical potential at each end. If V is the potential difference, an ion of charge qe will increase its energy by an amount $\Delta T = qeV$ in traversing the tube.

For many applications, the energy per nucleon T_n, rather than total energy T, is the decisive quantity; this is given by $T_n = T/A$, where A is the mass number. Hence a particle falling through a potential V will gain $\Delta T_n = (q/A)eV$.

Consequently, the charge to mass ratio of an ion is a most important quantity. The lower the value of q/A, the greater the voltage V required to reach a given T_n. Another consideration is that the deflection of particles by magnetic fields—important for bending and focusing—is proportional to the rigidity $B\rho$, which is given by momentum divided by charge: $eB\rho = Mv/q = (A/q)\gamma m_p v$. Here the particle mass $M \equiv A\gamma m_p$, where m_p is the proton mass, and γ is the relativistic mass increase. Of two particles having the same energy per amu but different q/A, the particle with the lower q/A will have a higher rigidity, requiring stronger magnetic fields for beam handling.

In principle, the acceleration process for heavy ions differs from protons only in that q/A is different. However, major differences occur in the ion sources, in the need for low velocity (low β) accelerator structures, and in the fact that heavy ions gain and lose electrons in charge-exchange processes.

The charge state is initially determined by the degree of ionization attained in the ion source, a device that typically contains a plasma from which charged particles are extracted. The extraction process forms the original beam and sets certain properties of it, such as emittance, brightness, and maximum macroscopic time structure. A fuller discussion of the parameters of ion sources is given in Section 4.1.

Beyond this starting phase of beam formation, the charge state at various stages of acceleration requires careful consideration. It can be increased by stripping (Section 2.1), usually reducing the intensity. Charge-exchange processes with the residual gas can cause a sudden change in $B\rho$ and in most situations means loss of the particle.

When q/A from the ion source is very low, as it often must be for very heavy ions, acceleration is relatively slow, causing problems with transverse-beam containment.

The inherent repulsion of charged particles with equal polarity gives rise to space-charge forces that in turn require compensation by focusing forces to contain them. At nonrelativistic energies, the space-charge forces per unit length are inversely proportional to β^2 (where $\beta = v/c$, the velocity of the particle v relative to the velocity of light c). Special low-β accelerating structures must be built in order to cope with this situation.

2.1   Charge Exchange

An ion passing through matter exchanges electrons with atoms in the medium. Two processes operate in competition—stripping of electrons from the moving ion and capture of electrons by it. The cross section for each of these processes varies with the nuclear charge z and the ion velocity β. At low velocities, capture is dominant, while at high β, stripping is more important. This is illustrated by Figure 1, which shows stripping and capture cross sections for iodine ions in nitrogen gas. At sufficiently high velocities, the ion will be fully stripped; at lower velocities an equilibrium is quickly reached, depending upon z, β, and the nature of the medium traversed.

Because of the statistical nature of the process, at equilibrium there will be a distribution of charge states about some mean charge $\bar{q}$. Due to the complexity of

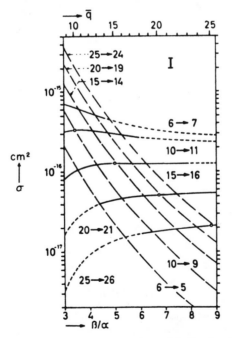

Figure 1   Calculated charge-exchange cross sections of iodine in air as a function of velocity. Velocity is plotted as β/α, where α is the fine-structure constant, equal to $1/137$. The upper scale on the graph is $\bar{q}$, the mean charge state of the stripped ions [from (12)].

the interactions involved, there is no theory that explains in satisfactory detail the observed stripping behavior; instead, a number of semi-empirical formulae have been relied upon in designing and operating heavy-ion accelerators. With these formulas, reviewed in a paper by Betz (13), it is possible to predict mean charge states to ± 1 in the areas of β and z where measurements exist. Large extrapolations are risky. Some indication of the use of stripping for heavy-ion accelerators can be obtained from an inspection of Figure 2, in which the empirical formulas were used to plot equilibrium charge states as a function of T_n (energy per nucleon) for dense-stripping media. The stripping region lies below a given z curve, the capture region above it.

The acceleration of argon by the Bevalac serves as an illustration. Starting with $q = 3$ from the ion source, the first linac section accelerates to $T_n = 1.2$ for the first stripping, yielding $\bar{q} = 12.4$. The next linac section accelerates to $T_n = 8.5$ where, after a second stripping, $\bar{q} = 17.3$. Definite charge states must be used for each stage of acceleration. Here, $q = 12$ is used after the first stripping, and $q = 18$ after the second stripping. After each stripping, some beam intensity is sacrificed because several charge states are created and only one can be used. The loss is 50–70% for argon at each stripping. For higher-z ions the losses are greater due to the wider charge-state distribution.

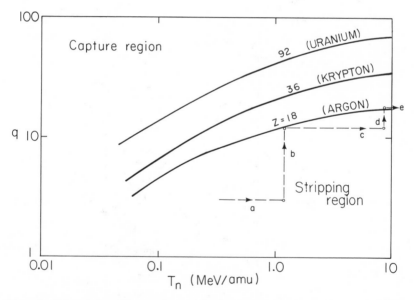

Figure 2   Calculated equilibrium charge states for dense strippers as a function of energy, for $Z = 18$, 36, and 92. The dotted lines indicate the processes of acceleration and stripping of argon ions in the Bevalac, which involves two stages of acceleration (a and c) and stripping twice (b and d). See Section 3.6.

2.2   Emittance and Brightness

The concepts of emittance and brightness are crucial because they determine beam size and particle flux that can be transmitted through a given ion optical system. Knowledge of linear dimensions of a beam and of physical boundaries like the vacuum chamber alone are inadequate, since any beam has a certain divergence, and strength and spacing of focusing elements must be taken into consideration. The concept of emittance is also helpful in understanding the minimum possible beam size at any point (for example at a target), and the relative importance of disturbances, such as electric and magnetic inhomogeneities, on beam quality.

Particles can be described by giving position relative to three orthogonal axes x, y, and z, and momenta p_x, p_y, and p_z along these axes. Beam particles form an ensemble in the six-dimensional phase space of coordinates and momenta. A fundamental theorem of statistical mechanics, known as Liouville's theorem, states that if the ensemble is Hamiltonian (i.e. if the equations of motion can be written in Hamiltonian form), then in the neighborhood of a given particle the density of particles in phase space remains constant in time (14). Charged particles acted upon by external electric and magnetic fields are Hamiltonian. Processes such as stripping are non-Hamiltonian, since knowledge of the particles' initial coordinates and momenta and the external forces acting upon them is not sufficient to describe beam behavior.

In some instances, the equations of motion are separable with respect to one or more coordinate axes. For example, motion in the x plane might be independent of the other coordinates; in this case, Liouville's theorem applies to the two-dimensional phase space x, p_x. This situation is often true in beam-transport systems. Considering such a case of one-dimensional motion, we can examine the beam at time t_0 and see that a curve E_0, in phase space will enclose all of the particles (see Figure 3). At a later time t, the particle distribution will be bounded by another curve E. It can be shown that, as a consequence of Liouville's theorem, the area enclosed by E_0 and E

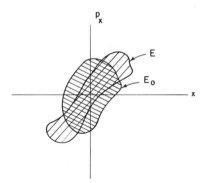

Figure 3   Representations of beam emittance in transverse phase space. x = transverse position, p_x = transverse momentum.

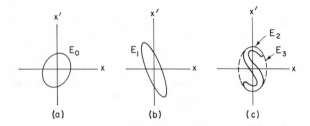

Figure 4 A beam of elliptical phase-space area E_0 in (*a*), when subjected to linear forces transforms to an ellipse of equal area E_1 in (*b*). However, when E_0 is subjected to non-linear forces, it translates to a nonelliptical shape E_2 in (*c*), which can only be enclosed by a larger ellipse E_3 for further linear transformations [after Banford (15)].

is the same and a particle within E_0 at time t_0 must be within E at time t. Taking z as the beam axis, the area in the x, p_x plane enclosed by E is called the normalized emittance. If the average beam momentum is p_0, and we transform to variables x, x', where $x' \equiv p_x/p_0$, the area occupied by the beam is called the emittance.

Beam intensity is also an important parameter. However, if a beam can be made more intense only at the expense of emittance, the amount of useful intensity might be no greater. To facilitate such comparison, beam brightness is defined as

$$B = I/E_x E_y, \tag{1}$$

where I is the beam current, and E_x and E_y are emittances of the transverse phase spaces occupied by the beam.

One might expect that beam brightness would be conserved, provided the acceptance of an accelerator is adequate to transmit emittance. This is rarely so, because loss of intensity can occur from charge exchange with residual gas, stripping to increase charge state, rf trapping, etc. Furthermore, emittance can be diluted in an accelerator or transport system even though, in a strict sense, Liouville's theorem is not being violated. This is illustrated with a common case in Figure 4. Beam emittance E_0, which is here taken to be elliptical in area, is shown in Figure 4a. After passing through a linear transport system, emittance is as in Figure 4b, an ellipse E_1 of different shape but the same area. However, if the transport system contains nonlinear fields, the result can be as is shown in Figure 4c, where the emittance E_0 has been transformed into E_2, a distorted shape with the same area. Unless stringent measures are taken, such as compensating nonlinearities, subsequent beam transport must accept the larger emittance figure enclosed by the dotted curve E_3. For further discussion of these questions see Banford (15).

3 TYPES OF HEAVY-ION ACCELERATORS

Electrostatic machines are in principle the simplest heavy-ion accelerators; operating on dc, they also inherently possess the highest duty factor. They have beams of low emittance and excellent energy resolution. They are restricted in maximum energy of ions, and, with the large tandem systems, are somewhat restricted in ion species because of the necessity for a negative ion source.

Linear accelerators can be built to any final energy. They can also have a high duty factor. However, the accelerator structure is used only once by the ions, so cost sets a limit on length of structure and hence energy. High duty factor is also limited because of the cost of rf power, of which only a small fraction will be transmitted to the beam, in typical applications.

Collective acceleration is still experimental, but is of interest because potentially it could be the most economical. The duty factor, however, is inherently low.

Cyclotrons economize on rf power by accelerating the particles repeatedly through the same structure. Maximum energy and therefore cost is set by the magnet dimensions. The duty factor is inherently high. When high energies, as well as variable ion species and energies are needed, changes in magnetic-field shape due to relativistic mass increases lead to sophisticated field-shaping techniques.

Synchrotrons economize on both rf power and magnets. With present technology, the synchrotron is the most economical machine to reach high energies. The duty

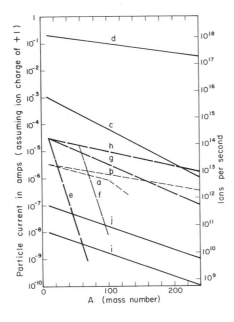

Figure 5 Approximate particle current vs mass number for heavy-ion accelerator types. (*a*) 25-MV tandem electrostatic accelerator; (*b*) tandem + cyclotron or linac booster; (*c*) existing linacs; (*d*) proposed high-intensity linac for heavy-ion fusion; (*e*) existing $K = 140$ cyclotron; (*f*) proposed $K = 500$ superconducting cyclotron; (*g*) proposed tandem cyclotrons, each of $K = 400$; (*h*) proposed tandem superconducting cyclotrons, $K = 500$ and $K = 800$; (*i*) linac and synchrotron (Bevalac, after completion of current improvement program); (*j*) state-of-the-art linac + synchrotron (20-Hz rep rate synchrotron).

362   GRUNDER & SELPH

factor is inherently low, but can be stretched to ~25–50% at the cost of average intensity. Vacuum requirements are severe because of the ion's long path length.

In principle, any acceleration scheme conceived and developed for protons can be adapted to the acceleration of heavy ions, and most of the necessary modifications to variable-energy cyclotrons, synchrotrons, and electrostatic accelerators can be readily accomplished even after the machine is built. Changes in linear accelerators must be considered at the design stage.

However, the existence of the fundamental technology has made possible rapid development of heavy-ion accelerators with rather well-defined specifications, including accelerator designs uniquely suited for the relatively low-q/A heavy ions. Some of the performance characteristics of the major heavy-ion accelerator types are shown in Figure 5, as intensity vs mass number, and in Figure 6, which shows energy per nucleon vs mass number. The reasons for the wide disparity in performance of the various types will become clear as each type is discussed in detail.

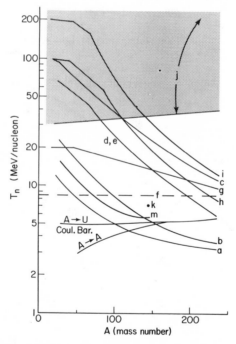

Figure 6 Energy per nucleon vs mass number for several accelerators. The descriptive symbols are explained in Section 5. (*a*) HHIRF tandem (T25); (*b*) HHIRF phase I (T25+K90); (*c*) HHIRF phase II (T25+K90+K400); (*d*) Chalk River (T13+K500); (*e*) Rochester (T13+K500); (*f*) SuperHILAC (D8.5); (*g*) UNILAC (D4.5+L23.8); (*h*) GANIL (K25+K400+K400); (*i*) MSU (K52+K500+K800); (*j*) Bevalac (D8.5+S2.6) (only lower limit is shown—upper is $T_n = 2.6$ GeV/amu for $A = 20$, and maximum $T_n = 1.3$ GeV/amu for $A = 238$); (*k*) Dubna JINR U200+U300; (*m*) JINR U400(K725).

3.1   Electrostatic Machines

Electrostatic accelerators utilize an insulated belt or chain to carry charge up to a terminal that is maintained at a high potential. Ions are accelerated through the potential difference between terminal and ground. Research and development on this type of accelerator was led by Van de Graaff for many years and the machine is frequently referred to by using his name.

The Van de Graaff accelerator is the only type that has achieved voltages above 10 MV. Another high voltage generator is the Cockcroft-Walton (16), and shunt-fed form of it, the Dynamatron (17). Voltages up to 4 MV with relatively high currents of many milliamperes make the Cockcroft-Walton generator a desirable preinjector for linacs.

For electrostatic accelerators above about 1 MV, pressurized vessels are used to contain the high-voltage components. This permits work with much higher electric fields and thus reduces the physical size of the apparatus, which is limited by breakdown and corona phenomena. Sulfur-hexafluoride alone or mixtures with nitrogen and carbon dioxide are used as insulating gas with pressures up to 250 psi.

The accelerating column, of course must have a high vacuum to avoid unwanted charge exchange for heavy ions. With electric-field gradients in the accelerating tubes of 2–3 MV/m, a conditioning process is necessary for steady, reliable operation. Conditioning a large electrostatic accelerator, a sensitive task involving, it seems, both folklore and technical skill, requires many hours to accomplish. The fundamental objective in conditioning is to eliminate by controlled discharges small (often microscopic) irregularities on the high-voltage surfaces.

A modern development has been the "tandem," in which negatively charged ions are accelerated up to the high-voltage terminal, where they are stripped, then accelerated to ground (18) (see Figure 7). New charging techniques have been devised, which replace the traditional charging belt with metal charge carriers called pellet chains. With these chains, there is a reduced chance of spark damage and frictional loss to the gas is low, giving modest power input and eliminating cooling requirements (19).

An example of the latest tandem technology is a 25-MV machine by National Electrostatics Corporation (NEC) now under construction at Oak Ridge National Laboratory (ORNL) (Figure 8). The tank is oriented vertically with the terminal at

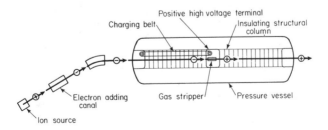

Figure 7   Schematic of a tandem electrostatic accelerator [after Van de Graaff (18)].

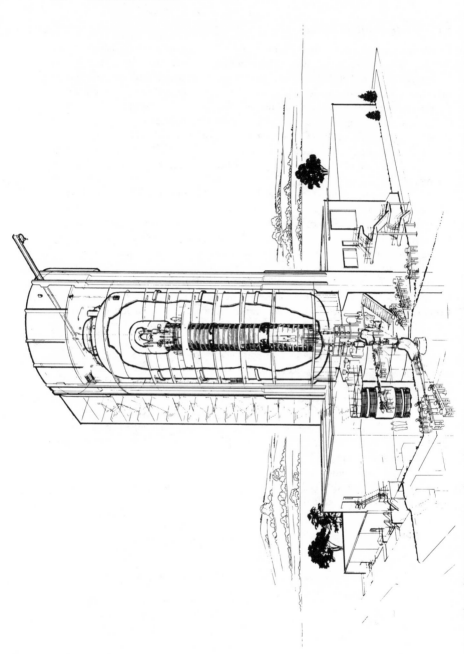

Figure 8   View of HHIRF 25-MV tandem built by NEC. A negatively charged beam from the ion source is bent 90° and accelerated upward to the terminal where after stripping it is bent 180° and accelerated downward through the same potential. A 90° magnet is arranged to select

the top. A 180° bending magnet is placed in the terminal, and the same column is used both for the negative- and positive-ion beams. This folded arrangement saves substantially in both tank and building height. The ion source is located at the base of the accelerator for convenience and building economy (20).

Another large tandem facility is under construction at Daresbury in England. Initial operation at 20 MV, with later upgrading to 30 MV is planned. This machine is also placed in a vertical tower, but with the ion source at the top (21).

Electrons or ions released from surfaces in the accelerating tube are suppressed by installing diaphragms with reduced apertures at intervals along the tube. Most ions or electrons originating between diaphragms are stopped at the next diaphragm. A few proceed to the second diaphragm, where they are stopped. Thus, positive-negative ion exchange, which is responsible for "electron loading," is depressed in a simple and elegant fashion. Other means are the introduction of magnetic fields or inclined electric fields.

Quadrupole lenses are appropriately located for beam control; with beam-profile monitors and steering elements, beam transmission from source to target is straightforward. A change in beam energy or in charge state is easily and quickly accomplished by a skilled operator.

Beams from an electrostatic accelerator can be injected into a cyclotron or a linear-accelerator booster to extend energies to 8 MeV/amu or more. The 25-MV Oak Ridge tandem will inject into the cyclotron "ORIC" to provide ions to energies above 9 MeV per nucleon for mass 100. Planning is also well advanced on an open-sector cyclotron booster to provide energies above 10 MeV per nucleon for all ions (22) (see Figure 9). A similar arrangement of lower energy is being constructed at the Hahn Meitner Institute in Berlin.

At Argonne National Laboratory, plans are under way to utilize an existing tandem for injection into a superconducting linac (23). Similar arrangements are being pursued at other laboratories.

Two attractive features of electrostatic machines are dc operation and modest power requirements. To reach higher energies, the problem of ion exchange on the surface of the accelerating tube must be overcome. Experiments show that clean surfaces of titanium or stainless steel have much higher thresholds for discharge than these metals when coated with condensed vapors. As tube length is increased, vapors migrate out more slowly and conditioning becomes more difficult. Another effect sets in. Condensed molecules dislodged by impinging ions temporarily form a gas, can be ionized, and can increase discharge, which when fed by vapor from condensed layers can spread to disruptive proportions. An all-metal and ceramic bakeable tube is very helpful for minimizing discharge problems, but recent difficulties have shown that vapors introduced inadvertently can persist over a long period.

3.2 Heavy-Ion Linear Accelerators

Although static potentials of electrostatic accelerators now under construction at Oak Ridge and Daresbury are expected to operate in the 25–30 MV range, a limit is set by voltage breakdown between the two terminals.

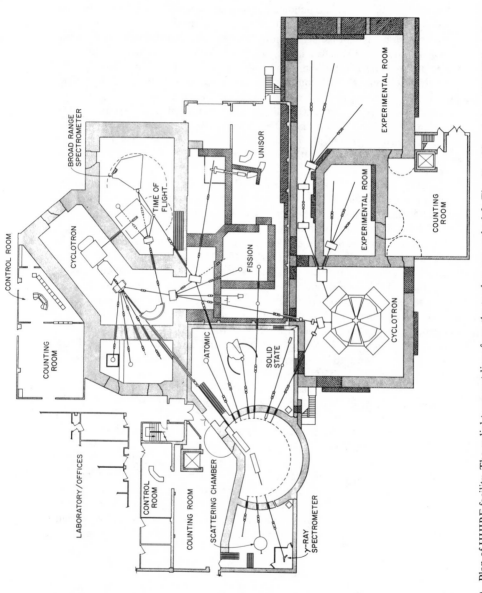

Figure 9   Plan of HHIRF facility. The radial transport lines from the tandem are at the left. The $K = 90$ cyclotron (ORIC) is at the top. The

A much higher effective potential can be obtained by arranging the accelerating tube so that a more modest accelerating potential is repeatedly applied to the ions in transit. Such an accelerator, or linac, will consist of many short accelerating stations arranged in sequence; in principle, there is no upper limit to the ion energy that can be achieved. Although a variety of linac structures have been employed to accelerate charged particles, the ones that so far have proved most practical as heavy-ion accelerators fall into the category of drift-tube linacs.

A drift tube is a cylindrical metal (usually copper) shell, closed at each end except for an axial hole for passage of the ions. A number of drift tubes are arranged in sequence, with gaps between (Figure 10), and an electric field is established in the gaps. The field originates with a rf oscillator, so that it periodically changes direction. Ions are accelerated in the gaps during the period when the field has the proper polarity; during the part of the cycle when the field is reversed, the ions are shielded from fields by the drift tube.

The behavior of a beam of ions traversing a linac can best be seen by first visualizing what happens to a reference particle, usually called a synchronous particle, as it is accelerated. It will gain exactly the right amount of energy at each gap by remaining (usually) at a constant phase angle relative to peak rf voltage. As the ion gains energy, its velocity will increase, so if rf phase is to remain constant, the separation of gaps must increase. This leads to drift-tube spacings that are determined by synchronous velocity. Once the linac is built, the spacings are fixed and the synchronous velocity at a given point cannot be changed. The situation is different if the gaps are electrically independent of one another, allowing rf phase at each gap to be varied at will. In such a single-gap linac, the synchronous velocity profile can be varied in an arbitrary manner, so long as it remains a continuous function, by proper programming of the rf gap phase.

A beam will be successfully accelerated if the particles remain near to the synchronous particle in both longitudinal and transverse phase space. The conditions

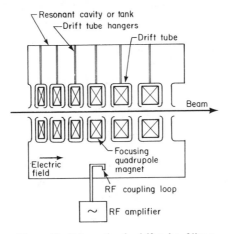

Figure 10   Schematic of a drift-tube rf linac.

for longitudinal phase stability in a linac require the electric field to be rising at the time the synchronous particle crosses the gap, because particles arriving at a gap earlier than the synchronous particle will then receive less energy, and consequently arrive later at the next gap. Particles arriving later than the synchronous particle gain more energy, hence arrive at the next gap earlier. Electric gap fields also exert transverse radial forces on the particles; if the rf phase is correct for longitudinal stability, these radial forces will be directed away from the axis, that is, they will defocus. Therefore, additional positive restoring forces must act on the particle to keep the bunch together during linac transit. The usual means for accomplishing this is to add focusing elements such as magnetic quadrupoles, either inside the drift tubes or in gaps between drift tube structures. Radial defocusing forces are inversely proportional to β^2, and because of the difficulties in offsetting them, these forces limit the use of linac structures to accelerate very-low-velocity ions. The lower β limit varies with the type of structure. With presently operating machines it is about $\beta = 0.005$ for the Wideröe structure and $\beta = 0.015$ for the Alvarez structure.

The structure diagrammed in Figure 10 is an Alvarez linac. In this machine, a cylindrical tank is made to resonate to the desired frequency in the T_{010} mode, giving an axial electric field with a maximum along the center. The drift tubes are supported by hangers attached to the tank wall. The cell length, or gap-to-gap distance advanced by the ions each n rf cycles is just $\beta\lambda$, where λ is the rf wavelength. The rf phase advance in one cell is $2n\pi$. In principle, n can be any integer, but in practice the accelerating field becomes inefficient when n is greater than 1. A typical Alvarez heavy-ion linac is the SuperHILAC (24), with rf of 70 MHz, $\lambda = 4.3$ m. The tank diameter, set by the requirement for resonance, is 3.0 m, about $3\lambda/4$.

For $\beta \ll 0.015$, the cell length $\beta\lambda$ becomes too short for adequate quadrupole focusing unless λ is made larger, which requires a resonant tank of larger diameter and hence greater cost. The difficulty can be avoided by using a Wideröe structure (sometimes called a Sloan-Lawrence structure), which does not rely upon a resonant cavity, but instead delivers rf power to the drift tubes by means of two conductors connected alternately to the drift tubes. Thus adjacent drift tubes have opposite polarity, and particles advance odd multiples of π in rf phase between one gap and the next, with the distance traveled being $n\beta\lambda/2$, where n is an integer. Figure 11 is a drawing of a Wideröe linac similar to the UNILAC design. The drift tubes are arranged in a $\pi - 3\pi$ sequence to allow room for quadrupoles in every other drift tube. The short drift tubes are connected to and supported by a metal pipe, which in turn is connected to three rf tuning stubs. The long drift tubes are suspended from the tank wall, which serves as the other conductor.

Recalling that the energy gained by an ion in a linear accelerator depends upon the charge number q in addition to the effective accelerating potential difference, we see that to reach high energies it is important to have q as high as possible. Ion sources show a strong intensity decrease with increasing q, so that in a practical linear accelerator, one faces a compromise between energy and intensity. This compromise is partially offset by the fact that once a moderate velocity has been reached, stripping can produce a higher q than could be achieved ordinarily in an ion source. Consequently, a typical heavy-ion linear accelerator design will usually

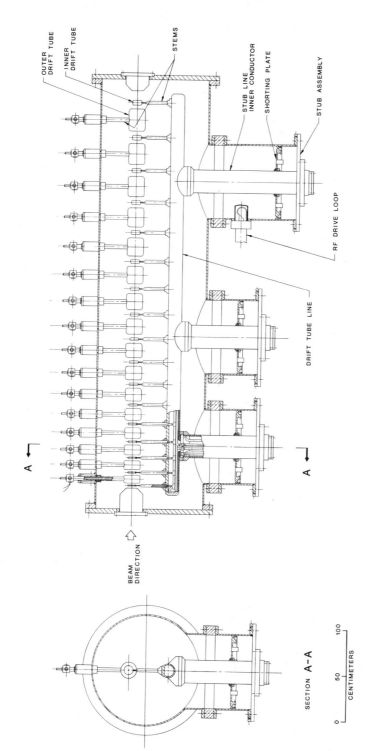

Figure 11   Drawing of a Wideröe drift-tube linac.

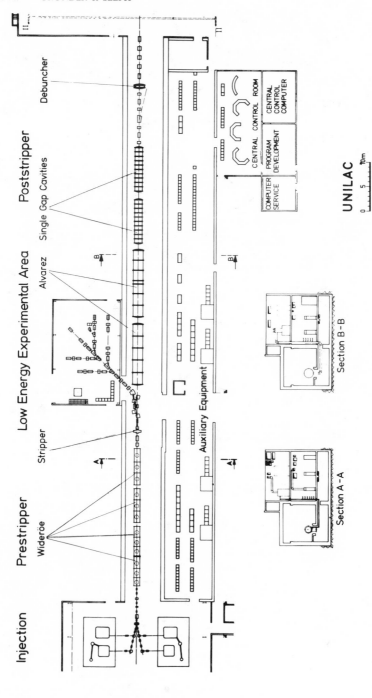

Figure 12 Plan of the UNILAC accelerator at Darmstadt.

have an ion source providing ions of moderate q injected in a dc accelerator that reaches $\beta = 0.005$ to $\beta = 0.015$, then a linac that accelerates to a modest energy, after which the ions are stripped to a q that is 2–4 times the injected q. This is followed by another linac stage to reach the final energy. Stripping does exact a penalty in loss of two thirds or more of original beam intensity, due to creation of many charge states, only one of which can be usefully accelerated in most cases.

The state of the art in heavy-ion linacs is typified by the UNILAC, built and operated by GSI at Darmstadt, Germany (25) (Figure 12). Two dc injectors of 320 kV, each containing two ion sources, are used to assure continuous operation. A

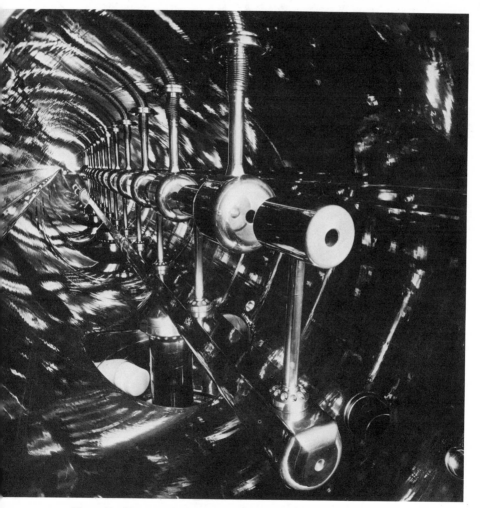

Figure 13   Photograph looking inside a Wideröe tank at the UNILAC.

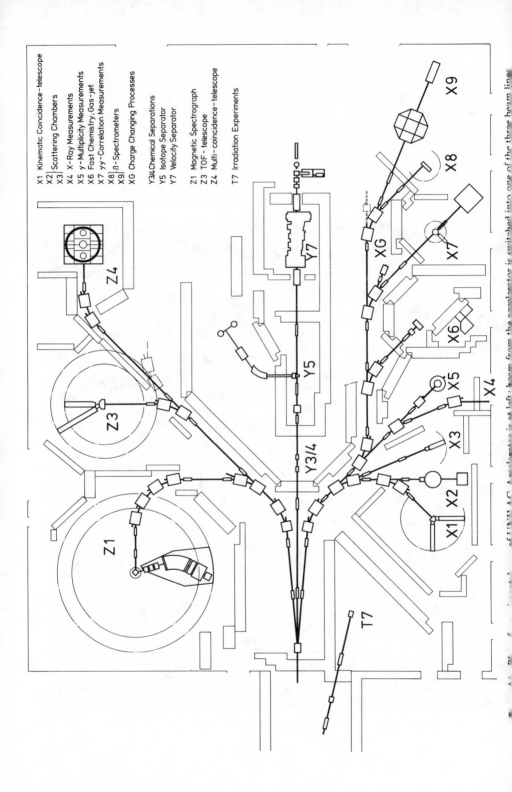

Wideröe linac operating at 27 MHz provides the first stage of acceleration, from $\beta = 0.005$ to $\beta = 0.055$ ($T_n = 0.012$ to 1.4 MeV/amu). A view of the drift tube structure of the Wideröe is shown in Figure 13.

After being accelerated to 1.4 MeV/amu in the Wideröe linac, the beam is stripped before being injected into the next section, an Alvarez linac. Note that the 1.4-MeV/amu beam is taken through a siding before being further accelerated (Figure 12). The purpose of this siding is to analyze charge states resulting from the stripping process, permitting only one charge state to enter the poststripper. A small experimental area has been installed to utilize rejected charge states. After being accelerated to 4.5 MeV/amu in the Alvarez section, the ions enter a chain of 20 resonant single-gap cavities, each one cell long. This permits each cavity to be driven independently of its neighbor, allowing arbitrary phase differences. The synchronous-particle profile may then be chosen at will. Variable-energy operation is easily accomplished, including the possibility of deceleration. At the end of the machine, neon energies from 2 to 15.4 MeV/amu and uranium energies from 2 to 8.5 MeV/amu can be achieved. A plan of the experimental area is shown in Figure 14. Figure 15 is a view from above of a portion of this area.

Improved linac structures for heavy ions are being developed in a number of laboratories. Figure 16 shows the Argonne split ring superconducting cavity, a unit of the tandem postaccelerator (23). At Heidelberg, helix resonators (26), and at Munich, an interdigital drift-tube structure (27), have been built as tandem post-accelerators. At Stanford University, substantial experience with superconducting niobium cavities has been gained in the design of an electron accelerator. Efforts are under way to make this technology available to heavy-ion acceleration (28).

3.3 Collective Acceleration

In an early paper, Veksler (29) suggested forming a bunch of electrons with some ions imbedded in it, then finding some way to accelerate the entire system. Electrons, being much lighter than ions and hence having a very large q/A, can be accelerated more readily. The ions will be trapped by Coulomb forces and will be carried along with the electron bunch. This idea is known today as collective acceleration. Successful acceleration of protons and other ions has been achieved experimentally by a number of collective methods.

The electron bunch can be produced in the form of a ring [electron-ring accelerator (ERA)] (30) or an intense burst [intense relativistic electron beam (IREB)] (31). Both methods are capable of short bursts of ion beams after the carrier electrons are separated from the embedded ions. The IREB may well emerge as a useful heavy-ion source. Typical IREBs have an energy from 100 keV to 10 MeV and currents of 10 kA–1 MA with pulse length from 10–100 nsec. The electron density achieved in IREBs is 10^{11}–10^{13} cm^{-3}. IREB development is being conducted at a number of laboratories. However, in most cases heavy-ion production is not the main goal of such work, but an incidental by-product. Groups are actively pursuing various ERA collective techniques at the University of Maryland (USA), the Max Planck Institut in Garching (West Germany), the Joint Institute for Nuclear Research at Dubna, and the Institute for Theoretical and Experimental Physics at Moscow (USSR).

Keefe has reviewed progress in the field of collective acceleration in a recent review article (32).

3.4   *Isochronous Cyclotrons*

In a cyclotron, ions are made to describe quasi-circular orbits in a magnetic field. This magnetic field is usually established between the poles of a large circular electromagnet, with the particle orbits constrained to the midplane. An ion of

Figure 15   View from above of a portion of the UNILAC experimental area. Just visible in the upper right hand corner are the three beam lines of Figure 14; the circular structure at left center is the scattering chamber labeled X2 in Figure 14.

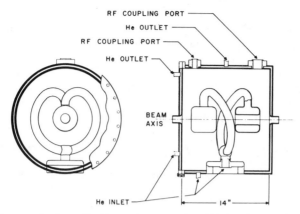

RF COUPLING PORT
He OUTLET
RF COUPLING PORT
He OUTLET

BEAM
AXIS

He INLET

14"

Figure 16   Superconducting split-ring accelerator cavity. This is one unit of the Argonne postaccelerator.

charge q and mass M will have an orbital frequency given by

$$f_{\mathrm{p}} = \frac{q\bar{B}e}{2\pi M},$$

(2)

where $\bar{B}$ is the average value of the magnetic field. Acceleration is accomplished by inserting D-shaped electrode plates, called dees, above and below the orbital plane, and attaching them to a rf resonator so that voltage of the appropriate frequency is seen by the particles. In a heavy-ion machine, the range of q/M values that must be accommodated can be large, requiring a variable-frequency rf system. In addition, it is often necessary to further extend the range by employing harmonic acceleration, in which the rf frequency f_{HF} is given by $f_{\mathrm{HF}} = hf_{\mathrm{p}}$, where h is a positive integer. The values of h that can be used in a given cyclotron will depend upon dee geometry (single dee or multiple symmetric dees) and the manner in which they are phased relative to one another. The necessary condition to satisfy, regardless of the number of dees or the harmonic number, is that the rf voltage be attractive as the ion enters the dee, and repulsive as it leaves the dee, so that the ion is accelerated both upon entering and leaving. Figure 17 shows a section through the midplane of the LBL 88-in machine. The dee is 180° in extent, filling the upper half plane in the drawing.

A useful cyclotron must provide some means for axial focusing of the ions. Early cyclotrons relied upon the weak-focusing principle, in which the magnetic field decreased with radius. In such a cyclotron, however, the energy ions can attain is limited by their relativistic mass increase. The ion mass can be written as $M = m_0 \gamma A$, where m_0 = nucleon mass, A = mass number, and γ = total energy divided by rest energy. Since m_0 and A are constants, we see that in Equation 2 that all values are constants except $\bar{B}$ and γ. If $\bar{B}$ decreases, as it must in a weak-focusing machine, the revolution frequency will decrease and the ion will slip in phase relative to the rf. As the ion gains energy, γ will increase, causing further slipping.

In an isochronous machine, $\bar{B}$ is made to increase with radius such that $\bar{B}/\gamma$

remains constant, hence the revolution frequency remains constant. Axial focusing with these machines is obtained by azimuthal variation of the magnetic field (the term AVF cyclotron for Azimuthally Varying Field cyclotron is often used). The magnetic field is divided into several (usually three or four) sectors, each consisting of a narrow iron gap followed by a wider gap, so that the ion sees a field that fluctuates in amplitude. This produces a scalloped orbit, with abnormal crossing of field boundaries, resulting in an axial focusing component at each crossing. Additional axial focusing is obtained if the field boundaries, instead of being radially straight, are made to curve at a constantly increasing angle away from the radial. The machine shown in Figure 17 has three sectors, with moderate spiral boundaries. For orbital stability, at least three sectors are required, because isochronism requires that the radial betatron frequency v_r (i.e. the number of radial oscillations about the equilibrium orbit per revolution) be approximately equal to γ, so that $v_r \simeq 1$ at the center of the machine, and greater than 1 as the ions gain energy. If the number of sectors $N = 2$, there will be a betatron-oscillation phase advance of π per sector, a condition for which the focusing forces are unstable (known as the π stopband). With $N = 3, 4 \ldots$ this is avoided. As ions gain energy, another stopband will be reached when $\gamma \simeq N/2$; this energy for $N = 3$ is 469 MeV/amu. For further details on isochronous cyclotron design, see Ref. (33).

In an isochronous cyclotron, since particles of different energy have the same revolution frequency, there is no longitudinal phase stability such as exists for a linac or synchrotron. If the field shape is exactly isochronous, particles that are

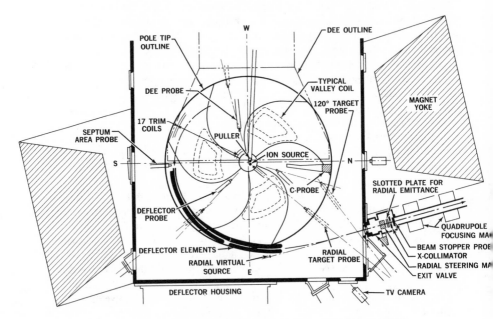

Figure 17   Plan of an isochronous cyclotron (the 88-in, at LBL).

injected into the center at a particular rf phase remain at this phase until maximum energy is reached. Any departure from the isochronous field, however, will cause particles to slip in rf phase, and with slippage beyond $\pm\pi$, the particles will be lost to further acceleration. This condition limits the allowable magnetic-field departure from isochronism. Typically, field tolerances must be held to one part in 10^{-5} or better.

It can be shown that there is no field contour that will satisfy Equation 2 for all radii as q/A is allowed to vary. This means that for a variable-ion machine, some means must be incorporated for appropriately shaping the field for the ion being accelerated. Usually, trim coils are employed, although movable shims also have been used. An accurate field mapping is necessary for various field levels and effects of trim coils or shims, so that field level and profile can be set to the required accuracy for a particular ion.

A number of methods are used for injecting ions into the innermost orbit. The simplest is to insert an ion source into the cyclotron. The diameter of the source is kept small, and the dee voltage is made large enough so that the first orbit clears the back of the ion source. Another method is to locate the ion source outside the magnet, with an axial transport system, terminating in a 90° inflector to inject into the first orbit. This can be combined with stripping. Still another means is employed with large separated sector cyclotrons. By preaccelerating the ions in another cyclotron or tandem-Van de Graaff, they have substantial energy at injection, yielding a larger first orbit.

Extraction is usually accomplished by exciting a betatron-oscillation resonance

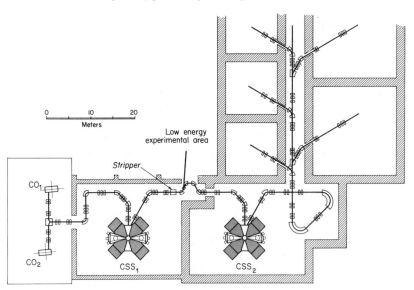

Figure 18   Plan of the GANIL facility. The two large cyclotrons are connected in tandem, with a stripper between them.

and thus increasing the radial position of the particle orbit in one revolution at a specified azimuth sufficient to insert a septum of an electrostatic or magnetic channel. This channel in turn will counteract the guide field and let the particle escape in a controlled fashion.

The maximum energy that can be attained in a cyclotron depends upon q/A and is given by $T_n \cong Kq^2/A^2$, where K is a constant for a particular accelerator. This expression is exact in the nonrelativistic limit. Since present heavy-ion cyclotrons seldom exceed $\gamma = 1.02$, the factor K is useful for comparing the capabilities of cyclotrons. Figure 18 shows a plan of the Grand Accélérateur National d'Ions Lourds (GANIL) project, now under construction in France, which will incorporate four cyclotrons, two (labeled CO_1 and CO_2) of $K = 25$ and two (labeled CSS_1 and CSS_2) of $K = 400$ (34).

3.5 Superconducting Cyclotrons

A new type of cyclotron, the so-called superconducting cyclotron, is a focus of major development effort. It combines high-field superconducting-magnet techniques with the basic isochronous cyclotron technology. Using superconducting main coils, the cyclotron's magnetic field can be raised by a factor of approximately three, to the 50-kG level. This high magnetic field leads to a major reduction in size and cost, even when the added expenses of cryogenic equipment and special conductor are included. (A superconducting cyclotron as estimated by Blosser is roughly one half the cost of a conventional isochronous cyclotron of the same energy and one third the cost of a separated-sector isochronous cyclotron.)

The anticipated properties of such a cyclotron make it a very attractive component for a number of heavy-ion accelerator systems. Figure 6d, e, and i compare estimated performance curves of superconducting cyclotrons with other heavy-ion accelerators. The expected high operating energies (above the constraints of the Coulomb repulsion, surface interactions, and normal density), will permit qualitatively new aspects of nuclei (such as high density, supersonic, and coherent mesic phenomena) to become the subject of sensitive investigation.

The superconducting cyclotron was first explored by a group at Michigan State University (35) in the early 1960s, but the idea was laid aside because of then underdeveloped conductor technology. A group at Chalk River (under Fraser) reconsidered the idea (36) in 1973, concluding that superconducting-coil technology had reached a level of development that not only established feasibility, but also offered a major economic breakthrough, lowering costs for both construction and operation. Today, three laboratories are actively exploring superconducting-cyclotron technology: the Atomic Energy of Canada Laboratory at Chalk River, Ontario, the University of Milan Cyclotron Laboratory at Milan, Italy, and the Michigan State University Cyclotron Laboratory at East Lansing, Michigan. The long-range goals of all three involve the pairing of two accelerators. At Chalk River and Milan, the first-stage accelerator is envisaged as a large tandem electrostatic accelerator, with the second stage a superconducting cyclotron of $K = 500$. In contrast, the MSU plan involves two superconducting cyclotrons, a $K = 500$ machine that injects into a $K = 800$ machine.

All three groups are presently involved in constructing prototype superconducting magnets; the prototypes at Chalk River (37) and MSU (38) are full-scale, and the one at Milan (39) is 1/6th scale. Also, all expect to have their prototype magnets in operation in 1977. Successful operation of these magnets will substantially advance the overall feasibility of superconducting cyclotrons, since all other major components operate at room temperature and involve established technologies. Major design constraints are set by the large attractive force (approximately 1000 tons) between the upper and lower halves of the coil, the need for good thermal insulation of the main coil, and the need for a sharp magnetic-field edge to facilitate extraction. All three projects plan a solenoidal-like main superconducting coil with a small median plane gap for injected and extracted beams and control hardware. The winding is further divided into two major sections, so that shifting

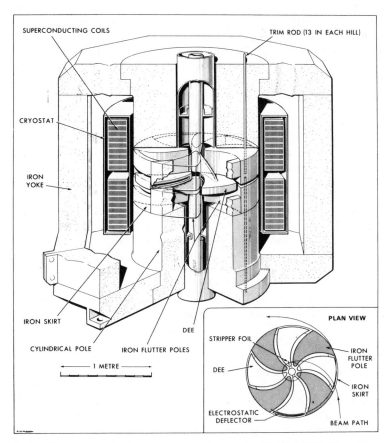

Figure 19 The superconducting cyclotron under construction at Chalk River, Canada. The insert drawing shows the method of external injection. A stripping foil placed near the center lowers the rigidity of the beam so that it is trapped in the first orbit.

of ampere turns from one coil section to the other will approximately match the variation in field shape required by particles of different mass. Major features of the structure planned at Chalk River are shown as an example in Figure 19.

The main-coil superconductor is a NbTi and Cu composite. In the MSU design, this coil is tightly packed onto a large stainless steel spool, with a picket-fence lattice between radial layers to allow helium cooling. All three projects use a massive magnet yoke, more or less completely encasing the coil. The major access to internal components will be from top and bottom by raising or lowering the appropriate elements of the yoke with a system of precision jacks.

Ion-source design also differs among the three projects. The Chalk River design omits entirely the use of a central ion source since the cyclotron is always intended to be a second-stage accelerator. For testing purposes, Milan provides a central ion source in its second-stage superconducting cyclotron and will then shift to use of the electrostatic first-stage accelerator. The MSU prototype magnet includes an interchangeable central geometry to allow the K=500 magnet to be set up as a conventional cyclotron with a central ion source, or alternatively used to study central-region design of a second-stage booster accelerator.

Beam quality for the superconducting cyclotrons is expected to be generally similar to that of present heavy-ion isochronous cyclotrons, with some improvement in intensity as a result of the much better vacuum systems provided in the ·design. The new machines are expected to produce beams of moderate emittance and energy spread in the microampere range and low emittance and energy spread at reduced intensity, characteristics that are well matched to contemplated experimental programs in heavy-ion physics.

3.6 Heavy-Ion Synchrotrons

To attain energies above 100 MeV/amu, the synchrotron becomes economically attractive as a heavy-ion accelerator. The synchrotron consists of a ring of guide field magnets whose apertures need only accommodate a single orbiting beam. The particles complete 10^4–10^6 revolutions with a modest energy gain for each revolution. This is in contrast to a linac, in which particles traverse the accelerating rf system only once, and to the cyclotron, with its guide-field magnet large enough to enclose a platter comprising orbits all the way from injection to extraction energies. In addition to guide-field magnets, the synchrotron has straight sections for injection, extraction, rf cavities, vacuum-pumping stations, and auxiliary magnets of various kinds.

The magnetic field is programmed to rise in a predetermined manner, usually either linearly or sinusoidally. As the field rises, the rf must change to keep in step with the particles, and the energy gained by the particles must be just sufficient to keep them within the magnetic guide field. Particles will oscillate about a synchronous phase angle, as in the linac, causing the beam to be bunched. If the rf is the same as the orbital frequency, there is only one bunch, but if, for practical reasons, the rf is chosen as a multiple of the orbital frequency, then the number of bunches depends upon the harmonic number. Beam control is usually accomplished by feedback systems in which the beam bunches themselves are used to generate radius and phase information and to control the position of the beam.

Such positive-feedback systems depend on sufficient beam intensity for favorable signal-to-noise ratio. Heavy-ion beams do not always possess sufficient intensity, or very low intensity may be desired, so in these cases the rf must be preprogrammed. But establishing the correct rf settings may require an accuracy on the order of one part in 10^6, a severe test of component reproducibility. One way to solve this problem is to run a light tracer ion with adequate intensity and a q/A as close as possible to the desired heavy-ion beam to find the correct rf settings, which are then recorded in a computer. For a discussion of the theory of synchrotrons, see (40) and (41).

Because of the relatively large cross section for charge-exchange processes with the residual gas and a long path during acceleration, a good vacuum is needed in heavy-ion synchrotrons. Any change of charge state will cause immediate loss, because the magnetic rigidity is inversely proportional to ionic charge.

Depending on length of acceleration, ion species, charge state, and tolerable losses, a vacuum of 10^{-9}–10^{-10} torr may be required. The vacuum requirement is less severe for fully stripped ions because their energy at injection is high enough to render the capture cross sections much smaller than the stripping cross sections (see Section 2.1). Conversely, if fully stripped ions are desired, then the minimum injection energy is determined. Contrary to the proton synchrotrons, space-charge effects for heavy ions to date are small. Hence, the magnetic field at injection must only be high enough to avoid problems with remnant fields. This establishes a minimum of $B\rho$ at injection and, therefore, the injection energy for a given q/A. The maximum energy attainable will then depend upon the ratio of maximum to minimum magnetic field. This ratio will usually fall in the interval 20–40.

To obtain intense beams of proper emittance, energy spread, and charge state,

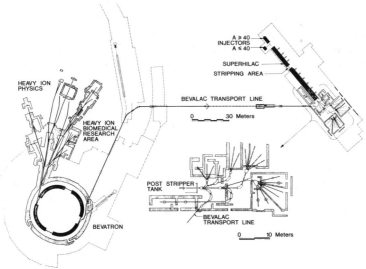

Figure 20 Plan of the Bevalac facility at LBL. The synchrotron is in the lower left-hand corner. The SuperHILAC injector is at the upper right.

an air-insulated Cockcroft-Walton dc accelerator and an rf linac with short pulses are the favored components of an injection system.

Following the shutdown of the Princeton-Pennsylvania Accelerator, there remains only one operating heavy-ion synchrotron, the Bevalac facility located at Lawrence Berkeley Laboratory, Figure 20 (42). Since the Bevatron itself was originally built to accelerate protons, its parameters had to be adapted for heavy-ion acceleration; it can now achieve maximum particle rigidities of 229 kG-m (2.6 GeV/amu for fully stripped particle with $q/A = \frac{1}{2}$). Its injector system consists of a 2.5-MV pressurized and a 750-kV air insulated Cockcroft-Walton dc accelerator, either of which can inject into the SuperHILAC, an Alvarez-type linac of 8.5 MeV/amu (43). The high-intensity beam from this machine is then transported via a 250-m long transfer line for injection into the Bevatron synchrotron. With an operating vacuum of $2 \cdot 10^{-7}$ torr in the Bevatron, only fully stripped beams can be accelerated. The heaviest particle accelerated to date is ^{56}Fe, which at 8.5 MeV/amu from the linac yields about 5% of 26+ (fully stripped) with a 400 μg cm^{-2} carbon stripping foil.

In accelerating the ion beam, the intensity was too low for operation of the feedback loops, so $^{15}N^{+7}$ ($q/A = 0.4667$) was used as a tracer particle for $^{56}Fe^{+26}$ ($q/A = 0.4643$). An improvement program is planned that will permit acceleration of very heavy ($A \approx 200$) and partially stripped ions.

4   IMPORTANT ACCELERATOR COMPONENTS

In this section we attempt to round out the sketch that has been given of each accelerator type with a discussion of some vital components common to all, with emphasis on those features that are of most importance for the heavy-ion machines.

4.1   Ion Sources for Heavy-Ion Accelerators

The ideal ion source for a modern heavy-ion accelerator should provide beams of all atomic species at high intensity, good emittance, and long lifetime. The source should be easily accessible for maintenance. For accelerators that require positive ions from the source, high-charge states are desirable. Cyclotron energy is proportional to charge-squared, and linac length can be reduced by using ions with highest-charge states. For tandem electrostatic accelerators, the charge state is −1 (only a few low-intensity ion species have been produced with −2 charge).

High-charge-state ion beams for positive-ion accelerators can be produced by electron bombardment of atoms and ions in a plasma or by stripping of fast ions. For electron-bombardment sources, the product of flux density and ion-confinement time must be sufficient to produce the desired charge state. Electron energies should be tens of eV up to hundreds or several thousand eV, depending upon the degree of ionization to be achieved. For a more complete discussion of heavy ion sources, see (44).

The traditional heavy-ion source for cyclotrons and linacs is the Philips Ion Gauge (PIG) (45, 46) (see Figure 21). Gas is introduced into the vertical tube, where the arc is struck between the two cathodes. The arc follows the direction of the magnetic field. Electrons for the arc are supplied by the cathodes. Ions are extracted through an aperture in the side or one end of the discharge tube (side-extracted or axially extracted PIG). Solid materials can be fed into the source by placing them in a cathode or tangentially to the bore in the anode, or by use of an oven.

Another type of source that has been used for heavy-ion production is the duoplasmatron, which also establishes a plasma in a magnetic field, with the ions

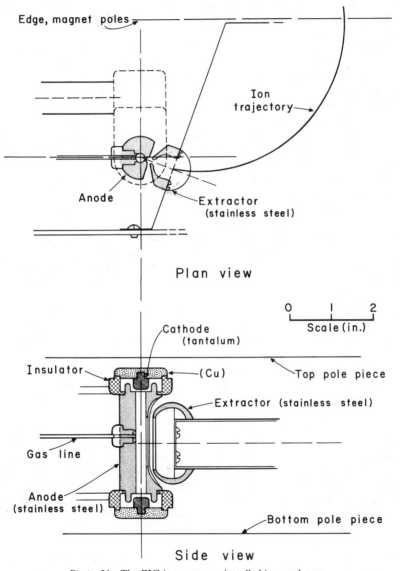

Figure 21 The PIG ion source as installed in a cyclotron.

extracted axially (47). It has undergone many years of development for high-intensity (hundreds of mA), high-quality pulsed beams of protons at high-energy synchrotron laboratories, and has recently been used for heavy ions at the UNILAC heavy-ion linac.

The duoplasmatron has higher intensity than the PIG, better emittance and lifetime, but lower average charge states. For heavy-ion linac requirements (e.g. UNILAC) of $q/A \geq 0.04$, the duoplasmatron is preferred over the PIG for ions lighter than xenon.

Several other types of sources are under development for high-charge states. One is the electron cyclotron resonance, or ECR source. Here the high-energy electrons used to strip ions are produced from the magnetically confined plasma by feeding in microwave energy at the cyclotron resonance frequency of the electrons in the applied magnetic field (48).

Another is the electron-beam ion source (49). In EBIS, a usually superconducting solenoid of 1-m length confines an electron beam from a gun placed at one end. The magnetic field keeps the beam limited to a radius of a few millimeters as it drifts down the solenoid axis to the collector at the other end. A pulse of gas injected at the gun end is ionized and confined radially by the potential well of the electron beam. A raised field at either end confines the ions longitudinally for typically 10–100 msec, until the desired charge state is reached. Then the barrier is lowered at the solenoid exit, providing beam extraction. The output current averaged over long times is orders of magnitude less than for the other sources, but the unique high-charge states would make possible significantly higher energies from cyclotrons than would the other sources. Also, the pulsed nature of the source, and the delivery of fully stripped ions (presently up to neon), make it a good match to synchrotron requirements.

Other sources that have great potential for high currents of low-charge-state ions are the multi-aperture sources developed for ion propulsion (50) and for injection of multi-ampere beams of hydrogen into thermonuclear-fusion reactors (51).

Tandem electrostatic accelerators must rely upon negative ion sources. The recently developed cesium-sputter ion source (52) is able to deliver greatly improved yields compared with earlier sources, and can be used with a much greater range of ion masses.

In summary, a large amount of work is being done at many different laboratories on new types of ion sources to achieve greater intensity and higher-charge states. The reason for this intense effort is that the design of accelerator systems depends critically on ion sources. A doubling of the charge state, for example, will cut in half the electric field required for acceleration, and also reduce the magnetic rigidity by one half. At the present time, however, the PIG and duoplasmatron sources are almost exclusively relied upon for heavy-ion accelerators.

4.2   Control Systems

An adequate control system will have some means for adjusting accelerator parameters to achieve optimal output. All heavy-ion accelerators described in this paper are sufficiently complicated to require many adjustable parameters in order to change ions, energy, etc. Some means should also exist for recording parameters for future use. Analog control systems have been used in the past, but great progress has been made recently in implementing computer-control systems for accelerators (53). They will perform the duties mentioned above admirably, and have other useful features.

One of the great advantages of a computer-control system is its potential for retaining a great amount of data, which can be presented to the operator on a selective basis. This permits the use of sophisticated instrumentation where necessary, while at the same time reducing the number of meters, displays, recorders, etc, competing for the operator's attention. Figure 22 shows schematically the computer-control system installed at the SuperHILAC. A central processor is used to transmit and receive information from operator consoles, and is also linked to standard peripheral devices. Two auxiliary computers are used for real-time control of accelerator hardware, while a third is used for graphic displays of beam data. This system employs four minicomputers for accelerator control because the complexity of operation at the SuperHILAC requires it; at other accelerator installations, a single minicomputer could perform all functions adequately.

The diagram in Figure 22 shows two operator-control consoles employed to run the accelerator in the "timeshare" mode, in which two different ions—argon and xenon, for example—are transmitted through the machine on successive pulses and delivered to separate experimental areas (43). Each console is devoted to one of the ions and an operator at that console tunes his ion beam independently of the other beam. This type of operation illustrates how accelerator capabilities may be extended with the imaginative use of control computers.

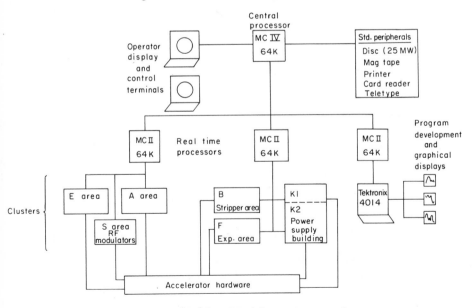

Figure 22   Schematic of SuperHILAC computer-control system.

4.3 Beam Instrumentation

For a heavy-ion accelerator to operate properly, it is necessary to measure beam intensity, energy, charge state, mass number, emittance, etc. In spite of the great importance of such instrumentation, space does not permit a discussion in this paper. The interested reader is referred to instrumentation journals, accelerator conference reports, and internal laboratory memoranda.

4.4 Beam-Transport Systems

Few accelerators could be utilized effectively without a beam-transport system of some kind, either to carry a beam from the source to the accelerator, or from the accelerator to a target area. Design of beam-transport systems has been described by several authors (15, 54). Here only some basic considerations are touched upon.

1. With heavy-ion beams, a good vacuum is necessary, particularly at low β, to prevent loss due to charge exchange.
2. In transporting heavy-ion beams, charge-state analysis, and sometimes isotopic analysis, can be very helpful. For example, for elements with several naturally occurring isotopes, only one of which is desired for acceleration, the mass separation can best be done in the injection line.
3. Clearly, a transport system should be able to transmit source or accelerator emittance without loss. However, as a safety factor, the transport system should be designed to transmit a somewhat larger emittance, say by a factor of 2, to minimize time and beam loss in correcting small errors in parameters that inevitably occur.

5 SUMMARY OF HEAVY-ION ACCELERATOR FACILITIES

Table 1 lists existing and proposed heavy-ion facilities. The second column indicates the accelerator type by a prefix followed by a number from which the maximum energy can be determined. A plus sign indicates accelerators linked in series. Facilities in boldface type are operating or are under construction; others are proposed. Meaning of symbols: Dn = drift linac, maximum energy n MeV/amu; En = simple electrostatic accelerator, maximum operating voltage n MV; Kn = isochronous cyclotron, maximum energy nq^2/A^2 MeV/amu; Ln = linear accelerator with independently phased resonators with maximum total accelerating voltage n MV, maximum energy nq/A MeV/amu; Sn = synchrotron with maximum energy n GeV/amu for ion of $q/A = 0.5$; and Tn = tandem electrostatic accelerator, maximum operating voltage n MV.

6 FUTURE PROSPECTS FOR HEAVY-ION ACCELERATORS

The many plans for expansion of existing heavy-ion accelerator facilities and for construction of new facilities (see Section 5) are an indication of the keen interest in these machines as tools for nuclear physics. Many of these projects are for isochronous cyclotrons; the next largest number involve electrostatic accelerators. There is a

Table 1   Existing and proposed heavy-ion facilities[a]

Location	Type and energy	Remarks	Ref.
Argentina			
Buenos Aires	T20		
Australia			
Canberra (Aust. Nat'l U)	**T14**	NEC 14UD	(55), p. 227
Belgium			
University of Louvain	K70 + **K110**	CYCLONE	(56), p. 80, 618
Brazil			
University of Sao Paulo	**E4 + T8**	NEC 4U, 8UD	(55), p. 213
Canada			
Chalk River	**T13 + K520**	superconducting cyclotron	(37, 57)
Quebec	K28 + K500	superconducting cyclotron	(56), p. 580
Denmark			
University of Aarhus	**T7**	EN tandem	(58)
Copenhagen	**T9**	FN tandem	(58)
France			
Caen	**K25 + K400 + K400**	GANIL	(34)
Grenoble	**K90**		(56), p. 614
Orleans	**K50**	constructed by CGR-MeV	(56), p. 275
Orsay	**D1.2 + K75**	ALICE	(59)
Orsay	**T12**	MP tandem	
Strasbourg	**T12**	MP tandem	
Saclay	**D5 + S1.2**	EBIS source development	
Germany			
Berlin	**E6 + K120**	VICKSI	(60)
Darmstadt	D4.5 + **L23.8** + S1.0	UNILAC	(25)
Heidelberg	**T13 + L3 + L9**	helix resonator postaccelerator	(26)
Munich	**T13 + L5**	interdigital H-type postaccelerator	(27)
India			
Calcutta	**K140**		(56), p. 84
Israel			
Weizmann Institute	**T14**	NEC 14UD	
Italy			
Catania, Sicily	T13		(58)
Padua	**T16 + K540**	XTU, SSC developed at University of Milan	(61)

Table 1—continued

Location	Type and energy	Remarks	Ref.
Japan			
Osaka University (RCNP)	**K120**		(56), p. 95
Kyushu University	**E5+T10**		(62)
Saitama (IPCR)	**L16**	RILAC	(63)
Saitama (IPCR)	**K77**	WFC	(82)
Tohoku University	**K50**	CGR-MeV, Sumitomo collaboration	(56), p. 79
Tokai (JAERI)	**T20**	NEC 20 UR	(58)
Tokyo University (INS)	**K68**		(56), p. 103
Tokyo University	D10+S1.4	numatron (INS study)	(81)
Tsukuba University	T12	NEC 12UD	
Netherlands			
University of Groningen	**K160**		(83)
Poland			
Warsaw University	**K178**	modified U-200	(56), p. 107
South Africa			
Cape Town	K200		(56), p. 117
Sweden			
University of Uppsala	K200	exist, synchrocyclotron convertor	(56), p. 127
Switzerland			
Villagen	**K135**	SIN injector	(64), p. 257
United Kingdom			
Daresbury	**T20** (T30)		(21)
Harwell	**K86**	VEC	
Oxford	**E8+T7**		
USA			
Argonne	**T10+L15**	SSRR postaccelerator	(23)
Berkeley	**K140**	88-inch	(76), p. 265
Berkeley	**D8.5+S2.6**	SuperHILAC, Bevalac	(24, 42, 43)
Brookhaven	**T8+T12**	MP tandems	(66)
U.C. Davis	**K90**	NHIP	
Indiana University	**K16+K220**	NHIP	(67)
University of Maryland	**K180**	MUSIC	(68)
University of Minnesota	T12		(69)
Michigan State University	**K50+K500**+K800	K500 & K800 are superconducting	(38)
Los Alamos	**E8+T9**	FN tandem	(70)

Table 1—continued

Location	Type and energy	Remarks	Ref.
Oak Ridge	**T7**	EN tandem	
Oak Ridge	**T25 + K90 + K400**	HHIRF, K90 is called ORIC	(20, 22)
University of Rochester	**T11 + K500**		(71)
Stonybrook	**T9 + L12**	SSRR postaccelerator	(72)
Texas A & M University	**K147**		(73)
University of Washington	**T7 + T9 + L**	FN tandems	(74)
Yale University	**T13**	MP tandem	(75)
USSR			
Dubna (JINR)	**K156 + K250**	U-200, U-300 (WFC)	(76)
Dubna (JINR)	**D4 + S4.6**		
Dubna (JINR)	**K725**	U-400	(77)
Kazakhstan	**K50**		(56), p. 630
Kiev	**K140**	U-240	(56), p. 326
Leningrad	K140	U-240	(78), p. 692
Moscow, Kurchatov Institute	**K60**		(56), p. 205
Moscow, Radiotechnical Institute	L102		(79)
Kiev	T20	EGP-20	(80)

trend to higher energies, as evidenced by the number of projects using two accelerators, some with a tandem injecting into a cyclotron, some with two cyclotrons in series. One (the GANIL project) plans to use three cyclotrons in series.

Linear accelerators have not been used as often as tandems or cyclotrons for nuclear science because of their greater cost. Interest in linacs remains high, however, because they potentially have fewer limitations in intensity, particle species, and maximum energy. They are also superior as a synchrotron injector. Work in several laboratories is devoted to the development of more efficient linacs. The prospect for heavy-ion synchrotrons for nuclear science in the near future is promising. Present activity indicates that at least two, and perhaps as many as four, will be started in the next five years.

There is growing interest from radiotherapists in using dedicated heavy-ion machines for the treatment of cancer. The ion mass used would be in the range $A = 12$–40, with energy around 500 MeV/amu. Duty factor could be low, intensity modest ($\sim 10^9$ pps). A synchrotron is the likely choice to fit the requirements.

The use of heavy ions as projectiles to trigger a thermonuclear reaction for fusion power generation has recently emerged as an attractive possibility. The requirement is for beams with $A \sim 200$, $T_n \sim 200$ MeV/amu, with $\sim 10^{15}$ W of peak power in a 10^{-8}-sec burst. There is no existing accelerator with these capabilities, but they seem to be within a reasonable extrapolation of existing accelerator technology. Studies are under way at several national laboratories to establish feasibility.

ACKNOWLEDGMENTS

The authors wish to thank Drs. R. Herb and H. Blosser, who have contributed the major parts of the Van de Graaff and superconducting cyclotron section, respectively. We are indebted to Dr. D. Clark, who contributed substantially to the ion-source section, as well as being an excellent consultant on many other sections. We received valuable display material and data from Drs. R. Bock, GSI Darmstadt, and J. Ball; and C. Jones, HHIRF Oak Ridge. We also wish to acknowledge help we have received from many others too numerous to list.

It is a pleasure to acknowledge the great help on editing we received from C. Weber and R. Hendrickson. We are thankful to Dr. Ch. Leemann for proofreading the final paper. Recognition has to go to ERDA and NSF for their support of many US heavy-ion facilities. ERDA Program Director for Nuclear Physics, Dr. G. Rogosa deserves our deepest appreciation for his unfailing enthusiasm and support in this field.

Literature Cited

1. Livingston, M. S., Blewett, J. P. *Particle Accelerators.* New York: McGraw-Hill
2. Hubbard, E. L. et al 1961. *Rev. Sci. Instrum.* 32:621–34
3. Bromley, D. A. 1974. *Nucl. Instrum. Methods* 122:1–34
4. Middleton, R. 1974. *Nucl. Instrum. Methods* 122:35–43
5. Isaila, M. V. et al 1972. *Int. Conf. Heavy Ion Sources, IEEE NS* 19(2):204–7
6. Grunder, H. A. et al 1971. *Int. Conf. High Energy Accelerators,* Geneva; p. 574
7. Stock, R. 1977. *Heavy Ion Collisions,* North Holland, ed. R. Bock, Vol. 1. To be published
8. *Biological and Medical Research with Accelerated Heavy Ions at the Bevalac 1974–1977,* LBL 5610 (1977)
9. Bakish, R., ed 1976. *Proc. Symp. Electron Ion Beam Sci. Tech., 7th Int. Conf.,* pp. 482–532
10. Brown, W. 1977. *Particle Accelerator Conf. Proc., IEEE NS.* In press
11. McHargue, C. J. 1975. *Particle Accelerator Conf., IEEE NS* 22(3):1743–48
12. Leischner, E. 1966. *UNILAC Ber.* No. 1–66. Univ. Heidelberg
13. Betz, H. D. 1971. *Particle Accelerator Conf., IEEE NS* 18(3):1110–14
14. Tolman, R. 1938. *The Principles of Statistical Mechanics.* Oxford
15. Banford, A. P. 1968. *The Transport of Charged Particle Beams.* E. & F. N. Spon Limited
16. Reinhold, G., Truempy, K. 1969. *IEEE NS* 16(3):117–18
17. Hanley, P. R. et al 1969. *IEEE NS* 16(3):90–95

18. Van de Graaff, R. J. 1960. *Nucl. Instrum. Methods* 8 : 195–202
19. Herb, R. 1974. *Nucl. Instrum Methods* 122 : 267–76
20. Blair, J. K. et al 1975. *Particle Accelerator Conf. IEEE NS* 22(3): 1655–58
21. Aitken, T. W. et al 1974. *Nucl. Instrum. Methods* 122 : 235–65
22. Martin, J. A. et al 1977. *Particle Accelerator Conf. Proc., IEEE NS.* In press
23. Bollinger, L. M. et al 1976. *Proton Linear Accelerator Conf., Chalk River* AECL-5677: 95–101
24. Main, R. M. 1971. *Nucl. Instrum. Methods* 97 : 51–64
25. Böhne, D. 1976. *Proton Linear Accelerator Conf., Chalk River* AECL-5677: 2–11
26. Jaeschke, E. et al 1977. *Particle Accelerator Conf. Proc., IEEE NS.* In press
27. Nolte, E. et al 1977. *Particle Accelerator Conf. Proc., IEEE NS.* In press
28. Sokolowski, J. S. et al 1977. *IEEE NS.* In press
29. Veksler, V. I., Tsytovich, V. N. 1959. *Int. Conf. High Energy Accel., Geneva*
30. Lambertson, G. R. 1974. *Int. Conf. High Energy Accelerators, Stanford*, pp. 214–17
31. Olson, O. L. 1975. *Particle Accelerator Conf., IEEE NS* 22(3): 962–69
32. Keefe, D. 1976. *Proton Linear Accelerator Conf., Chalk River* AECL-5677: 352–57
33. Livingood, J. J. 1961. *Principles of Cyclic Particle Accelerators.* New York: Van Nostrand
34. GANIL Study Group 1975. *Particle Accel. Conf., IEEE NS* 22(3): 1651–54
35. Berg, R. 1963. *Magnetic Coil Design for a Superconducting Air-Cored 40 MeV Cyclotron*, MSUCP-14
36. Bigham, C. B., Fraser, J. S., Schneider, H. R. 1973. *Superconducting Heavy Ion Cyclotron*, CRNL Rep. AECL-4654
37. Fraser, J. S., Tunnicliffe, P. R. 1975. *A Study of a Superconducting Heavy Ion Cyclotron as a Post Accelerator for the CRNL MP Tandem.* AECL-4913
38. 1976. *Proposal for a National Facility for Research with Heavy Ions Using Coupled Superconducting Cyclotrons.* MSUCL-222
39. 1976. *Studio del progetto di un ciclotrone superconduttore per ioni pesanti.* Instituto nazionale di fisica nucleare. (Sezione di milano) Milan, Italy
40. Green, G. K., Courant, E. D. 1959. *The Proton Synchrotron* in Handbuch der Physik, pp. 218–340. Berlin: Springer (Berlin)
41. Bruck, H. 1966. *Accel. Circulaires de*

Particules. Univ. France Press. 1966. Transl. 1976 as *Circular Particle Accelerators.* LASL LA-TR-72-10 Rev.
42. Barale, J. et al 1975. *Particle Accelerator Conf., IEEE NS* 22(3): 1672–74
43. Grunder, H. A., Selph, F. B. 1976. *Proton Linear Accelerator Conf., Chalk River* AECL-5677, pp. 54–61
44. Clark, D. 1977. *Particle Accelerator Conf. Proc. IEEE NS.* In press
45. Makov, B. N. 1976. *Int. Conf. Heavy Ion Sources, IEEE NS* 23(2): 1035–41
46. Schulte, H. et al 1976. *Int. Conf. Heavy Ion Sources, IEEE NS* 23(2): 1042–48
47. Keller, R., Muller, M. 1976. *Int. Conf. Heavy Ion Sources*, 23(2): 1049–52
48. Geller, R. 1976. *Int. Conf. Heavy Ion Sources, IEEE NS* 23(2): 904–12
49. Donets, E. D. 1976. *Int. Conf. Heavy Ion Sources, IEEE NS* 23(2): 897–903
50. Stuhlinger, E. 1971. *Proc. Symp. Ion Sources Form. Ion Beams*, BNL 50310: 47–60
51. Ehlers, K. W. et al 1976. *Proc. Symp. Fusion Technol., 9th*, ed. Garmish. New York: Pergamon (LBL-4471)
52. Middleton, R. 1975. *Int. Conf. Heavy Ion Sources, IEEE NS* 23(2): 1098–1103
53. Belshe, R. A., Elisher, V. P., Jacobson, V. 1975. *Particle Accel. Conf., IEEE NS* 22(3): 1036–40
54. Steffen, K. G. 1965. *High Energy Beam Optics.* New York: Wiley-Interscience
55. Electrostatic Accelerator Issue 1974. *Nucl. Instrum. Methods*, Vol. 122
56. 1975. *Int. Conf. Cyclotrons Appl., 7th.* Zurich, Basel & Stuttgart: Birkhauser
57. Omrod, J. H. et al 1977. *Particle Accel. Conf. Proc., IEEE NS.* In press
58. Jones, C., private communication
59. Cabrespine, A., Lefort, M. 1971. *Nucl. Instrum. Methods* 97 : 29–40
60. Maier, K. H. 1977. *Part. Accel. Conf. Proc., IEEE NS.* In press
61. Acerbi, E. et al 1977. *Part. Accel. Conf. Proc., IEEE NS.* In press
62. Isoya, A. 1973. *Int. Conf. Tech. Electrostat. Accel.* Daresbury DNPL/NSF/R5: 89–99
63. Odera, M. 1976. *Proton Linear Accel. Conf., Chalk River* AECL-5677: 62–66
64. Baan, A. et al 1973. *Part. Accel. Conf., IEEE NS* 20(3): 257–59
65. Clark, D. J. et al 1972. See Ref. 78
66. Thieberger, P. 1977. *Part. Accel. Conf. Proc., IEEE NS.* In press
67. Pollock, R. 1977. *Part. Accel. Conf. Proc., IEEE NS.* In press
68. Johnson, W. P. 1971. *Part. Accel. Conf., IEEE NS* 18(3): 268–71
69. Broadhurst, J. H., Blair, J. M. 1974.

Nucl. Instrum. Methods 122:143–46
70. Woods, R., McKibben, J. L., Hinkel, R. L. 1974. Nucl. Instrum. Methods 122:81–97
71. Purser, K. H., Gove, H. E., Lund, T. S. 1974. Nucl. Instrum. Methods 122:159–77
72. Noe, J. W. et al 1977. Part. Accel. Conf. Proc., IEEE NS. In press
73. McFarlin, W. A., Goerz, D. J. Jr. 1966. IEEE NS 13(4):401–10
74. Weitkamp, W. G., Schmidt, F. H. 1974. Nucl. Instrum. Methods 122:65–79
75. Sato, K. et al 1974. Nucl. Instrum. Methods 122:129–42
76. Shelaev, I. A. et al 1972. See Ref. 78

77. Flerov, G. N. 1976. J. Phys. Paris Colloq. 37:Suppl. II, pp. 233–35
78. 1972. Int. Cyclotron Conf., Vancouver, 6th. AIP Proc. No. 9
79. Murin, B. P. 1976. Proton Linear Accel. Conf., Chalk River AECL-5677:22
80. Hochberg, B. M., Mikhailov, V. D., Romanov, V. A. 1974. Nucl. Instrum. Methods 122:119–28
81. Hirao, Y., private communication
82. Kohno, I. et al 1969. Proc. Int. Cyclotron Conf., 5th, Oxford, pp. 487–98
83. Van Kranenburg, A. A., Wierts, D., Hagedoorn, H. L. 1966. Int. Conf. Isochron. Cyclotrons, Gatlinburg IEEE NS 13(4):447

Ann. Rev. Nucl. Sci. 1977. 27:393–464
Copyright © 1977 by Annual Reviews Inc. All rights reserved

PSIONIC MATTER[1] **✲5589**

W. Chinowsky

Department of Physics and Lawrence Berkeley Laboratory, University of California, Berkeley, California 94720

CONTENTS

[1] Work supported by the Energy Research and Development Administration, under the auspices of the Division of Physical Research.

INTRODUCTION

Before November of 1974, heavy hadrons that decay to hadrons were neatly classified into two groups, distinguished by their decay rates. Those are, very roughly, either of the order of 10^{-9} sec or 10^{-23} sec (1). Strangeness conservation distinguishes the longer-lived group from the shorter-lived one. The strikingly different decay rate of the J/ψ,[2] orders of magnitude different from either of these values, indicates the operation of a quite different kind of dynamics. With further developments, particularly the observation in e^+e^- annihilation of heavier, directly formed vector states as well as others produced in e^+e^- annihilations to multiparticle states, interpretation has narrowed. Arguments in favor of a new property of matter and a new conservation law for strong interactions have become very convincing, if not yet entirely compelling. Indeed, all observations now fit very comfortably into a theoretical framework originating with the introduction of a fourth quark precisely of the character proposed earlier, and for quite different reasons, by Glashow and co-workers (2, 3). In this article, I review the phenomenology of psion production and decay, and also charmed-hadron production and decay, as relevant to a broader conception of the subject. The simple model of the ψ states as composites of a charmed quark and its antiparticle dominates the interpretive discussion (4, 5).

1.1   Old Vector Mesons

It is inappropriate to give an extensive review of properties of the older hadronic resonant states, but still useful to summarize some characteristics of production and decay to contrast with the ψ-family characteristics. Only those vector mesons that couple directly to the photon, i.e. ρ^0, ω^0, and ϕ^0, are discussed. Table 1 shows selected decay parameters of these vector mesons (1). Values of full width, Γ, and partial width, Γ_{ee}, for decay to e^+e^- pairs, have been extracted from measurements of cross sections for e^+e^- annihilation to particular hadronic final states, $\pi^+\pi^-$ for ρ^0, $\pi^+\pi^-\pi^0$ for ω^0, and $K\bar{K}$ for ϕ (6). The annihilation cross section in the vicinity of a resonance varies with center of mass (c.m.) energy E, according to the Breit-Wigner expression

$$\sigma_f(E) = \frac{(2J+1)\pi}{E^2} \frac{\Gamma_{ee}\Gamma_f}{(E-M)^2 + \Gamma^2/4},$$

Table 1   Parameters of vector mesons ρ^0, ω^0, and ϕ^0

Meson	Mass (MeV/c^2)	Γ(MeV)	Γ_{ee}(keV)	$f_v^2/4\pi$
ρ^0	773 ± 3	152 ± 3	6.5 ± 0.7	2.1 ± 0.2
ω^0	782.7 ± 0.3	10.0 ± 0.4	0.76 ± 0.17	18.3 ± 4.0
ϕ^0	1019.7 ± 0.3	4.1 ± 0.2	1.31 ± 0.08	13.8 ± 0.8

[2] From now on, I eschew the typographically awkward "J/ψ" and use ψ or ψ (3095) for the lightest psion.

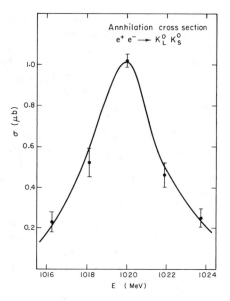

Figure 1 Cross section for e^+e^- annihilation to $K_S^0 K_L^0$ as a function of c.m. energy, showing ϕ^0-meson resonance excitation; data from (7).

where Γ_f and Γ_{ee} are the partial widths for decay to the final states f and e^+e^- respectively, and Γ is the full width. Suitable modification of this simple behavior must be made to take account of initial-state radiation in order to obtain the resonance widths. An exemplary excitation curve, for $K_L^0 K_S^0$ production (7), is shown in Figure 1. The full width and the product $\Gamma_{ee}\Gamma_f$ are determined from the annihilation cross sections near the resonance. Values of Γ_f are obtained from independent sources, thus allowing determination of electron-pair partial widths. A useful parameterization is in terms of the coupling of the vector meson to the virtual photon, indicated in the vector-dominance (8) diagram of Figure 2. The coupling constant f_v, defined by

$$G_v \equiv M_v^2/f_v,$$

is related to Γ_{ee} by (9)

$$f_v^2/4\pi = \frac{1}{3}\frac{\alpha^2 M_v}{\Gamma_{ee}}.$$

1.1

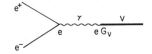

Figure 2 Direct coupling of a vector meson to a virtual photon.

Values of Γ in Table 1 are averages of values obtained from measurements of resonance production in hadron-hadron interactions, as well as the e^+e^- annihilation results. As has been remarked often, it is noteworthy that the width of the ϕ^0 is only one half that of the ω^0 and more remarkable still that the partial width for ϕ^0 decay to $\pi^+\pi^-\pi^0$ is just 0.67 ± 0.07 MeV, while the partial width is 9.0 ± 0.4 MeV for decay of the lighter ω^0 to $\pi^+\pi^-\pi^0$. Indeed the principal decay mode of the ϕ^0 meson is $K\bar{K}$, with 82% branching fraction, in spite of the relatively small phase space that is available to that state.

Meson resonances of the old style are produced in hadron-hadron collisions, restricted only by requirements of the strong-interaction conservation laws. Typical exclusive reaction cross sections, e.g. for $\pi p \to \rho p$, are in the range 0.1–10 mb (10), values typical of strong interactions. Data on inclusive vector-meson production in hadron-hadron interactions are sparse and suffer poor statistical accuracy. Inclusive ρ^0 production rates in π-p and p-p collisions have been determined from observations of interactions in bubble chambers (11). Cross sections increase smoothly from ~ 2 mb at 4.9 GeV c.m. energy to ~ 10 mb at 19 GeV. Inclusive production of ϕ^0 appears to be smaller by about a factor of ten, judging from measured values of 0.16 ± 0.04 mb at 6.8 GeV (12) and 0.6 ± 0.2 mb at 16.8 GeV c.m. energy (13). The first result is again from a pp bubble-chamber exposure. The other is from an experiment that measured rates for muon pair production in π^+ and proton collisions with Be nuclei. It is described in some detail below.

In the context of vector-meson dominance (8), vector-meson photoproduction (14) and scattering are related according to (15)

$$\frac{d\sigma}{dt}(\gamma p \to Vp) = \frac{4\pi}{f_v^2} \alpha \frac{d\sigma}{dt}(Vp \to Vp) \qquad 1.2$$

where t is the momentum transfer. With the optical theorem, neglecting the real part of the forward scattering amplitude,

$$\frac{d\sigma}{dt}(\gamma p \to Vp) = \frac{\alpha}{16\pi} \frac{4\pi}{f_v^2} \sigma_T^2(Vp \to Vp) \qquad 1.3$$

at $t = 0$. The value of the total vector-meson-nucleon interaction cross section, σ_T, is then obtained with f_v determined from the e^+e^- partial width Γ_{ee}.

Values of $d\sigma/dt$ at $t = 0$ for ρ^0 photoproduction on protons are, to within $\sim 15\%$, constant at 115 μb/GeV^2 for 3.6 GeV $< (s)^{1/2} < 5.9$ GeV (16). Using the vector-

Table 2   Properties of quarks

Flavor	Spin	I	I_3	Q/e	B	S	Y	C
u	1/2	1/2	1/2	2/3	1/3	0	1/3	0
d	1/2	1/2	$-1/2$	$-1/3$	1/3	0	1/3	0
s	1/2	0	0	$-1/3$	1/3	-1	$-2/3$	0
c	1/2	0	0	2/3	1/3	0	0	1

dominance equation (1.3) with $f_\rho^2/4\pi$ determined from a weighted average of measured $e^\pm$ and $\mu^\pm$ partial widths (1), we obtain the total cross section $\sigma(\rho\text{-nucleon}) = 24$ mb in that range of c.m. energy. Results of measurements of ϕ^0 photoproduction (17) at various energies in the range 2.2 GeV $< (s)^{1/2} < 4.2$ GeV are compatible with a constant value for the differential cross section at $t = 0$, $d\sigma/dt = 2.49 \pm 0.15\ \mu b\ \text{GeV}^{-2}$. Vector dominance then yields $\sigma(\phi\text{-nucleon}) = 8.7 \pm 0.5$ mb (17). Those values for ρ^0 and for ϕ^0 interaction cross sections are typical of strong interactions.

1.2   Quark-Model Classification

Hadron states group into multiplets specified by spin, parity, and baryon numbers indicative of the exact SU(2) symmetry and approximate SU(3) symmetry of the strong interactions. Baryons populate SU(3) multiplets of dimensionality ten and eight. Mesons fit into octet and singlet representations. The observed states can be considered composite structures whose basic constituents are the three spin-$\frac{1}{2}$ quarks of the 3 representation of SU(3). Their properties are listed in Table 2. In this model, baryons are systems of three quarks, and mesons are quark-antiquark composites. Properties of meson states are given in terms of the orbital and spin angular momentum of the $q\bar{q}$ system and its isospin, namely spin $J = |\mathbf{L}+\mathbf{S}|$, parity $P = (-1)^{L+1}$, charge conjugation $C = (-1)^{L+S}$ and G-parity $G = C(-1)^I$. Members of the lowest-lying meson multiplets are s states of the $q\bar{q}$ system. The quark contents assigned to the "old" vector mesons are

$$\rho^0 = (2)^{-1/2}\left[u\bar{u} - d\bar{d}\right]$$

$$\omega^0 = (2)^{-1/2}\left[u\bar{u} + d\bar{d}\right]$$

$$\phi^0 = s\bar{s}.$$

An essentially ad hoc rule based on this composition and those of the pseudo-scalar mesons, the Okubo (18)-Zweig (19)-Iizuka (20) (OZI) rule, has been introduced to explain the small width and dominant $K\bar{K}$ decay mode of the ϕ. This rule proclaims that a process is inhibited if its duality diagram contains disconnected quark lines. Such lines can be isolated in a duality diagram by drawing a line that crosses no quark lines. Allowed and forbidden decays of the ϕ-meson are illustrated in Figure 3. Some attempts have been made to put the OZI rule on a firmer dynamical foundation based on dual models (21) or asymptotically free gauge theories (22), but only semiquantitative success has yet been achieved.

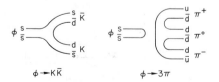

Figure 3   Duality (quark-line) diagrams for $K\bar{K}$, OZI-allowed; and 3π, OZI-forbidden, decays of the ϕ meson.

2   PRODUCTION OF $\psi(3095)$ AND $\psi(3684)$

The existence of the narrow state $\psi(3095)$ was first established independently, with quite different experimental techniques, by Aubert et al (23) and Augustin et al (24).

The Brookhaven National Laboratory-Massachussets Institute of Technology (BNL-MIT) (23) group measured the effective mass spectrum of e^+e^- pairs produced in the interaction of 28.5 GeV/c protons with beryllium, $p + \text{Be} \rightarrow e^+ + e^- + X$. In this experiment, a double-arm magnetic spectrometer, designed to have maximum detection efficiency for heavy, unstable particles produced at rest in the center of mass, produced the mass spectrum shown in Figure 4. A narrow resonance was definitively established at mass 3.112 GeV/c^2 and width less than the 5-MeV/c^2 experimental resolution of the apparatus. More details of this experiment and subsequent hadronic-production experiments are given below.

The Lawrence Berkeley Laboratory-Stanford Linear Accelerator Center (LBL-SLAC) (24) group measured the total cross section for e^+e^- annihilation as a function of energy. Measurements were made in small energy intervals near 3100 MeV with a

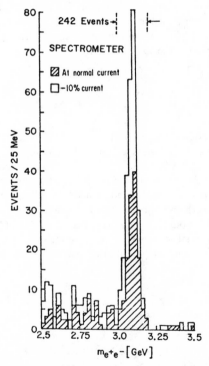

Figure 4   Invariant-mass spectrum of e^+e^- pairs produced in collisions of 28-GeV/c protons with beryllium showing J-meson production; from (23).

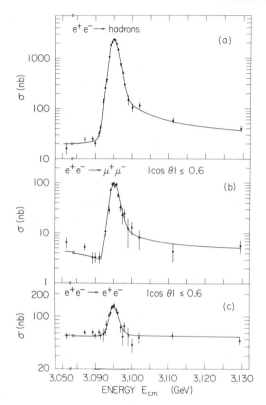

Figure 5 Cross sections for e^+e^- annihilation to (*a*) hadrons, (*b*) muon pairs, and (*c*) electron pairs, showing $\psi(3095)$-resonance excitation; from (43).

cylindrical detector array in an axially directed solenoidal magnetic field. Yields were measured for annihilation to e^+e^- pairs, $\mu^+\mu^-$ pairs, and multiparticle hadronic states. Results, shown in Figure 5, indicate a resonance at 3095 MeV, coupled to e^+e^-, $\mu^+\mu^-$, and hadronic states. Its width is less than 2 MeV. With the same technique, the LBL-SLAC group observed a second resonance at 3684 MeV, as indicated by the data of Figure 6 (25). Again the hadronic and lepton-pair decay modes appear, with the latter less prominent than at 3095 MeV.

The ψ resonance peak has been observed in e^+e^- annihilations at the storage ring ADONE (26); both ψ and ψ' have been observed at the higher-energy e^+e^- storage rings DORIS (27). Those results confirmed the existence of the two states and generally agree with the SLAC-LBL measurements of energy and decay widths of the resonances.

Further analyses of the shapes of these yield curves, particularly by the LBL-SLAC group, taking account of broadening due to finite beam-energy spread and initial-state radiation, led to determination of the full widths 69 ± 15 keV and

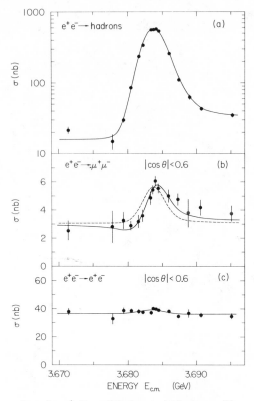

Figure 6   Cross sections for e^+e^- annihilation to (*a*) hadrons, (*b*) muon pairs, and (*c*) electron pairs, showing $\psi(3684)$ resonance excitation; from (51).

228 ± 56 keV for the lower- and higher-mass resonances, respectively. Compared with previously established resonances at considerably smaller energies, these widths are startlingly small, indicating the effect of some new mechanism inhibiting their decays.

2.1   *Production of ψ and ψ' in Hadron-Hadron Collisions*

All such experiments have until now searched for narrow peaks in mass distributions of pairs of oppositely charged leptons as a signature of decays of heavy particles produced in the primary hadronic reactions. Severe experimental difficulties in detecting decay products in heavy backgrounds of hadronic debris have been overcome with sophisticated, complex experimental arrangements with special capabilities (28, 29). These include $e^\pm$ or $\mu^\pm$ identification and rejection of $\pi^\pm$, $K^\pm$, $p^\pm$; momentum- and angle-measurement precision sufficient to yield mass resolution of order 10–100 MeV/c^2; multicoincident counting of particles to reduce accidental rates to acceptable levels; good multitrack efficiency and position resolution; rejection of photon and neutron backgrounds associated with incident beams and from secondary, diffuse sources; large angle and large mass acceptance; and, since rates are very low, reliability over long periods of experimental operation. Instrumental limitations restrict the detected regions of production and decay phase space, and further introduce mass dependence in the acceptance of the apparatus. To obtain absolute yields, corrections must be made for these geometric deficiencies, as well as detection inefficiencies. Calculations are made with Monte Carlo simulations of detector response with an assumed dependence on production and decay kinematic variables.

2.1.1   28.5-Gev/c AND 20-Gev/c PROTONS ON BERYLLIUM   Aubert et al (BNL-MIT) (30) used an arrangement designed for maximum acceptance of heavy particles produced at rest in the center of mass in the reaction $p + \text{Be} \to e^+ e^- + X$ (X signifies all other particles in the final state). The experiment was made with incident protons of 28.5 GeV/c momentum. Schematic views of the detector are shown in Figure 7. To extract a ψ production cross section from the observed peak in the mass distribution, they assumed the differential cross section

$$\frac{d^3\sigma}{dP_{\parallel}^* \, dP_{\perp}^{*2}} \propto \frac{\exp(-6P_{\perp}^*)}{E^*}$$

for production of a ψ with c.m. longitudinal momentum $P_{\parallel}^*$ and transverse momentum $P_{\perp}^*$ and allowed the ψ decay to be isotropic at rest. They obtained a

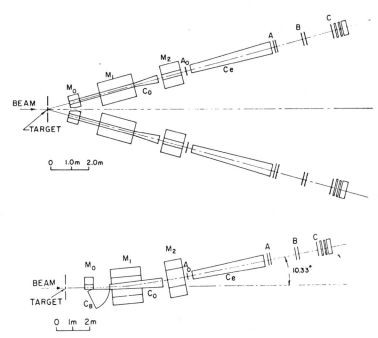

Figure 7   Spectrometer detection apparatus of the MIT-BNL group (23, 28–30) at Brookhaven National Laboratory.

production cross section $\sigma \sim 10^{-34}$ cm^2/nucleon for e^+e^- pairs with mass in the peak. With 20-GeV/c incident protons, the cross section is ten times smaller. No production of heavier particles was observed, again at 28.5 GeV/c, yielding an estimated upper limit $\sigma \sim 10^{-36}$ cm^2, with 95% confidence, for e^+e^- pairs of mass 3.2 GeV/$c^2 < M < 4.0$ GeV/c^2.

2.1.2 300-GeV NEUTRONS ON NUCLEI   Knapp et al (31) and later, Binkley et al (32, 33), measured the yield of muon pairs in interactions of neutrons with Be, Al, Cu, and Pb nuclei. A primary objective was to determine the A dependence of resonance production so that cross sections for single nucleon interaction could be determined by extrapolation. Neutrons were produced by 300-GeV/c protons in the earlier experiments and by 400-GeV/c protons in the later runs. The resultant neutron-energy spectra peak at ~ 250 GeV and ~ 300 GeV for the lower and higher proton energy, respectively. Binkley et al (32) made fits to the yields of dileptons of momentum $p < 75$ GeV/c with the power law A^γ, obtaining the best-fit value $\gamma = 0.93 \pm 0.04$ for $\psi(3095)$ production. In contrast, they find $\gamma = 0.62 \pm 0.03$ for ρ^0 and ω^0. Fits were made to the observed distributions in the dimuons' $P_\perp$ and $P_\parallel$ (Figure 8), taking account of acceptance limitations and resolution, using assumed production dependences on $x \, (\equiv P_\parallel^* / P_{\max}^*)$ and $P_\perp$ of the form

$$E \frac{d^3\sigma}{dP^3} = C(1-x)^\alpha \exp(-bP_\perp) \qquad \text{for} \quad x_0 < x < 1,$$

and

$$E \frac{d^3\sigma}{dP^3} = C(1-x_0)^\alpha \exp(-bP_\perp) \qquad \text{for} \quad 0 < x < x_0.$$

For dimuon decay products of $\psi(3095)$, they obtain values $\alpha = 5.2 \pm 0.5$, $b = (1.6 \pm 0.2) \, c$ GeV^{-1} with $x_0 = 0.3$. With these, they calculated an integrated yield of ψ-

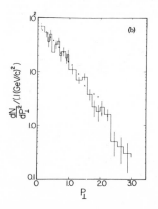
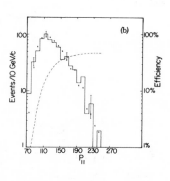

Figure 8   Differential yield of ψ mass dimuons as functions of transverse ($P_\perp$) and longitudinal ($P_\parallel$) momentum from interactions of ~ 300-GeV neutrons with nuclei; $\times$'s are results of a Monte Carlo calculation; from (33).

mass $\mu^{\pm}$ pairs

$$\int_{0.24}^{1.0} B_{\mu\mu}\frac{d\sigma}{dx}\,dx \simeq 3.5 \times 10^{-33}\,cm^2/nucleon.$$

With the further assumption $d\sigma/dy = $ constant ($y \equiv$ c.m. rapidity) in the region $0 < x < 0.3$, extrapolation into the unobservable region $x < 0.24$ yields the estimate for product of branching ratio, $B_{\mu\mu}$, and total inclusive cross section, $B_{\mu\mu}\sigma = 22 \times 10^{-33}\,cm^2/nucleon$ for 300-GeV neutrons. With no evidence for ψ' production, they conclude

$$B_{\mu\mu}(\psi')\sigma(\psi')/B_{\mu\mu}(\psi)\sigma(\psi) < 0.015.$$

2.1.3   150-GeV/c π^+ AND PROTONS ON BERYLLIUM   An extensive series of measurements of inclusive $\mu^{\pm}$ production by 150-GeV/c protons and positive pions on Be has been carried out by Anderson et al (34, 35). Muon pairs with mass 0.456 GeV/$c^2 < M < 3.5$ GeV/c^2 were detected efficiently, so that ρ^0 (or ω^0) and ϕ^0 decays were observed, as well as ψ decays. Their apparatus is shown in Figure 9. Muons are specified by the requirement that they penetrate a total of 4.7-m iron absorber. The observed dependence of dimuon yield on x and transverse momentum $P_\perp$ of the dimuon was fit with a parameterization of the invariant differential cross section, also of the form

$$E\frac{d^3\sigma}{dP^3} = C(1-x)^\alpha \exp(-bP_\perp)$$

after appropriate corrections for detection efficiency. Events with $x > 0.15$ and $P_\perp > 200$ MeV/c were included in the fits. For the dependence of $\psi(3095)$ yield on x, they find $\alpha = 2.9 \pm 0.3$ for production by protons and $\alpha = 1.7 \pm 0.4$ for production by pions. The steeper x dependence with incident protons than with pions agrees with earlier results of Blanar et al (36) for π^- and proton interactions with iron. The parameters of the $P_\perp$ dependence obtained are $b = 2.1 \pm 0.3$ c GeV^{-1} and $b = 2.6 \pm 0.4$ c GeV^{-1} for incident protons and pions, respectively. It is worth noting that the falloff with $P_\perp$ is significantly slower than that observed for ρ^0 (or ω^0) or ϕ^0 production, for which the corresponding parameter $b \sim 4$ for protons or pions. Integrated inclusive cross sections for production on single nucleons,

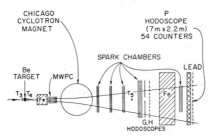

Figure 9 Spectrometer detection apparatus of Anderson et al (34) at Fermi National Accelerator Laboratory.

obtained by extrapolating the fit of $d\sigma/dx$ to $x = 0$ and using the dependence on A of (32), are $\sigma_p B_{\mu\mu} = (3.3 \pm 1.1) \times 10^{-33}$ cm²/nucleon for incident protons and $\sigma_\pi B_{\mu\mu} = (6.5 \pm 2.2) \times 10^{-33}$ cm²/nucleon for incident pions, for forward, $x > 0$, psions.

2.1.4 70-GeV/c PROTONS ON BERYLLIUM An arrangement similar to the above was used by Antipov et al (37) to measure $\mu^\pm$ pair yields from 70 GeV/c protons on beryllium. They fit the x distribution for ψ production with $\exp[-(6.0 \pm 1.2)x]$ for $x > 0.3$ and constant at smaller x. The $P_\perp$ dependence has the form

$$d\sigma/dP_\perp^2 \sim \exp[-(1.8 \pm 0.03)P_\perp^2].$$

For the inclusive $\psi(3095)$ production yield, they obtained

$$\sigma B_{\mu\mu} = (9.5 \pm 2.5) \times 10^{-33} \text{ cm}^2/\text{nucleus},$$

and, with the $A^{0.93}$ dependence of Binkley et al,

$$\sigma B_{\mu\mu} = (1.2 \pm 0.3) \times 10^{-33} \text{ cm}^2/\text{nucleon},$$

2.1.5 400-GeV PROTONS ON Be Hom et al (38, 40) and Snyder et al (39) have measured both $e^\pm$ and $\mu^\pm$ pair yields from interactions of 400-GeV protons with beryllium. Their spectrometer acceptance limits the detectable phase space to the small x interval $-0.06 < x < 0.08$ and to $P_\perp < 2$ GeV/c. They observed a transverse-momentum dependence well represented by either

$$E\frac{d^3\sigma}{dP^3} = (1.7 \pm 0.4) \times 10^{-32} \exp[-(1.1 \pm 0.4)P_\perp^2] \text{ cm}^2/\text{nucleus}$$

or

$$E\frac{d^3\sigma}{dP^3} = (2.5 \pm 0.6) \times 10^{-32} \exp[-(1.6 \pm 0.4)P_\perp] \text{ cm}^2/\text{nucleus}.$$

That result, extrapolated away from $x \simeq 0$ with the functional form

$$E\frac{d^3\sigma}{dP^3} = C(1-x)^{4.3} \exp(-1.6\,P_\perp),$$

yields an inclusive cross-section branching-ratio product

$$\sigma B = (1.0 \pm 0.3) \times 10^{-31} \text{ cm}^2/\text{nucleus},$$

which gives, with $A^{0.93}$ dependence,

$$\sigma B = (1.3 \pm 0.3) \times 10^{-32} \text{ cm}^2/\text{nucleon}.$$

A small peak at a mass consistent with ψ' yields an estimate of the ratio of production cross sections $\sigma(\psi')/\sigma(\psi) = (10 \pm 3)\%$.

2.1.6 26-GeV PROTONS ON 26-GeV PROTONS In p-p collisions at the CERN Intersecting Storage Rings (ISR), production of ψ has been observed, with ψ identified by e^+e^- (41) and $\mu^+\mu^-$ (42) decays. From the $\mu^+\mu^-$ yield at c.m. energy $(s)^{1/2} = 52$

GeV, there results an estimate of total cross section $\sigma B = (4.2 \pm 1.9) \times 10^{-32}$ cm², based on detection of eleven dimuon events above background.

2.1.7 SUMMARY   Dividing the measured values of dilepton yields by $B_{\mu\mu} = 0.069$, the branching ratio for $\psi(3095)$ decay to lepton pairs obtained by the LBL-SLAC group (43), one obtains the various ψ-production total cross-section estimates plotted as a function of c.m. energy in Figure 10. This shows a rise of more than two orders of magnitude from the lowest energy, $(s)^{1/2} = 7.4$ GeV, to the highest, $(s)^{1/2} = 52$ GeV. Cross sections for ϕ production are smaller than those for ρ by about a factor of ten. Production of the ψ's is more grossly inhibited; their rates are a factor of 1000 smaller still. Average transverse momenta of the ψ's are about 1 GeV/c, roughly twice as large as those of ρ^0 and ϕ^0.

2.2 Photoproduction of ψ and ψ'

Measurement of the ψ differential photoproduction cross section, extrapolated to $t = 0$, with application of the vector-meson dominance-model prescription, yields

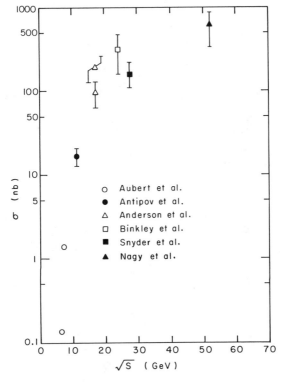

Figure 10 Compendium of results on the energy dependence of the cross section for inclusive $\psi(3095)$ production in hadron-hadron collisions; data from (30, 32–39, 42).

estimates of the ψ-nucleon total cross section (see Section 1.1) and so is important for determination of the strength of ψ coupling to hadrons.

Knapp et al (44) first observed ψ photoproduction and its diffractive character. Photons were produced by 300-GeV protons interacting with beryllium and had a continuous energy distribution to a maximum of $\sim$200 GeV. The photon beam bombarded a beryllium target, and produced muon pairs were detected. Camerini et al (45) observed production of electron pairs and muon pairs, made by bremsstrahlung photons of various energies from 13 GeV to 21 GeV, interacting with deuterium. Gittelman et al (46) measured ψ production by 11.8-GeV bremsstrahlung photons on a beryllium target, detecting electron-pair decays. Nash et al (47) measured electron-pair yields from bombardment of deuterium by photons of mean energy 55 GeV, in a range from 31–80 GeV. A typical yield curve is in Figure 11, showing the electron-pair effective mass distribution produced in the experiment of Gittelman et al. Figure 12, the momentum-transfer distribution of dimuon pairs with mass near 3100 MeV/c^2, the results of Knapp et al, shows the characteristic

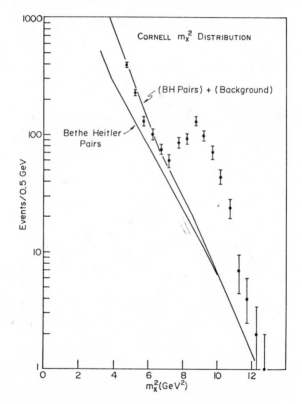

Figure 11 Invariant-mass distribution of dielectrons photoproduced on beryllium at 11 GeV, showing ψ(3095) yield; from (46).

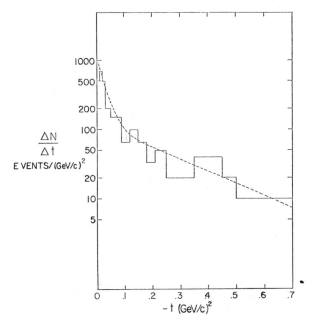

Figure 12   Differential yield of ψ(3095) photoproduction on beryllium as a function of momentum transfer. The superimposed curve is the calculated t distribution of the form $A^2 \exp{(40t)} + A \exp{(4t)}$ after correction for experimental resolution and detector acceptance; from (44).

exponential behavior near zero momentum transfer. In a subsequent experiment at SLAC, Camerini et al (45) detected single-muon decay products of ψ(3095) produced by photons on beryllium and tantalum. They provide a measurement of photoproduction rates for, among others, 9.5-GeV incident energy, the lowest energy at which ψ(3095) photoproduction has been observed and find, in fact, the smallest differential cross section, 0.18 ± 0.08 nb $(\text{GeV}/c)^{-2}$ at the minimum momentum transfer. We have applied the vector-dominance relation (Equation 1.3) to calculate ψ-nucleon total cross sections from reported values of $d\sigma/dt$ at $t = 0$, neglecting the real part of the forward-scattering amplitude and using the lepton-pair partial width $\Gamma_{ee} = 4.8$ keV (43). These values are shown in Figure 13 as a function of c.m. energy. The experiment of Camerini et al (45) has provided a direct measurement of the ψ-nucleon cross section, inferred from the ratio of yields in Be and Ta. From results with 20-GeV incident photons, they find $\sigma(\psi N) = 2.8 \pm 0.9$ mb, where only statistical errors are included in the uncertainty. The systematic error was estimated at $\sim \pm 0.5$ mb. The first directly measured ψ-interaction cross section agrees with values extracted from the photoproduction cross sections with vector-dominance model.

It appears that the ψN total cross section approaches a value $\sigma \simeq 1$ mb, small, but within the range typical of strong-interaction processes. It is noteworthy that

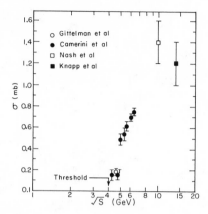

Figure 13   Compendium of results on the energy dependence of the ψ(3095)-nucleon total-interaction cross section obtained from photoproduction data of (44–48) and the vector-dominance model.

this ψ-nucleon cross section is smaller by a factor of ten than the ϕ-nucleon total cross section. Also the energy dependence has a shape somewhat suggestive of a threshold for an inelastic ψ-nucleon reaction at an energy near 5 GeV in the ψ-nucleon c.m. system. Candidates for such reactions are the processes yielding pairs of charmed mesons, implicit in the four-quark view of hadron microstructure.

The yield of ψ' in photoproduction is considerably smaller than that of ψ. Only Camerini et al (45) report a positive result. With 21-GeV photons, they measure $d\sigma/dt = (2.1 \pm 0.8) \times 10^{-33}$ cm^2 (GeV/c)$^{-2}$ at minimum momentum transfer, just about one seventh of the value for ψ photoproduction at the same energy and its minimum momentum transfer.

3   PROPERTIES OF ψ(3095) AND ψ(3684)

Essentially all information on the characteristics of the resonant states has come from analysis of e^+e^- annihilations and mostly from the SLAC-LBL group at SPEAR and the Double Arm Spectrometer (DASP) group at DORIS (27). The LBL-SLAC magnetic detector (49) is a cylindrical arrangement of counters and spark chambers arrayed around the direction of the colliding beams as the axis, all in an axial magnetic field of ~ 4000 G uniform to a few percent over its ~ 3-m-diameter-by-~ 3-m-length volume. Charged particles and photons are detected over about 65% of the full solid angle. A schematic view of a vertical cross section through the detector axis is shown in Figure 14. The DASP apparatus (50), Figure 15, is a symmetric pair of magnetic spectrometers providing precise momentum and time-of-flight measurements of charged particles produced in a solid angle of $\sim 0.1 \times 4\pi$ sr. The nonmagnetic inner detector covers $\sim 0.7 \times 4\pi$ sr. Both the SLAC-LBL and DASP apparatuses detect photons, DASP with better efficiency, energy, and position resolution.

3.1 Masses and Widths

The resonant-state energies, full widths, and partial widths were determined from data obtained at SPEAR on the energy dependence of the annihilation cross section shown in Figures 5 and 6 (43, 51). For these data the incident flux was obtained from measurement of Bhabha scattering at ~25 mrad. Hadron yields were corrected for the solid-angle and momentum acceptance of the apparatus, using a Monte Carlo simulation of the detector response and a phase-space production model (52). That calculation yields a net detection efficiency $\varepsilon = 0.4$ with $\pm 15\%$ estimated uncertainty. Lepton-pair cross sections refer only to the angular range $|\cos \theta| \leq 0.6$, where θ is the angle between the final-state leptons' and initial

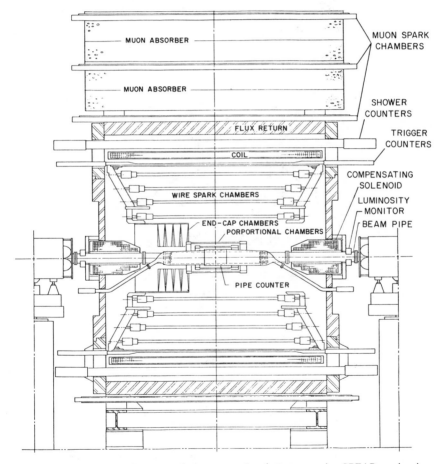

Figure 14 The SLAC-LBL magnetic detector at the e^+e^- storage ring SPEAR; a view in the vertical plane through the intersecting beams.

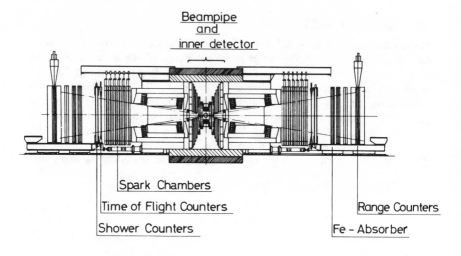

Beampipe
and
inner detector

Spark Chambers

Time of Flight Counters

Shower Counters

Range Counters

Fe - Absorber

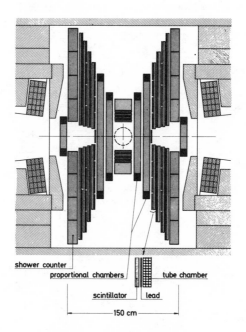

shower counter

proportional chambers

scintillator   lead

tube chamber

150 cm

DASP — Inner Detector

Figure 15 The Double Arm Spectrometer detector (DASP) at the e^+e^- storage rings DORIS.

Table 3   Parameters of ψ particles

Particle	Mass (MeV/c^2)	Γ (MeV)	Γ_{ee} (keV)	$\Gamma_{\mu\mu}$ (keV)	Γ_h (keV)	$\Gamma_{\gamma h}$ (keV)	$f_v^2/4\pi$
$\psi(3095)$[a]	3095 ± 4	0.069 ± 0.015	4.8 ± 0.6	4.8 ± 0.6	59 ± 15	12 ± 2	11.5 ± 1.4
$\psi(3684)$[b]	3684 ± 5	0.228 ± 0.056	2.1 ± 0.3	2.1 ± 0.3	224 ± 56	7 ± 1	31.1 ± 4.5
$\psi(4414)$[c]	4414 ± 7	33 ± 10.0	0.44 ± 0.14				178 ± 57

[a] LBL-SLAC values (43).
[b] LBL-SLAC values (51).
[c] Reference 101.

beams' directions. The observed excitation curves were each fit to a function consisting of a Breit-Wigner line shape for a resonance at energy M of full width Γ,

$$\sigma_\beta(W) = \frac{(2J+1)\pi}{W^2} \frac{\Gamma_{ee}\Gamma_\beta}{(W-M)^2 + \Gamma^2/4}, \qquad 3.1$$

convoluted with the energy-resolution function of the colliding beams. The parameters Γ_{ee} and Γ_β are the partial widths for decays to electron pairs and final state β, respectively. The observed cross section at nominal interaction energy W_0 (the storage-ring energy setting) is then

$$\sigma_\beta(W_0) = \int G_R(W_0 - W)\sigma_\beta(W)\,dW, \qquad 3.2$$

where initial-state radiation has been taken into account by replacing the resolution function $G(W_0 - W)$ by a radiatively corrected form (53)

$$G_R(W_0 - W) = t \int_0^{W_0/2} \left(\frac{2k}{W_0}\right)^t G(W_0 - W - k)\frac{dk}{k}, \qquad 3.3$$

where $t = 2(\alpha/\pi)[\ln(W_0/M_e)^2 - 1]$ and k is the energy of a radiated photon. Radiation causes a "tail" on the high-energy side of a resonance, decreasing the maximum observed cross section, but does not appreciably change the position of the maximum. Best-fit values obtained for M, Γ, Γ_{ee}, $\Gamma_{\mu\mu}$, and Γ_h, the hadron width, are listed in Table 3. It was assumed that no decay modes escaped detection, i.e. $\Gamma = \Gamma_h + \Gamma_{\mu\mu} + \Gamma_{ee}$. It is of some value to note here the more transparent relation between the resonance parameters and the area under the resonance peak

$$\iint \sigma_\beta(W_0)G(W_0 - W)\,dW\,dW_0 = \frac{2\pi^2}{M^2}(2J+1)\frac{\Gamma_{ee}\Gamma_\beta}{\Gamma}, \qquad 3.4$$

independent of the shape of the resolution function. Assuming no modes escape detection, the sum of the areas under the three resonance peaks shown determines the electron-pair partial width directly.

Included in Γ_h is a contribution $\Gamma_{\gamma h}$, from resonance decays via a second-order electromagnetic process, indicated in Figure 16b. With the usual assumption that

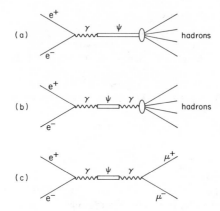

Figure 16 Diagrams of resonance production in lowest order in e^+e^- annihilation with decays to (*a*) hadrons, (*b*) hadrons via second-order electromagnetic interaction, and (*c*) muon pairs.

lepton pairs couple to the resonant state via an intermediate photon, it follows

$$\Gamma_{\gamma h} = \left[\frac{\sigma(e^+e^- \to \text{hadrons})}{\sigma(e^+e^- \to \mu^+\mu^-)} \right] \Gamma_{\mu\mu},$$

where the quantity in brackets is evaluated at an energy just below resonance. Values of the cross-section ratio measured by the SLAC-LBL group (52) were used to calculate the second-order electromagnetic widths given in the table. The fraction of indirect hadronic decays, $\Gamma_{\gamma h}/\Gamma$, is 0.17 ± 0.03 for $\psi(3095)$ and 0.031 ± 0.009 for $\psi(3684)$.

3.2   *Spin, Parity, and Charge Conjugation*

These quantum numbers are directly established to have values $J = 1$, $P = $ odd, and $C = $ odd from the observation of interference between single-photon–exchange and ψ-exchange amplitudes in the process $e^+e^- \to$ leptons near resonance, Figure 16*c* (43, 51). If the intermediate meson has the same quantum numbers as the photon, the muon-pair-annihilation differential cross section is

$$\frac{d\sigma}{d\Omega} = \frac{9}{16W^2} (1 + \cos^2 \theta) \left| \frac{-2\alpha}{3} + \frac{\Gamma_{\mu\mu}}{M - W - i\Gamma/2} \right|^2,$$

showing destructive interference below the resonant energy M. To minimize the effects of uncertainties in incident-flux determinations, the SLAC-LBL group examined the ratio of electron-pair to muon-pair cross sections, integrated over the angular interval $-0.6 < \cos \theta < 0.6$ covered by the detector. That ratio is shown as a function of energy in Figure 17, together with calculated curves, due account having been taken of radiative corrections. Both $\psi(3095)$ and $\psi(3684)$ data are consistent with expectations for spin-1 and exclude spin-0. Interference observations directly establish that both psions and the photon have the same value, -1,

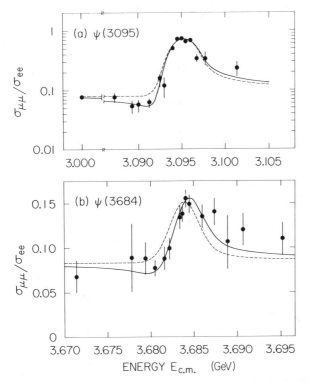

Figure 17   Ratio of cross sections for e^+e^- annihilation to muon pairs to that for annihilation to electron pairs at energies near (*a*) 3095 MeV and (*b*) 3684 MeV. The full and dashed curves show maximal and zero interference, respectively, between one photon and resonance amplitudes; from LBL-SLAC (43, 51).

of parity and charge-conjugation quantum numbers. Spin greater than one is not excluded a fortiori by an interference observation because the unit helicity projection of the intermediate ψ will produce interference effects in the energy variation of that fraction of the total cross section included in the limited angular range of detection. Since the QED amplitude is nonzero only for transitions between lepton-pair states of helicity ± 1, the interference term in the cross section is proportional to the overlap integral of the spin-rotation functions $d^{J_\psi}_{11}(\theta)$ and $d^1_{11}(\theta)$. It follows from the properties of the $d^J_{11}(\theta)$ functions that ψ-spin values two and three give rise to constructive interference at energies below resonance. Higher spins produce negligible interference. Spins greater than, as well as less than, one are thus excluded by the LBL-SLAC data.

3.3   *Hadronic Decays of $\psi(3095)$; Isospin and G-Parity; SU(3) Classification*

Table 4 presents information on hadronic-decay modes of $\psi(3095)$.

Table 4 Branching fractions for $\psi(3095)$ nonleptonic decays

Mode	Fraction (%)[a]	Ref.	Footnote(s)
$\pi^+\pi^-$	0.01 ± 0.007	54	b
$\pi^+\pi^-\pi^0$	1.6 ± 0.6	55	—
$\rho\pi$	1.3 ± 0.2	54, 55	c
$2\pi^+2\pi^-$	0.4 ± 0.1	55	d
$\pi^\pm A_2^\mp$	< 0.4	54	d
$2\pi^+2\pi^-\pi^0$	4 ± 1	55	—
$\omega\pi^+\pi^-$	0.7 ± 0.2	57	e
ωf	0.19 ± 0.08	57	e
ρA_2	0.8 ± 0.5	57	e
$\rho\pi\pi\pi$	1.2 ± 0.4	55	e
$3\pi^+3\pi^-$	0.4 ± 0.2	55	d
$3\pi^+3\pi^-\pi^0$	2.9 ± 0.7	55	—
$\omega 2\pi^+2\pi^-$	0.9 ± 0.3	57	f
$4\pi^+4\pi^-\pi^0$	0.9 ± 0.3	55	—
K^+K^-	0.02 ± 0.02	57	g
$K_S^0 K_L^0$	< 0.008	57	g
$K_S^0 K^\pm \pi^\mp$	0.26 ± 0.07	57	—
$K^0\bar{K}^{*0}$	0.27 ± 0.06	57	h, i, j
$K^\pm K^{*\mp}$	0.32 ± 0.06	57	h, j
$K^0\bar{K}^{**0}$	< 0.2	57	h, i, j
$K^\pm K^{**\mp}$	< 0.15	57	h, j
$K^+K^-\pi^+\pi^-$	0.7 ± 0.2	57	—
$K^{*0}\bar{K}^{*0}$	< 0.05	57	g, h, k
$K^{*0}\bar{K}^{**0}$	0.7 ± 0.3	57	h, i, k
$K^{**0}\bar{K}^{**0}$	< 0.3	57	g, h, k
$\phi\pi^+\pi^-$	0.14 ± 0.06	57	k
ϕf	< 0.04	57	—
$K^+K^-K^+K^-$	0.07 ± 0.03	57	—
ϕK^+K^-	0.09 ± 0.04	57	l
$\phi f'$	0.08 ± 0.05	57	l
$K^+K^-\pi^+\pi^-\pi^0$	1.2 ± 0.3	57	—
ωK^+K^-	0.08 ± 0.05	57	m
$\omega f'$	< 0.02	57	m
$\phi\eta$	0.10 ± 0.06	57	m
$\phi\eta'$	< 0.1	57	—
$K^+K^-2\pi^+2\pi^-$	0.3 ± 0.1	57	n
$\phi 2\pi^+2\pi^-$	< 0.15	57	—
$\bar{p}p$	0.22 ± 0.02	54, 56	o
$\bar{N}N\pi$	0.4 ± 0.2	56	p
$\Lambda\bar{\Lambda}$	0.16 ± 0.08	56	o
$\gamma\gamma$	< 0.05	74	q
$\gamma\pi^0$	< 0.055	74	—
$\gamma\eta$	0.14 ± 0.04	50, 73	—
$\gamma\eta'$	< 0.7	50, 73	—
$\gamma X(2800)$	< 1.7	68	—
$\gamma X(\to 3\gamma)$	~ 0.015	50, 73	—

Table 4 Continued

[a] All upper limits refer to the 90% confidence level.
[b] Two events detected with background estimated at 0.24 events. An isospin-nonconserving, second-order electromagnetic mode.
[c] $\rho^0\pi^0 + \rho^\pm\pi^\mp$. Included in $\pi^+\pi^-\pi^0$ grouping.
[d] Proceeds via second-order electromagnetic interaction.
[e] One-π^0 mode included in $2\pi^+2\pi^-\pi^0$.
[f] One-π^0 mode included in $3\pi^+3\pi^-\pi^0$.
[g] Forbidden for SU(3) singlet.
[h] $K^* \equiv K^*(892)$, $K^{**} \equiv K^*(1420)$.
[i] Includes charge-conjugate state.
[j] $K_S^0K^\pm\pi^\mp$ fraction included in $K_S^0K^\pm\pi^\mp$ grouping.
[k] $K^+K^-\pi^+\pi^-$ fraction included in $K^+K^-\pi^+\pi^-$ grouping.
[l] $K^+K^-K^+K^-$ fraction included in $K^+K^-K^+K^-$ grouping.
[m] $K^+K^-\pi^+\pi^-\pi^0$ fraction included in $K^+K^-\pi^+\pi^-\pi^0$ grouping.
[n] $K^+K^-2\pi^+2\pi^-$ fraction included in $K^+K^-2\pi^+2\pi^-$ grouping.
[o] Angular distribution $1+\cos^2\theta$ assumed, consistent with data.
[p] Includes $p\bar{p}\pi^0$, $p\bar{n}\pi^-$, and $n\bar{p}\pi^+$.
[q] Forbidden for angular momentum $J = 1$.

Two-body branching ratios have been determined by Braunschweig et al (54) from data obtained with the double arm spectrometer DASP. Two-body final states were selected, using criteria of collinearity, momentum, energy loss, and range appropriate to the particular pair. Evidence for the modes $\pi^\pm\rho^\mp$ and $K^\pm K^{*\mp}$ is presented in the missing-mass spectra of Figure 18, showing the mass recoiling against single pions and kaons.

Multibody decay rates were measured by the SLAC-LBL group (55–57) using kinematical fits of multiprong events in which the detected prongs have zero total charge. Events with no missing particles were selected by requiring that the missing momentum be less than 100 MeV/c. As an example, the distribution in total energy of the detected prongs in four-prong events is shown in Figure 19. Four-particle

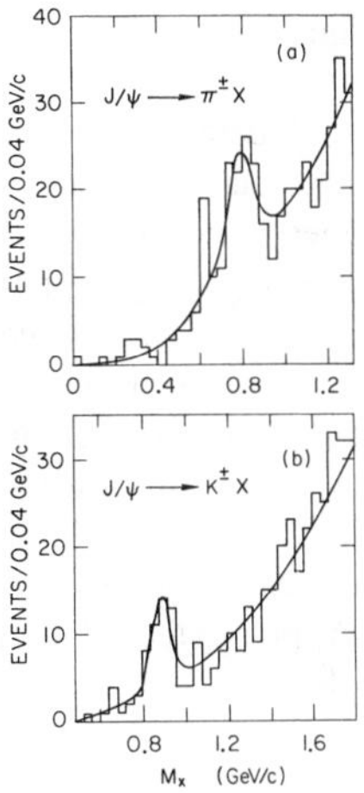

Figure 18 Distribution of mass recoiling against (*a*) single pions and (*b*) single kaons in multihadron decays of $\psi(3095)$ produced in e^+e^- annihilations; from DASP (54).

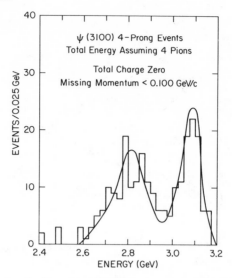

Figure 19 Total energy in four-prong ψ(3095) decay events with missing momentum less than 100 MeV/c. The solid curve is a Monte Carlo fit to the data; from SLAC-LBL (55, 56).

states, presumed to be $\pi^+\pi^-\pi^+\pi^-$, are indicated by the presence of the peak near 3100 MeV in total detected energy. Events with momentum unbalance greater than 200 MeV/c exhibit the missing-mass spectrum of Figure 20 with clear evidence for the $2\pi^+2\pi^-\pi^0$ decay mode of the ψ and, for contrast, its relative absence at the nonresonant energy 3.0 GeV. Similar analysis techniques were used by the LBL-SLAC group to identify samples of the various final states of Table 4 and, after correction for losses due to detection inefficiency, to determine the branching ratios shown.

Application of these results to the determination of strong-interaction properties of the ψ requires isolation of the direct-decay contribution (Figure 16a) from that of the second-order electromagnetic decays (Figure 16b). That is done by comparing the relative branching ratio $\sigma_\beta/\sigma_{\mu\mu}$ for a final state β at resonance to the same quantity at 3.0 GeV, where the resonance contribution is negligible. Jean-Marie et al (55) show that this ratio, measured at resonance, equals the nonresonant value for states with even numbers of pions. States with odd numbers of pions have significantly larger cross-section ratios at resonance than below resonance, leading to the conclusion that ψ couples directly to states with odd numbers of pions and not to states with even numbers of pions. It follows that the ψ is a hadronic state of odd G-parity. Applying the relation between isospin and G-parity for a state of arbitrary number of pions,

$$G = C(-1)^I = (-1)^{I+1}$$

it follows that the ψ is a state of even isospin and that direct decays of the ψ follow the hadronic rule of isospin conservation.

The $\pi^+\pi^-\pi^0$ event class of Jean-Marie et al (55) is dominated by the quasi-two-body $\rho\pi$ state, as indicated in the Dalitz plot of Figure 21. From the distribution of events, they concluded that for ψ decay,

$$\frac{\Gamma_{\rho^0\pi^0}}{\Gamma_{\rho^+\pi^-}+\Gamma_{\rho^-\pi^+}} = 0.59\pm0.17.$$

Comparison with the values 0.5 and 2.0, appropriate to isospin-conserving decays of objects of isospin zero and two, respectively, uniquely establishes $I = 0$ for the ψ. Confirmation of this result came with observation of the $\bar{p}p$ mode by the DASP and LBL-SLAC groups. As pointed out by Braunschweig et al (54), the measured ratio $\Gamma_{\bar{p}p}/\Gamma_{\mu\mu}$ is much larger than would be obtained with any plausible extrapolation of the nucleon form factor, so that the $\bar{p}p$ mode must represent a direct decay, rather than a second-order electromagnetic process. That state can occur only with $I = 0$ or $I = 1$, and the latter is excluded by inference from the existence of decays to odd numbers of pions. Lastly, the LBL-SLAC team has reported a

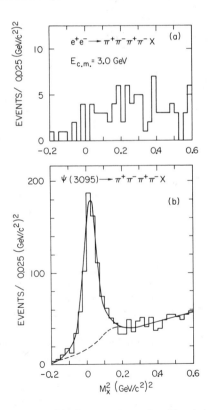

Figure 20 Missing mass-squared in four-prong events with missing momentum greater than 200 MeV/c produced in e^+e^- annihilations at (*a*) 3.0 GeV and (*b*) $\psi(3095)$ resonance. The solid curve is a Monte Carlo fit to the data; from LBL-SLAC (55, 56).

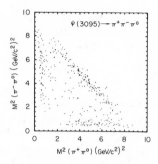

Figure 21   Dalitz plot for ψ(3095) decays to $\pi^+\pi^-\pi^0$; from SLAC-LBL (55, 56).

measurement of the $\Lambda\bar{\Lambda}$ branching ratio, $\Gamma_{\Lambda\bar{\Lambda}}/\Gamma = 0.0016 \pm 0.0008$, based on twenty well-identified examples (56). None of those is consistent with primary $\Sigma^0\bar{\Lambda}^0$ ($\bar{\Sigma}^0\Lambda^0$) events. Since second-order electromagnetic decays populate $\Sigma\bar{\Lambda}$ states six times more frequently than $\Lambda\bar{\Lambda}$ states (58), it follows that the $\Lambda\bar{\Lambda}$ pairs are direct decay products. This observation provides another argument for assigning $I = 0$ to the ψ(3095).

Two-body-decay branching ratios are especially useful in considering the applicability to ψ-decay interactions of the broader SU(3) strong-interaction symmetry (58, 59). Decay of an SU(3) singlet is constrained severely because only one amplitude determines the rates for all decays $\psi \to m_a + m_b$, where m_a and m_b represent any two members of meson octets. If m_a and m_b are the charged members of an $I = 1$ submultiplet, $m_a + m_b$ has even G-parity and is not an allowed final state for ψ(3095) decay. By extension, decay to any two members of the same octet is forbidden. If m_a and m_b are the $I = 1$, $I_3 = 0$, $Y = 0$ members of different octets, a and b, and the charge-conjugation quantum numbers of m_a and m_b are equal, then m_a and m_b have the same G-parity, so the state $m_a + m_b$ also has even G-parity. Decay of an odd G-parity, SU(3) singlet to that state is again forbidden. Symmetry under SU(3) then forbids ψ(3095) decay to a state of any two members of such octets a and b. Allowed decay modes are, for example, $\rho\pi$, $\eta\phi$, and $\bar{K}K^*(890)$, while K^+K^-, πA_2, $\bar{K}K^*(1420)$ are forbidden. Inspection of the branching fractions listed in Table 4 shows that the SU(3)-singlet-allowed modes $K\bar{K}^*(890)$ and $K^*(890)\bar{K}^*(1420)$ occur, and with branching fractions at least an order of magnitude larger than those for $\pi^+\pi^-$ and K^+K^-. Other forbidden modes, $\bar{K}K^*(1420)$, $K^*\bar{K}^*$, etc have not appeared, but present upper limits on their relative rates are not very convincingly small.

Exact SU(3) symmetry leads to the further prediction of equal branching ratios for allowed singlet decays to any two members, m_a and m_b, of any pair of octets a and b, e.g. to $\pi^0\rho^0$, $\pi^+\rho^-$, $\pi^-\rho^+$, $K^-K^{*+}(892)$, $K^+K^{*-}(892)$ etc. The relevant experimental ratios are

$$[\Gamma(\pi^+\rho^-)+\Gamma(\pi^-\rho^+)+\Gamma(\pi^0\rho^0)]/[\Gamma(K^+K^{*-})+\Gamma(K^-K^{*+})]/[\Gamma(K^0\bar{K}^{*0})$$
$$+ \Gamma(\bar{K}^0K^{*0})] = (1.3 \pm 0.2)/(0.34 \pm 0.06)/(0.27 \pm 0.05),$$

not quite in agreement with the predicted 3/2/2, indicating some SU(3) breaking in the decay interaction.

In summary, ψ decays appear to have properties in agreement with the strictures of SU(3) symmetry appropriate to a singlet state, although there are some discrepancies.

3.4 Hadronic Decays of $\psi(3684)$; Isospin and G-Parity

The dominant hadronic decay mode of $\psi(3684)$ is the cascade decay to $\psi(3095)$, $\psi' \to \psi\pi\pi$. The modes $\psi\pi^+\pi^-$, $\psi\pi^0\pi^0$, together with $\psi' \to \psi\eta$ and $\psi' \to \psi\gamma\gamma$, account for just over one-half of all decays. Table 5 lists branching ratios for all identified hadronic modes. Data from the DASP (50) and LBL-SLAC (60, 61) groups were obtained using the detection apparatus and analysis techniques described above. Hilger et al (62), at SPEAR, used two identical collinear spectrometers, each equipped with a large NaI crystal for photon and electron detection, an array of multiwire proportional chambers for track-trajectory sampling, and a magnetized iron block

Table 5   Branching fractions for $\psi(3684)$ nonleptonic decays

Mode	Fraction (%)[a]	Ref.	Footnote(s)
$\psi(3095)\pi^+\pi^-$	33 ± 3	50, 60–62	—
$\psi\pi^0\pi^0$	17 ± 3	50, 60–62	—
$\psi\eta$	4.2 ± 0.7	50, 61	—
$\psi\gamma + \psi\pi^0$	<0.15	61	b, c
$\pi^+\pi^-$	<0.04	54	d
$2\pi^+2\pi^-\pi^0$	0.35 ± 0.15	56	—
$\rho^0\pi^0$	<0.1	56	—
K^+K^-	<0.14	54	e
$p\bar{p}$	<0.05	54	f
$\gamma\gamma$	<0.5	69	g
$\gamma\pi^0$	<0.7	69	—
$\gamma\eta$	<0.13	50	—
$\gamma X(2800)$	<1	65, 68	—
$\gamma X(2800) \to 3\gamma$	<0.04	50	—
$\gamma\chi(3415)$	7 ± 2	65, 68	f
$\gamma\chi(3510)$	7 ± 2	68	h
$\gamma\chi(3550)$	7 ± 2	68	h
$\gamma\chi(3455)$	<2.5	68	—
$\gamma\chi(3455) \to \gamma\gamma\psi$	0.8 ± 0.4	65	—

[a] All upper limits refer to the 90% confidence level.
[b] $\psi\gamma$ forbidden by c invariance.
[c] $\psi\pi^0$ forbidden by isospin invariance.
[d] Isospin-nonconserving, second-order electromagnetic decay.
[e] Forbidden for an SU(3) singlet.
[f] Angular distribution $1 + \cos^2 \theta$ assumed.
[g] Forbidden for angular momentum 1.
[h] Isotropic angular distribution assumed. With $1 + \cos^2 \theta$ distribution, these branching fractions increase by a factor of 1.3.

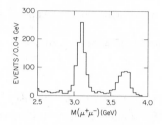

Figure 22 Inclusive distribution of muon-pair invariant masses from $\psi(3684)$ decays showing evidence of cascade decays to $\psi(3095)$; from LBL-SLAC (60).

for charged-particle momentum determination. The spectrometers were perpendicular to the direction of the colliding beams.

The LBL-SLAC evidence for the cascade decay $\psi(3684) \rightarrow \psi(3095) + $ anything (60) is presented in Figure 22. This shows the distribution in invariant mass of the two highest-momentum, oppositely charged prongs in the $\psi(3684)$ decays, calculated with the muon mass assumed for those two tracks. A peak at the ψ mass is apparent, as is the peak at 3684 MeV/c^2, due to direct decays of $\psi(3684)$ to muon pairs. Figure 23a shows the mass recoiling against all $\pi^+\pi^-$ pairs in all events, exhibiting again a clear peak at 3095 MeV/c^2. Figure 23b shows the recoil mass against those pion pairs in 4-prong events with zero total charge and, within experimental uncertainties, zero missing momentum, consistent with kinematics of the decay $\psi(3684) \rightarrow \mu^+\mu^-\pi^+\pi^-$. The peak results from that subset of $\psi(3684)$ cascade decays in which $\psi(3095)$ subsequently decays via its muon-pair mode. An example of a computer reconstruction of such an event is shown in Figure 24. The SLAC-LBL group identified $\psi' \rightarrow \psi\eta$ decays among the class of $\pi^+\pi^-\mu^+\mu^-$ events with missing momentum (61). The distribution in the square of the mass recoiling against the $\mu^+\mu^-$ pairs in Figure 25 shows a peak at the $\eta(\text{mass})^2$ whose

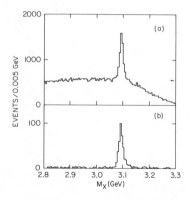

Figure 23 (a) Distribution of missing mass recoiling against all pairs of oppositely charged particles in $\psi(3684)$ decays to multihadron states. (b) same as (a), but for events consistent with zero missing energy and momentum; from SLAC-LBL (60).

width is consistent with that expected from experimental resolution folded with the natural width of the η meson. The $\psi\eta$ mode is also revealed in the LBL-SLAC group's mass spectrum of all-neutral states accompanying ψ decays via the $\mu^+\mu^-$

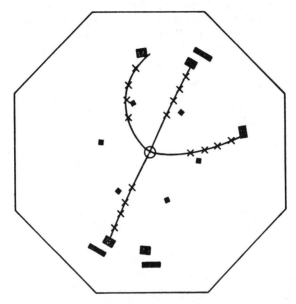

Figure 24   A computer reconstruction of a $\psi(3684) \rightarrow \pi^+\pi^-\psi(3095) \rightarrow \pi^+\pi^-e^+e^-$ event in the LBL-SLAC detector seen in projection on the plane perpendicular to the intersecting beams. The $\times$'s mark spark-chamber samplings. Black rectangles show counters. The straightest two tracks are the electron and positron; the others are the low-momentum pions; from SLAC-LBL (60).

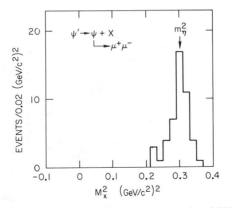

Figure 25   Distribution of square of recoil mass accompanying $\psi(3095)$-decay muons in $\psi' \rightarrow \psi\pi^+\pi^- \rightarrow \mu^+\mu^-\pi^+\pi^-$ events not consistent with kinematics of $\psi(3684) \rightarrow \psi(3095)\pi^+\pi^-$ decay; from LBL-SLAC (61).

mode. Figure 26 shows the spectrum of missing mass-squared in events of the type $\psi' \to \mu^+\mu^- + $neutrals, with a clear η peak. This is particularly prominent after subtracting a calculated contribution to the data expected from $\psi' \to \pi^0\pi^0\psi \to \pi^0\pi^0\mu^+\mu^-$, using the isospin-conservation prediction $\pi^0\pi^0/\pi^+\pi^- = \frac{1}{2}$ for a $\pi\pi$ system of zero isospin and the measured rate for $\psi' \to \psi\pi^+\pi^-$. Subsequently, direct observation was made by the DASP group of the decay $\psi' \to \psi\pi^0\pi^0$ (50), examples of which were reconstructed from events with ψ-decay muon pairs and detected γ rays. They measured a branching ratio

$$\Gamma(\psi' \to \psi\pi^0\pi^0)/\Gamma = 0.18 \pm 0.06.$$

The yield of events with muon pairs and charged particles provided data for their determination of the branching ratio

$$\Gamma(\psi' \to \psi\pi^+\pi^-)/\Gamma = 0.36 \pm 0.06,$$

in good agreement with the LBL-SLAC value 0.32 ± 0.04. The LBL-SLAC results yielded

$$\Gamma(\psi' \to \psi\pi^0\pi^0)/\Gamma = 0.17 \pm 0.04,$$

upon subtraction of the $\psi\eta$ (61) and $\psi\gamma\gamma$ (63) branching fractions from the directly measured value

$$\Gamma(\psi' \to \psi + \text{neutrals})/\Gamma = 0.25 \pm 0.04.$$

The LBL-SLAC and DASP values are in good agreement. Weighted averages are listed in Table 5, which summarizes the available data.

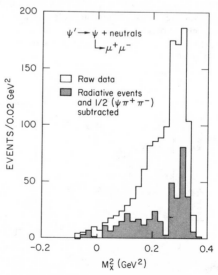

Figure 26 Distribution of square of missing mass accompanying $\psi(3095)$-decay muons in events containing $\mu^+\mu^-$ and undetected neutral particles; from SLAC-LBL (61).

Branching ratios to other exclusive hadronic states are relatively small. Comparison of the rates for ψ decay in Table 4 with those for ψ' in Table 5 reveals strong differences in the various branching ratios. In particular, the $\pi^+\pi^-\pi^+\pi^-\pi^0$ mode, prominent in $\psi(3095)$ decay, has a branching ratio smaller by an order of magnitude in $\psi(3684)$ decay.

Assignment of odd G-parity to $\psi(3684)$ follows directly from the existence of the $\psi\pi\pi$ mode and the value $G = -1$ for $\psi(3095)$ previously established.

Isospin is determined from the ratio of partial widths

$$\Gamma(\psi' \to \psi\pi^0\pi^0)/\Gamma(\psi' \to \psi\pi^+\pi^-)$$

which, corrected for the difference between the $\pi^+\pi^-$ and $\pi^0\pi^0$ phase space, has the predicted value 0.52 for $I = 0$, zero for $I = 1$, and 2.1 for $I = 2$. The measured value is

$$\Gamma(\psi' \to \psi\pi^0\pi^0)/\Gamma(\psi' \to \psi\pi^+\pi^-) = 0.5 \pm 0.1,$$

based on the combined results of the LBL-SLAC and DASP experiments. That conclusion appears to be consistent with the measurement of Hilger et al (62) of

$$\Gamma(\psi' \to \psi + \text{neutrals})/\Gamma(\psi' \to \psi\pi^+\pi^-) = 0.64 \pm 0.15,$$

considering that there is some unevaluated contribution from the $\psi\eta$ and $\psi\gamma\gamma$ modes to the $\psi +$ neutrals inclusive rate. Isospin conservation and the isosinglet character of ψ', as well as ψ, is established. This assignment of $I = 0$ is confirmed by the existence of the $\psi' \to \psi\eta$ decay mode.

3.5 Hadronic Properties of ψ and ψ'

In summary, the hadronic decays of the psions are characterized by strong-interaction conservation laws for isospin and G-parity. Both $\psi(3095)$ and $\psi(3684)$ have $I = 0$ and $G = -1$. SU(3) appears to be a symmetry of the decay of $\psi(3095)$, and that state has been provisionally classified as a singlet. Cascade to $\psi(3095) +$ pions dominates $\psi(3684)$ decays, with less suppression than other exclusive hadronic modes. These features of the decay dynamics, together with the estimates of $\psi(3095)$ interaction cross sections inferred from photoproduction, establish that the two lightest ψ particles are hadrons.

4   INTERMEDIATE STATES

No resonance peak at any energy between 3095 and 3684 MeV has been found in a series of measurements of e^+e^--annihilation cross sections at small energy intervals (25). Rather, such states have been observed in radiative decay, $\psi' \to \gamma\chi$, of the heavier ψ particle. Those states must thus have even charge-conjugation quantum numbers. The presence of such states has been revealed by the use of three distinct experimental techniques and analysis methods. First, the full cascade-decay chain $\psi' \to \gamma\chi \to \gamma\gamma\psi$ has been observed by detecting the lepton-pair decay products of the ψ with one or two coincident photons. Second, hadronic decay products of χ have been detected. Third, monochromatic photon lines have been observed in the inclusive energy distribution of photons produced in $\psi(3684)$ decays. For clarity of presentation, we use the generally agreed-upon symbol χ to denote the intermediate states of the psionic system.

4.1   Cascade Radiative Decays

Evidence for the existence of intermediate states was first obtained from the study of the $\psi' \to \psi\gamma\gamma$ decays by the DASP group (64) and later corroborated by the LBL-SLAC group (63). Both groups used the $\mu^+\mu^-$ decay mode as a signature of a ψ produced in ψ' decay. In one method of χ-mass determination, used by both groups, two photons were required to have been detected in shower counters. Since the energy resolution is poor, use was made only of the measured photon directions. Those, together with momentum and direction of the ψ, inferred from the $\mu^+\mu^-$ momentum, provided input for a two-constraint fit to the kinematics of the process $\psi' \to \psi\gamma\gamma$. It was estimated that the background from $\psi\pi^0\pi^0$ events is less than 10% of the number of events selected on the basis of quality of the kinematic fit. The DASP group found two clusters of points on a scatter plot of one photon energy versus the other. Since it is not known which of the two photons is the decay product of the χ, each group of photons yields two possible values of χ mass. Using central values of the energies of the two observed groups, they determined the masses to be (3.507 ± 0.007) or (3.258 ± 0.007) GeV/c^2 and (3.407 ± 0.008) or (3.351 ± 0.008) GeV/c^2, with decay widths consistent with zero.

To obtain better resolution in their χ-mass measurements, the LBL-SLAC group used the magnetic detector as a pair spectrometer to measure the energy of photons that had converted in the various pieces of the apparatus preceding the tracking

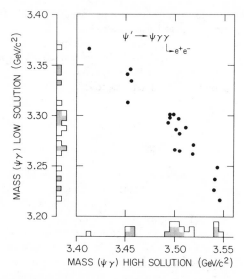

Figure 27   Scatter plot of $\psi\gamma_1$ invariant mass against $\psi\gamma_2$ invariant mass for $\psi(3684)$ decays to $\psi(3095)\gamma_1\gamma_2$; from LBL-SLAC (65).

chambers. The total conversion material has a thickness of some 5% of a radiation length. From an initial sample of 54 events with one converted photon, a subset of $\psi\gamma\gamma$ candidates was selected on the basis of consistency with zero missing mass and indication of the presence of a second photon, revealed by a shower-counter signal. With the elimination of events in which the inferred mass of the photon pair was consistent with the mass of the η, there remain 21 examples of $\psi' \rightarrow \psi\gamma\gamma$. One-constraint fits were made to the kinematics of $\psi' \rightarrow \psi\gamma\gamma$, in order to obtain values of both photon energies and, for each event, again two values of the mass of the parent of a $\psi\gamma$ pair. These masses are shown plotted against each other in Figure 27 (65). Three clusters are evident, at masses (3543 ± 10), (3504 ± 10), and (3454 ± 10) MeV/c^2. These, rather than the lower-mass alternatives, were chosen as χ masses because the spread of mass around the lower central values is in all cases consistent with that expected from Doppler broadening and the resolution of the apparatus. The one event with $M_\chi \simeq 3415$ MeV/c^2 appears to correspond to the two-event cluster in the DASP results, and the large clumps at $M_\chi = 3504$ MeV/c^2, it is presumed, are to be identified with the DASP events at $M_\chi = 3507$ MeV/c^2.

4.2   Hadronic Decays of Intermediate States

Hadronic decays of intermediate states were first identified by Feldman et al (66), who used the LBL-SLAC magnetic detector. Events were restricted to those con-

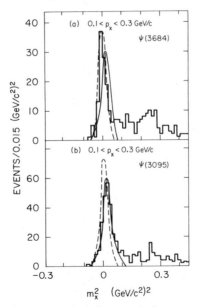

Figure 28   Spectrum of the square of missing mass in four-prong, zero-charge events with missing momentum between 100 and 300 MeV/c for (a) $\psi(3095)$ and (b) $\psi(3684)$ decays. The solid and dashed smooth curves show calculated resolution functions for a missing π^0 and missing photon, respectively; from SLAC-LBL (66).

426   CHINOWSKY

taining even numbers of detected particles with total charge zero and momentum unbalance between 100 and 300 MeV/c. This restriction to small missing momentum allows a good separation of events with undetected π^0 from those with undetected photons, based on the observed distribution of missing mass. An example is the missing-mass distribution for the selected four-track events shown in Figure 28, which contrasts the results obtained for the $\psi(3095)$ and $\psi(3684)$ decays. The low-mass peak in the $\psi(3684)$ event sample is consistent with that expected for missing photons and inconsistent with that resulting from undetected neutral pions. Just the opposite conclusion is drawn from the $\psi(3095)$ sample. Adjusted values of particle momenta were obtained from one-constraint fits to the kinematics of $\psi' \rightarrow \gamma + \pi^+ \pi^- \pi^+ \pi^-$ and $\psi' \rightarrow \gamma + \pi^+ \pi^- K^+ K^-$. In subsequent analyses of a larger sample, the SLAC-LBL group used time-of-flight measurements, as well as the quality of the kinematic fit, as criteria for separating events with $K^+ K^-$ and $\bar{p}p$ pairs from those

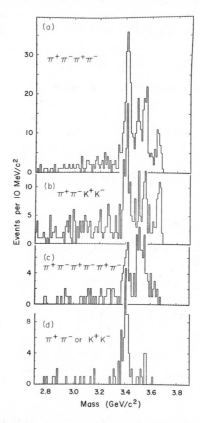

Figure 29   Spectra of invariant masses of states χ fit to the kinematics of the radiative decay $\psi(3684) \rightarrow \gamma\chi$ for χ constituent particles identified as indicated in the legends; from LBL-SLAC (67).

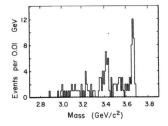

Figure 30   Invariant-mass spectrum of $\pi^+\pi^-p\bar{p}$ products of decays $\psi(3684) \to \gamma\pi^+\pi^-p\bar{p}$ identified with the aid of kinematic fits; from LBL-SLAC (67).

with pions only (67). Similar methods were applied to the analysis of two-prong and six-prong events. Results indicating the presence of decays of distinct states with small decay widths are shown in the mass distributions of Figures 29 and 30. All show prominent peaks at (3415 ± 10) MeV/c^2. Other peaks present at various levels of significance among the spectra are at (3500 ± 10) and (3550 ± 10) MeV/c^2. Noticeably lacking are two-body decays of the state at 3500 MeV/c^2. There is no indication of hadronic decay modes of the proposed state near 3450 MeV/c^2 inferred from results of analysis of the $\psi(3684) \to \psi(3095)\gamma\gamma$ events.

4.3   Monochromatic Photons

Certainly, the conceptually simplest and most direct evidence for distinct intermediate states would be the presence of monoenergetic lines in the spectrum of photon-decay products of $\psi(3684)$. Measurements of the yield of photons in the radiative decays of $\psi(3684)$ to the various χ states are essential for the determination of the $\psi' \to \gamma\chi$ and χ decay branching ratios.

The LBL-SLAC group measured the photon-energy spectra using only those photons that had converted to detected electron-positron pairs and whose momenta were then determined from measurements of reconstructed tracks sampled in the track chambers of the detector (65). Energy spectra of photons from $\psi(3684)$ decay and from $\psi(3095)$ are shown in Figure 31. The yield of photons from $\psi(3095)$ decays varies smoothly with photon energy. The shape of the $\psi(3684)$ distribution is similar, but for the evident peak at an energy determined to be (261 ± 10) MeV after correction for energy lost by the electron and positron by ionization. The observed width of the peak is consistent with a true width of zero, broadened by resolution. The inferred mass of the χ state accompanying the photon is (3413 ± 11) MeV/c^2. With appropriate corrections for detection inefficiencies and the assumption of a $1 + \cos^2\theta$ dependence of the photon angular distribution, the LBL-SLAC group finds a branching fraction of 0.075 ± 0.026 for radiative decay to this state. It is reasonably certain that this state is to be identified with the state at (3415 ± 10) MeV/c^2 whose hadronic decays have been observed.

Biddick et al (68) used an arrangement of proportional counters and NaI crystals arrayed about the other interaction region at SPEAR. Superior photon efficiency and energy resolution were achieved with that apparatus. Their results are shown in

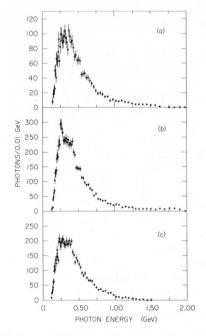

Figure 31 Inclusive energy spectra of photons from decays of (*a*) $\psi(3095)$ and (*b*) $\psi(3684)$; (*c*) a Monte-Carlo-simulated spectrum of π^0 decay photons in multihadron decays of $\psi(3095)$; from SLAC-LBL (67).

Figure 32, containing the spectrum of photons from $\psi(3684)$ decays and, for contrast, that from $\psi(3095)$ decays. In addition to the line at (260.6 ± 2.9) MeV, peaks above the continuum are evident at energies (169.2 ± 1.4) and (120.9 ± 1.3) MeV. The corresponding values of M_χ, 3413, 3511, and 3561 MeV/c^2, may be identified with the states at 3415, 3500, and 3550 MeV/c^2 discussed above, all consistent with the uncertainties in the various mass determinations. Branching ratios obtained by Biddick et al from these yields, corrected for detector acceptance, are

$$\Gamma[\psi' \to \gamma\chi(3415)]/\Gamma(\psi') = 0.072 \pm 0.023,$$

$$\Gamma[\psi' \to \gamma\chi(3510)]/\Gamma(\psi') = 0.071 \pm 0.019,$$

and

$$\Gamma[\psi' \to \gamma\chi(3550)]/\Gamma(\psi') = 0.070 \pm 0.020.$$

The first agrees well with the LBL-SLAC result, to within the substantial errors.

With values of the radiative-decay branching ratios and the branching-ratio products $\Gamma(\psi' \to \gamma\chi)/\Gamma(\psi') \times \Gamma(\chi \to f)/\Gamma(\chi)$ determined from the measured yields corrected for detection inefficiency, we obtain branching fractions for the various

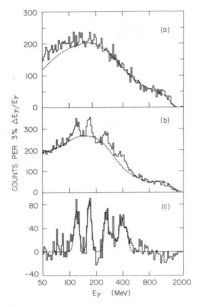

Figure 32   Inclusive energy spectra of photons from (a) $\psi(3095)$ and (b) $\psi(3684)$; (c) continuum subtracted from data of (b); from Biddick et al (68) at SPEAR.

detected decay modes of the χ states. The values in Tables 6–8 are to be considered estimates, accurate to approximately a factor of two.

Table 6   Branching fractions for $\chi(3415)$ decays

Mode	Branching fraction (χ) × branching fraction ($\psi' \rightarrow \gamma\chi$) (%)	Branching fraction (χ) (%)[a]	Ref.
$\pi^+\pi^-$	0.07 ± 0.02	1.0	67
K^+K^-	0.07 ± 0.02	1.0	67
$\pi^+\pi^-\pi^+\pi^-$	0.32 ± 0.06	4.6	67
$\pi^+\pi^-K^+K^-$	0.27 ± 0.07	3.9	67
$\pi^+\pi^-\bar{p}p$	0.04 ± 0.013	0.6	67
$\pi^+\pi^-\pi^+\pi^-\pi^+\pi^-$	0.14 ± 0.05	2.0	67
$\gamma\psi$	$\sim 1^b$	14.0	50
	0.2 ± 0.2^c	3.0	65
	3.3 ± 1.7^d	47.0	68

[a] Calculated from the branching-ratio product and 0.07 for the $\psi' \rightarrow \gamma\chi$ fraction. Estimates, with large uncertainties.
[b] Based on two events.
[c] Based on one event.
[d] Based on a fit to the inclusive energy spectrum of photon-decay products of $\psi(3684)$.

Table 7 Branching fractions for $\chi(3500)$ decay

Mode	Branching fraction (χ) × branching fraction $(\psi' \to \gamma\chi)$ (%)	Branching fraction (χ) (%)[a]	Ref.
$\pi^+\pi^-$ and K^+K^-	<0.015	<0.2	67
$\pi^+\pi^-\pi^+\pi^-$	0.11 ± 0.04	1.6	67
$\pi^+\pi^- K^+K^-$	0.06 ± 0.03	0.9	67
$\pi^+\pi^- \bar{p}p$	0.01 ± 0.008	0.1	67
$\pi^+\pi^-\pi^+\pi^-\pi^+\pi^-$	0.17 ± 0.06	2.4	67
$\gamma\psi$	4 ± 2	57	50
	2.4 ± 0.8	34	65
	5.0 ± 1.5[b]	71	68
$\gamma\gamma$[c]	<0.0013	<0.02	50

[a] Calculated from the branching-ratio products and 0.07 for the $\psi' \to \gamma\chi$ fraction. Estimates, with large uncertainties.
[b] Based on a fit to the inclusive energy spectrum of photon decay products of $\psi(3684)$.
[c] Forbidden for spin-1.

Table 8 Branching fractions for $\chi(3550)$ decays

Mode	Branching fraction (χ) × branching fraction $(\psi' \to \gamma\chi)$ (%)	Branching fraction (χ) (%)[a]	Ref.
$\pi^+\pi^-$ and K^+K^-	0.02 ± 0.01	0.3	67
$\pi^+\pi^-\pi^+\pi^-$	0.16 ± 0.04	2.3	67
$\pi^+\pi^- K^+K^-$	0.14 ± 0.04	2.0	67
$\pi^+\pi^- \bar{p}p$	0.02 ± 0.01	0.3	67
$\pi^+\pi^-\pi^+\pi^-\pi^+\pi^-$	0.08 ± 0.05	1.1	67
$\gamma\psi$	1.0 ± 0.6	14	65
	2.2 ± 1.0[b]	28	68

[a] Calculated from the branching ratio products and 0.07 for the $\psi' \to \gamma\chi$ fraction. Estimates, with large uncertainties.
[b] Based on a fit to the inclusive energy spectrum of photon decay products of $\psi(3684)$.

4.4 Properties of the Intermediate States. J, P, C, G

It follows directly from the observation of the radiative decays $\psi' \to \gamma\chi$ that the various χ states have even charge-conjugation quantum numbers. Information needed for assignment of parity and G-parity is available only for those three intermediate states whose decays to hadrons have been observed. Each decays to one or another state of even numbers of pions and thus has even G-parity. The states $\chi(3415)$ and $\chi(3550)$ decay to $\pi^+\pi^-$ pairs and so must be of natural spin and parity 0^+, 1^-, 2^+, etc. Of these, only the even spin, even parity assignments are compatible with the even charge conjugation of the χ states.

Angular correlations (70–72) among the various particles in the cascade-decay chains $e^+e^- \to \psi' \to \gamma\chi$, $\chi \to \gamma\psi$ or $\chi \to$ hadrons depend upon the spin of the state

χ, but are uniquely determined only in the case of zero spin. For instance, photons are produced in lowest-order e^+e^- annihilation with an angular distribution of the form

$$W(\theta_{\gamma_1}) = 1 + \alpha \cos^2 \theta_{\gamma_1},$$

where θ_{γ_1} is the angle between the photon and the initial e^+ (or e^-) direction. In the case that the χ spin is zero, $\alpha = 1$. For other spins, α cannot be specified further than by the general limit $|\alpha| < 1$ without knowledge of the strengths of the various multipole contributions to the transition amplitude. Although, in contrast to the situation with nuclear transitions, there are no strong arguments justifying neglect of higher multipoles, should it happen that $E1$ amplitudes are dominant, then the magnitude of the coefficient in the angular distribution is fixed by the χ spin, i.e. $\alpha = -1/3$ for $S_\chi = 1$ and $\alpha = +1/13$ for $S_\chi = 2$.

Figures 33 and 34 show the LBL-SLAC data on angular distributions (67) of the photons produced with those χ states that have substantial branching fractions to hadronic final states. Values of the inferred anisotropy coefficients for the three χ states are $\alpha = 0.3 \pm 0.4$ for $\chi(3550)$, $\alpha = 0.1 \pm 0.4$ for $\chi(3500)$, and $\alpha = 1.4 \pm 0.4$ for $\chi(3415)$. These results indicate spin-zero for $\chi(3415)$, and at best only weakly exclude spin-zero for the other two.

For those χ states decaying to two particles, as in, for example, $\psi' \to \gamma\chi \to \gamma\pi^+\pi^-$, the distribution in angle θ' between the photon and a decay particle, in the χ rest frame, is described by a polynomial in $\cos^2 \theta'$. For the decay into two pseudoscalars, this angular correlation function is isotropic for spin-zero and contains terms in $\cos^2 \theta'$ and $\cos^4 \theta'$ for spin-two and becomes increasingly complex for higher spins. The data of the SLAC-LBL group are shown in Figure 34a for $\chi(3415)$ decays to pion and kaon pairs. Poor statistical accuracy precludes the

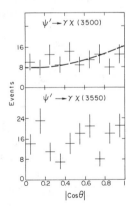

Figure 33 Distributions in the angle between the initial-state colliding beams' direction and the photons produced in $e^+e^- \to \psi(3684) \to \gamma\gamma$. The χ masses are indicated in the legends. The smooth dashed curve gives the distribution for χ's of zero spin; from LBL-SLAC (67).

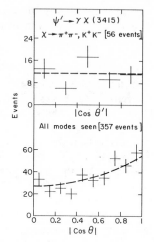

Figure 34 Angular correlations in radiative transitions $\psi(3684) \to \gamma\chi(3415)$ of psions created in e^+e^- annihilations. θ' is the angle between the photon and the dimeson direction in $\pi^+\pi^-$ or K^+K^- decays. θ is the angle between the photon and the colliding beams. The dashed curves are distributions for χ's of zero spin; from SLAC-LBL (67).

possibility of making definitive conclusions, but no deviation from isotropy is indicated, consistent with the expected distribution of the decay products of a zero-spin object.

As yet, there have been no reported measurements of angular correlations between the two photons in the cascade process $\psi' \to \gamma_1\chi \to \gamma_1\gamma_2\psi$. In this case, decays of states with angular momentum greater than zero may produce isotropic angular distributions. Lack of γ_1-γ_2 correlation is not a specific signature of zero spin. Further information about spin is contained in the correlation between γ_2 and lepton decay products of ψ in the processes $\psi' \to \gamma_1\gamma_2\psi \to \gamma_1\gamma_2 l^+l^-$. For zero spin, that correlation function has the simple form

$$W(\theta_1, \theta_2, \phi_2, \theta_l, \phi_l) = (1 + \cos^2\theta_1)(1 + \cos^2\theta_l),$$

where θ_l and ϕ_l are the polar and azimuthal angle of a decay lepton measured from an axis defined by the γ_2 direction in the rest frame of the ψ. Expressions for decay of states with greater spin are complicated, and their forms depend again on the multipolarity of the χ transitions.

In sum, the study of decay angular distributions has yielded one conclusive result, $S_\chi = 0$ for $\chi(3415)$ and tentative suggestions that $\chi(3500)$ and $\chi(3550)$ have greater spin. Nothing is known of the spin of $\chi(3455)$, observed so far only via its $\psi\gamma$ decay mode.

4.5 *Lower-Mass States*

Searches have not uncovered odd-C states formed in e^+e^- annihilations in the energy range 1.9–3.1 GeV (see Section 5). Evidence has been presented for existence of a state with mass near 2800 MeV/c^2, revealed in three-photon decays of $\psi(3095)$ produced in e^+e^- annihilations. Both the DASP group (50) and the Deutsches Electronen Synchrotron DESY-Heidelberg group (73, 74) measured only production angles of the three photons and determined their energies from the kinematics of the process $\psi(3095) \rightarrow 3\gamma$. The DASP data have been presented as the plot, reproduced in Figure 35, of the smallest against the largest opening angle of any pair of photons in each event. Decays of an intermediate state of unique mass in the sequence

$$\psi(3095) \rightarrow \gamma_1 X \rightarrow \gamma_1 \gamma_2 \gamma_3$$

produce points common to a locus determined by the mass of X. Figure 35 shows such curves for π^0, η, η' and $M_X = 2800$ MeV/c^2. An η signal is clear. A cluster of points is also shown in the region of the plot populated by events due to two-photon decays of a particle of mass ~ 2800 MeV/c^2 produced in radiative decay of $\psi(3095)$. Calculations of background from three-photon events produced according to electrodynamics indicate smooth variation of density with mass in this region of the plot. It was concluded that a new narrow resonance had been observed with production-decay branching-ratio product

$$B[\psi \rightarrow \gamma X(2800)] \times B(X \rightarrow 2\gamma) \simeq 1.5 \times 10^{-4}.$$

Similar conclusions have been reached by the DESY-Heidelberg group, who detected three-photon events with the apparatus shown in Figure 36. The large array of NaI and lead-glass counters provides good photon-detection efficiency over $\sim 60\%$ of the full-production solid angle. Their results are shown in Figure 37, a Dalitz plot of the lowest vs the highest γ-γ pair mass in each event. Data from

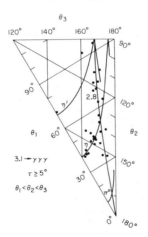

Figure 35 Scatter plot of the largest opening angle, θ_3, versus the smallest opening angle, θ_1, between any two photons in decays of $\psi(3095)$ to three photons. Loci of constant $\gamma\gamma\gamma$ invariant mass for η and η' and 2.8 GeV/c^2 are shown. Results from DASP (50) at DORIS.

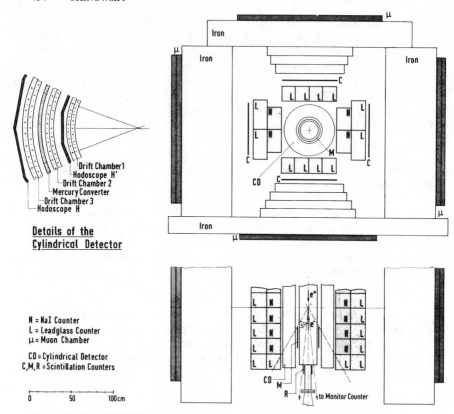

Details of the
Cylindrical Detector

Drift Chamber 1
Hodoscope H'
Drift Chamber 2
Mercury Converter
Drift Chamber 3
Hodoscope H

N = NaI Counter
L = Leadglass Counter
μ = Muon Chamber

CD = Cylindrical Detector
C,M,R = Scintillation Counters

0   50   100 cm

Figure 36 The DESY-Heidelberg detector apparatus at the e^+e^- intersecting storage rings DORIS (73).

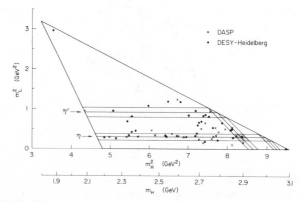

Figure 37 Dalitz plot of the smallest $\gamma\gamma$ invariant mass-squared vs the largest $\gamma\gamma$ mass-squared in $\psi(3095)$ decays to three photons. Data from DASP and DESY-Heidelberg (73).

Table 9 Properties of the intermediate states

State	Mass[a]	J^{PC} [b]	Ref.	Footnote(s)
$\chi(3415)$	3412 ± 3	0^{++}	50, 65, 67, 68	c, d
$\chi(3455)$	3454 ± 10	0^{-+}	65	e, f
$\chi(3500)$	3508 ± 4	1^{++}	50, 65, 67, 68	c
$\chi(3550)$	3553 ± 5	2^{++}	65, 67, 68	c, d

[a] Weighted averages of values quoted in the cited references.
[b] Even charge conjugation established by the radiative decay $\psi(3684) \rightarrow \psi(3095) + \gamma$.
[c] Preferred assignments of spin and parity consistent with observed angular correlations.
[d] Assignment of 0^- or 1^+ excluded by decays to $\pi^+\pi^-$ and K^+K^-.
[e] Four $\chi \rightarrow \gamma\psi$ decays observed; no evidence for hadron modes.
[f] Speculative assignment of spin and parity, neither supported nor contradicted by experimental evidence.

the DASP experiment are included. Again, the excess of events at mass near 2750 MeV/c^2 was attributed to $\gamma\gamma$ decay of a resonant state with production-decay branching-ratio product comparable to that due to η radiative production followed by $\gamma\gamma$ decay.

4.6 Summary

Five additional members of the ψ family have been established experimentally, albeit with reliability varying from firm [for $\chi(3415)$, $\chi(3500)$, and $\chi(3550)$] to mushy [for $X(2800)$ and $\chi(3455)$]. They are connected to the $\psi(3684)$, or $\psi(3095)$ via radiative transitions. Prodded by the suggestive level structure of a hypothesized bound quark-antiquark system, charmonium, these states have been assigned spectroscopic labels consistent with experimental results: $^3P_0$, $^3P_1$, and $^3P_2$ for the first three, and 1^1S_0 and 2^1S_0 for the others, respectively. Decays $\psi(3684) \rightarrow \chi(^3P) + \gamma$ occur with comparable branching fractions, $B(\psi' \rightarrow \gamma\chi) \simeq 0.1$, for each of the ^3P states. Properties of these states are listed in Table 9.

5   OTHER NARROW STATES, 1900 MeV/$c^2 < M < 8000$ MeV/c^2

A search for other narrow states coupled to lepton pairs was made by the LBL-SLAC group, which measured e^+e^--annihilation cross sections in small energy intervals from 3.2 to 7.6 GeV (75, 76). As demonstrated by the results plotted in Figure 38, none was discovered. Those results establish upper limits on the electron-pair decay widths of a resonance whose full width is much smaller than ~ 2 MeV (see Equation 3.1 with $\Gamma_h \simeq \Gamma$). Those limits are summarized in Table 10. Similar systematic searches at the storage ring ADONE, operated at energies between 1.9 and 3.1 GeV, also failed to unearth any new resonances. Limits on Γ_{ee} were established at $\sim 5\%$ of the ψe^+e^- width (77–81). Those limits are included in Table 10. Measurements at SPEAR of e^+e^--annihilation rates in coarser energy steps revealed the striking behavior near 4 GeV shown in Figure 39. At energies above the second ψ, the character of the energy dependence changes markedly.

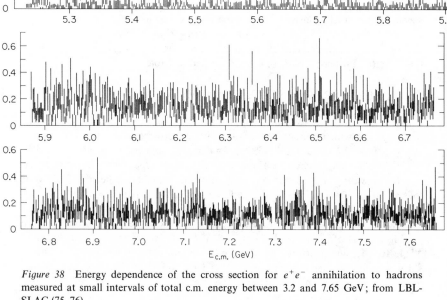

Figure 38   Energy dependence of the cross section for e^+e^- annihilation to hadrons measured at small intervals of total c.m. energy between 3.2 and 7.65 GeV; from LBL-SLAC (75, 76).

Table 10   Upper limits on parameters of narrow resonances other than $\psi(3095)$ and $\psi(3684)$

Mass range (MeV/c^2).	$\int \sigma_{\text{Res}}(E)\,dE^a$ (nb MeV)	Γ_{ee} (keV)^a	Ref.
1910–2200	1150	0.21	77
2200–2545	800	0.20	77
2520–2595	690	0.20	81
2600–2694	1400	0.42	81
2700–2799	800	0.26	81
2800–2899	800	0.28	81
2900–3000	1600	0.62	81
2970–3090	800	0.32	78
3200–3500	970	0.47	75
3500–3680	780	0.44	75
3720–4000	850	0.55	75
4000–4400	620	0.47	75
4400–4900	580	0.54	75
4900–5400	780	0.90	76
5400–5900	800	1.11	76
5900–7600	450	0.87	76

^{a} The results from ADONE (77, 78, 81) have been presented as ratios of upper limits on the integrated resonance cross section to the $\psi(3095)$ integrated cross section measured with the same apparatus. To eliminate effects of different detection efficiencies, we have used the SLAC-LBL value of the $\psi(3095)$ lepton-pair partial width (Table 3) in calculating the limits on the integrals and Γ_{ee} listed here (see Equation 3.4).

Below 4 GeV, the ratio of nonresonant hadronic to muon-pair cross sections, R, is essentially constant at $\simeq 2.5$. Above 4 GeV, R is also constant, but larger: $R \simeq 5.5$. In the transition interval, there appears to be a series of resonant structures that, however many there are, have widths more comparable to those of the lighter vector mesons than the ψ particles at 3095 and 3684 MeV/c^2.

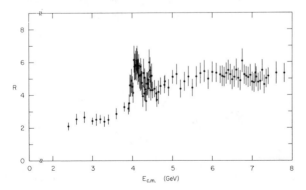

Figure 39   Energy dependence of the ratio R of the cross section for e^+e^- annihilation to hadrons to that for annihilation to muon pairs; from SLAC-LBL (76).

6 THEORETICAL ORIENTATION

Confronted with the necessity of understanding the mechanisms inhibiting the hadronic decay of the strongly interacting psions, appeal was made to new, hypothetical hadronic degrees of freedom. It had already been diversely suggested that the strong-interaction symmetry group is larger than SU(3). Certain suggestions, variously motivated, had been made to introduce a new additive quantum number (2, 82–85), charm (2), to be carried by a fourth quark, thus extending SU(3) to SU(4). Greenberg (86) and Han & Nambu (87) had argued for SU(3)' × SU(3)'' as the proper symmetry group. In the first version cited (86), quarks are fractionally charged, while the Han-Nambu quarks have integral charges. Those schemes have nine quarks, each carrying one of three SU(3)'' "colors" as well as one of the three usual SU(3)' "flavors." Ordinary hadrons must be SU(3)'' singlets. With SU(4) × SU(3)'' symmetry, three colors, charm, and three other flavors equip the basic constituents of hadrons.

The fundamental hadronic entities now are the four flavored u, d, s, and c (charmed) quarks, whose properties are listed in Table 2. For these quarks, the Gell-Mann–Nishijuma formula now reads $Q = I_3 + (B + S + C)/2$. Each flavored quark has, in addition, the threefold color degree of freedom. The simple quark-parton model for e^+e^- annihilation to hadrons via the diagram of Figure 40, which predicts constant $R = \Sigma_j q_j^2$, the sum of squares of quark-parton charges, yields $R = 2$ below the threshold for annihilation to charmed quarks and $R = 3\frac{1}{3}$ above. The rough agreement with observation supports both the charm and color hypotheses of fractionally charged quarks.

It was suggested early that the narrow widths of ψ and ψ' could be accounted for with a color quantum number that is conserved in strong interactions (88, 89). In general, SU(3) color models of the Han-Nambu type suffer however, from the basic difficulty that radiative decays to ordinary hadrons conserve color, as do, in some versions, the cascade-decay processes $\psi' \rightarrow \psi\pi\pi$ and $\psi' \rightarrow \psi\eta$. Ad hoc suppression mechanisms must be imposed to obtain agreement with observation. Furthermore, since higher-dimensionality representations of color SU(3) are populated, there must exist many other colored states, for example colored ρ mesons.

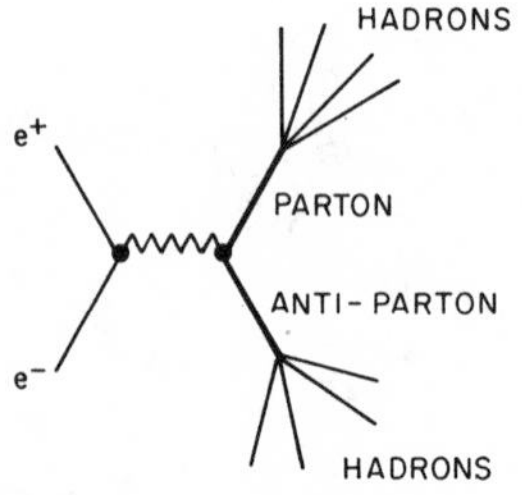

Figure 40   Feynman diagram for e^+e^- annihilation to hadrons via intermediate-quark–parton-antiparton pairs.

Table 11 Quark composition and properties of singly charmed hadrons

Particle[a]	Quark components	I, I_z	C	S
D^+, D^0	$c\bar{d}, c\bar{u}$	$(\frac{1}{2}, \frac{1}{2}), (\frac{1}{2}, -\frac{1}{2})$	$+1$	0
$D^-, \bar{D}^0$	$\bar{c}d, \bar{c}u$	$(\frac{1}{2}, -\frac{1}{2}), (\frac{1}{2}, \frac{1}{2})$	-1	0
F^+	$c\bar{s}$	0	$+1$	$+1$
F^-	$\bar{c}s$	0	-1	-1
Λ_c^+	cud	0	$+1$	0
Σ_c^0	cdd	$1, -1$	$+1$	0
Σ_c^+	cud	$1, 0$	$+1$	0
Σ_c^{2+}	cuu	$1, 1$	$+1$	0
Ξ_c^0	cds	$\frac{1}{2}, -\frac{1}{2}$	$+1$	-1
Ξ_c^+	cus	$\frac{1}{2}, \frac{1}{2}$	$+1$	-1
Ω_c^0	css	0	$+1$	-2

[a] This nomenclature is for states with lowest angular momentum, $J = 0$ for mesons, $J = \frac{1}{2}$ for baryons.

The decay $\psi(3684) \to \rho^c + \pi$ should, in some models (88), occur with a branching fraction comparable to that for decay to $\psi\pi\pi$. No evidence for such a particle has been revealed in the inclusive momentum distribution of pions from $\psi(3684)$ decay (50). Color models, at least in their simpler forms, are in serious contradiction with observed ψ and ψ' decay properties, and in any case have no place for the recently discovered mesons of mass ~ 1860 MeV/c^2 (see Section 7).

A generally successful description of the psionic states is obtained with a theoretical structure in which hadrons are colorless composites of the four fractionally charged, flavored quarks of Table 2, including charmed quarks. Strong interactions are mediated by exchanges of color gluons that serve to confine the quarks. An extended spectrum of states now occurs. Mesons still have $q\bar{q}$ content. Meson nonets grow to sixteen-dimensional multiplets, with the addition of new mesons containing one or more charmed quarks. Of these, one, $c\bar{c}$, has zero, or hidden, charm; the other six have exposed charm, $C = +1$ or -1. Table 11 lists these additional meson states, with designations D, F, etc introduced by Gaillard, Lee & Rosner (90).

Physically realized baryons, still with qqq content, are now grouped into two nonequivalent representations of dimensionality 20, one containing spin-$\frac{1}{2}$ and another spin-$\frac{3}{2}$ baryons. The additional states are charmed, with values $C = +1$, $+2$, or $+3$. Table 11 lists the charmed, spin-$\frac{1}{2}$ baryons that occur as composites of three quarks.

6.1 Charmonium

States found in e^+e^- annihilation, directly coupled to photons, are zero-charm vector mesons, and so in this picture, have $c\bar{c}$ quark content. Presuming that the χ states are reached in ordinary electromagnetic transitions from the heavier ψ states, the intermediate states and $X(2800)$ are also to be specified as $c\bar{c}$ composites.

(a)   (b)

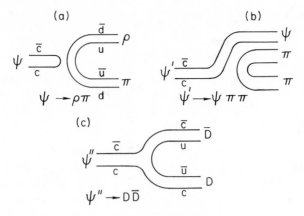

(c)

$$\psi'' \longrightarrow D\bar{D}$$

Figure 41 Duality diagrams of OZI-forbidden psion decays to (*a*) ordinary hadrons and (*b*) $\psi\pi\pi$; (*c*) OZI-allowed decays to charmed-meson pairs.

The empirical OZI rule must be invoked to provide a means of inhibiting the decays of the $c\bar{c}$ states to ordinary mesons composed only of *u*, *d*, or *s* quark-antiquark pairs. Strong, charm-conserving decays can proceed only to states containing *D* mesons, as depicted in the duality diagrams of Figure 41. Kinematics is invoked to prohibit decay to charmed mesons, whose masses must then be greater than 1842 MeV/c^2. Calculations have been made by De Rujula et al (91) of hadron masses in a general theory of quark confinement. Their estimates of the

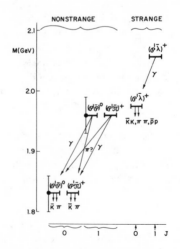

Figure 42 Spectrum of masses of lowest-lying charmed mesons as predicted by De Rujula, Georgi & Glashow (91). Their notation for the quark constituents may be readily translated to *u*, *d*, *s*, and *c*.

masses of charmed S-state quark-antiquark bound systems are shown as the energy level diagram of Figure 42. The results, admittedly not very precise, support the experimental requirement that $\psi(3684)$ be still below threshold for decay to charmed mesons. It should be noted, in this connection, that the energy dependence of the e^+e^--annihilation cross section (see Figure 39) behaves very much as if there were a threshold for pair production of particles of mass between 1.85 and 2.0 GeV/c^2. Also, the photoproduction results of Figure 13 invite a speculative interpretation in terms of a threshold at ~ 5 GeV energy for the inelastic, charm-conserving, OZI-allowed reaction $\psi N \to D\bar{D}N$, producing a pair of charmed mesons.

Identification of the psions with $c\bar{c}$ bound states requires comparison with the complete spectrum of such states. Even before the first ψ discovery, Appelquist & Politzer (5), reasoning from ideas of asymptotic freedom, expounded the view that this system of bound massive quarks could be described as a nonrelativistic atomic system, charmonium, that is analogous to positronium. Properties of charmonium states are determined by the combined effects of short-range $c\bar{c}$ interactions via exchange of massless vector gluons and some long-range confining potential. The pattern of charmonium energy levels that emerges, labeled with spectroscopic notation, is shown schematically in Figure 43. The pattern of observed energy levels is plotted suggestively in Figure 44, with $\psi(3095)$ the lowest-lying $^3S_1$ state and $\psi(3684)$ the first radially excited $^3S_1$ state. Remarkable qualitative agreement with the model predictions is apparent. That statement must be tempered with the reminder that the identities of the intermediate states have not been firmly established. Spins and parities of the intermediate χ states have been tentatively assigned values consistent with experimental data. Some theoretical arguments of Chanowitz & Gilman (92), based on quantitative predictions of radiative decay widths, have been brought to bear in assigning $\chi(3510)$, rather than $\chi(3445)$, the $^3P_1$ classification.

Numerous more or less quantitative predictions of level spacings, hadronic decay rates, leptonic decay rates, and radiative transition rates are available for comparison with the data. We briefly consider a selection from among the results of those calculations.

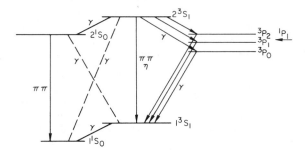

Figure 43   Schematic diagram of the energy levels of charmonium, the bound system of a charmed quark and its antiparticle. Some transition modes are indicated. See Appelquist & Politzer (5), for example.

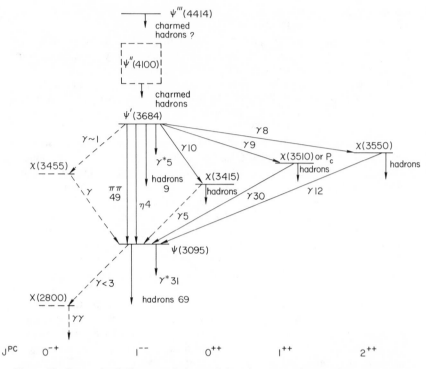

Figure 44 Energy-level diagram of observed ψ and χ states shown with spectroscopic notation corresponding to the levels of charmonium. Dashed lines indicate uncertain states. Numbers are now-outdated branching fractions. See text for current values.

6.1.1 LEVEL SPACINGS De Rujula, Georgi & Glashow (91) represent the quark-confinement interaction by a long-range, spin-independent potential common, in fact, to all quarks, ordinary as well as charmed ones. They argue that hadron mass splittings are determined by additional short-range perturbing forces resulting from exchange of a single, massless vector gluon. The corresponding interaction Hamiltonian has the form of the Fermi-Breit potential for the positronium atom (93), with the electrodynamic $\alpha = e^2/4\pi$ replaced by an effective hadronic coupling constant. Treating the Fermi-Breit interaction as a first-order perturbation, they obtain expressions for charmonium S- and P-state energy eigenvalues in terms of known hadron masses and the mass of the charmed quark. Applying their mass formula to $\psi(3095)$, the charmed-quark mass is estimated to be ~ 1650 MeV/c^2. With that, all other S- and P-state masses are determined. They obtain S-state hyperfine splittings

$$M(^3S) - M(^1S) \simeq 27 \text{ MeV}/c^2$$

for both the ground and first radially excited state, $\psi(3684)$; singlet P-state mass

$M(^1P) \simeq 3650 \text{ MeV}/c^2$;

triplet P-state mass

$M(^3P_0) \simeq 3640 \text{ MeV}/c^2$;

and hyperfine splittings

$$[M(^3P_1) - M(^3P_0)] \simeq 7.8 \text{ MeV}/c^2$$

and

$$[M(^3P_2) - M(\ P_1)] \simeq 6.2 \text{ MeV}/c^2.$$

These last values differ by roughly an order of magnitude from measured mass differences of the observed states given the charmonium classification shown in Figure 44. An alternative description, introduced by Eichten et al (94), assigns the $c\bar{c}$ system a potential that is partially Coulombic, $1/r$, and partially linearly increasing in its dependence on quark-antiquark separation. The linearly increasing part is meant to provide the long-range quark-confining force. They find a splitting of 230 MeV between the center of gravity of the 1^3P and the 2^3S levels, in reasonable agreement with observations. With that potential taken as the Fourier transform of $V(k^2)$ in a modified gluon propagator $g_{\mu\nu}V(k^2)$, Pumplin et al (95) and Schnitzer (96) arrive at hyperfine splitting

$$M(^3S) - M(^1S) \simeq 100 \text{ MeV}/c^2,$$

still about three times smaller than is observed with the current hesitant assignment of states. Better agreement is obtained with this prescription, but still, the agreement between model predictions of masses and observed values can hardly be characterized as better than at a semiquantitative level.

Further radial excitations, n^3S_1 bound states of charmonium, occur at energies determined by the characteristics of the $c\bar{c}$ binding potential and the quark mass. Using the 1^3S_1 and 2^3S_1 masses as inputs to fix the strength of a linear potential and the quark mass, Harrington et al (97) calculate eigenvalues

$$E(3^3S_1) = 4.18 \text{ GeV},$$

$$E(4^3S_1) = 4.61 \text{ GeV},$$

$$E(5^3S_1) = 5.00 \text{ GeV},$$

for the next three charmonium states accessible to e^+e^- annihilation in lowest order.

6.1.2 LEVEL WIDTHS Since annihilations to hadrons or photons occur at zero quark-antiquark separation, it may be expected that conditions are more favorable for validity of a perturbative, weak-coupling treatment of $c\bar{c}$ annihilation (22). Annihilations to lepton-pairs proceed via $c\bar{c}$ single-photon transitions, as indicated in Figure 45a; hadronic decays of $^3S_1$ proceed through the three-gluon intermediary of Figure 45b; $^1S_0$, $^3P_0$, and $^3P_2$ states decay to hadrons via two gluons,

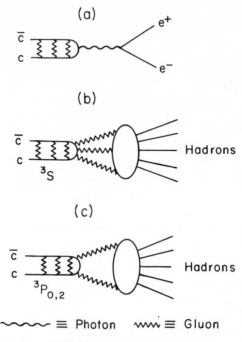

$\text{\small \raisebox{0pt}{$\sim\!\sim\!\sim$}} \equiv$ Photon $\text{\small \raisebox{0pt}{$\wedge\!\wedge\!\wedge$}} \equiv$ Gluon

Figure 45 Diagrams for annihilation of charmonium states to: (*a*) electron pairs via intermediate photons; (*b*) and (*c*) hadrons via intermediate gluons.

shown in Figure 45*c*. Radiative decays are represented in Figure 45*b* and *c*, with gluons replaced by photons, as in positronium annihilation. The rate for each of these decay processes is proportional to $|\psi(0)|^2$, the square of the $c\bar{c}$ bound-state wave function at the origin, while the hadronic rates also depend upon α_s, the "running" gluon coupling constant. Radiative decay rates transpose directly to hadronic rates with the replacement of α by α_s. Formulas for these decay widths are (22)

$$\Gamma(\psi \to e^+ e^-) = 4(Q_c \alpha)^2 \frac{1}{m_c^2} |\psi(0)|^2$$

$$\Gamma(\psi \to \text{hadrons}) = \frac{16}{9\pi}(\pi^2 - 9) \frac{5}{18} \alpha_s^3 \frac{1}{m_c^2} |\psi(0)|^2,$$

where Q_c is the charge of the *c*-quark and m_c its mass. The ratio of these two partial widths determines the coupling constant $\alpha_s \simeq 0.2$ for $q^2 = (3095)^2$. Only the small variation in α_s between 3095 and 3684 MeV distinguishes the ratio $\Gamma[\psi(3684) \to e^+ e^-]/\Gamma[\psi(3684) \to \text{hadrons}]$ from $\Gamma[\psi(3095) \to e^+ e^-]/\Gamma[\psi(3095) \to \text{hadrons}]$. Neglecting the energy dependence of α_s, we obtain

$$\Gamma[\psi(3684) \to \text{hadrons}] = 2.1/4.8 \times 47 = 21 \text{ keV.}$$

The measured branching fractions are $\sim 53\%$ for $\psi\pi\pi + \psi\eta$, $\sim 3\%$ second-order electromagnetic, $\sim 30\%$ for $\gamma\psi$, and $\sim 2\%$ for lepton pairs, totaling 88%, so that $\Gamma[\psi(3684) \to \text{hadrons}] \simeq 27$ keV, perhaps fortuitously in rather good agreement with the prediction.

6.1.3 TRANSITIONS BETWEEN LEVELS   Radiative transition rates provide another test of the bound-state model. Transitions from $\psi(3684)$, the 2^3S_1 level, to the 2^3P_2 states should be dominantly electric dipole in nature. With that presumption, the ratios of partial widths should be simply related to the transition energies

$$\Gamma[\psi(3684) \to {}^3P_2]/\Gamma[\psi(3684) \to {}^3P_1]/\Gamma[\psi(3684) \to {}^3P_0] \simeq 5k_2^3/3k_1^3/k_0^3,$$

where the k's are the photon momenta and the numerical factors are the statistical weights of the P states. The measured photon momenta give ratios 1.0/1.3/1.5, to be compared with the ratios of measured rates 1/1.1/1.25, the latter ratios with substantial errors. Early calculations with potential models, e.g. those of Eichten et al (94), produced a total width $\Gamma[\psi(3684) \to \gamma\chi] = 215$ keV for transitions to the three ^3P states, accounting for a major fraction of $\psi(3684)$ and in gross disagreement with the facts. Incorrect calculated values were used for the transition energies, however. Using the observed rather than the calculated values of photon momenta, we obtain the corrected values $\Gamma[\psi(3684) \to \gamma\chi_2] \simeq 23$ keV, $\Gamma[\psi(3684) \to \gamma\chi_1] \simeq 28$ keV, $\Gamma[\psi(3684) \to \gamma\chi_0] \simeq 36$ keV, which are in rather more acceptable agreement with observation. Later estimates, obtained with a modified calculation by Eichten et al (98), are also in satisfactory agreement with the data. It is not clear to what extent the improvement results from the more sophisticated, extended model, since experimental values of transition energies were used as input to the calculation. It appears that the $\psi(3684) \to \chi$ radiative transition rates can be accommodated within the context of the bound-state charmonium model and indeed, provide some support for the chosen spectroscopic classification.

Further predictions await definitive experimental confrontation. Among them we cite the rates for radiative transition from the ^3P states to $^3S_1$ ground state (98), $\Gamma(^3P \to {}^3S) \simeq 100-300$ keV; magnetic-dipole transition partial width (94) $\Gamma(2^3S_1 \to 1^1S_0) \simeq 1$ keV; ^3P-state hadronic decay widths (99) $\Gamma(^3P_2 \to \text{hadrons})/\Gamma(^3P_0 \to \text{hadrons}) = 4/15$, obtained with a two-gluon annihilation process; and $\Gamma(^3P_1 \to \text{hadrons})/\Gamma(^3P_2 \to \text{hadrons}) \simeq 1/4$, gotten from the dominant gluon plus quark-pair annihilation (100).

6.2 Charmed-Particle Decays

It is necessary to depart from a strict delineation of the subject matter to consider properties of those ingredients crucial to the expanded hadronic substructure, states with exposed charm. Glashow, Iliopoulis & Maiani (3) recognized that the fourth quark, introduced to bring a symmetry to the fundamental entities of the weak interactions, provided the extra degree of freedom necessary for the natural proscription of strangeness-changing neutral currents. In fact, the prescribed neutral hadronic weak current

$$J^0_{h_v} = \bar{u}\gamma_v(1+\gamma_5)u + \bar{c}\gamma_v(1+\gamma_5)c - \bar{d}\gamma_v(1+\gamma_5)d - \bar{s}\gamma_v(1+\gamma_5)s$$

Table 12 Selection rules for weak decays of charmed particles

Decay-amplitude θ_c dependence	Selection rules	Examples
$\cos \theta_c$	$\Delta S = \Delta C = \Delta Q = \pm 1, \lvert \Delta I \rvert = 0$	$D^0 \to K^- \mu^+ \nu$ $F^+ \to \mu^+ \nu$ $\Lambda_c^+ \to \Lambda^0 e^+ \nu$
$\sin \theta_c$	$\Delta S = 0, \Delta C = \Delta Q = \pm 1, \lvert \Delta I \rvert = \frac{1}{2}$	$D^+ \to \mu^+ \nu$ $F^+ \to K^0 \mu^+ \nu$ $\Sigma_c^{2+} \to p e^+ \nu$
$\cos^2 \theta_c$	$\Delta S = \Delta C = \pm 1, \lvert \Delta I \rvert = 1$	$D^0 \to K^- \pi^+$ $D^\pm \to K^\mp \pi^\pm \pi^\pm$ $F^+ \to K^+ K^- \pi^+$ $\Lambda_c^+ \to \Lambda^0 \pi^+ \pi^+ \pi^-$
$\cos \theta_c \sin \theta_c$	$\Delta S = 0, \Delta C = \pm 1, \lvert \Delta I \rvert = \frac{1}{2}, \frac{3}{2}$	$D^0 \to \pi^+ \pi^-$ $F^+ \to K^0 \pi^+$ $\Lambda_c^+ \to n \pi^+$
$\sin^2 \theta_c$	$\Delta S = -\Delta C = \pm 1, \lvert \Delta I \rvert = 0, 1$	$D^0 \to K^+ \pi^-$ $D^\pm \to K^\pm \pi^- \pi^+$ $F^+ \to K^+ K^+ \pi^-$ $\Xi_c^0 \to p \pi^-$

is charm- as well as strangeness-conserving. Decay selection rules, or rather, ordering of amplitude strengths, are implicit in the form of the charged hadronic current

$$J_{h_\nu} = \bar{u} \gamma_\nu (1 + \gamma_5)[d \cos \theta_c + s \sin \theta_c] + \bar{c} \gamma_\nu (1 + \gamma_5)[-d \sin \theta_c + s \cos \theta_c],$$

where θ_c is the Cabibbo rotation angle ($\simeq 0.23$ radians). Decays involving quark-constituent transformations $c \to s$ are favored because of the presence of $\cos \theta_c$ in their transition amplitudes. We list in Table 12 selection rules corresponding to the θ_c dependence of the decay amplitudes and some selected examples of charmed-particle decay modes that obey those rules. Meson decay rates have been estimated to be $\sim 10^9$ sec^{-1} for leptonic, $\sim 10^{12}$ for semileptonic, and $\sim 10^{13}$ sec^{-1} for hadronic modes (90). There do not exist reliable estimates for partial widths for decays to the various allowed final states of hadrons, although naive phase-space considerations lead to the expectation of relatively low average multiplicity. Such simple-minded arguments may require significant modification on consideration of the role of resonant states as decay products.

7   STATES HEAVIER THAN 4000 MeV/c^2

The LBL-SLAC group has reported results of detailed measurements of the $e^+ e^-$-annihilation cross section in the energy interval 3.9–4.6 GeV (101). Detection efficiency determined from a Monte Carlo simulation of the apparatus' response varied from 0.53 at the lowest energy to 0.57 at the highest. Possible systematic errors in absolute scale are estimated to be $\sim \pm 15\%$, the result of uncertainties in the efficiency calculation. Those errors should introduce smaller relative varia-

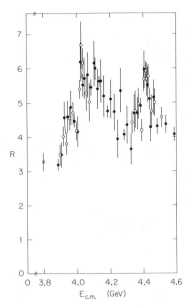

Figure 46   Energy dependence of the ratio $R = \sigma(e^+e^- \to \text{hadrons})/\sigma(e^+e^- \to \mu^+\mu^-)$ at c.m. energies between 3.8 and 4.6 GeV; from LBL-SLAC (101).

tions, less than 5%, among the values at the various energies. Figure 46 shows the data, plotted again as the ratio of hadronic-annihilation cross section to the muon-pair cross section, with statistical errors indicated. At the very least, two peaks are evident. It is, however, not possible to determine details of the structure in the variation of R with energy between 3.9 and 4.3 GeV. There appear to be a number of resonances; there may be thresholds for production of new particles. Interferences among resonant amplitudes may also account for some of the violent appearance of this substructure.

The data in the interval $\sim 4.3 - \sim 4.6$ GeV appear more consistent with the presence of a single, isolated resonant state, as seen in Figure 47. A fit was made

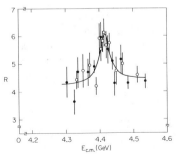

Figure 47   Resonance fit to the energy dependence of R near 4.4 GeV. See text for parameters of the state; from SLAC-LBL (101).

to a Breit-Wigner line shape, including radiative corrections, with a noninterfering background. Resonance parameters determined from the fit are shown in Table 3, together with those for $\psi(3095)$ and $\psi(3684)$, for comparison. The striking feature of this resonance is the magnitude of its full width, $\Gamma(4414) = 33 \pm 10$ MeV, more characteristic of strong decay than of the inhibited decay of $\psi(3095)$ and $\psi(3684)$. The electron width $\Gamma_{ee} = 0.44 \pm 0.14$ keV is one fifth that of $\psi(3684)$. Branching ratios to particular distinct hadronic final states have not as yet been determined.

A resonance at ~ 4.4 GeV is nicely qualitatively compatible with an interpretation as an anticipated higher excitation of the charmonium structure. Indeed, as discussed above, two $^3S_1$ states may be expected in the energy interval 3.9–4.5 GeV. The smaller lepton-pair partial width of $\psi(4414)$ needs to be accounted for with a decreased modulus-squared of the wave function at zero separation, $|\psi(0)|^2$, of the higher-energy bound eigenstate. Unfortunately, all $^3S_1$ $c\bar{c}$ eigenfunctions have the same value of $|\psi(0)|^2$, independent of principal quantum number, in the simplest model of a long-range binding potential, linearly rising with distance. Some "tuning" of the shape of the potential is required if the charmonium model is to give more than a qualitative description of the psionic states.

Two orders of magnitude increase in total width, compared to $\psi(3684)$, indicate that the OZI suppression mechanism has ceased to operate on $\psi(4414)$ decays. If the hidden charm interpretation of psions is to be maintained, it must be that this resonance rest energy is above the threshold for decay to pairs of particles bearing exposed charm.

8   CHARMED PARTICLES

Following some years of unsuccessful searches for charmed particles produced in hadron-hadron collisions (102) and in e^+e^- annihilations (103), all but conclusive evidence for their existence has been obtained by the LBL-SLAC group (104, 105), who observed decays of charmed mesons produced in e^+e^- annihilations, and by Knapp et al (106), who detected decays of photoproduced charmed baryons.

8.1   Mesons

In the LBL-SLAC experiment, evidence for $K\pi$ and $K\pi\pi\pi$ decay modes of charmed mesons was found in a study of 29,000 multiprong events produced in e^+e^- annihilations at c.m. energy between 3.90 and 4.60 GeV. The data are summarized succinctly in the effective-mass plots of Figure 48. A significant signal is evident near 1900 MeV/c^2 in the effective-mass distributions of all neutral combinations of two charged particles to which kaon mass and pion mass were arbitrarily assigned. Independent measurements of momentum and time of flight are made with the detection apparatus, but the timing resolution, 0.4 nsec, is insufficient to allow a distinction to be made between pions and kaons whose flight times in these events typically differ by 0.5 nsec. To make use of the time-of-flight information, the alternate mass assignments to each track were assigned relative weights appropriate to a Gaussian probability distribution with $\sigma = 0.4$ nsec:

$$W_i = \exp\left[-(t_i - t_T)^2/2\sigma^2\right],$$

where t_i is the flight time for mass m_i, determined from measured track parameters, and t_T is the directly measured time. The sum of track weights is made equal to unity and each mass combination, $\pi\pi$, πK, and KK, assigned a weight equal to the product of the track weights. With this method, the sum of weights assigned to all mass combinations just equals the total number of two-particle combinations in each event, so that there is no double counting. The second row of Figure 48 shows the invariant-mass spectra. A clear peak is evident in the $K^\pm\pi^\mp$ mass distribution, and the small residual signals in $\pi\pi$ and KK are understood to result from misidentification of true $K^\pm\pi^\mp$ events expected from inaccurate time-of-flight measurements. In the third row of Figure 48 are similarly weighted distributions, showing a peak in $K\pi\pi\pi$ at essentially the same mass as the $K\pi$ combinations. No significant signals were found in corresponding doubly charged two- or four-particle mass spectra.

Fits to the data with assumed Gaussian-shaped peaks above smoothly varying backgrounds yielded central values of mass 1870 and 1860 MeV/c^2 for $K\pi$ and $K\pi\pi\pi$, respectively. With consideration of the effects of systematic and random errors, it was concluded that both mass distributions are consistent with having a common source, decays of a particle of mass $M(D^0) = 1865 \pm 15$ MeV/c^2. The

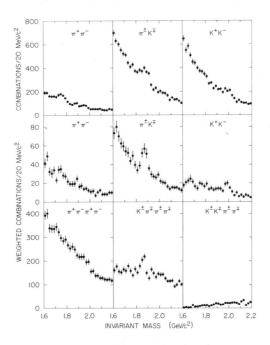

Figure 48 Inclusive neutral two- and four-particle invariant-mass spectra in multi-hadronic e^+e^- annihilation at c.m. energies 3.9–4.6 GeV. Top row, indiscriminate-mass assignments; middle row, mass assignments with time-of-flight weighting; bottom row, four-body mass assignments with time-of-flight weighting; from LBL-SLAC (104).

observed widths of the mass peaks are those expected from experimental resolution alone, 25 MeV/c^2 (rms) for the $K\pi$ system and 13 MeV/c^2 for the $K3\pi$ system, but a finite width of the state was not definitively excluded. A 90% confidence limit of 40 MeV/c^2 was deduced for the resonance decay width, based on quality of fits of a Breit-Wigner resonance shape convoluted with a Gaussian resolution function.

Peruzzi et al (105) report similar observations of peaks in effective-mass distribution of $K^-\pi^+\pi^+$ and $K^+\pi^-\pi^-$ combinations selected from multihadron e^+e^- annihilations at a c.m. energy of 4.03 GeV/c. Those data were obtained in a run at SPEAR with the LBL-SLAC magnetic detector, stimulated by the observation of the D^0 decays. Invariant mass spectra are shown in Figure 49 for three-body mass combinations, with $K\pi\pi$ mass assigned and each event weighted according to the prescription discussed above. Events with $\pi^+\pi^-$ pairs in the final state show no evidence of resonance peaks in their mass distributions, while a clear peak appears, at $M(D^\pm) = 1876 \pm 15$ MeV/c^2, in the exotic $K^-\pi^+\pi^+$ and $K^+\pi^-\pi^-$ states. That peak again has the appearance of resulting from decays of a resonance state whose width is a creature of merely experimental resolution. As with the neutral state, the 90% confidence-level upper limit to the width of the resonance is 40 MeV/c^2. Charged and neutral masses are equal within errors, but the measurements allow of a mass splitting of some 10 MeV/c^2, not atypical of splittings within isospin multiplets. It is natural and sensible then to identify the charged $K^\pm\pi^\mp\pi^\mp$ events as members of an isospin multiplet whose neutral partners are the D^0 mesons whose decays were revealed in $K^\pm\pi^\mp$ and $K^\pm\pi^\mp\pi^+\pi^-$ modes.

Charmed mesons must be produced in particle-antiparticle pairs in the annihilation of uncharmed positrons on electrons. Two pieces of experimental evidence

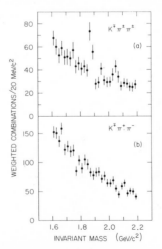

Figure 49 Inclusive mass spectra of charged three-particle combinations in multihadron products of e^+e^- annihilations at 4.03-GeV c.m. energy. Mass assigned according to time-of-flight weights in (a) exotic and (b) nonexotic systems; from SLAC-LBL (105).

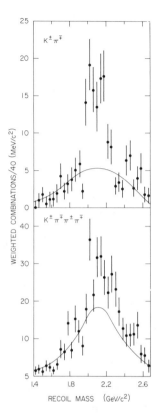

Figure 50   Distribution of masses accompanying $K^{\pm}\pi^{\mp}$ and $K^{\pm}\pi^{\mp}\pi^{+}\pi^{-}$ in events with $K\pi$, $K3\pi$-invariant masses in the peaks near 1865 MeV/c^2 seen in Figure 48. The smooth curves show background shapes obtained from events with $K\pi$ and $K3\pi$ masses adjacent to the peaks; from LBL-SLAC (104).

have been brought to bear on the question of the process by which the D^0 and $D^{\pm}$ mesons are produced. First, the observed spectra of masses recoiling against the $(K\pi)^0$, $(K3\pi)^0$, and $(K\pi\pi)^{\pm}$ systems with masses in the resonance region give evidence that the D mesons are produced in association with systems whose masses are greater than ~ 1870 MeV/c^2. Figures 50 and 51 show those recoil-mass spectra for the D^0 and $D^{\pm}$ peaks, respectively, both with estimated background indicated. All three spectra show evidence of a recoil system with well-defined mass near 2000 MeV/c^2. A second peak near 2200 MeV/c^2 appears in the spectra of masses accompanying the neutral $K\pi$ and $K3\pi$ combinations. Since there is little signal above background near 1870 MeV/c^2 in any of the recoil-mass distributions, $\bar{D}D$-associated production must be a minor contributing process. Although the data are hardly conclusive, the results suggest that some D^0's and $D^{\pm}$ whose decays are observed are produced in the cascade decay of a heavier, narrow-width object.

It had previously been suggested (90, 91) that the $^3S_1$ vector charmed meson D^* is only some 100 MeV/c^2 heavier than the $^1S_0$ pseudoscalar, D, charmed meson. It is natural then to interpret the observations as indicating that major sources of the observed D mesons are the cascade decay of D^* mesons produced in the associated production processes $e^+e^- \to D\bar{D}^*$, $\bar{D}^*D$ and $e^+e^- \to D^*\bar{D}^*$. Comparable rates for these two production reactions is implied by the data. In addition to these results, which may be said to be direct observation of associated production, the LBL-SLAC group has searched for such $(K\pi)^0$, $(K3\pi)^0$ and $(K\pi\pi)^\pm$ mass peaks among the debris of e^+e^- annihilations at the ψ resonant energies, 3095 and 3684 MeV. Some 150,000 ψ and 300,000 ψ' decays were examined. Of those, some 72,000 are in fact decays via second-order electromagnetic decays of $\psi(3095)$. Since no statistically significant peaks were found near 1870 MeV/c^2 in the mass spectra of these states, it may be taken as established that the threshold energy is more than 3095 MeV for production of D mesons in e^+e^- annihilation via the one-photon intermediate state. Cross-section branching-ratio products were estimated to be 0.20 ± 0.05 nb for $K\pi$, 0.7 ± 0.1 nb for $K3\pi$, and 0.3 ± 0.15 nb for charged-$K\pi\pi$ modes. Allowing for numerous alternative decay modes, comparison with the total annihilation cross section at 4.03 GeV, $\sigma \simeq 30$ nb, makes it clear that D-meson production accounts for a substantial fraction of all annihilations at this energy.

These LBL-SLAC results indicate the existence of an isospin multiplet of heavy mesons of mass ~ 1870 MeV/c^2, which are produced, not in pairs, but very likely as decay products of somewhat more massive particles produced in pairs, or in association with the heavier particles. The mesons have, most likely, small widths as measured against typical or expected hadronic widths of the order 100 MeV/c^2. The characteristics of the charged decays show extreme departures from ordinary hadronic-decay norms. Only the exotic $K^\pm \pi^\mp \pi^\mp$-particle combinations occur; the isospin of the source, if conserved, is at least $\frac{3}{2}$, but the isospin siblings with charge 2 or more do not appear. While this behavior would be exceptionally aberrant in ordinary hadrons, it precisely conforms to the rules for decays of charmed mesons

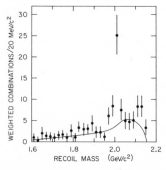

Figure 51   Spectrum of masses accompanying $K^\pm \pi^\mp \pi^\mp$ exotic combinations with mass in the peak near 1875 MeV/c^2 seen in Figure 49. The smooth curve is a background estimate based on the shape of the recoil-mass spectrum with nonexotic $K^\pm \pi^+ \pi^-$ combinations; from SLAC-LBL (105).

in the Glashow-Iliopoulos-Maiani form of weak interaction theory with a charmed quark included among the four fundamental hadronic components of nature. Specifically,

$$D^0(c\bar{u}) \rightarrow K^-(s\bar{u}) + \pi^+(u\bar{d})$$

$$D^0(c\bar{u}) \rightarrow K^-(s\bar{u}) + \pi^+(u\bar{d}) + \pi^-(\bar{u}d) + \pi^+(u\bar{d})$$

$$D^+(c\bar{d}) \rightarrow K^-(s\bar{u}) + \pi^+(u\bar{d}) + \pi^+(u\bar{d})$$

are favored decays, with rates proportional to $\cos^4\theta_c$, as are the corresponding antiparticle decays.

8.1.1 PARITY VIOLATION These already strong indications of the existence of the charmed quark are greatly reinforced by the observation that parity is not conserved in the decay of the charged D mesons. Unless very fundamental theoretical bases are to be overturned, parity violation is to be accepted as proof that decays of D mesons to hadrons proceed via weak interaction, violating the charm-conservation rule of the strong interaction. Wiss et al (107) have presented a discussion reminiscent of the τ-θ puzzle of 20 years ago, whose resolution came

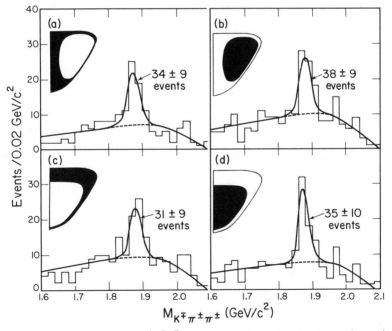

Figure 52 Invariant masses of $K^\pm\pi^\mp\pi^\mp$ combinations in multihadron states from e^+e^- annihilations at 3.9–4.25 GeV. Events restricted to those with recoil mass in the range 1.96–2.04 GeV/c^2. Distributions are for events that populate the dark regions of the Dalitz plots. Regions in (*a*) and (*b*) [(*c*) and (*d*)] are separated by contours of constant amplitude for decay of a 1^- [2^+] particle; from LBL-SLAC (107).

with the recognition of parity violation in the weak interaction. A sample of $K\pi\pi$ decays of $D^{\pm}$ was chosen from events produced in $e^{\pm}$ annihilations at c.m. energy between 3.9 and 4.25 GeV, restricted to those $K\pi\pi$ combinations accompanied by missing mass between 1.96 and 2.04 GeV, a procedure that produced 70 events in a narrow peak above a background of some 50 events. A Dalitz plot for those $K\pi\pi$ combinations with mass between 1860 and 1920 MeV/c^2 has density indistinguishable from uniform. Figure 52 shows the data divided into Dalitz-plot regions chosen to be equally populated by decays of 0^- mesons, but populated in the ratio $\sim 1/8$ for spin-parity 1^- and $\sim 1/6$ for spin-parity 2^+ meson decays. In both cases, the observed population ratio is unity to within errors; $J^P = 1^-$ is excluded to a confidence level 2×10^{-5}, 2^+ is ruled out with a confidence level of 0.002. Since three pseudoscalars cannot be in a zero-spin, even-parity state, only 0^- is a possible $D^{\pm}$ spin-parity assignment, unless the spin is more than two. However, the $K\pi$ state has natural spin parity. Presuming $D^{\pm}$ and D^0 to be members of an isospin multiplet and so to have the same parity, the observations show that the decay proceeds with comparable rates to states of opposite parity, and it follows that the decay interaction is parity-violating, another property of charmed mesons.

Presented with this assemblage of properties congruent to those of the predicted particles, it makes sense to end the reticence maintained by the experimentalists and recognize the discovery of charmed mesons.

8.2 Baryons

One other member of the charmed-particle family has been established with reasonable certainty, a charmed baryon. Knapp et al (106) have reported observation of a sharp peak in the mass spectrum of $\bar{\Lambda}\pi^+\pi^-\pi^-$ produced by photons interacting with beryllium. Some details of the experimental arrangement have been presented above in the discussion of ψ production by this group at Fermilab. Identification of Λ^0 and $\bar{\Lambda}^0$ decays is beyond question, as shown in the mass plots of Figure 53. Evidence of a negatively-charged narrow state decaying to $\bar{\Lambda}\pi^+\pi^-\pi^-$ is shown in the invariant-mass plots of Figure 54. Also shown there is the nonexistence of a positively charged partner, $\bar{\Lambda}\pi^+\pi^+\pi^-$. The state's mass was determined to be 2260 ± 10 MeV/c^2 and the measured full width of the peak is (40 ± 20) MeV/c^2. That width is consistent with the estimated experimental mass resolution of a zero-width state. Some evidence was found for a cascade decay of a heavier particle of mass ~ 2500 MeV/c^2 to the $\Lambda(3\pi)^-$ and a positive pion, as shown in Figure 55. There is a clear peak in the distribution of the difference in mass between the $\bar{\Lambda}(4\pi)^0$ combinations and those $\Lambda(3\pi)^-$ combinations with mass near the 2260 peak value. Those data indicate a cascade process

$$\bar{\Lambda}\pi^-\pi^-\pi^+\pi^+ (\sim 2.5 \text{ GeV}/c^2) \rightarrow \bar{\Lambda}\pi^-\pi^-\pi^+ (2.26 \text{ GeV}/c^2) + \pi^+.$$

A charmed antibaryon with charge -1 has quark components $\bar{c}\bar{u}\bar{d}$, i.e. one charmed antiquark and two uncharmed antiquarks. It is impossible to construct a charmed, positively charged antibaryon with a charmed antiquark and two ordinary antiquarks. It is natural to identify and classify the particle decaying to the

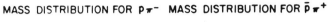

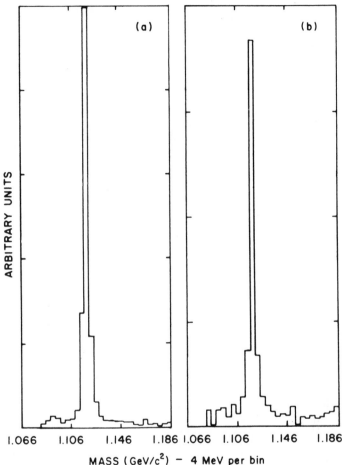

Figure 53   Invariant mass distributions of (*a*) $p\pi^-$ and (*b*) $\bar{p}\pi^+ V^0$'s in multihadron states photoproduced at ~ 250 GeV; from Knapp et al (106) at Fermilab.

$\Lambda\pi^+\pi^-\pi^-$ system at 2260 MeV/c^2 as the charmed antibaryon $\overline{\Lambda}_c(\overline{cud})$. Its higher-mass parent is then recognized as either $\overline{\Sigma}_c^0(\overline{cdd})$, a spin-$\frac{1}{2}$, $I = 1$ charmed baryon or $\overline{\Sigma}_c^{*0}(\overline{cdd})$ of spin $\frac{3}{2}$, isospin 1.

A single example of just such a production-decay sequence had previously been found among events initiated by neutrino interactions in a hydrogen bubble chamber (108). The event, shown in Figure 56, is interpreted as an example of the reaction $\nu p \rightarrow \mu^- \Lambda^0\pi^+\pi^+\pi^+\pi^-$. With those assignments of particle identities, the mass of the $(\Lambda 4\pi)^{2+}$ system is calculated to be 2426 ± 12 MeV/c^2 and the three possible

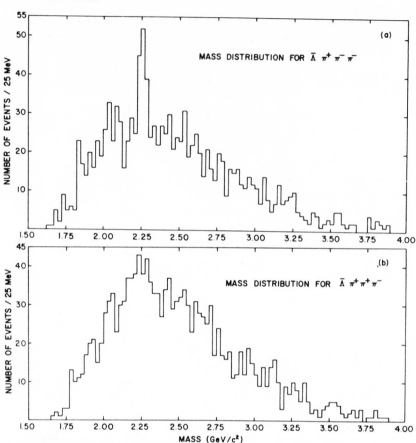

Figure 54 Distribution of invariant masses for (*a*) $\overline{\Lambda}(3\pi)^-$ and (*b*) $\overline{\Lambda}(3\pi)^+$ combinations in multihadron states photoproduced at ~250 GeV; from Knapp et al (106).

$(\Lambda\pi^+\pi^+\pi^-)$ combinations are found to be smaller in mass by 166 ± 15 MeV/c^2, 338 ± 12 MeV/c^2, and 327 ± 12 MeV/c^2. Noting that this reaction violates the selection rule $\Delta S = \Delta Q$ for ordinary hadrons, it was suggested that this event is an example of the sequential reactions, beginning with weak production of a charmed baryon

$$vp \to \Sigma_c^{2+} (cuu)\mu^- \left[\text{or } \Sigma_c^{*2+}(cuu)\mu^- \right],$$

Figure 55 Evidence for the cascade process $\overline{\Lambda}(4\pi)^0 \to \overline{\Lambda}(3\pi)^-\pi^+$ in multihadron states $\rightarrow$ photoproduced at 250 GeV. (*a*) $\overline{\Lambda}(4\pi)^0$ invariant-mass spectrum; (*b*) $\overline{\Lambda}(3\pi)^-$ invariant-mass spectrum; (*c*) difference in mass between $\overline{\Lambda}(4\pi)^0$ and $\overline{\Lambda}(3\pi)^-$ combinations shown as the solid histogram for $\overline{\Lambda}(3\pi)^-$ masses in the peak at ~2.25 GeV/c^2 and as the dashed histogram for $\overline{\Lambda}(3\pi)^-$ just outside the peak; from Knapp et al (106).

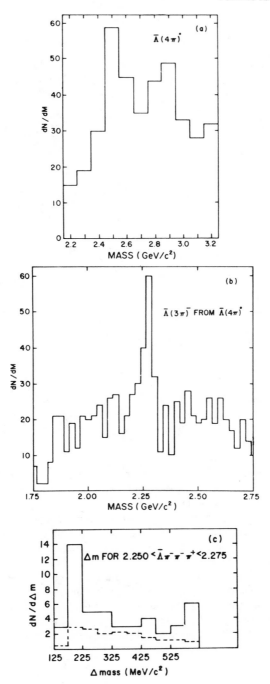

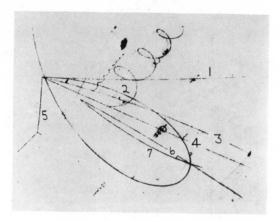

Figure 56   View of an event interpreted as an example of $vp \to \mu^- \Lambda^0 \pi^+ \pi^+ \pi^+ \pi^-$ produced in the BNL liquid-hydrogen bubble chamber. The Λ^0 is particularly conspicuous as tracks 6 and 7; observation of Cazzoli et al (108) at Brookhaven National Laboratory.

followed by strong decay

$$\Sigma_c^{2+} \ (\text{or} \ \Sigma_c^{*2+}) \to \Lambda_c^+ (cud) \pi^+,$$

and the succeeding, favored weak decay

$$\Lambda_c^+ \to \Lambda^0 \pi^+ \pi^+ \pi^-.$$

The notation here is the same as above, indicating quark contents and spins. Particularly noteworthy are the remarkable coincidences between the masses of these objects and those predicted by De Rujula et al (91), i.e. 2250 MeV/c^2 for Λ_c, 2410 MeV/c^2 for Σ_c, and 2480 MeV/c^2 for Σ_c^*.

8.3 Miscellany

Other observations have provided more indirect evidence of the presence of charmed-quark components of hadrons. Events with two muons produced in high-energy neutrino-nucleon interactions have been interpreted in terms of production of new particles and their subsequent weak leptonic or semileptonic decays (109, 110). From the measured characteristics of the events, Benvenuti et al (109) deduce that the new particle, if a hadron, has mass between ~ 2 and ~ 4 GeV/c^2 and has a lifetime less than 10^{-10} sec.

Events with a negative muon, positron, and neutral strange particle in the final state have been observed as products of neutrino-initiated reactions in bubble chambers. Von Krogh et al (111) found four examples of the reaction $v_\mu N \to \mu^- e^+ K_s^0 X$ in an exposure of the Fermi National Accelerator Laboratory FNAL 15-ft bubble chamber filled with a neon-hydrogen mixture. Blietschau et al (112) have reported observation of three examples of $v_\mu N \to \mu^- e^+ V^0 (K^0 \ \text{or} \ \Lambda^0) X$ in the freon-filled bubble chamber Gargamelle exposed to a neutrino beam at CERN.

One may surmise (both groups have) that these events show cascade production and semileptonic decay of charmed hadrons of mass about 2000 MeV/c^2.

Semileptonic decays of charmed hadrons have been invoked to explain the single-electron-plus-hadrons final states made in e^+e^- annihilations at c.m. energies between 4.0 and 4.2 GeV and detected in the DASP apparatus (113). For improved electron detection in this experiment, the apparatus of Figure 15 was supplemented with threshold Cerenkov counters on each side of the intersection region at DORIS. Of 87 electron + hadron events, 28 survived cuts on shower-counter pulse height, electron momentum, and vertex position. Of these, 22 had hadron multiplicity of four or greater. Calculated rates for conventional sources of these events are too low to account for any but a small fraction of those detected. In data taken at c.m. energies less than 3.7 GeV, events of this type did not occur with greater than calculated background rates. The electron-momentum distribution reproduced in Figure 57 disagrees with that expected of electron-decay products of a heavy lepton of mass 1.9 GeV/c^2, a conceivable source of the events. The posited new hadrons whose semileptonic decays are presumed to have been observed are to have mass between 1.8 and 2.1 GeV/c^2 and to be supplied with a new quantum number (charm, perhaps) conserved in strong and electromagnetic processes.

Rates for e^+e^- annihilations to states with a neutral K meson and an electron have been measured as a function of c.m. energy by Burmester et al (114) with the magnetic detector PLUTO at DORIS. That apparatus consists of fourteen cylin-

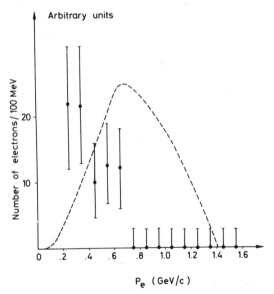

Figure 57   Inclusive momentum distribution of electrons in multiparticle e^+e^- annihilations at energies of 4.0–4.2 GeV. Events with multiplicity of four or greater are included. The dashed smooth curve shows a calculated distribution of electrons from decay of a 1.9-GeV/c^2 heavy lepton; from DASP (113).

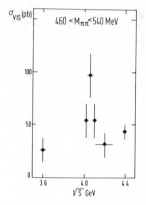

Figure 58 Energy dependence of the yield of detected correlated $K_s^0 e^{\pm}$ combinations in multiparticle states made in e^+e^- annihilations; from PLUTO (114) at DORIS.

drical proportional wire chambers arrayed coaxially about the direction of the intersecting beams. Two cylindrical lead converters, 0.44 and 1.7 radiation lengths thick, with following wire chambers provide for detection of electron-initiated electromagnetic showers. A super conducting coil surrounding the track-sampling chambers produces a 20,000-G magnetic field directed parallel to the axis of the detector array. Clear evidence for K^0 production was found in distributions of invariant mass of oppositely charged tracks in multiparticle events. Figure 58 shows the observed energy dependence of the detected yield of $e^{\pm} K_s^0 (\to \pi^+ \pi^-)$ events, the K_s^{0}'s having been chosen as $\pi^+ \pi^-$ pairs with mass in the observed peaks between 460 and 540 MeV/c^2. After background subtraction and correction for deficiencies, Burmester et al arrive at an estimated inclusive cross section at the maximum, $\sigma(e^+ e^- \to e^{\pm} K^0 + \text{anything}) \simeq 3$ nb, accurate to about a factor of 2.

Table 13 Properties of observed narrow states with classifications as charmed hadrons

State	Mass (MeV/c^2)	Decay modes	Ref.	Footnote(s)
$D^0, \bar{D}^0$	1865 ± 15	$K^{\pm}\pi^{\mp}, K^{\pm}\pi^{\mp}\pi^+\pi^-$	104	a
$D^{\pm}$	1876 ± 15	$K^{\pm}\pi^{\mp}\pi^{\mp}$	105	a
Λ_c^-	2260 ± 10	$\bar{\Lambda}\pi^-\pi^-\pi^+$	106	—
Λ_c^+	2260 ± 15	$\Lambda\pi^+\pi^+\pi^-$	108	b, c
$\bar{\Sigma}_c^0$	~ 2500	$\bar{\Lambda}_c^-\pi^+$	106	d
Σ_c^{2+}	2426 ± 12	$\Lambda_c^+\pi^+$	108	b, d

ª Presuming these to be isospin partners, parity violation in their decays has been demonstrated (107).
ᵇ One event with a cascade decay.
ᶜ Mass not uniquely determined. One of three possible $\Lambda\pi^+\pi^+\pi^-$ from decay of the Σ_c^{2+} (see text for others).
ᵈ Not uniquely classified. May be excited state Σ_c^*.

These data are proffered as evidence for the presence of charmed mesons whose semileptonic decays, $D^{\pm} \to K^0 e^{\pm} v$ have been detected. Boldly accepting that interpretation, and presuming ~ 10 nb (of the 30 nb total) to be the cross section for production of charged D-meson pairs in $e^+ e^-$ annihilations at 4.05 GeV, we may make the rough estimate $\sim (3/2)/10 = 15\%$, for the Kev decay branching ratio. That value is not incompatible with crude theoretical estimates (90) of charmed-meson decay rates.

8.4 Summary

New mesons and baryons of long lifetime have been observed. Their currently established masses, modes of production, and decay characteristics, as listed in Table 13, correspond without exception to those of charmed hadrons manifesting a basic structure that includes the fourth, charmed quark.

9 SUMMARY AND CONCLUSIONS

In somewhat less than two years of intense experimental activity, at least 13 new unstable hadrons have been discovered. They fall into two separate but related classes. One, the psion family, includes, first, the new vector mesons $\psi(3095)$, $\psi(3684)$, and $\psi(4414)$; second, the intermediate states $\chi(3415)$, $\chi(3500)$, and $\chi(3550)$; and third, the less well-established $\chi(3455)$ and $X(2800)$. The more recent assembly of new states includes the neutral and charged D mesons, D^0 (and presumably $\bar{D}^0$) at 1865 MeV/c^2 and D^+, D^- at 1876 MeV/c^2, together with the first new baryons, Λ_c at 2260 MeV/c^2 and Σ_c at 2450 MeV/c^2. All the characteristics of these new particles, their quantum numbers, the pattern of masses, and the variety of decay modes and widths, are in close correspondence with those of states that occur in a four-quark model of hadronic substructure.

The body of experimental evidence presented in this review strongly supports the notion that this small number of fundamental objects is responsible for the structure and interactions of the hadrons; however, there are gaps and quantitative discrepancies. Charmonium spectroscopy is still in a relatively primitive state; the pseudoscalar $^1S_0$ states have only a tenuous connection with reality; properties of the intermediate states are not sufficiently well-determined to permit definitive identification as ^3P levels; radiative transition rates are poorly known and not in good agreement with theoretical values; the structure of the higher radially excited ^3S states is very confused and their decay rates, particularly to states with charmed mesons, are essentially unknown. Only a small fraction of the charmed hadrons required by the model has yet been discovered. One may, in fact, also still feel uneasy about the value of R for $e^+ e^-$ annihilations at energies above the charmed-quark threshold.

The new discoveries have provided some basis for optimism that a correct basic theoretical framework has been invented. It remains to be seen, and no doubt will be seen in experiments in the near future, whether that structure is complete, or, at least, needs expansion to include still more, heavier quarks.

ACKNOWLEDGMENT

Much of what I know of the subject of this review has been learned from my colleagues, past and present, in the SLAC-LBL collaboration. They are: G. S. Abrams, M. S. Alam, J.-E. Augustin, A. M. Boyarski, M. Breidenbach, D. D. Briggs, F. Bulos, W. C. Carithers, S. Cooper, J. T. Dakin, R. G. DeVoe, J. M. Dorfan, G. E. Fischer, C. E. Friedberg, D. Fryberger, G. Goldhaber, D. Hitlin, G. Hanson, R. J. Hollebeek, B. Jean-Marie, J. Jaros, A. D. Johnson, J. A. Kadyk, R. R. Larsen, A. M. Litke, D. Lüke, B. A. Lulu, V. Lüth, H. L. Lynch, D. Lyon, R. J. Madaras, C. C. Morehouse, H. K. Nguyen, J. M. Paterson, M. L. Perl, F. M. Pierre, I. Peruzzi, M. Piccolo, T. Pun, P. Rapidis, B. Richter, B. Sadoulet, R. Schindler, R. F. Schwitters, J. Siegrist, W. M. Tanenbaum, G. H. Trilling, F. Vannucci, J. S. Whitaker, F. C. Winkelmann, J. Weiss, J. E. Wiss, and J. E. Zipse. I am greatly indebted to them for their efforts.

Particular thanks go to S. Cooper, M. Mandelkern, S. Shannon, M. Suzuki, and J. E. Wiss for corrections and clarifications that resulted from their careful reading of the text.

I am grateful for the hospitality and support of the Conselho Nacional de Pesquisas and the Universidade do Brasilia, Brazil where part of this review was prepared while I was a visiting professor in the Department of Physics.

Literature Cited

1. Particle Data Group, Trippe, T. G. et al 1976. Review of particle properties. *Rev. Mod. Phys.* 48:S1
2. Bjorken, J. D., Glashow, S. L. 1965. *Phys. Lett.* 11:225
3. Glashow, S. L., Iliopoulos, J., Maiani, L. 1970. *Phys. Rev. D* 2:1285
4. Carlson, C. E., Freund, P. G. O. 1972. *Phys. Lett. B* 39:349
5. Appelquist, T., Politzer, H. D. 1975. *Phys. Rev. Lett.* 34:43
6. Schwitters, R. F., Strauch, K. 1976. The physics of e^+e^- collisions. *Ann. Rev. Nucl. Sci.* 26:89
7. Cosme, G. et al 1974. *Phys. Lett. B* 48:159
8. Sakurai, J. J. 1969. *Currents and Mesons.* Chicago: Univ. Chicago Press
9. Nambu, Y., Sakurai, J. J. 1962. *Phys. Rev. Lett.* 8:79
10. High-Energy Reactions Analysis Group, Bracci, E. et al 1972. Compilation of cross sections $I - \pi^-$ and π^+ induced reactions. *CERN/HERA 72-1*, CERN, Geneva, Switzerland
11. Whitmore, J. 1976. *Phys. Rep. C* 27:188
12. Blobel, V. et al 1975. *Phys. Lett. B* 59:88
13. Anderson, K. J. et al 1976. *Phys. Rev. Lett.* 37:799
14. Silverman, A. 1975. *Proc. 1975 Int. Symp. Lepton Photon Interactions High Energies*, Stanford, Calif., p. 355
15. Ross, M., Stodolsky, L. 1966. *Phys. Rev.* 149:1172
16. Anderson, R. et al 1970. *Phys. Rev. D* 1:27
17. Behrend, H. J. et al 1975. *Phys. Lett. B* 56:408
18. Okubo, S. 1963. *Phys. Lett.* 5:165
19. Zweig, G. 1964. *CERN TH401, CERN TH412*, CERN, Geneva, Switzerland
20. Iizuka, J. 1966. *Suppl. Prog. Theor. Phys.* 37–38:21
21. Rosenzweig, C. 1976. *Phys. Rev. D* 13:3080
22. Appelquist, T., Politzer, H. D. 1975. *Phys. Rev. D* 12:1404
23. Aubert, J. J. et al 1974. *Phys. Rev. Lett.* 33:1404
24. Augustin, J.-E. et al 1974. *Phys. Rev. Lett.* 33:1406
25. Abrams, G. S. et al 1974. *Phys. Rev. Lett.* 33:1453
26. Bacci, C. et al 1974. *Phys. Rev. Lett.* 33:1408
27. Criegee, L. et al 1975. *Phys. Lett. B* 53:489
28. Becker, U. 1975. *New Directions in Hadron Spectroscopy, Proc. Summer*

Symp. Argonne Natl. Lab., July 7–10, 1975. Argonne, Ill., p. 209

29. Ting, S. C. C. 1975. See Ref. 14, p. 155
30. Aubert, J. J. et al 1975. Nucl. Phys. B 89:1
31. Knapp, B. et al 1975. Phys. Rev. Lett. 34:1044
32. Binkley, M. et al 1976. Phys. Rev. Lett. 37:571
33. Binkley, M. et al 1976. Phys. Rev. Lett. 37:574
34. Anderson, K. J. et al 1976. Phys. Rev. Lett. 36:237
35. Anderson, K. J. et al 1976. Phys. Rev. Lett. 37:799
36. Blanar, G. J. et al 1975. Phys. Rev. Lett. 35:346
37. Antipov, Y. M. et al 1976. Phys. Lett. B 60:309
38. Hom, D. C. et al 1976. Phys. Rev. Lett. 36:1236
39. Snyder, H. D. et al 1976. Phys. Rev. Lett. 36:1415
40. Hom, D. C. et al 1976. Phys. Rev. Lett. 37:1374
41. Büsser, F. W. et al 1975. Phys. Lett. B 56:482
42. Nagy, E. et al 1975. Phys. Lett. B 60:96
43. Boyarski, A. M. et al. 1975. Phys. Rev. Lett. 34:1357
44. Knapp, B. et al 1975. Phys. Rev. Lett. 34:1040
45. Camerini, U. et al 1975. Phys. Rev. Lett. 35:483
46. Gittelman, B. et al 1975. Phys. Rev. Lett. 35:1616
47. Nash, T. et al 1976. Phys. Rev. Lett. 36:1233
48. Ritson, D. M. 1976. Particle Searches and Discoveries—1976, Vanderbilt. AIP Conf. Proc. No. 30, Part. Fields Subser. No. 11, p. 75
49. Augustin, J.-E. et al 1975. Phys. Rev. Lett. 34:233
50. Wiik, B. H. 1975. See Ref. 14, p. 69
51. Lüth, V. et al 1975. Phys. Rev. Lett. 35:1124
52. Augustin, J.-E. et al 1975. Phys. Rev. Lett. 34:764
53. Jackson, J. D., Scharre, D. L. 1975. Nucl. Instrum. Methods 128:13
54. Braunschweig, W. et al 1976. Phys. Lett B 63:487
55. Jean-Marie, B. et al 1976. Phys. Rev. Lett. 36:291
56. Abrams, G. S. 1975. See Ref. 14, p. 25
57. Vannucci, F. et al 1977. Phys. Rev. D 15:1814
58. Gilman, F. J. 1975. High Energy Physics and Nuclear Structure, Los Alamos, 9–13 June 1975. AIP Conf. Proc. No. 26,

p. 331
59. Gupta, V., Kögerler, R. 1975. Phys. Lett. B 56:473
60. Abrams, G. S. et al 1975. Phys. Rev. Lett. 34:1181
61. Tanenbaum, W. et al 1976. Phys. Rev. Lett. 36:402
62. Hilger, E. et al 1975. Phys. Rev. Lett. 35:625
63. Tanenbaum, W. et al 1975. Phys. Rev. Lett. 35:1323
64. Braunschweig. W. et al 1975. Phys. Lett. B 57:407
65. Whitaker, J. S. et al 1976. Phys. Rev. Lett. 37:1596
66. Feldman, G. et al 1975. Phys. Rev. Lett. 35:821
67. Trilling, G. H. 1976. Lawrence Berkeley Lab. Rep. LBL-5535
68. Biddick, C. J. et al 1977. Phys. Rev. Lett. 38:1324
69. Hughes, E. B. et al 1976. Phys. Rev. Lett. 36:76
70. Brown, L. S., Cahn, R. N. 1976. Phys. Rev. D 13:1195
71. Karl, G. et al 1976. Phys. Rev. D 13:1203
72. Kabir, P. K., Hey, A. J. G. 1976. Phys. Rev. D 13:3161
73. Heintze, J. 1975. See Ref. 14, p. 97
74. Bartel, W. et al 1976. DESY 76/65, DESY, Hamburg, Germany
75. Boyarski, A. M. et al 1975. Phys. Rev. Lett. 34:762
76. Schwitters, R. F. 1975. See Ref. 14, p. 5
77. Esposito, B. et al 1975. Phys. Lett. B 58:478
78. Bacci, C. et al 1975. Phys. Lett. B 58:481
79. Bacci, C. et al 1976. Phys. Lett. B 64:356
80. Barbiellini, G. et al 1976. Phys. Lett. B 64:359
81. Esposito, B. et al 1976. Phys. Lett. B 64:362
82. Tarjanne, P., Teplitz, V. L. 1963. Phys. Rev. Lett. 11:447
83. Amati, D. et al 1964. Phys. Lett. 11:190
84. Hara, Y. 1964. Phys. Rev. B 134:701
85. Maki, Z. et al 1964. Prog. Theor. Phys. Kyoto 32:144
86. Greenberg, O. W. 1964. Phys. Rev. Lett. 13:598
87. Han, M. Y., Nambu, Y. 1965. Phys. Rev. B 139:1006
88. Bars, I., Peccei, R. D. 1975. Phys. Rev. D 12:823
89. Feldman, G., Matthews, P. T. 1976. Nuovo Cimento A 31:447
90. Gaillard, M. K., Lee, B. W., Rosner, J. L. 1975. Rev. Mod. Phys. 47:277

91. De Rujula, A., Georgi, H., Glashow, S. L. 1975. *Phys. Rev. D* 12:147
92. Chanowitz, M., Gilman, F. J. 1976. *Phys. Lett. B* 63:178
93. Bethe, H. A., Salpeter, E. 1957. *Quantum Mechanics of One and Two Electron Atoms.* Berlin: Springer
94. Eichten, E. et al 1975. *Phys. Rev. Lett.* 34:369
95. Pumplin, J. et al 1975. *Phys. Rev. Lett.* 35:1538
96. Schnitzer, H. J. 1975. *Phys. Rev. Lett.* 35:1540
97. Harrington, B. J. et al 1975. *Phys. Rev. Lett.* 34:168
98. Eichten, E. et al 1976. *Phys. Rev. Lett.* 36:500
99. Barbieri, R. et al 1976. *Phys. Lett. B* 60:183
100. Barbieri, R. et al 1976. *Phys. Lett. B* 61:465
101. Siegrist, J. et al 1976. *Phys. Rev. Lett.* 36:700
102. Cester, R. et al 1976. *Phys. Rev. Lett.* 37:1178
103. Boyarski, A. M. et al 1975. *Phys. Rev. Lett.* 35:196
104. Goldhaber, G. et al 1976. *Phys. Rev. Lett.* 37:255
105. Peruzzi, I. et al 1976. *Phys. Rev. Lett.* 37:569
106. Knapp, B. et al 1976. *Phys. Rev. Lett.* 37:882
107. Wiss, J. et al 1976. *Phys. Rev. Lett.* 37:1531
108. Cazzoli, E. G. et al 1975. *Phys. Rev. Lett.* 34:1125
109. Benvenuti, A. et al 1975. *Phys. Rev. Lett.* 34:419
110. Barish, B. C. et al 1976. *Phys. Rev. Lett.* 36:939
111. Von Krogh, J. et al 1976. *Phys. Rev. Lett.* 36:710
112. Blietschau, J. et al 1976. *Phys. Lett. B* 60:207
113. Braunschweig, W. et al 1976. *Phys. Lett. B* 63:471
114. Burmester, J. et al 1976. *Phys. Lett. B* 64:369

Ann. Rev. Nucl. Sci. 1977. 27:465–547
Copyright © 1977 by Annual Reviews Inc. All rights reserved

DAMPED HEAVY-ION COLLISIONS[1]
      ×5590

W. U. Schröder[2] and J. R. Huizenga

Departments of Chemistry and Physics and Nuclear Structure Research Laboratory,[3]
University of Rochester, Rochester, New York 14627

CONTENTS

[1] This work was supported by the US Energy Research and Development Administration.
[2] Supported in part by a grant from the German Academic Exchange Service (Deutscher Akademischer Austausch-Dienst).
[3] Supported by the National Science Foundation.

465

1 INTRODUCTION

In this review we discuss the mechanisms of reactions between very heavy ions, with emphasis on a recently discovered process known by the various names of deep inelastic transfer, quasi-fission, strongly damped collisions and relaxed processes. Apart from the very early pioneering experiment (1) and a few studies in the early 1970s (2, 3), this new reaction process began to be clearly recognized in 1973–1974 (4–8). The characteristic features (4–37) of this new reaction mechanism for heavy ions are the following.

1. Damping of the initial relative kinetic energy of the target and projectile nuclei takes place, resulting in a range of binary-product kinetic energies down to the Coulomb energies for charge centers of highly deformed fragments. The events that are the most relaxed are reminiscent of nuclear fission. What fraction of the kinetic-energy loss initially goes into internal excitation energy rather than into collective degrees of freedom is an open question of current interest.
2. Nucleon transfer or diffusion occurs during the short time the two nuclei are in contact; and the magnitude of the nucleon exchange is correlated with the kinetic-energy dissipation (28). However, the reaction-product mass distributions are bimodal with centroids near the target and projectile masses.
3. The angular distributions for the products with masses in the vicinity of the projectile mass have features characteristic of a fast peripheral or direct reaction process occurring on a time scale of approximately 10^{-21} sec or less. Products near the projectile mass produced by reactions for which the relative incident energy above the Coulomb barrier, i.e. $(E_{c.m.} - E_{Coul})/E_{Coul}$, is small tend to have differential cross sections that are preferentially sideways-peaked at angles slightly forward of $\theta_{1/4}$ (the angle where the elastic-to-Rutherford cross-section ratio is 0.25). As the above parameter takes on larger values, a larger percentage of the fragments are emitted at forward angles, and some fragments may display orbiting (emission of the fragments at negative angles). The first angle-energy correlation plot for a heavy-ion reaction was made by Wilczyński (38).

In order to place this new reaction process in perspective, an overall classification of heavy-ion nuclear reactions (39a, 40), as shown in Figure 1, is discussed. The various terms in this figure, such as *distant* and *touching* collisions are to be understood in the context of a matter density of nuclei with diffuse surfaces. The radial dependence of the nuclear density ρ is approximated by a two-parameter Fermi distribution function,

$$\rho = \rho_0/\{1 + \exp[(r - C)/a]\}, \qquad\qquad 1.$$

where C is the nuclear matter half-density radius, a is a measure of the surface diffuseness, and ρ_0 is the central density. Distant collisions occur when the extreme tails of the nuclear density of each nucleus overlap, where the centers of the nuclei are separated by a distance greater than $C_T + C_P + S$ [where S varies slightly with ion charge or mass (see Section 3.6) and $S = \zeta(R_{SA})$]. At these large distances, only electromagnetic interactions, slightly modified by the extreme tail of the nuclear potential, occur. Touching or grazing collisions take place at nuclear separation distances equal to or less than $C_T + C_P + S$. At these distances the attractive nuclear potential is sufficiently strong (relative to the repulsive Coulomb potential) to lead to the onset of nuclear interaction. The flux of touching collisions comprises the total reaction cross section and is subdivided in Figure 1 into a number of continuously evolving reaction processes. As the penetration depth of the two ions increases, the energy damping and mass transfer, as well as the reaction time, increase.

The touching collisions for light-ion reactions at low and medium energies undergo both grazing (larger impact parameters) and solid-contact (smaller impact parameters) collisions. The latter collisions have a high probability for fusion,

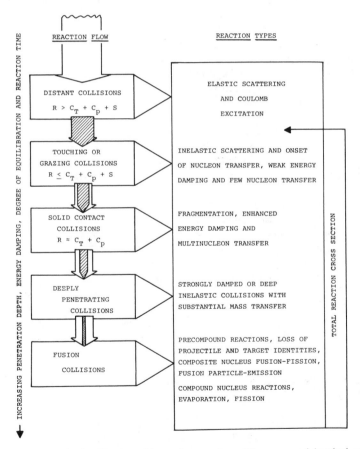

Figure 1   Schematic classification of heavy-ion reactions. The open and hatched arrows depict the reaction flow for light- and heavy-ion reactions, respectively.

which in turn may lead to trapping and compound-nucleus formation, although to some fraction, only a composite nucleus with partial equilibrium (preequilibrium decay) is formed. Hence, for light-ion reactions, the compound-nucleus mechanism dominates (as illustrated by the open arrows in Figure 1, a major part of the reaction flow proceeds to complete damping of the projectile-target system and equilibration of all degrees of freedom prior to decay).

The reaction flow (as illustrated in Figure 1) for very-heavy-ion reactions is remarkably different from that for light-ion reactions. Little, if any, cross section goes into the compound-nucleus process (fully equilibrated fusion) for very-heavy-ion reactions. Instead, a new reaction process occurs that comprises a dominant fraction of the total reaction cross section and has the striking properties enumerated previously. In terms of Figure 1, these very-heavy-ion reaction processes occur at penetration depths prior to the complete loss of projectile and target identities. The reaction processes observed correspond to a wide range of energy damping and mass transfer, depending on the target-projectile combination and the bombarding energy. It is sometimes inferred that the events from all very-heavy-ion reactions fall into two distinct classes, quasi-elastic events with energies slightly smaller than that of elastic scattering, and fully relaxed events with final kinetic energies corresponding to the Coulomb barrier of touching spheres. Although such an extreme view has some validity for special reactions and bombarding energies, it is necessary to adopt a broader point of view in order to understand heavy-ion reactions in general. Experimentally, one observes cross sections for events with a continuous range of energy damping and mass transfer as illustrated in Figure 1.

In Section 2, the macroscopic and microscopic approaches to heavy-ion reactions are discussed, with emphasis on the theoretical models that are more easily adaptable to the interpretation of present experiments. Some aspects of heavy-ion potentials and interaction radii are treated in Section 3, especially those aspects related to experiment. In Section 4, the characteristic properties of damped heavy-ion collisions are enumerated. The intercorrelation of the experimental properties of these damped collisions are discussed in some detail in Section 5. The emission of light particles and γ rays in strongly damped collisions is the topic of Section 6, and a summary is given in Section 7. Attention is called to the proceedings of two conferences on very-heavy-ion reactions, held during 1976, where review papers on several aspects of very-heavy-ion reactions were presented (36, 37a,b). In addition, Fluery & Alexander (37c) have surveyed several aspects of heavy-ion reactions through 1973. In this article, the new damped reaction mechanism is emphasized for which most of the literature has appeared since 1973.

2   MACROSCOPIC AND MICROSCOPIC APPROACHES TO HEAVY-ION REACTIONS

This section is intended to introduce the concepts and theoretical framework relevant to the discussion of heavy-ion reactions. It is beyond the scope of this article, which emphasizes the phenomenology of strongly damped reactions, to discuss in detail the different theoretical approaches. Therefore, only the main implications pertaining to current theories on heavy-ion reactions are outlined. Particular model calculations are discussed, along with the presentation of data whenever applicable.

2.1   Heavy Ions as Semiclassical Objects

Although quantum-mechanical calculations are performed for heavy-ion elastic scattering (41, 42) and few-nucleon transfer processes (42–45), the understanding of reactions between heavy ions is greatly facilitated by applying semiclassical concepts to these processes. The theory of classical and semiclassical approximations of the scattering or reaction amplitudes is rather involved (46–52). Therefore, only the most important results and conditions of applicability are mentioned in this section.

Semiclassical reaction amplitudes can be evaluated (47, 48) as a function of time, assuming the particle trajectory to be determined by classical dynamics, including Coulomb and real nuclear potentials. The method can be extended (48) to describe interference of different l waves due to strong nuclear attraction and absorption caused by the imaginary nuclear potential.

The most important parameter pertaining to semiclassical considerations is the reduced wave length λ of the system of two interacting heavy ions

$$\lambda(r) = k^{-1}(r) = \hbar\{2\mu[E_{c.m.} - V(r)]\}^{-1/2}. \qquad\qquad 2.$$

Here, μ is the reduced mass and $E_{c.m.}$ and $V(r)$ are the center-of-mass energy and the interaction potential, respectively. The classical approximation is generally valid if $|\,\text{grad}\,\lambda\,| \ll 1$, equivalent to the condition that the radius R of curvature of the trajectory be large: $R/\lambda \gg 1$. The more elaborate JWKB method is applicable if the potential varies slowly over the distance of one wavelength (46):

$$\hbar\mu\,|\,\text{grad}\,V\,|\,\{2\mu\,|\,E_{c.m.} - V\,|\}^{-3/2} \gg 1.$$

An approximate condition for classical behavior is given if $D/\lambda \gg 1$, where D is a typical dimension of the system and $\lambda = k^{-1}$ is the wavelength at infinity. For

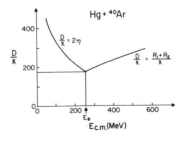

Figure 2   The ratio between the characteristic length D and the wavelength in the relative motion for argon projectiles on mercury as a function of bombarding energy in MeV. For energies below the Coulomb barrier ($E_B \approx 270$ MeV), D is taken to be the distance of closest approach in a head-on collision, while for $E > E_B$, D is taken to be the sum of the nuclear radii [from (47)].

Coulomb scattering, the relevant dimension is the distance of closest approach $2a = e^2 Z_1 Z_2 / E_{c.m.}$ and the above condition then reads

$$\eta = a/\lambda = e^2 Z_1 Z_2 / \{\hbar[2E_{c.m.}/\mu]^{1/2}\} \gg 1, \qquad\qquad 3.$$

where η is the Sommerfeld parameter. The high degree to which the classical condition is typically fulfilled in heavy-ion reactions is illustrated in Figure 2 for the system Hg + Ar (47). Nevertheless, deviations from classical behavior may be expected for large scattering angles and certain domains of impact parameters (42, 52).

Allowing for such reservations, classical trajectories may be defined for heavy ions, as shown in Figure 3 (42) for the system $^{18}O + ^{120}Sn$ at 100 MeV. Trajectories may be classified according to the degree of matter overlap to which they lead in the collision. The figure exhibits a pure Coulomb trajectory (1), the grazing trajectory (g), a skimming trajectory (2), indicating a slight overlap of the diffuse nuclear surfaces, and a plunging trajectory (3) leading to full matter-density saturation in a considerable part of the interaction region.

The deflection function $\theta(l)$ and the differential cross section $d\sigma_l/d\Omega$ as a function of angular momentum l are easily calculated in the classical limit,

$$\theta(l) = \pi - 2(l + \tfrac{1}{2})\hbar \int_{r_0}^{\infty} \frac{dr}{r^2} \left\{ 2\mu \left[E_{c.m.} - V(r) - \frac{\hbar^2(l+\tfrac{1}{2})^2}{2\mu r^2} \right] \right\}^{-1/2}, \qquad 4.$$

$$\frac{d\sigma_l}{d\Omega} = \frac{\lambda^2}{\sin\theta} \frac{l+\tfrac{1}{2}}{|d\theta/dl|}, \qquad\qquad 5.$$

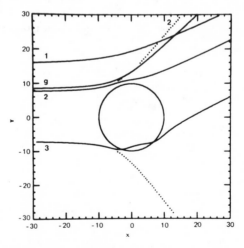

Figure 3 Four classical orbits for the $^{120}Sn + ^{18}O$ reaction at 100 MeV are plotted, three of which, having decreasing impact parameters 1, 2, and 3, scatter at the same angle θ. The orbit g is the grazing one. The circle marks the half-value of the Woods-Saxon nuclear potential $V = -40$ MeV, $r = 1.31$ fm, and $a = 0.45$ fm). Dotted lines show pure Coulomb orbits. The scale is in femtometers (fermis) [from (42)].

where r_0 denotes the classical turning point for the radial motion. In order to calculate the total reaction cross section σ_R from Equation 5, one has to know the transmission coefficients T_l for the given interaction potential. Integrating Equation 5, one obtains

$$\sigma_R = \pi \lambda^2 \sum_{l=0}^{\infty} (2l+1)T_l. \qquad 6.$$

The transmission coefficients in a purely classical picture, the sharp cutoff limit, are given by

$$T_l = \begin{cases} 1 & l \le l_{max} \\ 0 & l > l_{max} \end{cases}. \qquad 7.$$

Here, the maximum angular momentum leading to a reaction is determined by a critical distance between the two ions, the interaction radius R_{Int}:

$$l_{max} = \hbar^{-1} R_{Int} \{2\mu[E_{c.m.} - V(R_{Int})]\}^{1/2}. \qquad 8.$$

In the sharp cutoff limit of Equation 7, Equation 6 reduces to

$$\sigma_R = \pi \lambda^2 (l_{max}+1)^2, \qquad 9.$$

or, using the above definition of l_{max}:

$$\sigma_R = \pi R_{Int}^2 \{1 - [V(R_{Int})/E_{c.m.}]\}. \qquad 10.$$

The assumption given by Equation 7 is in good agreement with reality for relatively heavy systems, as is intuitively expected for nuclei where the surface diffuseness is small compared to the nuclear radius. As a consequence, the interaction radius attains fundamental importance in heavy-ion reactions.

The observation of diffraction effects in elastic heavy-ion scattering has led to an interpretation in terms of classical optics. It has been noticed by Frahn and collaborators (53–56) that the partial-wave expansion obtained in the strong-absorption model of Blair (57–59) reduces to formulas describing classical Fresnel diffraction in the limit of strong Coulomb fields and high energies. The occurrence of Fresnel rather than Fraunhofer scattering appears to be owing to a "diverging lens effect" caused by the Coulomb field. The ratio of the elastic-to-Rutherford cross section is given in this model by

$$\sigma_{el}(\theta)/\sigma_{Ruth}(\theta) = \tfrac{1}{2}\{[\tfrac{1}{2} - S(y)]^2 + [\tfrac{1}{2} - C(y)]^2\}, \qquad 11.$$

where $S(y)$ and $C(y)$ are the Fresnel sine and cosine integrals of argument

$$y = \left(\frac{2\eta}{\pi}\right)^{1/2} \csc\left(\tfrac{1}{2}\theta_{1/4}\right)\sin\left[\tfrac{1}{2}(\theta - \theta_{1/4})\right]. \qquad 12.$$

Here, $\theta_{1/4}$ is the "quarter-point" angle, where $\sigma_{el}(\theta)/\sigma_{Ruth}(\theta)$ assumes the value 0.25. The angular momentum l_{max} and the interaction radius are related to this angle according to

$$l_{max} = \eta \cot\left(\tfrac{1}{2}\theta_{1/4}\right), \qquad 13.$$

$$R_{Int} = \eta\lambda[1 + \csc(\tfrac{1}{2}\theta_{1/4})].$$ 14.

A similar relation had already been found empirically by Blair (57), establishing the "quarter-point technique" now commonly used in heavy-ion physics to determine l_{max} and R_{Int}. The applicability of the Fresnel model is subject to the conditions

$$l_{max} \gg 1 \quad \text{and} \quad l_{max} \sin\theta_{1/4} \gtrsim 1,$$ 15.

criteria which are easily met for most heavy-ion reactions at moderate bombarding energies.

In concluding this section, it should be emphasized that the conditions discussed above for the validity of various semiclassical approximations to heavy-ion elastic scattering and reactions are only necessary ones. They indicate that for trajectories exploring the tails of the nuclear interaction, classical dynamics holds, to a good approximation. However, nothing can be inferred for trajectories, such as orbit 3 in Figure 3, that plunge deeply into the nuclear matter distribution, permitting large transfers of mass and energy to occur between the two interacting ions. Examination of the validity of semiclassical approaches in general must resort to the study of the quantum-mechanical S matrix and deflection function (42, 47, 48).

2.2   Concepts of Energy Dissipation in Heavy-Ion Collisions

As is discussed in the experimental sections, heavy-ion reactions at energies several MeV/amu above the Coulomb energy are distinctly characterized by a large loss of kinetic energy of the relative motion (38), which is dissipated among intrinsic and shape degrees of freedom. It is therefore of considerable significance, for explaining the features of these reactions, to explore in detail the importance of various degrees of freedom and their coupling to the relative motion of the colliding heavy ions. Because of the large number of degrees of freedom that could conceivably be relevant to dissipative phenomena, statistical and macroscopic concepts such as friction and viscosity have received particular interest in recent theoretical work on heavy-ion reactions and nuclear fission.

Attempts to quantize dissipative systems (60–65), in order to describe the motion of quantum-mechanical wave-packets under the influence of friction, have revealed that such a quantization is not unique. So far, studies have been mostly carried out for rather simple physical situations, although progress has been made (64, 66–68) in applying the formalism to problems encountered in heavy-ion reactions.

In view of these difficulties, current theoretical models treat the relative motion classically or semiclassically. They calculate dissipative effects either microscopically or assume phenomenological friction forces to parameterize data. Although Gross (69) has demonstrated for the simple case of a wall moving with constant velocity through a Fermi gas of nucleons (piston model) that the interaction of intrinsic with collective degrees of freedom is describable by a classical, velocity-proportional friction force, such a result is not generally expected from realistic calculations (70). It turns out that it is, in fact, rather problematic to separate collective from intrinsic motion (71–74).

Microscopic evaluations of dissipation effects start with a total Hamiltonian of the system

$$\mathscr{H} = \mathscr{H}_{rel} + \mathscr{H}_{intr} + V, \qquad\qquad 16.$$

where $\mathscr{H}_{intr}$ describes the intrinsic degrees of freedom, $\mathscr{H}_{rel}$ is the relative (collective) motion, and V is an interaction potential. Different assumptions are made for various models as to the nature of the constituents of $\mathscr{H}$ and their dependence on the coordinates. For example, $\mathscr{H}_{intr}$ may be either the sum of unperturbed intrinsic Hamiltonians of the two nuclei (69, 75, 76), or a two-center model Hamiltonian depending on some collective variables (72, 77). Correspondingly, V represents either the residual one- or two-body interaction leading to intrinsic excitation of the system or an average Hartree-Fock–type field, which might be included in $\mathscr{H}_{rel}$. Dissipative effects are then studied either by solving the time-dependent, many-body Schrödinger equation (71, 72, 77), in perturbation theory (69, 78), in the formalism of the linear-response theory (75, 76), or in other statistical approximations (71, 79–83). The various approaches are also characterized by different model spaces they consider, e.g. particle-hole (69, 76–78, 83, 84) or quasi-particle excitations (71–73).

By way of analogy to the piston model, Gross et al (69, 78, 84) and Hofmann et al (75, 76, 83) considered one-body dissipation and assumed for V the single-particle potential of the combined system. Because of the dependence of V on the separation **r**, this variable of relative motion determines the rate of excitation. This scattering of nucleons off the wall of the single-particle potential has to be distinguished from two-body dissipation, where intrinsic excitation is due to collisions between two nucleons (72, 85, 86). The calculations indicate that there are dissipative and nondissipative terms arising from the coupling of the intrinsic degrees of freedom to the relative motion. Constructing effective equations of motion, it is seen that the former contributions give rise to macroscopic forces **F** with components

$$F_\nu(t) = \sum_\mu C_{\nu\mu}(t)\dot{r}_\mu(t), \qquad\qquad 17.$$

i.e. classical friction forces. In the simplest case of an isotropic friction tensor independent of time, $C_{\nu\mu}(t) = C\delta_{\nu\mu}$, the friction force (Equation 17) leads to a rate of energy dissipation proportional to the kinetic energy at a given time. Generally, however, $C_{\nu\mu}$ will be anisotropic and time-dependent because of changes in both the excitation energy and the spectrum of intrinsic eigenstates during the course of the reaction. Explicit numerical calculations (72, 87), which consider the effect of level crossing induced by the interpenetration of the two nuclei and excitations due to residual pairing forces, indicate that the dissipation rate is highest at the initial stages of a heavy-ion reaction, followed by a weak time dependence at later stages. It is instructive to notice (67, 69, 72, 88) that dissipation in heavy-ion reactions is related to the imaginary optical potential.

As has been emphasized above, the force (Equation 17) is typical of a one-body dissipation process. The velocity dependence of a two-body force generated by the two-body collisions of nucleons in the interaction region is generally rather complicated, as has been shown by Albrecht & Stocker (86). Only in a rough approximation may such a force reduce to a velocity-proportional force of the Tsang-Swiatecki type (85, 89, 90):

$$\mathbf{F} = -k \int d^3 r \rho_1 \rho_2 \dot{\mathbf{r}}, \qquad 18.$$

where ρ_1 and ρ_2 are the density distributions of the two nuclei, k is the friction constant, and the integral is to be taken over the overlap region.

Apart from the above energy-dissipation mechanisms associated with particle-hole excitation, kinetic energy is also lost in nucleon exchange between target and projectile (91, 92). In a simple version of a one-body nucleon-exchange process or window model (39b) the transferred nucleon is assumed to be loosely bound and at rest with respect to the donor nucleus (93). In a transfer, the nucleon with mass m deposits a relative momentum $\Delta p = m |\dot{\mathbf{r}}|$, which has to be cancelled by a recoil momentum experienced by the system. The loss of total kinetic energy associated with the exchange of a single nucleon δE_{ex} is then proportional to the kinetic energy $E_{av} = E_0 - E_B - E_{loss}$ available prior to the transfer:

$$\delta E_{ex} = \frac{m}{\mu}(E_0 - E_B - E_{loss}) = \frac{m}{\mu} E_{av}. \qquad 19.$$

Here, E_0 and E_B are the incident c.m. kinetic and the Coulomb energies, respectively, and μ is the reduced mass. The energy dissipation given by Equation 19 is equivalent to that resulting from a one-body friction force $F_{ex} = -k_{ex} |\dot{\mathbf{r}}|$. Its strength is directly given by the flux $j = dN/dt$ of nucleons exchanged between target and projectile according to $k_{ex} = jm/2$. Denoting the friction force constant due to one-body dissipation processes such as particle-hole excitation without nucleon exchange by k_{nex}, the total one-body friction force can be written

$$F = -k |\dot{\mathbf{r}}| = -(k_{nex} + k_{ex}) |\dot{\mathbf{r}}| \qquad 20.$$

resulting in an energy-dissipation rate

$$\frac{dE_{av}}{dt} = -2 \frac{k_{nex} + k_{ex}}{\mu} (E_0 - E_B - E_{loss}). \qquad 21.$$

It may be expected that the rates of one-body processes without and with nucleon exchange represented by the friction coefficients k_{nex} and k_{ex}, respectively, are somewhat different in magnitude. However, since both processes are induced by the same time-dependent single-particle potential, the two friction coefficients are expected to have the same time dependence. This is equivalent with both friction forces having the same spatial form factor. Under this condition, the energy loss rates due to particle-hole excitation and nucleon exchange are proportional to each other. The microscopic time scale corresponding to the nucleon exchange mechanism may then be used to calculate the total energy lost during the time needed for one nucleon exchange. Consequently, this total energy loss associated with a single exchanged nucleon, δE, is given by

$$\delta E = -\frac{dE_{av}}{dN} = \frac{m}{\mu}(1 + k_{nex}/k_{ex})(E_0 - E_B - E_{loss}). \qquad 22.$$

Since microscopic calculations of dissipative features of heavy ion reactions are numerically quite elaborate, most models describe such effects by phenomenological friction forces. Among the simplest possibilities for including friction effects in the description of heavy-ion reactions are assumptions of a sudden onset of infinite radial or tangential friction at a critical separation distance in the entrance channel (94–97) or velocity-proportional friction forces characterized by a strength parameter and a range (96, 97).

The one-body dissipation mechanism suggests a velocity-proportional friction force given by an anisotropic friction tensor as used by Gross et al (98).

$$C_r(r) = C_r \cdot g(r), \qquad C_t(r) = C_t \cdot g(r), \qquad\qquad 23.$$

where the subscripts r and t refer to the radial and tangential parts, respectively, and the radial form factor g is proportional to the square of the gradient of the nuclear ion-ion interaction potential U_N

$$g(r) = [\vec{\nabla} U_N(r)]^2. \qquad\qquad 24.$$

More recent studies (99, 100) using a different nuclear potential employ a form factor proportional to the fourth power of the potential gradient.

The rate of one-body energy dissipation has been calculated by Randrup (101) in the proximity formalism. The result for a grazing collision where the velocity is almost exclusively tangential during the interaction is

$$\frac{dE}{dt} = 4\pi \frac{n_0}{\mu} \frac{C_T C_P}{C_T + C_P} b\psi(\zeta_0)E \qquad\qquad 25.$$

Here, $n_0 (\approx 2.5 \cdot 10^{-23}$ MeV sec fm^{-4}) is the transfer flux density, C and b are the nuclear half-density radius and the diffuseness, respectively. $\psi(\zeta_0)$ is a universal proximity flux function at a surface separation distance ζ_0. The applicability of this formalism was first realized by Bass (102).

The above approaches all assume adiabaticity, i.e. an internal response time short compared to that in which the relative coordinates change, as well as frozen nuclear matter distributions. However, recently Broglia et al (103) have argued that characteristic times over which the interaction matrix elements change may in fact be of the order of 10^{-23}–10^{-22} sec in contradiction to the adiabaticity assumption. This leads to a strong coupling of collective excitation modes, e.g. giant resonances, to the relative motion and their subsequent fast damping into particle-hole degrees of freedom for relatively low values of angular momentum. Similarly, there are suggestions (104, 105) that rapid changes of the shape degree of freedom may also cause nonadiabatic damping of the relative motion. This must be seen in contrast to dynamical deformations of the interacting ions which evolve slowly in time. The importance of this latter mechanism of energy loss, which is not readily describable by nuclear friction, has been emphasized by several authors (38, 90, 96, 106, 107), particularly for the exit channel.

2.3   Microscopic Transport Theories of Nucleon Exchange

The microscopic statistical treatment of the coupling of internal degrees of freedom to those of the relative motion generally leads to coupled equations, one of which is the classical Newton equation for average values of the coordinates **r** and **p** of relative motion in terms of potentials including friction forces. The other one describes the time evolution of macroscopic probabilities. Dissipation and fluctuations are due to the same phenomenon, as can be illustrated by considering the motion of Brownian particles experiencing random forces caused by a heat bath (108). Much theoretical work (75, 79, 81, 81a, 83, 109–116) has been devoted to establishing a statistical description of nuclear reactions in terms of transport equations. Of particular importance is the derivation of master equations for the dynamics of heavy-ion reactions (81, 81a, 83, 110) initiated by the pioneering work of Nörenberg (109).

Starting from the Liouville equation for the density matrix ρ of the system constructed from asymptotic scattering channel wave functions, Nörenberg (110) introduces occupation-probability distributions $P(\mathbf{x}, t)$ for a set $\mathbf{x}$ of macroscopic observables by a "coarse-graining" method. Assuming a random distribution of the matrix elements of a single-particle potential $V(t)$, a generalized master equation is derived that includes correlations and memory effects. The typical memory time τ_{mem} during which phase coherence between different channels $\mathbf{x}$ and $\mathbf{x}'$ is experienced, is given by

$$\tau_{mem}(\mathbf{x}, \mathbf{x}'; t) = \left[\frac{2\pi\hbar^2}{\langle V^2(t)\rangle_{\mathbf{x}} + \langle V^2(t)\rangle_{\mathbf{x}'}} \right]^{1/2}. \qquad 26.$$

Here $\langle V^2\rangle_{\mathbf{x}}$ denotes the average expectation value for a fixed macroscopic variable $\mathbf{x}$, including a sum over all channels connected by V to those belonging to $\mathbf{x}$. This time is closely related to the response time mentioned in the preceding section. If τ_{mem} is short compared to the relaxation time τ_{rel} during which the macroscopic variables [i.e. $P(\mathbf{x}, t)$] change, memory effects can be neglected. Such irreversible Markov processes are governed by a master equation

$$\frac{\partial}{\partial t} P(\mathbf{x}, t) = \int d\mathbf{x}' w(\mathbf{x}, \mathbf{x}'; t)[\rho(\mathbf{x})P(\mathbf{x}', t) - \rho(\mathbf{x}')P(\mathbf{x}, t)], \qquad 27.$$

where ρ denotes the density of channels and $w(\mathbf{x}, \mathbf{x}'; t)$ is the transition probability. In dealing with an equilibration process of a finite system, it has also to be assured that the Poincaré recurrence time $\tau_{rec} = 2\pi\hbar\rho_{int}$ is long compared to τ_{rel} and the interaction time τ_{int}. This time determined by the internal level density ρ_{int} is the time needed for the system to return to its initial state. Applicability of the master-equation approach is then granted under the conditions

$$\tau_{mem} \ll \tau_{rel}, \tau_{int} \ll \tau_{rec}. \qquad 28.$$

Rough estimates using elastic-transfer matrix elements lead to memory times of the order 10^{-23}–10^{-22} sec (110). If the transition probabilities $w(\mathbf{x}, \mathbf{x}'; t)$ are essential only for $\mathbf{x} \approx \mathbf{x}'$, a second-order expansion of the master equation around $\mathbf{x}$ is appropriate, leading to the Fokker-Planck equation

$$\frac{\partial}{\partial t} P(\mathbf{x}, t) = -\sum_i \frac{\partial}{\partial x_i}[v_i(\mathbf{x}, t)P(\mathbf{x}, t)] + \sum_{i,j} \frac{\partial^2}{\partial x_i \partial x_j}[D_{ij}(\mathbf{x}, t)P(\mathbf{x}, t)], \qquad 29.$$

where the x_i are the components of $\mathbf{x}$. Drift and diffusion coefficients are given by

$$v_i(\mathbf{x}, t) = \frac{1}{\rho(\mathbf{x})} \sum_j \frac{\partial}{\partial x_j} [\rho(\mathbf{x}) D_{ij}(\mathbf{x}, t)],$$

$$D_{ij}(\mathbf{x}, t) = \rho(\mathbf{x}) \mu_{ij}(\mathbf{x}, t) = D_{ji}(\mathbf{x}, t),$$

30.

respectively. The μ_{ij} are second moments of the probability w. In its simplest case, with one observable x and constant coefficients v and D, Equation 29 reduces to

$$\frac{\partial}{\partial t} P(x, t) = -v \frac{\partial}{\partial x} P(x, t) + D \frac{\partial^2}{\partial x^2} P(x, t),$$

31.

having Gaussian solutions

$$P(x, t) = (4\pi Dt)^{-1/2} \exp\left[-\frac{(x - vt)^2}{4Dt} \right].$$

32.

The centroid $x_0 = vt$ and the variance $\sigma^2 = 2Dt$ are linear functions of the interaction time t, as is illustrated in Figure 4 for an initial δ distribution for P. Ayik et al (111, 112) have evaluated transport coefficients for mass-fragmentation and excitation energy using a single-particle model for the intrinsic system. They obtain the approximate relations

$$v_A = -\frac{1}{T} D_A \frac{\partial}{\partial A_1} U_l(A_1),$$

33.

$$v_E = \frac{1}{T} D_E$$

34.

where U_l denotes the ground-state energy of the combined system with relative angular momentum l and fragmentation A_1. The local temperature T is determined by the excitation energy. Equation 33 is the well known Einstein relation, while Equation 34 represents the fluctuation-dissipation theorem as can be realized by noticing that the average drift coefficient $\langle v_E \rangle$ describes the rate of energy loss.

A generalized master equation has also been derived in the framework of the linear-response theory (83), assuming for the internal system a canonical ensemble described by a temperature. A Fokker-Planck equation can be deduced from this in the limit of high temperature. This temperature concept has been criticized by Agassi et al (81), who employ a random matrix model, including angular momentum,

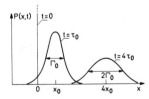

Figure 4 Illustration of the solution of the Fokker-Planck equation with constant drift and diffusion coefficients. Γ is the FWHM [from (110)].

to deduce a master equation, but are as yet not able to generally describe mass exchange in heavy-ion reactions. In addition, several phenomenological parameterizations of the quantities entering the Fokker-Planck equation (Equation 29) have been proposed (33, 117–119). There are also attempts (120) to simulate nucleon exchange in heavy-ion reactions by Monte Carlo methods, solving the classical equations of motion consistently.

2.4   Reaction Models for Heavy-Ion Collisions

Classical dynamical models comprise the main body of theories available to date describing the features of strongly damped and fusion reactions between heavy ions. As has been discussed above, classical calculations are expected to reveal the important average properties of these reactions for systems sufficiently heavy and for energies well above the Coulomb barrier. Many theories (38, 85, 90, 94–100, 121–133) emphasize the role of the ion-ion potential and friction in the reaction dynamics. Elementary models neglect dynamical distortions of the reaction partners and the effect of nucleon exchange. In this picture, two spherical nuclei with sharp or diffuse surfaces approach each other on Coulomb trajectories that are modified by the action of friction forces and nuclear attraction. The models differ with respect to the interaction potentials used (see Section 3) and the relative strengths and form factors of the friction forces. Because of the nuclear attraction, the system in its contact configuration is allowed to rotate through a certain angle as a whole before it either fuses or breaks apart. The importance of friction in the reaction is illustrated in Figure 5, indicating that both weak and strong friction inhibits fusion of the system in this picture of a frozen configuration.

The classical dynamical variables describing the system of two interacting nuclei 1 and 2 can be specified as the radial and polar coordinates of the nuclear centers, r and θ, respectively, and the angles θ_1 and θ_2 denoting the orientation of the two nuclei (123). In order to diagonalize the friction tensor (cf Equation 17), generalized coordinates q_i are defined

$$q_1 = r, \quad q_2 = R_1(\theta_1 - \theta) + R_2(\theta_2 - \theta), \quad q_3 = \theta_2 - \theta, \quad q_4 = \theta, \qquad 35.$$

where R_1 and R_2 are the nuclear radii. The Lagrangian $L = E - V$ is then expressed in terms of these coordinates. where E is the kinetic energy and V the interaction potential $V = V_{\text{nucl}} + V_{\text{Coul}}$, consisting of a nuclear and a Coulomb part. The Lagrangian equations

$$\frac{\mathrm{d}}{\mathrm{d}t}\left(\frac{\partial L}{\partial \dot{q}_i}\right) - \frac{\partial L}{\partial q_i} = C_i q_i \qquad i = 1,\ldots,4 \qquad 36.$$

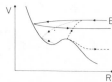

Figure 5   A schematic diagram of the effective potential V and the total kinetic energy E for three different frictional strengths, all given as a function of r [from (129)].

that determine the equations of motion include friction forces of the form given by Equation 17. The friction form factors C_i are usually taken to depend on $q_1 = r$ only. Using a friction force of the form of Equation 18, one arrives at the following equations of motion (85, 90):

$$\mu \frac{d^2 r}{dt^2} = \frac{e^2 Z_1 Z_2}{r^2} - \frac{d}{dr} V_{\text{nucl}} + \frac{l^3}{\mu r^3} - k \frac{dr}{dt} \int \rho_1 \rho_2 \, d^3 \tau, \qquad\qquad 37.$$

$$\frac{dI_1}{dt} = k \int \rho_1 \rho_2 \, d^3 \tau [(\dot{\theta} R - \dot{\theta}_1 R_1' - \dot{\theta}_2 R_2') R_1' - g^2(\dot{\theta}_1 - \dot{\theta}_2)], \qquad\qquad 38.$$

$$\frac{dI_2}{dt} = k \int \rho_1 \rho_2 \, d^3 \tau [(\dot{\theta} R - \dot{\theta}_1 R_1' - \dot{\theta}_2 R_2') R_2' - g^2(\dot{\theta}_2 - \dot{\theta}_1)], \qquad\qquad 39.$$

$$\frac{dl}{dt} = -\left(\frac{dI_1}{dt} + \frac{dI_2}{dt} \right). \qquad\qquad 40.$$

Here, l is the relative angular momentum, I_1 and I_2 are the intrinsic spins of the two ions, R_1' and R_2' are the distances from the nuclear center to the centroid of the overlap region, and g is an effective radius of gyration of the overlap region around its centroid.

The first relation (Equation 37) describes the radial motion under the influence of Coulomb, nuclear, centrifugal, and friction forces. Equation 40 relates to the conservation of total angular momentum $L = l + I_1 + I_2$. The changes in the intrinsic spins I_1 and I_2 of the two nuclei are due to tangential-sliding and rolling friction, corresponding to the two terms in brackets in Equations 38 and 39. In the limiting case $\dot{\theta} = \dot{\theta}_1 = \dot{\theta}_2$, the two nuclei stick together and rotate as one rigid body. The maximum value Δl of angular momentum transformed into intrinsic spin can be written in terms of the moments of inertia $\mathscr{I}_{\text{NS}}$ and $\mathscr{I}_{\text{S}}$ for the nonsticking and sticking conditions

$$\mathscr{I}_{\text{NS}} = \mu r^2, \qquad\qquad 41.$$

$$\mathscr{I}_{\text{S}} = \mathscr{I}_{\text{NS}} + \tfrac{2}{5}(M_1 R_1^2 + M_2 R_2^2), \qquad\qquad 42.$$

where M_1 and M_2 are the nuclear masses, as

$$\Delta l = l_i(\mathscr{I}_{\text{S}} - \mathscr{I}_{\text{NS}})/\mathscr{I}_{\text{S}}. \qquad\qquad 43.$$

In Equation 43, l_i is the initial value of angular momentum. The system of coupled differential equations (Equations 37–40) can be solved numerically. The projectile-target system is regarded as having fused if, for $t \to \infty$, r stays always smaller than some arbitrarily chosen distance R_{min}. These and all other trajectories experiencing nuclear and friction forces contribute to the reaction cross section. The deflection function (Equation 4), the differential cross section (Equation 5), and the energy spectrum of the fragments are then obtained. Although model calculations (90, 123) indicate that the deflection function depends on details of the friction tensor, the effect of tangential friction is too weak to account for an appreciable energy loss (90, 123, 134) due to the large moment of inertia. Tangential friction has been reported, however, to be important for l values approaching the grazing angular momentum (125–128). Most commonly, the rolling-friction term is neglected and the friction tensor is assumed to be of the form of Equation 23.

The models discussed so far are unsatisfactory in that dynamical deformation effects, such as neck formation and elongation of the system during the interaction, are not properly taken into account. Several theories account partly for such effects by phenomenologically introducing either different ranges of friction forces in the entrance and exit channels (96, 124) or modifying the exit-channel nuclear potential (90). Most recently (106, 107), reaction models have been proposed that explicitly treat axially symmetric deformations as collective variables of motion. The most important results from these calculations indicate that slightly oblate nuclear deformations occur in the entrance channel, whereas during the time of mutual overlap of the two nuclei, prolate deformations develop. Consequently, the entrance-channel interaction barrier is decreased and the interaction time increased. These effects lead to an increase of the strongly damped cross section at the expense of fusion. These calculations further indicate that tangential friction is important for the reaction dynamics and is of the same strength as radial friction, in contrast to most other calculations.

It is obvious that the transfer of mass and charge between the reaction partners can influence the reaction kinematics appreciably. So far, little attention has been paid to this effect by the various theoretical approaches. A simple diffusion model has been suggested by Bondorf et al (123) to describe approximately such transfer processes, which are simulated by altering the conservative potential. It has also been suggested (90) that ingoing and outgoing trajectories be matched at an arbitrary distance R_{min}, where transfer is supposed to occur, taking into account recoil effects (91, 92). Another deficiency of most current models is the neglect of the influence of statistical fluctuations on the reaction dynamics (cf Section 2.3). Most recently, fluctuations and mass transfer are incorporated in dynamical calculations employing either the linear-response theory (134, 135) or Monte Carlo methods (120).

2.5   Time-Dependent Hartree-Fock and Fluid-Dynamical Approximations

These two approaches represent very recent theoretical developments and hence it seems premature to review them here in any detail. However, these theories are mentioned here for the sake of completeness.

The difficulty of distinguishing unambiguously between intrinsic and collective degrees of freedom in heavy-ion reactions has renewed interest (136) in the self-consistent, time-dependent Hartree-Fock approximation (TDHF). This approximation basically assumes that at all times the many-body wave function representing the nucleons of projectile and target is given by a single Slater determinant Ψ, which can be determined by the condition that $\langle \Psi \, | \, i\hbar \partial/\partial t - \mathcal{H} \, | \, \Psi \rangle$ be a minimum. The Hamiltonian $\mathcal{H}$ is given by the sum of the one-body kinetic-energy operator and a two-body effective interaction. Using a simplified geometry, an initial treatment (136) resulted in a variety of collision phenomena, including damped reactions associated with high losses of kinetic energy. Recent studies (137–142) employ more realistic assumptions. However, owing to their computational complexity, such calculations are so far manageable only for comparatively light systems or for slightly heavier systems if the dimensions and symmetries of the nuclei are restricted. A quantitative description of data has not yet been achieved. However, TDHF calculations provide the opportunity for studying interesting effects pertaining to damped reactions, such as neck formation, density oscillations, and static fragment deformations, as well as energy and angular-momentum dissipation.

A possibility of bypassing some practical problems encountered in TDHF calculations, but retaining essential quantum features, is represented by fluid-dynamical theories. Such an approach is based on the observation (143–145) that the time-dependent Schrödinger equation is equivalent to a continuity equation and one analogous to the Euler equation describing classical fluid dynamics. Wong et al (146–148) derived fluid-dynamical equations from the many-body quantum kinetic equation and the TDHF equations. They have also estimated (147) the time and length scales for which a completely macroscopic fluid-dynamical description may be valid. Effects of viscosity may be treated in terms of the Navier-Stokes equations.

The macroscopic dynamics of viscous nuclear drops have been studied by Alonso (149). Nuclear fluid-dynamical concepts have been applied to nuclear fission and damped heavy-ion reactions by Nix and collaborators (150–152).

3 HEAVY-ION POTENTIALS AND INTERACTION RADII

The determination of the potential acting between two complex nuclei from measurements of the elastic-scattering and inelastic-reaction processes is a goal of experimentalists. This interaction potential is sometimes assumed to be a function of the separation of the centers of the two nuclei alone, where all the internal degrees of freedom are frozen. Such potentials develop a repulsion when the densities of the two ions overlap appreciably. This is due both to the Pauli exclusion principle and to the saturation of the nucleon-nucleon force at the normal nuclear density (42, 153). In contrast to the above sudden approximation, some interaction potentials have been computed with an adiabatic approximation, where the nuclear shapes are allowed to adjust so as to minimize the potential energy at each distance of separation. Due to the continual change in shape, the adiabatic procedure does not lead to a short-range repulsion. There are actually many adiabatic potentials, each corresponding to a different degree of excitation of the intrinsic states of the deformed system (42).

Most information about the interaction potential comes from analyses of elastic scattering data, which are sensitive to the nuclear potential in the vicinity of the strong-absorption radius R_{SA} (154). In addition, important information comes from inelastic-reaction processes, which are sensitive to the potential at radial distances smaller than R_{SA}.

3.1 Phenomenological Potentials

The basic assumption of the phenomenological models, e.g. the optical model, is that heavy-ion scattering is describable by a simple, complex potential that depends only upon the separation distance r of the centers of mass of the two heavy ions.

The radial dependence of the diffuse-surface potential energy is usually of the Saxon-Woods type (42, 90, 94, 95, 123, 154–163). Several of these potentials used for heavy-ion trajectory calculations are quite different (164).

The parameterization of the optical potential is subject to considerable ambiguities when applied to the scattering of strongly absorbed particles such as heavy ions (154, 161, 165). This leads to a wide choice of potentials, each of which may give an equivalent fit to the elastic-scattering data. However, various sets of real potentials that give these equivalent fits to the elastic-scattering data tend to give a common value of the real potential near the strong-absorption radius. This is also true for very heavy ions like Ar (159), Kr (159, 162), and Xe (21, 30). However, with these very heavy ions, the measured energy widths (FWHM) of the elastic spectra are in the vicinity of 1–1.5% of the beam energy. Hence, near $\theta_{1/4}$, it is difficult to effect a good separation of elastic and inelastic events. This problem leads to additional uncertainty in the real potential near R_{SA} (163).

Elastic scattering data for the ^{28}Si + ^{16}O reaction have been analyzed for energies ranging from 33–215 MeV, in an attempt to find a common, energy-independent Saxon-Woods optical potential (166). This is the most extensive set of heavy-ion elastic-scattering data giving information on the slope of the real potential near R_{SA}. Even so, it is not possible to determine V_N very far inside R_{SA}.

If one assumes a nuclear potential of the Saxon-Woods form, the derivative of such a potential with respect to the radial distance r has a maximum given by

$$(dV_N/dr)_{max} = -V_0/4a,$$

44.

where V_0 is the central depth and a is the diffuseness parameter. If this force is equated to the liquid-drop model force for two touching spherical nuclei (167), the Saxon-Woods parameters can be evaluated if the potential is known at a second radial distance (90, 159, 164) (e.g. at R_{SA} from elastic-scattering data). Of course, such an approximation of the nuclear potential has no validity at distances less than the touching distance.

3.2   Folding Potentials

There are two basic folding models; the single-folding and double-folding models. In the single-folding model, the nuclear energy is obtained by folding an empirical nucleon-nucleus-1 optical potential into the nuclear-matter density distribution of nucleus-2 (47, 168–170). The double-folding model requires an effective nucleon-nucleon interaction, that is folded into the density of both nuclei (41, 154, 171–175). Although folding potentials have been used extensively to fit heavy-ion elastic-scattering data, these potentials are probably too deep (153), especially at smaller radial distances. For example, the folding potential deduced from the ^{28}Si + ^{16}O elastic-scattering data produces "pockets" for $l \leqq 45$ (175). Although inelastic reaction channels are sensitive to the nuclear potential at smaller distances, reactions in which one or two nucleons are transferred do not offer a way to distinguish between most potentials (175). Determination of the number of l waves leading to fusion is a more sensitive test of the real potential at radial distances inside the strong-absorption radius R_{SA} (163). In general, the folding potential can be justified only if all antisymmetrization effects between the two nuclei are negligible and the changes of the nuclear structure due to the interaction can be neglected (176). This corresponds to the sudden-approximation or frozen-configuration assumption.

3.3 Proximity Potential

The proximity potential (177) is based on a very general model that expresses the force between two gently curved leptodermous surfaces as a product of a geometrical factor and a universal function equal to the interaction energy per unit area between two parallel surfaces. The proximity potential is given by (177)

$$V_{prox} = 4\pi\gamma \, \frac{C_T C_P}{C_T + C_P} \, b\Phi(\zeta). \qquad\qquad 45.$$

The proximity potential function $\Phi(\zeta)$ is rather well approximated by the expressions (39a, 177, 178)

$$\Phi(\zeta < 1.2511) = -\tfrac{1}{2}(\zeta - 2.54)^2 - 0.0852(\zeta - 2.54)^3, \qquad\qquad 46.$$

$$\Phi(\zeta \geq 1.2511) = -3.437 \exp(-\zeta/0.75), \qquad\qquad 47.$$

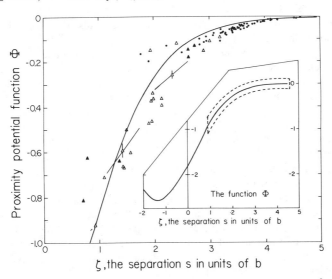

Figure 6 A comparison of the universal proximity potential (177) $\Phi(\zeta)$ as a function of the dimensionless quantity ζ (*solid line*) with experimental data (*points*). The analysis of elastic-scattering data produces the circles and the analysis of inelastic reaction data produces the triangles. Comparisons made by (177 and 180). The open triangles are based on excitation function measurements of fusion cross sections by counter-telescope measurements of evaporation residuals and/or fission fragments, while the filled triangles rely on excitation functions based on summing measured partial fusion cross sections. The latter values are subject to the largest errors. The region enclosed by the dashed line in the insert shows the part of the proximity potential tested by experimental data. Errors in ζ are reflected also in Φ (see text) as illustrated for two fusion points.

where $\zeta = s/b$, $s = r - C_{\text{T}} - C_{\text{P}}$ and $b = 1$ fm. The surface-energy coefficient $\gamma = 0.9517 \, (1 - 1.7826 \, I^2)$ MeV fm^{-2} where $I = (N - Z)/A$, and N, Z, and A refer to the combined system of the two interacting nuclei. The half-density matter radius C is calculated from the effective sharp radius R by the formula (169, 177)

$$R = 1.28 \, A^{1/3} - 0.76 + 0.8 \, A^{-1/3} \qquad\qquad 48.$$

and

$$C = R\{1 - (b^2/R^2) + \cdots\}. \qquad\qquad 49.$$

The theoretical proximity function $\Phi(\zeta)$ in the extreme tail region ($\zeta \geq 2.5$) has been compared (177) with nuclear potentials deduced from an analysis of elastic-scattering data (160). These values of the dimensionless function $\Phi(\zeta)$, ranging from 0 to -0.16, are reproduced as a function of ζ in Figure 6 as circles. The "experimental" values of $\Phi(\zeta)$ on this figure are obtained by dividing the evaluated potential at some fixed distance r_{exp} by $4\pi\gamma b C_{\text{T}} - C_{\text{P}}/(C_{\text{T}} + C_{\text{P}})$. The corresponding separation distance ζ_{exp} (in units of b) is equal to $(r_{\text{exp}} - C_{\text{T}} C_{\text{P}})/b$. The theoretical and experimental values of $\Phi(\zeta)$ from the elastic-scattering data are in reasonable agreement (177), although the elastic-scattering data test the potential over only a small interval in $\Phi(\zeta)$ at very large values of ζ (i.e. radial distances near the strong absorption radius R_{SA}).

As stated previously, the inelastic-reaction processes test the nuclear potential to smaller radial values considerably farther inside R_{SA}. It is possible to calculate the fusion radius R_{B} and the fusion barrier $V_{\text{B}}(R_{\text{B}})$ from an expression for the fusion cross section that is analogous to Equation 10:

$$\sigma_{\text{fus}} = \pi R_{\text{B}}^2 [1 - (V_{\text{B}}/E_{\text{c.m.}})]. \qquad\qquad 50.$$

The values of R_{B} and $V_{\text{N}}(R_{\text{B}})$, where $V_{\text{N}}(R_{\text{B}}) = V_{\text{B}}(R_{\text{B}}) - V_{\text{C}}(R_{\text{B}})$, determined from a plot (179) of the fusion cross section as a function of $1/E_{\text{c.m.}}$ are used to determine $\Phi(\zeta)$ and ζ. These "experimental" values of $\Phi(\zeta)$, ranging from -0.1 to -1.0, are plotted as a function of ζ (where $0.7 \leq \zeta \leq 3.2$) on Figure 6 as triangles (180). These reaction data test the proximity potential over a range six times larger than the extreme value from the elastic scattering data. The theoretical proximity potential is in excellent agreement with the data over the entire range of radial distances ζ. It is somewhat surprising that the proximity potential fits the data so well for ζ values as small as 1, where the combined nuclear density of the overlapping nuclei is approximately 60% of the central density.

An error in the experimental barrier radius alters $V_{\text{N}}(R_{\text{B}})$ through the strong radial dependence of the Coulomb potential and causes a fusion data point in Figure 6 to move roughly parallel to the theoretical curve for $\Phi(\zeta)$. Thus the agreement between experiment and the proximity potential is not markedly changed by errors in R_{B}. However, it is important to emphasize that the analysis of the fusion data as displayed in Figure 6 was performed by neglecting friction. In a simple analysis of the effect of including friction, the magnitude of the nuclear potential is increased, moving the fusion points below the theoretical curve in Figure 6, although the exact shifts in the points are at present uncertain. Currently, one of the important challenges is to find the value of ζ where the neck between the two nuclei begins to form and the sudden approximation fails. A nucleus-nucleus potential similar to the proximity potential has been derived by Bass (181). In Figure 7, the radial dependence of the effective total real potential $V = V_C + V_l + V_N$ is plotted for the ^{209}Bi + ^{84}Kr and ^{208}Pb + ^{16}O reactions where V_N is approximated by the proximity potential. The most important feature displayed in this figure is the presence of deep minima or pockets in the total potentials for the ^{208}Pb + ^{16}O reaction and the absence of deep minima for the ^{209}Bi + ^{84}Kr reaction. Based on the information shown in Figure 6, the proximity potential is expected to give a rather good estimate of V_N, at least up to a radial distance equal to R_B. Furthermore, the l waves with pockets in Figure 7 give estimates of the fusion cross sections that agree with experiment.

3.4 Self-Consistent Microscopic-Model Potentials

A general treatment of the collision between heavy ions requires the solution of Hartree-Fock–type equations for the projectile-target system at each point of the trajectory. Potential energies have been evaluated with microscopic models that are based on the Skyrme interaction and Hartree-Fock or time-dependent Hartree-Fock approximations (136, 137, 182, 183). The complexity of the many-body problem is greatly reduced by a statistical approach (184). The use of such a statistical theory of finite nuclei based on nuclear-matter calculations to evaluate the interaction energy between the complex nuclei has recently been further explored (185). The potential based on this energy density formalism is usually evaluated for the sudden approximation and resembles rather closely the proximity potential.

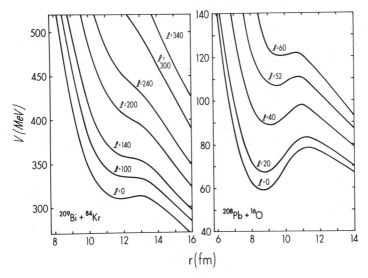

Figure 7   Total real potentials ($V = V_C + V_l + V_N$), based on the nuclear proximity potential.

Table 1 Comparison of parameters deduced from the Fresnel and optical models for several heavy-ion reactions (163)

Reaction	Energy	η	$\theta_{1/4}^{c.m.}$	Optical			Fresnel		
				$l_{1/2}$	R_{SA}(fm)	σ_R(mb)	l_{max}	R_{Int}(fm)	σ_R(mb)
^{209}Bi + ^{40}Ar	340	80.63	47°	186	13.30	2480	185	13.21	2382
	286	87.91	60°	150	13.35	1926	151	13.43	1887
^{238}U + ^{40}Ar	340	89.37	51°	187	13.60	2410	187	13.59	2336
	286	97.44	68°	147	13.68	1800	144	13.56	1648
^{209}Bi + ^{84}Kr	714	161.4	49.5°	353	14.35	2710	350	14.31	2650
	712	161.6	50.5°	346	14.25	2606	343	14.16	2533
	600	175.9	66°	272	14.28	1922	270	14.24	1880
^{209}Bi + ^{136}Xe	1130	244.8	54°	484	15.21	2780	481	15.10	2700

3.5   Liquid-Drop Model (Macroscopic-Microscopic) Potentials

Potential-energy surfaces have been calculated with modified liquid-drop models for a number of systems as a function of mass asymmetry and distance between mass centers for various shape configurations (186–195). In a number of calculations, the energies from microscopic shell and pairing corrections are added to the nuclear macroscopic potential energy, which takes into account the finite range of the nuclear force. The topography or multidimensional map of the potential-energy surface for heavy-ion reactions is essential for understanding the close-contact collisions. The potential for the mass-transfer degree of freedom has been investigated for contact configurations (191) and more deeply penetrating collisions (192, 194).

3.6   Interaction and Barrier Radii

The strong-absorption radius for the optical model is defined by

$$R_{SA} = \lambda\eta + \lambda[\eta^2 + l_{1/2}(l_{1/2} + 1)]^{1/2}, \qquad\qquad 51.$$

where R_{SA} represents the distance of closest approach for the classical Rutherford orbit of angular momentum $l = l_{1/2}$, for which the transmission coefficient $T_l = 0.5$. The inclusion of the nuclear potential changes R_{SA} by only a very small amount. The magnitude of R_{SA} for a large number of very-heavy-ion reactions has been found to be nearly the same as that for R_{Int} (Equation 14), based on the Fresnel model (159, 162, 163) (see Table 1). Hence, we use these two quantities interchangeably. The interaction radius may be parameterized in terms of the expression

$$R_{Int} = R_{SA} = C_T + C_P + \zeta(R_{SA}), \qquad\qquad 52.$$

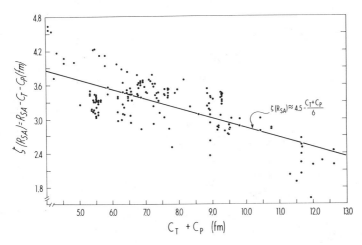

Figure 8   Plot of $\zeta(R_{SA})$, where $\zeta(R_{SA}) = R_{Int} - C_T - C_P$, as a function of $C_T + C_P$. The interaction radius R_{Int} is deduced from elastic scattering data (from J. R. Birkelund, J. R. Huizenga. 1977. Private communication).

where C_T and C_P are the target and projectile half-density matter radii of Equation 49, and ζ represents the distance between the half-density points. The value of $\zeta(R_{SA}) = (R_{Int} - C_T - C_P)$ is plotted in Figure 8 as a function of $C_T + C_P$ for a large number of heavy-ion reactions, where R_{Int} has been derived from elastic-scattering data. The value of $\zeta(R_{SA})$ decreases with increasing values of $C_T + C_P$ and is given approximately by (J. R. Birkelund, J. R. Huizenga. 1977. Private communication)

$$\zeta(R_{SA}) \approx 4.5 - (C_T + C_P)/6. \qquad\qquad 53.$$

If the radial dependence of the nuclear density is approximately a two-parameter Fermi distribution function with the surface diffuseness parameter $a = 0.551$ fm, the matter density of each nucleus in the overlap region increases from 4.0% of its central density for $\zeta = 3.5$ to 9.4% of its central density for $\zeta = 2.5$.

Information on the radial distance of the barrier R_B, defined by the condition $(dV_C/dr) = -(dV_N/dr)$, is summarized in Figure 9 for a number of heavy-ion reactions. The solid line represents a theoretical calculation using the proximity potential. The experimental points are based on the measured energy dependence of the fusion cross section (for the difference between the fusion and compound-nucleus cross section, see Section 1), from which it is possible to calculate V_B and R_B from Equation 50. The data points are for the same reactions as those represented by the triangles in Figure 6. The agreement between theory and experiment is reasonable, although the experimental values of R_B tend to be slightly smaller than the theoretical values. The above comparison is a rather stringent test of the proximity potential (see also Section 3.3).

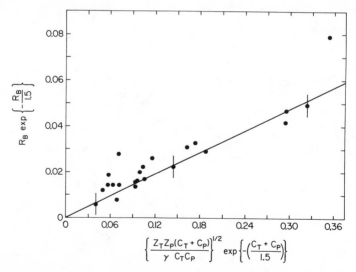

Figure 9 Plot of $R_B \exp (-R_B/1.5)$ as a function of $\{Z_T Z_P(C_T + C_P)/(\gamma C_T C_P)\}^{1/2}$ $\exp [-(C_T + C_P)/1.5]$. The experimental fusion-barrier radii are given as points. The solid line represents the theoretical fission-barrier radii for a nuclear proximity potential (from J. R. Birkelund, J. R. Huizenga. 1977. Private communication).

3.7 Reaction Cross Sections

The total reaction cross section, calculated from an optical-model analysis of the elastic-scattering angular distribution, is given by Equation 6, and the corresponding Fresnel total-reaction cross section is given by Equation 7.

A particular optical-model fit to the elastic scattering of 712-MeV ^{84}Kr on ^{209}Bi leads to a value of $l_{1/2} = 346$, and the value of T_l falls from 0.9 to 0.1 between l values of 333 and 362. The corresponding value of l_{max} from the Fresnel model is 343. The comparison of parameters, including the total-reaction cross sections, deduced from the optical and Fresnel models, is presented in Table 1 for several very-heavy-ion reactions (163). On the basis of the good agreement in σ_R for the two models, we conclude that the simple Fresnel model is extremely useful for estimating the total-reaction cross section between two very heavy ions by the following procedure: (a) estimate R_{Int} from Figure 8, (b) calculate l_{max} from the relation $l_{max} = (\lambda)^{-1} R_{Int}(1 - 2\eta\lambda R_{Int}^{-1})^{1/2}$, and (c) calculate the total-reaction cross section by Equation 9.

3.8 Classical Deflection Functions

For very-heavy-ion reactions, the large number of l waves insures that the deflection function (Equation 4) is sampled at very small intervals, assuring that the classical description becomes useful (42). Although classical deflection functions are not quantitatively valid, they are of instructive value in the classification of different types of orbits, e.g. distant, skimming, and plunging orbits, as illustrated in Figure 3. Analysis of elastic-scattering data for the ^{208}Pb + ^{16}O reaction at $E_{lab} = 107.7$ MeV gives $l_{1/2} = 52\hbar$ (196). The first plunging orbit, produced when $E_{c.m.}$ coincides

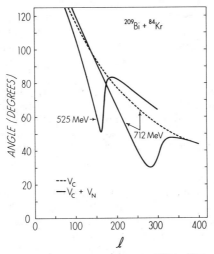

Figure 10   Classical elastic deflection functions for the ^{209}Bi + ^{84}Kr reaction. The nuclear proximity potential is employed for the attractive potential. The dashed line is the Rutherford deflection function [from (163)].

with the top of an l barrier, occurs for $l = 42\hbar$ when the proximity potential is employed in a deflection-function calculation (163). The latter l value divides the cross section into compound and noncompound contributions in approximately the same ratio as the experimental ratio (196). Comparison of the classical deflection functions at two energies for the $^{209}\text{Bi} + {}^{84}\text{Kr}$ reaction is shown in Figure 10 for a proximity potential (163). Such a nuclear potential produces two "rainbows." Other potentials produce different deflection functions (23, 98, 107).

4   CHARACTERISTIC PROPERTIES OF STRONGLY DAMPED HEAVY-ION COLLISIONS

In this section, experimental evidence is provided for the characteristic reaction patterns of strongly damped collisions. Apart from their basic two-body nature, other gross properties of these processes, such as angular distributions, cross sections, mass and energy distributions, are discussed, with emphasis on their dependence on the projectile-target system and the bombarding energy. Specific correlations of experimental quantities lending insight to particular properties of these reactions are discussed in Section 5.

4.1   The Binary Reaction Mode

In strongly damped reactions between heavy ions, in which several hundred MeV of kinetic energy can be lost, a large number of reaction channels is energetically permitted. Hence, there is no a priori obvious preference of the system for simply breaking up into two final fragments. This is indeed a very distinct feature of damped heavy-ion reactions to which experiments lend some support. The definition of a binary reaction mode, as it is understood here, implies the production of only two massive fragments in the exit channel, but allows for the emission of light particles in the course of the reaction. The subsequent decay of the highly excited primary fragments by fission or emission of particles and γ rays is a secondary process that will not conflict with the above restricted definition of the reaction mode.

There is some indirect evidence indicating a two-body nature of strongly damped reactions. The mean kinetic energies of products from reactions induced by relatively light projectiles are close to the Coulomb energies of the corresponding fragments (8, 197). Furthermore, energy, angular, and mass distributions of light fragments are reflected in those of their heavy partners (198–201). However, a direct proof of the binary nature must rely on experiments where two reaction fragments are measured in coincidence.

A technique used frequently to study binary processes consists of measuring kinematical coincidences of the correlated fragments (202, 203). This method has been applied to damped heavy-ion reactions induced by ions ranging from ^{40}Ar to ^{136}Xe for various energies (5, 6, 9, 14, 198, 204). Here, the reaction angles and energies of two fragments are measured in coincidence, from which the two masses and the c.m. energy are inferred, accounting for particle evaporation from the excited fragments. Typical examples of in-plane correlations of fragments from the

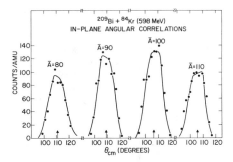

Figure 11 In-plane angular correlation of fragments from the reaction ^{209}Bi + ^{84}Kr at 600 MeV, where the light fragment of mass $\bar{A}$ was measured at $\theta_L = 48.5°$. The average emission angles expected from two-body kinematics are indicated by arrows. The curves represent an evaporation calculation [from (205)].

reaction ^{209}Bi + ^{84}Kr at 600 MeV (205) are displayed in Figure 11. The results are consistent with a binary fragmentation of the intermediate system followed by neutron evaporation, as indicated by model calculations represented by the curves in Figure 11.

However, this method has inherent experimental limitations, e.g. emission of particles during the reaction is not readily detectable. More stringent methods employ the identification of one or two fragments with respect to their mass or charge, in addition to the angle-energy coincidence measurement. Such studies have been performed for reactions induced by ^{12}C (206), ^{16}O (207), ^{20}Ne (208), ^{32}S (209, 210), ^{35}Cl (211), ^{40}Ar (212, 213), and ^{63}Cu (35) ions. With the reservations made above with respect to light-particle emission, all these experiments lend support to the two-body nature of strongly damped reactions. Data on such processes involving very heavy targets are scarce, but there are indications that sequential fission of the heavy fragment is an important decay mode (27, 35, 214, 215).

4.2 Angular Distributions

It is well known from reactions with relatively light ions that the angular distribution $d\sigma(\theta)/d\Omega$ of the reaction cross section carries a signature of the reaction mechanism. For example, compound-nuclear reactions associated with interaction times long compared to typical rotation times of the system and complete equilibration of all degrees of freedom are characterized by a c.m. angular distribution that is symmetric around 90° and in the classical limit of the form $d\sigma(\theta)/d\Omega = 1/\sin\theta$.

At the other extreme, one-nucleon transfer reactions exhibit a variety of angular distributions: at relatively low bombarding energies, the angular distribution is dominated by a "grazing peak" arising from l waves near the maximum of the deflection function (42, 48, 216), which moves forward in angle with increasing energy. This is illustrated by the reaction ^{103}Rh(^{16}O, ^{15}O)^{104}Rh (1) shown in Figure 12. Such reactions are characterized by very short interaction times and

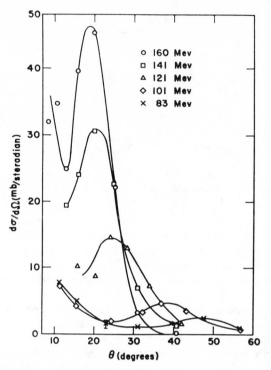

Figure 12 Differential cross section for the reaction $^{103}Rh(^{16}O, ^{15}O)^{104}Rh$ at various bombarding energies [from (1)].

small mutual mass overlap. The small range $\Delta\zeta$ of interpenetration depths that contribute to these reactions can be estimated semiclassically from the angular width $\Delta\theta \approx (2)^{1/2}/(k\Delta\zeta)$, which is, however, modified by dynamical dispersion effects (43, 217). It results in a broadening of the angular distribution with decreasing bombarding energy (cf Figure 12). The angular distributions also become progressively smoother (12, 44, 216) as the energy and the degree of kinematical mismatch are increased, finally approaching exponentially forward-rising distributions. Examples for this behavior are presented in Figure 13, where, among others, the angular distributions of damped one-proton transfer processes for the reaction $^{40}Ca + ^{40}Ca$ (11) are shown. Such reactions involve complex multistep processes and are hence associated with deeper interpenetration, and interaction times which comprise a significant fraction of a rotation period. The semiclassical model for surface reactions developed by Strutinski (217) explains the evolution from sideways-peaked (cf Figure 12) to forward-peaked (cf Figure 13) angular distributions in terms of a mean classical trajectory and a range Δl of l waves contributing to the reaction. However, the latter form of the angular distribution is approached only in the quantal limit of the model. Since classicality is not in general realized for the lighter systems (42), an interpretation of such angular distributions in terms of the orbiting picture (38) has to be taken with caution. On the other hand, the study of heavy systems may help to interpret the effects encountered with light systems. In order to be able to apply safely classical concepts, damped heavy-ion reactions induced by ions heavier than $A = 40$ are discussed first, although they were historically discovered later.

Experimental studies of strongly damped reactions involving heavy projectiles with $A > 40$ have been performed using ^{52}Cr, ^{56}Fe, ^{63}Cu, ^{84,86}Kr, ^{129,132,136}Xe, and ^{238}U beams and targets ranging from ^{56}Fe to ^{238}U. In Table 2, the available measurements of concern to this subject are listed, corresponding to the different projectile-target systems. Techniques applied and experimental quantities extracted from the data are indicated.

There are a number of observations that should be made when discussing experimental angular distributions. Due to the rather poor energy resolution of typically 1-2% (159, 162) for heavy projectile-target combinations, it is difficult to separate experimentally quasi-elastic and damped processes. Therefore, most of the angular distributions for heavy systems include both components. The interpretation of

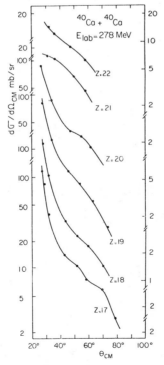

Figure 13 Angular distributions of damped events from the reaction ^{40}Ca + ^{40}Ca. Curves are drawn through the data points and labelled by the fragment Z [from (11)].

Table 2 Survey of experiments on damped heavy-ion reactions. The interaction radius R_{Int} is evaluated using the prescription discussed in Section 3. The Coulomb energy V_{Coul} is calculated for two spheres at a separation distance equal to R_{Int}. The classification parameter ξ is calculated according to $\xi = (E_{c.m.}/V_{Coul} - 1)\hbar l_{max}/\mathscr{I}_{Ns}$ and given in units of 10^{21} rad sec^{-1}.

Projectile	Target	E_{Lab}(MeV)	Measurement	Technique	Ref.	R_{Int}(fm)	$E_{c.m.}/V_{Coul}$	ξ
[11]B	[232]Th	87	f	B,C	32	12.11	1.55	1.1
[12]C	[27]Al	120	c,g	F	1	8.81	6.52	25.3
	[63]Cu	120	c,g	F	1	9.86	3.97	11.45
	[89]Y	107,197	a,c,d,f,g	B	206	10.39	2.91,5.35	6.2,21.3
	[92]Mo	90	c,d,i,j	B,E	222	10.44	2.29	3.5
	[98]Mo	107,197	a,c,d,f,g	B	206	10.54	2.77,5.10	5.6,19.6
	[116]Sn	107,197	a,c,d,f,g	B	206	10.83	2.43,4.47	4.2,15.9
	[107,109]Ag	86,107,109	a,c,d,f,g	B	3,206,218	10.69–10.72	2.04,2.53,2.58	2.6,4.6,4.8
	[197]Au	107–166	j	B,H	206,219	11.83	1.75–2.71	1.7–5.9
[13]C	Ni	105	j	B,E	220	9.95	3.59	8.7
[14]N	[27]Al	65–107	c,d	H	221	8.94	2.92–4.81	5.2–14.6
	[52,53]Cr	64–95	c,d,i,j	B,E	222	9.73,9.76	2.03–3.03	2.2–6.1
	[63]Cu	140	c,g	F	1	10.00	3.92	11.1
	Sn	140	c,g	F	1	10.97–11.02	2.72–2.74	5.5
	[90]Zr	75	c,d,i,j	B,E	222	10.53	1.70	1.36
	[93]Nb	95,120	c,d,i,j	B,E	222	10.59	2.12,2.67	2.8,5.1
	[92-100]Mo	97	c,d,i,j	B,E	222	10.57–10.71	2.10–2.15	2.8–2.9
	[103]Rh	140	c,g	F	1	10.76	2.92	6.4
[16]O	[107,109]Ag	78–250	c,d,f,g	B	3,8,17,218	10.82–10.85	1.58–5.05	1.1–19.7
	[27]Al	90–165	a–g,i,j	B,F,G	1,2,22,223,224,305,313	9.06	3.42–6.27	7.4–23.8
	[48]Ca	56	a–e,g	B,D	225	9.74	1.78	1.4
	[51]V	126	e,f	B	2	9.82	3.56	8.6
	[58]Ni	60–96	d,e,g,i,j	B,D,E,G	12,13,207,226–228,306	10.00	1.46–2.33	0.7–3.4

Projectile	Target							
^{18}O	^{63}Cu	160	c,g	F	1	10.11	3.86	10.7
	^{93}Nb	131	e,f	B	2	10.71	2.53	4.4
	^{94}Zr	140,315	f	B	229	10.72	2.78,6.26	5.5,27.7
	^{103}Rh	160	c,g	F	1	10.87	2.91	6.2
	Sn	160	c,g	F	1	11.08	2.70	5.4
	^{197}Au	125–315	d,f,g	B,H	219,229,230	12.09	1.54–3.87	1.0–12.6
	^{208}Pb	140,315	f	B	229,231	12.20	1.68,3.78	1.5,12.0
	^{232}Th	112–315	f	B	229,230,232	12.43	1.26–3.53	0.3–10.6
^{19}F	^{27}Al	96	c,e–g	B,D	233	9.17	3.53	7.5
	^{63}Cu	190	c,g	F	1	10.28	4.0	11.1
	^{103}Rh	190	c,g	F	1	11.04	3.04	6.6
^{20}Ne	^{27}Al	85–206	a,d	B,G	29,208,223	9.27	2.42–5.86	3.4–21.3
	Ni	164,173	c,f,h	B	234	10.35	3.21,3.38	7.0,7.9
	Ag	175,250	a,c,f,g,i	B	8,24,314	11.18	2.44,3.49	4.1,9.2
^{22}Ne	^{12}C	173	f	B	32	8.59	≦6.07	≦21.0
	^{94}Zr	174	f,h	B	235	11.03	2.70	4.8
	^{232}Th	174	f	B	232	12.74	1.56	1.3
^{40}Ar	^{58}Ni	187–280	c,g,i,j	B,D,E,G	15,34,212,213,236–238	10.97	1.60–2.51	1.2–3.9
	^{64}Ni	280	f,g	B,D,E	237	11.11	2.64	4.3
	^{107,109}Ag	169–340	c,d,f,g	B	8,16,239	11.91,11.94	1.20–2.42	0.2–3.7
	^{121}Sb	169–340	f,g	B	239	12.13	1.17–2.34	0.1–3.4
	^{139}La	282	a,c,d,f	F,H	199	12.38	1.83	1.7
	^{140,142}Ce	282	a,c,d,f	F,H	199	12.39,12.42	1.81	1.6
	^{141}Pr	282	a,c,d,f	F,H	199	12.40	1.78	1.5
	^{142,145}Nd	282	a,c,d,f	F,H	199	12.42,12.45	1.76,1.77	1.5
	$^{144-154}$Sm	282	a,c,d,f	F,H	199	12.44–12.57	1.71–1.75	1.4
	^{151,153}Eu	282	a,c,d,f	F,H	199	12.53,12.56	1.71,1.72	1.3–1.4
	$^{154-160}$Gd	282	a,c,d,f	F,H	199	12.57–12.64	1.70–1.72	1.3–1.4
	^{159}Tb	282	a,c,d,f	F,H	199	12.63	1.69	1.3
	^{197}Au	159–340	f,h	D	25,244a	13.06	0.84–1.80	≦1.7

Table 2 *continued*

Projectile	Target	E_{Lab}(MeV)	Measurement	Technique	Ref.	R_{Int}(fm)	$E_{c.m.}/V_{Coul}$	ξ
^{40}Ca	^{232}Th	288, 379	h	B, C	4	13.41	1.41, 1.86	0.6, 1.9
	^{238}U	≦288	f	H	27	13.46	≦1.39	≦0.6
^{40}Ca	^{40}Ca	278, 288	c, e, f, g	B, C, D	11, 197, 240, 241, 241a	10.49	2.53, 2.62	4.2, 4.5
^{40}Ca	^{58}Ni	280	d, f–h	B, D	15	10.97	2.25	3.1
	^{64}Ni	180–280	c, f	B, E	237, 238	11.11	1.53–2.37	0.8–3.5
	^{208}Pb	288	c, d, f, g	B	197, 242	13.17	1.35	0.5
^{52}Cr	^{56}Fe	180–280	c, d, f	C, D	238	11.26	1.17–1.82	0.1–1.6
^{56}Fe	^{148}Sm	300, 377	c–e	F, H	243	12.93	1.21, 1.52	0.2–0.9
	^{197}Au	538	a, b	H	244	13.50	1.91	2.0
	^{238}U	538	a, b	H	244	13.90	1.76	1.5
^{63}Cu	^{63}Cu	441, 455	f	B, C, D	245	11.69	2.13, 2.20	2.6, 2.8
	^{93}Nb	280	a–d, f	D	246	12.28	1.20	0.2
	$^{94-96}$Zr	165–280	b	H	247	12.30	0.73–1.23	≦0.2
	^{124}Sn	441, 455	f	B, C, D	245	12.77	1.79, 1.85	1.5, 1.6
	^{148}Sm	342, 431	c–e	F, H	243	13.09	1.21, 1.53	0.2, 0.9
	^{197}Au	365, 443, 455	a–i	B, C, D, E	9, 19, 35, 245, 248	13.66	1.15, 1.39, 1.43	0.1, 0.6, 0.6
	^{232}Th	441, 455	f	B, C, D	245	14.01	1.29, 1.33	0.4, 0.4
^{84}Kr	^{65}Cu	494, 604	f, g	B	239	12.16	1.74, 2.13	1.3, 2.5
	^{72,74}Ge	100–300	b	H	249	12.31, 12.35	0.34–1.05	≦0.02
	^{107,109}Ag	600	c, d, f, g	B	8, 250	12.94, 12.98	1.79, 1.80	1.5
	^{144,154}Sm	470, 595, 720	a, c–h	B	257	13.47–13.60	1.30, 1.66, 1.97	0.3, 1.1, 2.0
^{84}Kr	^{165}Ho	450, 492, 720	a–f, h	B, E	14, 198, 205	13.73	1.18, 1.29, 1.89	0.2, 0.3, 1.8
	^{186}W	492, 500	b	E	14, 198	13.97	1.23, 1.25	0.2, 0.3
	^{208}Pb	494, 510, 718	a, c–e	A, B	23	14.20	1.18, 1.21, 1.71	0.2, 0.2, 1.3
	^{209}Bi	500, 525, 600, 720	a–h	B, D, E	5, 6, 14, 198, 251	14.21	1.18–1.70	0.2–1.3
	^{238}U	500, ≦605	a, b, f	E, H	7, 14	14.50	1.12, ≦1.36	0.1, ≦0.5

^{86}Kr	^{107,109}Ag	606, 620	c, d, f, g, i	B	250	12.98, 13.01	1.77, 1.79	1.4, 1.5
	^{159}Tb	620	f, g	B	118	13.70	1.64	1.1
	^{181}Ta	620	f, g	B	119	13.95	1.55	0.9
	^{197}Au	620	c, d, f, g, j	B	26, 252, 307	14.13	1.49	0.7
^{129}Xe	^{197}Au	761	f	H	253	14.81	$\leqq 1.11$	$\leqq 0.1$
^{132}Xe	^{56}Fe	779	c, f	B	254	12.72	1.46	0.7
	^{197}Au	779	f	H	253	14.85	$\leqq 1.13$	$\leqq 0.1$
^{136}Xe	^{109}Ag	1130	c, f	B	205	13.79	1.90	1.8
	^{159}Tb	979	f, g	B	118	14.48	1.51	0.8
	^{165}Ho	1130	a, c–h	B	205	14.55	1.73	1.3
	^{181}Ta	1120	a, c–e, h	B	30	14.73	1.66	1.1
	^{197}Au	979	f, g	B	215, 258	14.91	1.41	0.5
	^{208}Pb	1120	a, c–e, h	A, B	30	15.02	1.60	1.0
	^{209}Bi	1130	a, c–h	A, B	21, 205, 255, 256	15.03	1.59	1.0
	^{238}U	$\leqq 1150$	f	H	20	15.31	$\leqq 1.57$	$\leqq 0.9$

a σ_R.

b σ_{CF}/σ_R.

c $d\sigma/d\Omega$.

d $d\sigma/dE$.

e $d^2\sigma/d\Omega\, dE$

f $d\sigma/dA$, $d\sigma/dZ$.

g $d^2\sigma/d\Omega\, dA$, $d^2\sigma/d\Omega\, dZ$.

h $d^2\sigma/dE\, dA$, $d^2\sigma/dE\, dZ$.

i γ multiplicity.

j light particles.

The following are the techniques used in making the table measurements: A: single counter; B: ΔE-E telescope; C: magnetic spectrometer; D: time of flight; E: correlated fragments; F: recoil range; G: in-beam γ spectroscopy; and H: off-beam induced radioactivity.

angular distributions from single-counter measurements is sometimes complicated by contributions from other processes such as fusion-fission and sequential fission. Transformation of the data into the center of mass requires assumptions on the mass-to-charge ratio in cases where only the fragment charge is measured. It also depends on an accurate knowledge of the initial fragment energies, which are obscured by neutron emission and pulse-height defects (259). Neutron emission from the excited fragments also tends to broaden angular distributions, as well as mass distributions (260). Finally, two-body kinematics is always assumed, neglecting the possibility of particle emission during the reaction. The above effects have been accounted for in the analyses to various degrees of accuracy.

As examples for very heavy systems, the ^{136}Xe- and ^{84}Kr-induced reactions on ^{209}Bi and ^{165}Ho are discussed first. The energies range from 1.6 times the Coulomb barrier for the ^{209}Bi + ^{136}Xe reaction to 1.9 for the system ^{165}Ho + ^{84}Kr, where the Coulomb energy is calculated for two spheres at a separation distance equal to the interaction radius R_{Int} (cf Sections 2.1 and 3.6). The large values of the Sommerfeld parameter ($\eta = 245$–130) indicate that these systems are close to the classical limit (see Equation 3). The total-reaction cross sections are, within experimental accuracy, equal to $\sigma_R \approx 2.8$ b, most of which is due to the strongly damped mechanism. The energy and the charge product $Z_P \cdot Z_T$ determining the distance of closest approach and the character of the total-interaction potential are seen to vary substantially for the four reactions. As discussed in Section 3, the interaction potential for such heavy systems is likely to possess no minima at all or only for a limited number of low l values. However, the steepness of the potential, i.e. the repulsive force, is quite different for the four reactions. Therefore, the region of interpenetration depths or l values where strong nuclear attraction is felt becomes narrower with increasing charge product and decreasing energy.

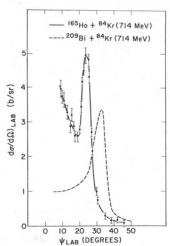

Figure 14 The laboratory angular distributions of light fragments from the reactions ^{165}Ho + ^{84}Kr (*solid curve*) and ^{209}Bi + ^{84}Kr (*dashed curve*) at $E_{\text{Lab}} = 714$ MeV [from (205)].

The angular distribution of the light fragments from the reaction ^{209}Bi + ^{136}Xe (21, 256) exhibits a sharp peak at $\theta_{Lab} = 29.8°$, slightly forward of the quarter-point angle $\theta_{1/4} = 33°$. The cross section drops rapidly and almost symmetrically on both sides of the peak with essentially no intensity at $\theta_{Lab} = 0°$.

A similarly sharp peak is observed (205) in the case of the ^{165}Ho + ^{136}Xe reaction, which, however, shows a slight forward asymmetry, much the same as that of the ^{209}Bi + ^{84}Kr reaction depicted in Figure 14. Although the mass and charge of the composite system are similar for the two reactions, the asymmetry is somewhat more pronounced for the ^{209}Bi + ^{84}Kr reaction. The angular distribution of the reaction ^{165}Ho + ^{84}Kr, also shown in Figure 14, is also dominated by a narrow peak. However, the forward asymmetry has now developed into a conspicuous forward rise of the angular distribution, suggesting that there is considerable cross section at zero degrees. On the other hand, preliminary measurements (205) on the system Ag + ^{136}Xe at 1124 MeV indicate that such a forward asymmetry may be very weak, although this system is well matched to the ^{165}Ho + ^{84}Kr system, with respect to charge product and energy. The sideways-peaking phenomenon has also been observed for reactions induced by 620-MeV Kr ions on targets of ^{197}Au (26) and ^{181}Ta (119), as well as for the somewhat lighter systems ^{197}Au + ^{63}Cu at 443 MeV (19, 35), Ag + ^{84}Kr at 620 MeV (250), and ^{93}Nb + ^{63}Cu at 280 MeV (246). The angular distributions of these reactions all show important forward-tending contributions, which are enhanced as the system becomes lighter.

When an attempt is made (19, 31, 250) to separate quasi-elastic from damped events, it is seen that the angular distribution of the former contributes mainly to the sharp peak. The angular distributions of the remaining damped events appear generally flatter. However, for the very heavy systems, damped events still exhibit angular distributions with well-defined peaks staying at roughly the same angle, though somewhat broader and reduced in height. The appearance of sharply peaked angular distributions, most strikingly observed in reactions induced by ^{84}Kr and ^{136}Xe ions, has been referred to as "angular focusing" (6, 132).

Since the type of the angular distribution for damped heavy-ion reactions depends so markedly on the charge product and therefore on the interpenetration depth accessible to the ions, it is crucial to examine also the evolution of the angular distribution with increasing bombarding energy. In Figure 15 are shown angular distributions of light fragments from the reaction ^{208}Pb + ^{84}Kr at bombarding energies ranging from 1.2 to 1.7 times the Coulomb barrier (23). Again, all distributions are sideways-peaked. The peak becomes narrower with increasing bombarding energy and shifts forward in angle, quite similar to the peripheral grazing reactions discussed earlier (see Figure 12). At the same time, a forward asymmetry develops at high bombarding energies. The latter behavior is also observed in experiments on lighter systems, such as ^{197}Au + ^{63}Cu (19, 35), although such measurements are scarce (see Table 2). Summarizing, it appears that sideways peaking of the angular distributions is most important for very heavy systems and incident energies not too high above the Coulomb barrier. At higher energies, and in particular for lighter systems, forward-tending continuous contributions gain significance.

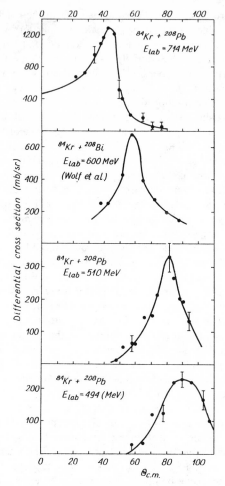

Figure 15   Angular distribution of the light fragments from the reaction ^{208}Pb + ^{84}Kr (23). Included are data of (6) for the similar system ^{209}Bi + ^{84}Kr [from (23)].

Apart from these general trends, a particularly interesting observation has been made for the reactions induced by 470- and 595-MeV ^{84,86}Kr ions on the two isotopes ^{144}Sm and ^{154}Sm (257). A strong forward asymmetry of the angular distribution is found, more prominent and forward-rising at the higher energy. However, the relative contribution of the smooth part appears to be always higher for the highly deformed and neutron-rich ^{154}Sm target, as compared to the nearly spherical and neutron-poor ^{144}Sm isotope. This behavior is most evident for the lowest energy.

The existence of the focusing effect observed in reactions between very heavy systems, which remains also for the strongly damped events, suggests immediately the conclusion that interaction times must be strongly l dependent. More than 400 l waves contribute to the damped-reaction cross section in the case of ^{209}Bi + ^{136}Xe. Despite the large variation in rotational velocity, these many l values all lead to reaction angles concentrated in a narrow angular range. Since the impact parameter is connected to the distance of closest approach, it is also possible to conclude that the interaction time increases with increasing interpenetration of the two ions.

The observed shift of the peak in the angular distribution with changing bombarding energy is roughly accounted for by the shift of the classical deflection function illustrated in Figure 10, which also demonstrates the angular confinement of the cross section. Depending on the bombarding energy, a band of intermediate l waves associated with deeply plunging orbits and longer interaction times may cause rotation of the intermediate system near to the grazing angle, beyond it, or even to the opposite side of the beam. The latter effect corresponds to the orbiting situation proposed by Wilczyński (38) for the ^{232}Th + ^{40}Ar system. It is consistent with this picture to expect orbiting to be more important for deformed or neutron-rich targets. For a prolate nucleus, the Coulomb potential at the nuclear surface is reduced at the tips with respect to the spherical case, whereas the nuclear potential is not, resulting in a deeper interpenetration and longer reaction times. A similar result may be produced by the charge-diluting effect of the extra neutrons.

An attempt to interpret the 714-MeV ^{165}Ho + ^{84}Kr angular distribution is shown in Figure 16 (261), where the more appropriate quantity $d\sigma/d\theta$ is plotted versus positive and "negative" angles. The separation of the two components has been achieved using a correlated-fragment technique applying energy and mass windows.

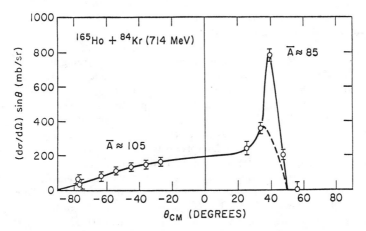

Figure 16   The differential cross section of light fragments from the reaction ^{165}Ho + ^{84}Kr at $E_L = 714$ MeV, decomposed into contributions from positive and negative angles. Indicated values for $\bar{A}$ denote average fragment mass numbers [from (261)].

It should be emphasized that this is only an interpretation of the data consistent with the orbiting picture. A more direct proof of this effect would require measurements at very forward angles or the determination of the fragment-spin polarization for positive- and negative-angle branches, e.g. by measuring the circular polarization of the deexcitation γ rays (38).

In turning now to the lighter systems, one notices that the situation is more complicated than with the heavier ones. Compound-nucleus formation is known to be an important reaction channel for low and medium l values (262), which may either lead to subsequent particle evaporation or fission. Damped reactions are due to an intermediate band of l waves, the highest l values contributing to quasi-elastic events. Depending on the asymmetry of the target-projectile system and the details of the reaction dynamics, fusion-fission events may complicate the analysis. On the other hand, higher experimental energy and mass resolution for the lighter ions permits a more detailed interpretation of data (15, 16).

Measurements on damped reactions induced by projectiles with $A \leq 40$ include projectiles from ^{11}B to ^{40}Ca and targets ranging from ^{27}Al to ^{238}U (see Table 2). Experiments by Artukh et al (4) on the reaction ^{232}Th + ^{40}Ar at 297 and 388 MeV, extensively studied both experimentally and theoretically, revealed sideways-peaked angular distributions with strong forward-rising continuous contributions. Similar observations were reported (232) for ^{22}Ne- and ^{16}O-induced reactions on ^{232}Th at 174 and 137 MeV, respectively. However, for these lighter projectiles it can be concluded (1, 3, 218, 232, 250) that the peak in the angular distribution is almost entirely due to weakly damped few-nucleon transfer processes, whereas for the heavy systems, it includes many more reaction channels. If such events are excluded from consideration, rather structureless angular distributions appear. They are substantially forward-peaked in the c.m. systems in clear excess of a $1/\sin \theta$ law (16, 18, 24) similar to the angular distributions shown in Figure 13. Some ^{14}N-, ^{20}Ne-, and ^{40}Ar-induced reactions on medium-weight targets (119) lead to angular distributions $d\sigma/d\Omega$ exhibiting an additional slight backward rise, indicating contributions from highly equilibrated processes such as fusion-fission.

The dependence of the angular distributions on the bombarding energy expected from that of heavy systems is reproduced by the ^{40}Ar-induced reactions on ^{232}Th (4): a forward shift of the side peak and enhancement of the forward-rising contribution with increasing energy. Quite surprisingly, the latter effect is not always seen. In fact, the reactions Ag + ^{14}N (17) and Ag + ^{20}Ne (24), measured at two energies, show the opposite behavior: a steeper forward rise at the lower energy. This result is possibly due to the interplay between the l-dependent driving forces governing nucleon transfer (18) and the energy dissipation in the exit channel.

On the average, light and heavy systems follow the same systematics. The more abrupt change in the character of the angular distributions when going from quasi-elastic to damped events for the lighter systems can be attributed to the existence of minima in the interaction potential for these systems (163). As illustrated in Figure 3 by trajectories 2 and 3, only a slight change in impact parameter leads from a skimming trajectory to a deeply plunging one. As an example, only three l waves have been estimated to contribute to the damped reaction ^{27}Al + ^{16}O at

90 MeV (22). Therefore, unless an l-wave decomposition is made, it is difficult to compare various reactions in detail.

The character of a reaction certainly depends on the delicate balance between nuclear, centrifugal, and Coulomb forces, as well as on dynamical and kinematical variables. It seems, therefore, difficult to conceive of a classification of damped reactions in terms of a single overall parameter. However, some importance may be attributed to the kinetic energy available at the Coulomb barrier determining the interpenetration depth and the angular velocity. Some features of the angular distribution may then be expected to be correlated with the parameter $\xi = [E_{c.m.}/V(R_{int}) - 1]\hbar l_{max}/\mathscr{I}_{NS}$, which represents a modified angular velocity and is also listed in Table 2. Generally, one observes sideways-peaked angular distributions for $\xi \lesssim 10^{21}$ rad sec^{-1}. With increasing values of ξ, the forward-rising contribution gains rapidly in importance. Recently, Galin (263) introduced a classification parameter η', defined as a ratio between Coulomb and velocity-proportional friction forces. Parameters like those discussed above may be of some value in the classification of heavy-ion reactions. However, in comparing reactions between light and very heavy ions, it is extremely important to remember the very different range of l values contributing to damped reactions. In light-ion reactions, a large fraction of lower l waves leads to fusion, whereas the whole range of l values goes into damped processes for very heavy systems.

Classical dynamical calculations of damped heavy-ion reactions have been performed for various light and heavy systems (89, 90, 98–100, 123, 132, 135). Gross et al

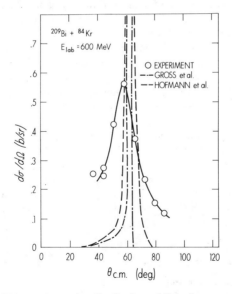

Figure 17 The experimental angular distribution of light fragments from the reaction ^{209}Bi + ^{84}Kr at 600 MeV (6) is compared to the calculations of Gross et al (100) and Hofmann et al (134).

(98–100) have solved Newton's equations for radial and polar coordinates (cf Equations 37–40), including friction forces defined by Equation 17. Using the friction form factor of Equation 23, they obtain values for the friction constants $C_r = 4 \cdot 10^{-23}$ sec MeV^{-1} and $C_t = 10^{-25}$ sec MeV^{-1}, i.e. dominant radial friction. As an example, Figure 17 shows the experimental and theoretical angular distributions of the reaction ^{209}Bi + ^{84}Kr at 600 MeV.

Although the position of the peak in the angular distribution is reasonably well reproduced by theory, there is a substantial disagreement between calculation and experiment with respect to its shape. This is not too surprising for a purely classical calculation neglecting nucleon transfer and dynamical fluctuations. The effect of fluctuations is to broaden the theoretical angular distribution, as demonstrated in Figure 17, by the calculation of Hofmann et al (134). A greatly improved description of data is achieved when mass exchange is taken into account (120, 135).

The character of the deflection function obtained in a model calculation depends crucially on the choice of the nuclear potential. In contrast to the deeper potentials used by Gross et al, the potential of Deubler & Dietrich (106, 107, 133) is close to the proximity potential (see Section 3.3) strongly supported by experiments. As for other cases, both models give comparably good descriptions also for the reaction ^{232}Th + ^{40}Ar. However, whereas the deeper potential leads to a deflection function reflecting the classical orbiting picture (38), the deflection function of Deubler & Dietrich (106, 107) exhibits two rainbow extrema. The "inner" rainbow located at small negative angles is then associated with strongly damped events.

In order to estimate the extent to which the predictive power of various models is tested by a comparison to experimental angular distributions, it is useful to reconsider the elastic deflection functions. Examples of these for the ^{209}Bi + ^{84}Kr reaction at two energies are displayed in Figure 10. First, one notices that the shift in the position of the peak in the angular distribution with changing bombarding energy reflects a similar shift of the elastic-scattering rainbow angle. Second, the deflection functions of heavy systems always lead to a confinement of the reaction cross section into a narrow angular range. The increasing effect of the Coulomb repulsion, as kinetic energy is lost in the reaction, results in a further flattening of the deflection function. The latter can be interpolated by elastic deflection functions at different energies as done, for example, by Vandenbosch et al (23). Hence, the angular focusing effect is a rather model-independent feature of damped reactions between very heavy ions. It is then the shape, rather than the peak position of the angular distribution, that discriminates between models. More stringent tests can be imposed on the calculations if energy loss and nucleon transfer are considered.

4.3 Fragment Mass and Charge Distributions

The complexity of reactions that occur in the bombardment of a very heavy target with heavy ions is impressively demonstrated by the mass distribution of the final fragments. Figure 18 shows the fragment mass distribution obtained from a thick-target experiment on the reaction ^{238}U + ^{40}Ar at 288 MeV (27), employing radio-

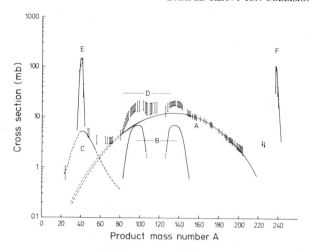

Figure 18 Total integrated mass yields (upper and lower limits are indicated at those mass numbers for which experimental data were obtained) and their decomposition into individual components: (A) complete fusion-fission, (B) transfer-induced fission, (C) strongly damped events, and (E) and (F) quasi-elastic transfer. The existence of products from the sequential fission of heavy fragments formed in damped collisions (D) is also indicated; however, a mass distribution for this component could not be deduced [from (27)].

chemical methods. The mass distribution is interpreted as a superposition of several components: apart from quasi-elastic, few-nucleon transfer products concentrated around projectile and target masses, compound-nucleus fission and transfer-induced fission are seen to be important reaction channels. The damped-reaction process leads to a mass distribution centered around the projectile mass. Due to the high excitation energy, the corresponding heavy fragments undergo sequential fission. Similar measurements have been performed for the systems ^{238}U + ^{84}Kr (7), ^{238}U + ^{136}Xe (20), and ^{197}Au + ^{129,132,136}Xe (253).

Since for very heavy systems, compound-nucleus formation is not a frequent process (204), the correlated-fragment technique (see Section 4.1) can be used to discriminate against events from sequential fission of the target-like fragment. Such a measurement is shown in Figure 19 for the ^{209}Bi + ^{84}Kr reaction at 600 MeV (6). Although considerable mass exchange occurs during the reaction, indicated by a large width of the mass distribution, the mass distribution exhibits two well-defined peaks near the masses of projectile and target. This has been confirmed by measurements of the Z distributions of light fragments from the damped reactions induced by ^{84,86}Kr or ^{136}Xe ions on targets of ^{181}Ta, ^{197}Au, ^{208}Pb, and ^{209}Bi at various bombarding energies (6, 21, 30, 118, 251, 256, 264).

Seemingly in contrast to these findings are observations made with somewhat lighter ions and targets, such as those for the systems Ag + ^{14}N (17) and Ag + ^{20}Ne (24) at 100 MeV and 175 MeV bombarding energy, respectively. The fragment Z distributions are characterized by a peak at very low Z. However, whereas the

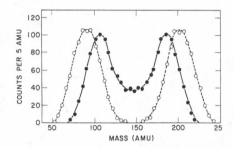

Figure 19 Mass distributions for the strongly damped reaction ^{209}Bi $+ ^{84}$Kr at 600 MeV, when the light fragment is detected at 59° lab (*solid curve*), and 34° lab (*dashed curve*) [from (6)]. The fragment coincidence technique biases the mass distributions due to a small percentage of sequential fission of the primary heavy fragments.

cross section for the former reaction decreases rapidly for higher Z, the charge distribution corresponding to the Ag $+ ^{20}$Ne reaction shows an additional contribution increasing with increasing Z. This effect becomes more apparent for the Ag $+ ^{40}$Ar (16) and ^{197}Au $+ ^{40}$Ar reactions (25). The Z distribution for the latter reaction also exhibits a peak near $Z = 18$ (Ar). Due to the importance of compound-nucleus fission in these reactions, nothing quantitative can be inferred from the continuous parts of the Z distributions. However, the latter process cannot explain the preference of low fragment Z in ^{14}N- and ^{20}Ne-induced reactions on Ag mentioned above.

The markedly different appearance of fragment charge distributions for relatively light and very heavy systems poses questions with regard to the driving forces governing the transfer of charge and mass between the constituents of the intermediate system. In a simple approach (164), such driving forces are determined by a potential of the form

$$U_l(3, 4) = B_3 + B_4 + V_C(3, 4) + V_l(3, 4) + V_N(3, 4) + \text{const.,} \qquad 54.$$

where B_3 and B_4 are the mass excesses of the outgoing fragments 3 and 4. V_C, V_l, and V_N are the Coulomb, centrifugal, and nuclear-interaction potentials for two touching spherical nuclei, respectively. The constant is adjusted so as to give $U_l = 0$ for the initial fragmentation. The masses B_3 and B_4 may be calculated in the droplet model (265), including shell corrections. However, at least in the limit of very high excitation energies $E^* > B^2 A/24$, where B denotes the single-particle gap of one fragment and A the combined mass, liquid-drop masses are appropriate because shell effects are believed (110) to disappear.

The mass-asymmetry potential (Equation 54), evaluated for $l = 0$ and liquid-drop masses, shows different features for light and heavy systems; whereas for the system Ag $+ ^{14}$N, U_0 increases steadily towards symmetric charge fragmentation (17), it develops a flat top many Z units wide for the Ag $+ ^{40}$Ar system (117). For even heavier systems, such as ^{209}Bi $+ ^{84}$Kr (264), U_0 exhibits a minimum at symmetry. As l is increased, a depression at symmetry develops in U_l, which is shallow for the lighter systems and steeper for the heavier ones. Hence, depending on the injection point given by the target-projectile combination and the angular momentum, the intermediate system will be subject to forces of different strength driving it in the direction of either symmetry or asymmetry.

Examining the potentials (Equation 54) for the reactions discussed above, one notices that the predicted direction of the driving force generally complies with what is experimentally observed: In both the ^{14}N- and ^{20}Ne-induced reactions on Ag, the injection point is on the low-Z side or near the potential peak for all l values (18). Therefore, if forces are exerted, they tend to drive the system toward asymmetry. For reactions induced by ^{40}Ar and heavier ions on heavy targets, liquid-drop driving forces are directed towards symmetry for most l values. Experimental results (9, 16, 21, 25, 256, 264) are consistent with the expected drift in Z, but only for very high excitation of the intermediate system. The energy-integrated Z distributions are still centered near the charges of projectile and target. A possible manifestation of shell effects in heavy-ion–induced multinucleon exchange is currently an interesting but as yet unresolved problem.

As discussed in Section 2.3, nucleon exchange in damped heavy-ion reactions may be viewed in terms of a diffusion mechanism proceeding during the interaction. The broad fragment mass and charge distributions, extending to both sides of the projectile mass and charge, signify such a mechanism. As is indicated by the Einstein relation (Equation 33), the drift coefficient is influenced less by the excitation energy but strongly dependent on the driving force $\partial U_l / \partial A_1$. It has been observed for several cases, such as the reactions ^{232}Th + ^{40}Ar (4) or ^{209}Bi + ^{136}Xe (21), that a charge drift is negligible, such that the charge distributions are characterized by a diffusion broadening. A detailed analysis of mass and charge distributions in terms of a diffusion model must employ l-dependent interaction times that can only be inferred from the "experimental" deflection function, as is discussed in Section 5.

Additional insight into the reaction mechanism can be gained by studying the correlation between neutron and proton numbers of the final fragments. These quantities are experimentally not easily accessible because of particle evaporation from the highly excited fragments after the breakup of the intermediate system (199, 260). The importance of this effect even for relatively light systems is demonstrated by measurements of the reaction ^{40}Ca + ^{40}Ca at 278 MeV (11). Since this is a symmetric system, the fragment charge distribution is expected to be symmetric around $Z = 20$, whereas the experimental Z distribution for damped events is found to be shifted down in Z by several units.

Masses and charges have been identified for fragments produced by several ^{40}Ar-, ^{40}Ca-, ^{63}Cu-, ^{56}Fe-, and ^{129,132,136}Xe-induced reactions on heavy targets (10, 199, 237, 245, 253). Table 3 compares some average or most probable N/Z ratios of the light final fragments with those of projectile, target, and combined system. It can be inferred from this table that the measured N/Z ratio is correlated with that of the combined system, although in some cases, it is closer to the projectile value. Comparing the four ^{40}Ar- and ^{40}Ca-induced reactions on Ni isotopes, one can conclude that neutron-rich fragments are produced when the

Table 3 Comparison of N/Z ratios of light fragments from damped reactions with those of target, projectile, and combined system

Reaction	E_{Lab}(MeV)	$\theta_{Lab}(°)$	N/Z target	projectile	combined	N/Z fragment	Ref.
$^{40}Ca + ^{40}Ca$	280	10	1.00	1.00	1.00	1.05	10
$^{58}Ni + ^{40}Ar$	280	25	1.07	1.22	1.13	1.09	237
$^{58}Ni + ^{40}Ca$	280	25	1.07	1.00	1.04	1.05	237
$^{64}Ni + ^{40}Ar$	280	25	1.29	1.22	1.26	1.14	237
$^{64}Ni + ^{40}Ca$	280	25	1.29	1.00	1.17	1.10	237
$^{63}Cu + ^{63}Cu$	455	10	1.17	1.17	1.17	1.15	245
$^{124}Sn + ^{65}Cu$	441	14	1.48	1.24	1.39	1.18	245
$^{141}Pr + ^{40}Ar$	282	0–60	1.39	1.22	1.35	1.36[a]	199
$^{144}Nd + ^{40}Ar$	282	0–60	1.40	1.22	1.36	1.37[a]	199
$^{148}Sm + ^{40}Ar$	282	0–60	1.39	1.22	1.35	1.36[a]	199
$^{197}Au + ^{63}Cu$	455	10	1.49	1.17	1.41	1.23	245
$^{232}Th + ^{40}Ar$	295	18	1.58	1.22	1.52	1.29	10
$^{232}Th + ^{65}Cu$	441	14	1.58	1.24	1.50	1.29	245

[a] Pre-evaporation value, deduced from measurement of a particular fragment, assuming a division of the total excitation energy according to the masses of the fragments.

combined system is neutron-rich. Analogously, neutron-deficient fragments result from a neutron-deficient combined system. This is further illustrated by Figure 20, where the most probable fragment N/Z ratios for the four above reactions are shown as a function of the fragment charge. Apart from the fact that the C, N, O, and F isotopes tend to be preferentially produced with N/Z values closer to their most stable ones, the fragment N/Z ratio is rather constant, but strongly dependent on that of the combined system. This indicates that, on the average, mass-to-charge equilibration is reached on a fast time scale by the initial nucleon-exchange processes in a damped heavy-ion reaction. Radiochemical studies of the fragment isotopic distributions (27, 253) arrive at a similar conclusion. Deviations of the fragment N/Z ratio from the equilibrium value, as well as odd-even effects observed (237) in the fragment Z distributions for neutron-deficient combined systems, can be understood by considering subsequent particle evaporation (237, 260).

In proceeding with the discussion of in-beam measurements of the fragment isotopic distributions, one must refer to the extensive work of the Dubna group (4, 32, 230, 232, 235, 266–268), which studied ^{11}B-, ^{15}N-, ^{16}O-, ^{22}Ne-, and ^{40}Ar-induced reactions on ^{232}Th, as well as the systems $^{12}C + ^{22}Ne$, $^{197}Au + ^{16}O$, and $^{94}Zr + ^{22}Ne$. Production cross sections are measured for various isotopes at a fixed reaction angle. Figure 21 shows data obtained for the $^{232}Th + ^{16}O$ system at $\theta_{Lab} = 40°$ (230). The differential cross section is plotted versus the ground-state Q value $Q_{gg} = B_P + B_T - B_3 - B_4$, where the B_i are the mass excesses of the initial and final ions.

As can be seen in Figure 21, the differential cross section $d\sigma/d\Omega$ exhibits an

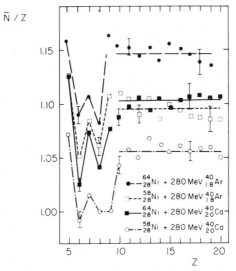

Figure 20 Variation of the most probable fragment N/Z ratio with Z of the fragment measured at $\theta = 25°$. Lines are drawn to guide the eye [from (237)].

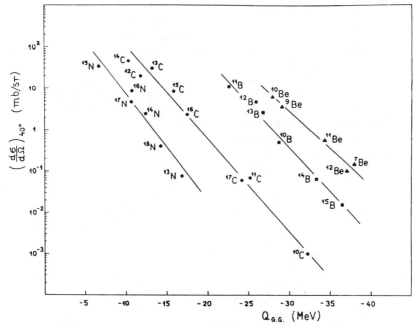

Figure 21 Differential cross sections $(d\sigma/d\Omega)_{40°}$ for the production of Be, B, C, and N isotopes in the ^{232}Th $+ \,^{16}$O reaction as a function of Q_{gg} [from (230)].

exponential dependence on Q_{gg} for the isotopes of a given element. The logarithmic lines have similar slopes, but are successively displaced along the Q_{gg} axis as Z is increased. The distances between the lines are larger when going from an odd Z to the higher even Z. These systematic features of the cross section are fairly commonly observed in the above reactions, although lines sometimes have different slopes and are disordered with respect to Z of the fragment. The O isotopes measured in the reaction $^{94}Zr + ^{22}Ne$ (235) do not follow the systematics at all. If, however, instead of Q_{gg}, a Q value is used which is corrected for the pairing energy of the transferred nucleons, the Q systematics is restored.

The strong dependence of the cross section on the ground-state Q value has to be taken as a hint to the statistical nature of the transfer mechanism (269). Using a simple phase-space argument, the transfer probability is determined by the level density $\rho = \rho_3 \rho_4$ of the final system. In a constant-temperature Fermi gas model, the intermediate system is then expected to populate different exit channels, according to

$$\sigma \sim \rho_3 \rho_4 \sim \exp(\Delta E^*/T). \qquad 55.$$

Here, T is the temperature, taken to be the same for both ions. It may deviate considerably from the equilibrium value. The gain ΔE^* in excitation energy may be defined as

$$\Delta E^* = Q_{gg} + \Delta E_C + \Delta E_{rot} - \delta(p) - \delta(n), \qquad 56.$$

where ΔE_C and ΔE_{rot} are the differences in Coulomb and rotational energy, respectively. The last two terms in Equation 56 denote the pairing energies of the transferred protons and neutrons, respectively. Noticing that for a heavy system, ΔE_{rot} is negligible with respect to the other quantities, one understands the experimentally observed exponential dependence of the cross section. Deviations from this Q-value systematics are observed (10, 32) for neutron-deficient isotopes and at angles where the principal contributing process is quasi-elastic transfer. It should be emphasized again that the experimental cross sections may contain substantial contributions from secondary evaporation processes, which will enhance in particular the cross section for the neutron-deficient final fragments (270). However, as indicated by optimum Q-value systematics (271, 235) and evaporation calculations (32), the observed Q_{gg} systematics of isotope formation are not due to deexcitation-evaporation of the final fragments. The information on detailed properties of the damped-reaction mechanism that such studies provide is somewhat limited as long as the dependence of the cross sections on other parameters such as the energy loss is not investigated. Only then is it possible to place the observed Q-value systematics in proper perspective with the diffusive features of the reaction mechanism.

The correlation of fragment N/Z ratios with that of the combined system can be taken as strong evidence for the existence of an intermediate double-nucleus system (237). During the interaction, partial equilibrium is achieved with respect to the kinetic energy dissipated and the mass- and charge-asymmetry degrees of freedom. Nevertheless, identities of target and projectile are approximately conserved in the reaction, indicating that statistical mass equilibrium is established only by the outer "valence" nucleons of each ion (32). This is in contrast to what seems to be implied by the term "deep inelastic transfer," frequently used to describe damped reactions.

4.4 Kinetic-Energy Damping

Certainly one of the most striking features of the damped-reaction mechanism is the large amount of kinetic energy dissipated during the reaction. This effect is clearly demonstrated by the laboratory energy spectra of the light fragments from the reaction ^{209}Bi + ^{136}Xe at 1130 MeV (256), shown in Figure 22. The spectrum taken at $\theta_{\text{Lab}} = 19.8°$ is dominated by an intensive elastic scattering line at $E_{\text{Lab}} = 1050$ MeV. At some 330 MeV lower in energy appears a distinctive symmetric peak of damped events. As the angle is increased, an enhanced yield of events appears in the energy region between the elastic and strongly damped events, until near the quarter-point angle, $\theta_{\text{Lab}} = 33°$, the intensity increases monotonically with energy from strongly damped events to quasi-elastic and elastic ones. At more backward angles, the strongly damped part of the spectrum is again well separated from the quasi-elastic peak, which has now developed into a rather broad and less intense bump. The continuous transition from elastic to strongly damped events near the quarter-point angle demonstrates the experimental difficulty in separating the different contributions from each other.

Similar energy spectra are obtained for many other light and heavy systems (6, 8, 13, 22–24, 205, 250). However, a peculiarity is observed in the energy spectra for the heavy systems ^{139}La + ^{84}Kr (31), ^{154}Sm + ^{84}Kr (257), and ^{165}Ho + ^{84}Kr (205), as well as for the reaction ^{232}Th + ^{40}Ar (4). At certain angles, these spectra exhibit two damped peaks, in addition to the quasi-elastic one, as is shown in

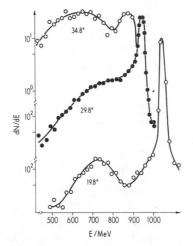

Figure 22   Laboratory energy spectra of light fragments from the reaction ^{209}Bi + ^{136}Xe at $E_{\text{Lab}} = 1130$ MeV for the indicated laboratory reaction angles [from (256)].

Figure 23 for the case of ^{165}Ho + ^{84}Kr. At the forwardmost angles, an asymmetric damped distribution is observed, which can be decomposed into two components. Whereas the high-energy bump moves toward the quasi-elastic peak as the angle approaches the quarter-point angle, the low-energy distribution moves continuously down in energy as the angle is increased. The assumption that the energy-loss mechanism operates continuously as the system rotates around its center of mass leads naturally to the occurrence of a second damped peak associated with negative-angle scattering for certain l values, bombarding energies, and observation angles. Decomposing the ^{165}Ho + ^{84}Kr energy spectra under this assumption leads to the orbiting-type angular distribution displayed in Figure 16.

In order to compare in a meaningful way different reactions with respect to their energy damping, the corresponding energy spectra have to be transformed into the c.m. systems and integrated over the reaction angle. Such a comparison is made in Figure 24 for four ^{84}Kr- and ^{136}Xe-induced reactions (28). Here the cross section is plotted as a function of energy loss. All spectra show a quasi-elastic peak at low energy loss and a continuous distribution of damped-reaction cross section extending over a range of several hundred MeV. A substantial amount of cross section is found with kinetic energies lower than the entrance-channel Coulomb energy evaluated for two spheres at a distance equal to the strong-absorption radius R_{SA}. These Coulomb energies are indicated by the arrows in Figure 24. Comparing the ^{136}Xe-induced reaction with the reaction ^{209}Bi + ^{84}Kr, the latter appears to be more strongly damped. The ^{165}Ho + ^{84}Kr data have to be taken

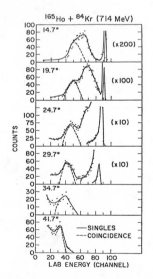

Figure 23   Laboratory energy spectra of light fragments from the reaction ^{165}Ho + ^{84}Kr at $E_{Lab} = 714$ MeV for the indicated laboratory reaction angles. The dashed curve represents the energy distribution corresponding to a mass window around $\bar{A} = 105$ [from (205)].

with caution because some cross section, presumably corresponding to high energy losses, is missing in the figure. The energy-loss spectra suggest that significance must be attributed to deformation effects in the exit channel.

Angle-integrated c.m. energy distributions are not available for the lighter systems, but observations similar to those made for the very heavy systems seem to apply. Generally, two components are exhibited in the energy spectra. The high-energy component can be attributed to quasi-elastic events. It converges with a broad lower-energy damped contribution. Mean energies of the latter distribution are often close to or lower than Coulomb energies for two touching spheres. This is illustrated by Figure 25 where the mean energies of fragments from the reaction Ag + ^{20}Ne (24) are compared with the Coulomb energies for two touching spheres or liquid-drop spheroids that are allowed to assume their equilibrium deformation. These data are not corrected for particle evaporation and are obscured by compound-nucleus fission contributions for $Z \geqq 15$ (239).

The damped energy distributions are generally bell-shaped for the lighter systems, but may exhibit low-energy asymmetries (11, 222, 272, 273) that are angle-dependent. For the ^{232}Th + ^{40}Ar reaction at 388 MeV (4), the damped distribution shows two well-separated energy peaks at angles backward from the grazing angle. This behavior can be understood in terms of the orbiting picture discussed above. At the lower energy of 297 MeV, this effect is not seen in the same reaction. Experi-

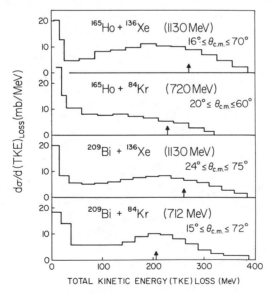

Figure 24   Differential cross section, $d\sigma/d(TKE)_{loss}$, for events of various degrees of kinetic-energy damping TKE_{loss}. As discussed in the text, some cross section is not accounted for in the ^{165}Ho + ^{84}Kr reaction. Each arrow corresponds to the kinetic-energy loss for an event with final kinetic energy equal to that of the Coulomb energy at the separation distance of the strong absorption radius R_{SA} [from (28)].

514   SCHRÖDER & HUIZENGA

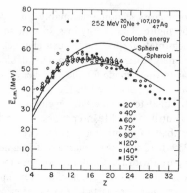

Figure 25 Average center-of-mass kinetic energies as a function of Z for various laboratory angles. The two lines are the calculated fragment energies arising from Coulomb repulsion for two spheres in contact and two spheroids in contact at equilibrium deformation [from (24)].

mentally, some caution has to be exercised in the case of heavier targets such as ^{197}Au, where secondary fission of the target-like fragments may result in a low-energy bump, in addition to the distribution of damped events. Such processes may, however, be recognized by their characteristic kinematics (215). It should also be kept in mind that nucleon evaporation may lead to a distortion of the spectra, resulting in a shift and a broadening (270).

The effect of an increased bombarding energy on the fragment energy spectra is a general broadening and a shift of the most probable and the centroid energies to higher values (4, 23, 24, 200, 274). For example, the spectrum of Mg-like fragments from the reaction ^{232}Th + ^{40}Ar (4) has a width of 40–50 MeV for a bombarding energy of $E_{Lab} = 288$ MeV, which increases to 80–90 MeV at $E_{Lab} = 379$ MeV. While the increase of the width with increasing bombarding energy is a common observation, for some systems the most probable fragment kinetic energy is reported to be rather independent of the bombarding energy (16, 17, 19, 25, 35, 119, 206). An example for this feature is the reaction ^{197}Au + ^{63}Cu, studied at $E_{Lab} = 365$ and 443 MeV (35), where an average fragment kinetic energy of $E_{c.m.} \approx 200$ MeV was observed for both bombarding energies. However, recent experiments (200) measuring particular fragments from ^{63}Cu- and ^{56}Fe-induced reactions on ^{148}Sm at similar bombarding energies arrive at a different conclusion.

The observation that the most-probable fragment energies for some systems are nearly independent of bombarding energy, unfortunately, has led to a somewhat narrow view of the damped-reaction mechanism and introduced some confusion in the literature. In this picture (117, 119, 275), a fully relaxed kinetic energy resulting from a dissipation mechanism proceeding on a time scale short compared to that on which transfer processes occur is assumed to be a characteristic feature of damped collisions. The appearance of two damped peaks in the fragment-energy distributions in some reactions referred to above is a clear disproof of the validity of this assumption. In addition, there is some evidence that energy spectra are entrance-channel-dependent. For example, the two combinations $^{165}_{67}Ho + ^{84}_{36}Kr$ and $^{107,109}_{47}Ag + ^{136}_{54}Xe$ lead to almost identical composite systems and similar charge products. Yet the resulting fragment energy distributions obtained for comparable c.m. bombarding energies reveal quite a different appearance. Whereas double-humped energy spectra are observed for the $^{165}Ho + ^{84}Kr$ reaction (cf Figure 23), the $Ag + ^{136}Xe$ fragment energy spectra are bell-shaped. These facts and the characteristic angular and mass distributions (cf Sections 4.2 and 4.3) indicate that damped reactions generally bear very little resemblance to the fission process, contrary to what is suggested by the term "quasi-fission" sometimes applied to these reactions.

The degree of kinetic-energy damping appears to be generally higher in reactions induced by lighter ions, particularly at bombarding energies close to the Coulomb barrier. However, it is evident that the damped-reaction mechanism leads to continuous kinetic-energy distributions connecting the domains of "completely relaxed" events with quasi-elastic ones. The final fragment kinetic energies are composed of Coulomb, radial, centrifugal, and internal rotation energy. It is presently not clear how to disentangle them unambiguously. Certainly, highly detailed multiparameter measurements are necessary to approach this goal. Whereas there is little hope of learning anything quantitative from a consideration of energy distributions alone, the energy-loss parameter appears to be a significant quantity since it samples the evolution of the damped reaction at least over a significant fraction of the total interaction time.

5 INTERCORRELATION OF EXPERIMENTAL PROPERTIES OF DAMPED COLLISIONS

After the discussion of the gross features of damped heavy-ion reactions presented in the preceding sections, this section is concerned with correlations between mass exchange, reaction angle and kinetic-energy loss. The limit of very large mass and energy transfer represented by compound-nucleus reactions is considered in Section 5.3. It should be pointed out again that in a comparison of the different systems, it is important to recognize that different bands of angular-momentum space contribute to different reactions. For a lighter system, a large fraction of the low-l waves leads to fusion, whereas most of the l waves of a very heavy system result in damped collisions. Hence, damped collisions arise from very different l windows for different systems. This difference sometimes obscures common features of the above correlations for reactions with different projectile-target combinations. It is the angular-momentum–dependent reaction time that is the underlying feature connecting the various correlations for the damped collisions.

Three degrees of freedom that are important for an understanding of damped collisions are the relative motion, mass asymmetry, and neutron-to-proton ratio. The relative time scales for the relaxation of these degrees of freedom are active topics of research and available information on these time scales is discussed in the following subsections.

5.1   Angular Distributions: Dependence on Mass (or Charge) Exchange and Energy Damping

The first evidence for the new damped-collision process was presented by Kaufmann & Wolfgang (1) in 1961. An example of their results for the ^{103}Rh + ^{16}O(E_{Lab} = 101 MeV) reaction is shown in Figure 26. The angular distribution for the product ^{15}O has a large peak near the quarter-point angle, a feature characteristic of single-nucleon transfer. The products ^{18}F, ^{13}N, and ^{11}C (and some fraction of the ^{15}O) have a very different angular distribution where the differential cross section rises continuously with decreasing angle. These latter products result from a collision in which the diffuse nuclear surfaces overlap to some degree. At least three types of processes (see Figure 1) are distinguishable for this reaction, which can be interpreted as follows: (a) The largest l waves lead to grazing collisions where the entrance and exit channels are strongly coupled and these reaction products have sideways-peaked angular distributions characteristic of a fast process. (b) The smallest l waves lead to fusion and compound-nucleus formation. This is the dominant reaction mode for the above reaction. (c) A band of intermediate l

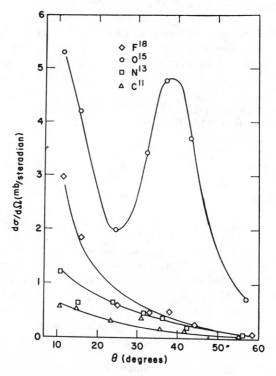

Figure 26   Differential cross sections with respect to solid angle are shown for products from the reaction of 101-MeV ^{16}O with a 7.35-mg/cm^2 Rh target [from (1)].

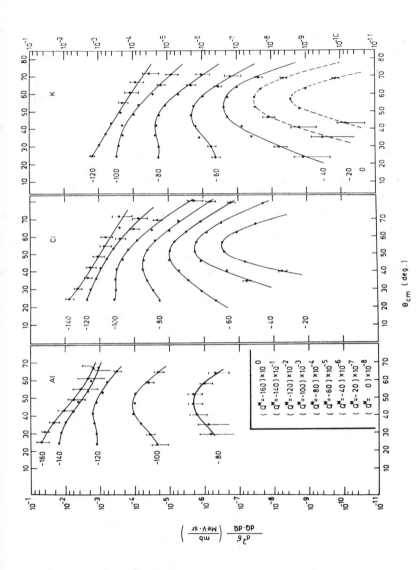

Figure 27 Angular distributions of isotopes of Al, Cl, and K (element identification only) from the ^{232}Th + ^{40}Ar ($E_{Lab} = 288$ MeV) reaction as a function of energy damping. The data for a given value of Q^* have been multiplied by the coefficient listed in the figure [from (4)].

waves penetrates sufficiently deeply for significant mass exchange to occur but decays prior to fusion. These l waves lead to processes with intermediate lifetimes and angular distributions of the type shown in Figure 26.

Artukh et al (4) made a detailed study of the ^{232}Th $+ ^{40}_{18}$Ar reactions at laboratory energies of 288 and 397 MeV, with both Z and energy analyses. The addition of the energy parameter was a significant advance that, however, was neglected in most of the subsequent studies. In Figure 27, the angular distributions for three elements from the above reaction ($E_{Lab} = 288$ MeV) are shown as a function of energy damping. The isotopes of $_{19}$K (one unit higher in Z than the projectile) have angular distributions that are sideways-peaked for large Q values (small energy damping) and forward-peaked for the smallest Q values. Similar angular distributions are observed for isotopes of $_{17}$Cl and $_{13}$Al, although the narrow sideways-peaking is beginning to disappear for $_{13}$Al. This figure demonstrates the rate of energy damping in the sequence of angular distributions for a particular element.

Contour diagrams of the double-differential cross section $d^2\sigma/dE\,d\theta$ were first published by Wilczyński (38) for the ^{232}Th(^{40}Ar, K)Ac reaction at $E_{Lab} = 388$ MeV, as shown in Figure 28. This plot contains two ridges in the θ vs E space. A major ridge is seen coming down in energy and moving to forward angles. This ridge develops into a peak at the most forward angles. In addition, a ridge is observed that starts at the most forward angle and moves to larger angles with a slight decrease in energy. The products in this forward peak have undergone considerable energy damping, and this peak supposedly arises from a crossing of the two above ridges. The major ridge corresponds to positive reaction angles, whereas the low-energy ridge represents its continuation to negative angles reflected at $\theta = 0°$. An energy spectrum at some fixed angles contains two peaks, a high-energy peak and one at an energy approximately equal to the Coulomb barrier, although it should be emphasized that events of a wide range of energy damping are present in the entire spectrum. The cross section for the damped (non–compound nucleus) events for the ^{232}Th $+ ^{40}$Ar reaction is substantially larger than for the previously discussed

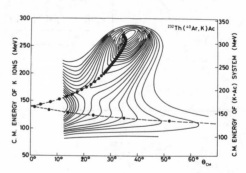

Figure 28 Contour diagram of the double differential cross section $d^2\sigma/dE\,d\theta$ for the ^{232}Th(^{40}Ar, K)Ac reaction at a laboratory energy of 388 MeV (38). The circles are based on a calculation with angular momentum values ranging from $l = 250\hbar$ to $l = 180\hbar$ [from (90)].

lighter-ion reactions. Such comparisons of cross sections depend on the product $Z_T Z_P$, but, in addition, on the bombarding energy via a factor like $(E_{c.m.} - E_C)/E_C$. Wilczyński interpreted the experimental data from the ^{232}Th$+^{40}$Ar reaction in terms of nuclear orbiting (see Figure 3, trajectory 3) combined with the dissipation of energy due to frictional forces. It is important to notice the continuous increase of energy damping as the l values decrease, as indicated by the points in Figure 28 representing a trajectory calculation (90). In this calculation, exit-channel deformation effects are accounted for by means of a phenomenological potential. Such phenomena support the existence of an intermediate double-nucleus system rotating in a dynamical-force equilibrium for which the decay time of the complex depends principally on dynamical quantities, including dissipative forces. The decay times for the complex produced by the ^{232}Th$+^{40}$Ar reaction range at least up to times necessary for one fourth of a full revolution (from $\theta_{1/4} \approx 37°$ to $-60°$, as seen in Figure 28). A quantitative determination of the reaction times requires a decomposition of the damped-reaction cross section into angular-momentum bins, as discussed below.

Initial experiments with the heavier projectile ^{84}Kr on a ^{209}Bi target produced sideways-peaked angular distributions; however, surprisingly little, if any, fusion (5, 6). Hence, almost the entire cross section is made up of damped collision events. However, the orbiting feature is not observed and the reaction mechanism was thought inititially to be different from the lighter-ion reactions (5). This strong focusing of the angular distribution is now understood as being due to the small Kr laboratory bombarding energies of 525 (5) and 600 (6) MeV (only about 1.2 and 1.6 times the Coulomb barrier, respectively, compared to 1.8 for the ^{232}Th$+$ ^{40}Ar reaction) and the small angular velocity of the heavy intermediate system. In the limit of very low energies, such strongly peaked angular distributions are similar in shape to those of one-nucleon transfer shown in Figure 26. However, there are several very important differences that need to be strongly emphasized. The first is that the differential cross section for the heavier systems, where strong focusing occurs, is orders of magnitude larger than for typical one-nucleon transfer reactions. Second, the events are strongly damped to various degrees in energy, as compared to very weak damping for one-nucleon transfer. Third, various degrees of mass exchange, up to many nucleons, occur compared to one-nucleon transfer. Finally, a large fraction of the l waves contributes to the observed peak, whereas only a small band of the highest l waves contributes to one-nucleon transfer. This means that the strongly focused peak contains events with a wide range of lifetimes, and it is not correct to assume that the side-peaking is due to very short lifetimes of the order of those associated with one-nucleon transfer. The smaller-impact parameters have the deeper penetrations, leading to larger energy damping and mass exchange corresponding to longer lifetimes. In the case of the ^{209}Bi$+ $^{84}Kr system at 600 MeV, for example, there is more mass transfer at the more backward angles, indicating a fractionation of events according to impact parameter, where the smaller impact parameters are decaying prior to reaching the $\theta_{1/4}$ angle (6).

Another feature of heavy-ion reactions at energies near the Coulomb barrier

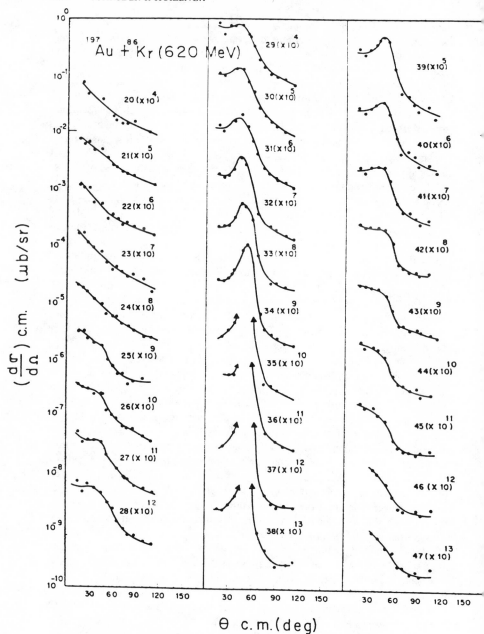

Figure 29 Angular distributions as a function of fragment atomic number for the $^{197}Au + {}^{86}Kr$ reaction at a bombarding energy of 620 MeV [from (26)].

is the much smaller penetration depth for equivalent l values as the energy is reduced. This results in shorter lifetimes (164), smaller energy damping (because $E_{c.m.} - E_C$ is small), and smaller mass exchange. Increasing the bombarding energy for a heavier projectile-target combination not only moves the peak in the cross section to smaller angles (see Figure 15), but also may add a forward-angle component, in analogy to results displayed in Figure 28 for a lighter projectile. The ^{165}Ho + ^{84}Kr ($E_{Lab} = 714$ MeV) reaction (205) is an example where $E - \theta$ contour plots for various elements show two cross-section ridges similar to the ^{232}Th + ^{40}Ar reaction (see Figure 28).

Angular distributions as a function of fragment atomic number are shown in Figure 29 for the ^{197}Au + ^{86}Kr reaction at $E_{Lab} = 620$ MeV (26). Events of all degrees of energy damping are included in such a plot. Strong sideways-peaking occurs for elements near the projectile. However, some six Z units away from the projectile, the angular distributions become smoother, with the differential cross section increasing monotonically as the angle decreases. The small net transfer of charge (and mass) alone does not insure sideways-peaking. The degree of energy damping is also important. For example, the angular distributions of fragments with $Z = 35$-37, produced by the ^{209}Bi + ^{84}Kr ($E_{Lab} = 712$ MeV) reaction, show sideways-peaking for small energy damping and a forward-rising differential cross section for large energy damping (264). This result is analogous to that for lighter heavy-ion reactions (see Figure 27). In general, fragments with Z and A near that of the projectile have, on the average, small kinetic-energy loss and angular distributions characteristic of very short lifetimes. If, however, fragments are selected with Z and A near the projectile (small net transfer) and with a large kinetic-energy loss, these fragments have angular distributions that correspond more nearly to those many Z units away from the projectile (264).

General properties of damped collisions initiated with still heavier projectiles like Xe are similar to those described above. Results from the ^{209}Bi + ^{136}Xe ($E_{Lab} = 1130$ MeV) reaction are shown in Figure 30 (21, 256). The entire cross section for this reaction is comprised of reaction processes damped to various degrees in energy. An interesting new feature is a cross-section ridge that moves down in energy at an almost constant angle. This is a striking example of strong angular focusing caused by a large fraction of the impact parameters producing an intermediate double nucleus which in turn decays at approximately the same angle, independent of angular momentum (see Sections 4.2 and 5.2). The angular distributions for fragments with $Z = 53$-55 are plotted in Figure 31 for the latter reaction as a function of energy damping. In contrast to the other reactions discussed previously, the angular distributions for elements near Xe continue to be sideways-peaked even for large energy damping, although the width does increase substantially with energy damping.

When the Xe projectile energy is lowered, the cross-section ridge moves initially to larger angles (see Ref. 118). This is because the interaction time for similar l waves decreases and the intermediate complex decays for a number of impact parameters before the system rotates to the grazing angle. Hence, depending on the values of the parameters $Z_T Z_P$ and $(E_{c.m.} - E_C)/E_C$, the cross-section ridge in an energy-angle contour plot (starting near the elastic energy and the grazing angle) may move forward (Figure 28), stay constant (Figure 30), or move backward in angle as the energy decreases.

The bombardment of rare-earth targets with 8.5-MeV/nucleon Kr leads to a forward moving cross-section ridge in an E-θ plot and some orbiting (31, 205, 257, 261), similar to that described previously for lighter projectiles. The movement of the ridge depends critically on the dynamical equilibrium created by the conservative and dissipative forces for each of the different impact parameters of the system. There is still a significant orbiting component for these targets when the energy is reduced to approximately 7.2 MeV/nucleon (257). In addition, orbiting is more pronounced for deformed and neutron-rich targets at very small energies (205, 257).

As demonstrated by Figure 28, classical dynamical calculations describe well the correlation between energy loss and reaction angle for the reaction ^{232}Th(^{40}Ar, K). However, for ^{84}Kr-induced reactions on ^{208}Pb (23) and ^{209}Bi (205), similar calculations (89, 276) do not reproduce fragment-mass–integrated data even if exit-channel deformations are roughly accounted for. Typically, the peak of the angular distribution is slightly missed (cf Figure 17) by theory and the rate of energy loss with changing angle is underestimated. In a recent extension of an earlier work (96) Bondorf et al (97) point out that the angular focusing effect observed for heavy systems, even for events with large energy damping, may be accounted for in trajectory calculations by employing l-dependent friction-force coefficients.

Natural silver has been bombarded with a sequence of projectiles N, Ne, Ar, and

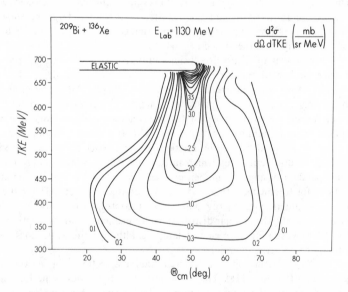

Figure 30   Contour diagram of the double-differential cross section $d^2\sigma/d\Omega\,d(TKE)$ for projectile-like products from the ^{209}Bi + ^{136}Xe reaction at a laboratory energy of 1130 MeV [from (256)].

Kr (see Table 2 for energies) to study the angular distributions and energy damping as a function of Z (16, 17, 24, 250). In addition, similar studies have been carried out for the Ag + ^{136}Xe reaction (205). This impressive sequence of reactions shows changes of the fragment angular distributions as a function of the projectile mass and indicates that the charge distributions depend upon the initial asymmetries. The cross sections are enhanced in the Z region below the projectile for Ag + ^{14}N and Ag + ^{136}Xe, in the Z region above the projectile for Ag + ^{40}Ar and Ag + ^{86}Kr, but distribute about equally above and below for the Ag + ^{20}Ne reaction. These findings are consistent with the respective potential-energy diagrams.

The angular distributions for the Ag + ^{14}N reaction vary considerably for Z smaller and larger than the projectile. The smaller Z have angular distributions that are highly forward-peaked, while for $Z > 13$ the angular distributions are essentially of a $1/\sin\theta$ type. These results are interpreted in terms of a diffusion-controlled evolution of an intermediate complex along the mass-asymmetry degree of freedom (18). The injection point in the asymmetry parameter on the potential surface favors the drift to smaller Z. Fast decay then produces the strong forward peaking. The higher Z values arise from the spreading associated with mass diffusion on a rising potential-energy surface. Larger Z values have, on the average, longer lifetimes and angular distributions approaching $1/\sin\theta$. A somewhat similar pattern is observed for the Ag + ^{20}Ne reaction. The pattern is, however, quite different for the Ag + ^{40}Ar reaction, where the injection point in the asymmetry parameter favors drift to larger Z. The magnitude of the forward peaking now decreases as one moves to Z lighter than the projectile. It should be stated that the kinetic-energy spectrum of the damped events could be interpreted as compound-nucleus

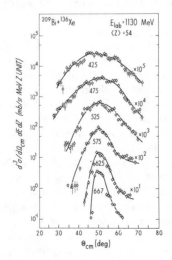

Figure 31   Angular distributions for fragments $Z = 53$–55 produced by the ^{209}Bi + ^{136}Xe ($E_{Lab} = 1130$ MeV) reaction as a function of energy damping. Each curve is identified by a center-of-mass energy that is the midpoint of a 50-MeV energy bin [from (256)].

decay products where the light particles are due to evaporation and the heavy particles are due to fission. The angular distributions, as discussed above, argue against the compound-nucleus mechanism for some Z, but not for those fragments displaying a $1/\sin\theta$ angular distribution. For example, in the case of the $Ag + {}^{40}Ar$ reaction, a substantial fraction of the fragments with $Z = 25$–40 may be due to fission rather than the damped-reaction process (239). The injection point for the $Ag + {}^{86}Kr$ system is not too far from the symmetry minimum. This system has sideways-peaking for a few Z values near Kr but otherwise forward-peaked angular distributions. In general, the angular distributions rise too rapidly to be associated with compound-nucleus reactions; however, some amount of fission cannot be ruled out. In the case of the $Ag + {}^{136}Xe$ reaction, the injection point is above the symmetry minimum. The energy damping for this reaction is less than for the similar mass system ${}^{165}Ho + {}^{84}Kr$ (at an energy where l_{max} is approximately the same).

An extensive set of experiments with intermediate projectiles of ${}^{40}Ar$, ${}^{40}Ca$, ${}^{52}Cr$, ${}^{56}Fe$, and ${}^{63}Cu$ has been performed by the Orsay group (see Table 2 for references), which bombarded a series of targets from ${}^{40}Ca$ to ${}^{238}U$ with energies up to 7 MeV/nucleon. The symmetric ${}^{40}Ca + {}^{40}Ca$ system has been studied extensively (11, 240, 241, 241a). The angular distributions are observed to change from a forward-peaked pattern for Z values near the projectile, to a $1/\sin\theta$ pattern for the lightest Z observed. The Z distributions show an odd-even effect that favors even Z, presumably due to Q-value effects and the sequential decay of primary fragments where the proton binding energies favor even Z. The contour plots of the cross section in the E, θ plane for a number of the above reactions show features that change systematically as the system becomes heavier. The ${}^{58}Ni + {}^{40}Ar$ (280-MeV) reaction (34) has a forward ridge for Z near the projectile as well as a ridge due to an orbiting component very similar to the ${}^{232}Th + {}^{40}Ar$ (388 MeV) reaction. As Z becomes progressively far from the projectile Z, the angular distributions become rather flat and are composed only of strongly damped events. One sees for this reaction a continuous evolution between very weak and very strong energy damping. The production of K isotopes has been studied also for the ${}^{58}Ni + {}^{40}Ar$ (280 MeV) reaction. As the reaction product requires more complex nuclear exchange processes, the intensity of the high-energy component decreases, as well as shifts in the direction of smaller angle and smaller energy. Such a sequence of products shows clearly the continuous progression in energy damping with reaction complexity. Lowering the bombarding energy changes the overall pattern of the E-θ contour plots. The ${}^{64}Ni + {}^{40}Ca$ (182-MeV) system has been studied (238) at an energy of 1.5 E_B, as compared to an energy of 2.5 E_B for the ${}^{64}Ni + {}^{40}Ar$ (280-MeV) (237) reaction. Due to the smaller energy difference between the entrance and the fully damped exit channel, the separation between the two cross-section ridges is almost completely masked for elements near the projectile. However, for elements $Z = 18$ to $Z = 12$, the transition from side-peaking to forward-peaking is evident as is enhanced energy damping.

In summary, the lighter systems resemble the heavier ones in all aspects except for a different division of the cross section into various degrees of energy damping and the stronger angular focusing of the heavy systems when compared at the same energy above the barrier. In making these comparisons, it is important to consider also the l windows involved for the light and heavy systems because they may be very different.

From the discussion of angular distributions presented above, it is evident that the kinetic-energy loss is intimately related to the interaction time. For example, small energy losses associated with very narrow angular distributions are characteristic of short interaction times. As the energy loss increases, processes with longer interaction times are sampled, resulting in increasingly broader angular distributions. In order to assess these interaction times quantitatively, an l-wave decomposition of the cross section has to be made to construct the "experimental" deflection function. Such a classical deflection function can only represent an average because fluctuations induced by the internal degrees of freedom are not accounted for.

The construction of an experimental deflection function discussed below (277) is based on the very general and natural assumption that the energy loss increases monotonically with decreasing values of l such that the energy axis of a Wilczyński-type plot (cf Figures 28 and 30) also represents an l scale. Employing, for simplicity, the sharp cutoff model of Equation 9, the experimental cross section $\Delta\sigma_{ij} = \sigma_j - \sigma_i$ belonging to an energy window (see Figure 24) $E_i \leq E \leq E_j$ can be related to the corresponding l window $l_i \leq l \leq l_j$ by

$$l_i = \left\{(l_j+1)^2 - \frac{\Delta\sigma_{ij}}{\pi\lambda^2}\right\}^{1/2} - 1.$$ 57.

Starting with $l_j = l_{max}$ and zero energy loss, Equation 57 allows the successive determination of l as a function of energy loss, for a range of higher l values. Since the mean deflection angle for a given energy window is an experimental quantity, one obtains immediately the average deflection function. An example of such a deflection function is shown as a solid line in Figure 32 for the $^{209}Bi + ^{136}Xe$ reaction at $E_{Lab} = 1130$ MeV. One observes that for this reaction the average emission angle θ_{exp}, where the cross section has a maximum, is independent of the energy loss (cf Figure 30) and hence also independent of l. The shape of this function depends on the reaction and the bombarding energy, e.g. for the $^{165}Ho + ^{84}Kr$ reaction, the experimental deflection angle of Kr-like fragments decreases as l decreases for a range of intermediate and high l values.

Figure 32 shows the Coulomb part of the deflection function $\theta_C(l)$, representing Rutherford scattering. The difference $\Delta\theta(l)$ between these two deflection functions directly gives the angle through which the intermediate system has rotated during the interaction time τ. The angular momentum–dependent interaction time can then be calculated with the expression (96, 164)

$$\tau(l) = \Delta\theta(l)\mathcal{I}(l)/\hbar l,$$ 58.

where $\mathcal{I}(l)$ is the moment of inertia of the intermediate system. The evaluation of $\tau(l)$ requires the adoption of a collision model. Calculations are presented for two rather different models, which are labeled nonsticking (NS) and sticking (S) collisions. A sticking collision is defined by rigid rotation of the double-nucleus system as

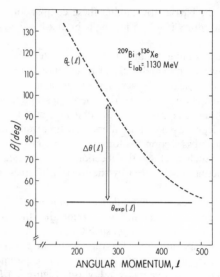

Figure 32   Deflection function for the ^{209}Bi + ^{136}Xe reaction. The dashed curve is the Coulomb part of the total deflection function, calculated for the NS condition by Equations 59 and 60. The curve $\theta_{\mathrm{exp}}(l)$ is the deflection function generated from experimental data (*solid line*). $\Delta\theta(l)$ is the angle through which the intermediate double-nucleus system rotates during the interaction [from (277)].

a whole. By a NS collision, we specify that the entrance- and exit-channel orbital angular momenta are the same ($l_f = l_i$). The moment of inertia for the orbital motion in both cases is given by Equation 41 with $r = R_{\mathrm{SA}}$. In the sticking case, the final angular momentum is reduced with respect to the initial one by an amount determined by Equation 43. In the calculations presented, any variation of $\mathscr{I}$ with angular momentum or time is neglected. The Coulomb deflection angle is estimated by (96)

$$\theta_{\mathrm{C}}(l) = 180° - \theta_1 - \theta_3, \qquad\qquad 59.$$

where the subscripts refer to the entrance and exit channels, respectively, and

$$\theta_j = +\arccos\frac{1}{\varepsilon_j} - \arccos\frac{K_j+R_j}{\varepsilon_j R_j} \quad \text{(for } j = 1 \text{ and 3).} \qquad 60.$$

In Equation 60, the parameters ε and K are determined by

$$\varepsilon = \left[1 + \frac{2El^2\hbar^2}{\mu(Z_{\mathrm{P}}Z_{\mathrm{T}}e^2)^2}\right]^{1/2}, \qquad\qquad 61.$$

$$K = \frac{l^2\hbar^2}{\mu Z_{\mathrm{P}}Z_{\mathrm{T}}e^2}. \qquad\qquad 62.$$

The angular momentum–dependent interaction times calculated by the above procedure are displayed in Figure 33 for three reactions. Multiplication of the angular

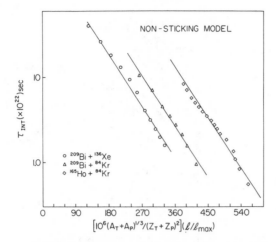

Figure 33 Plot of the interaction time as a function of angular momentum l for a nonsticking model. The laboratory bombarding energy is 8.5 MeV/amu for each reaction. The solid lines all have the same slope. The interaction times depend on the contact radius which has been assumed to be the strong-absorption radius, R_{SA} [from (278)].

momentum l by the factor $[10^6 (A_T + A_P)^{1/3} / (Z_T + Z_P)^2]$ $(1/l_{max})$ produces straight lines on a semilog plot of approximately the same slope for the three reactions (278). The angular momentum range for the ^{209}Bi + ^{136}Xe reaction shown in Figure 33 is $120 \leq l \leq 430$, for which the interaction time decreases by a factor of 33. The sticking model gives interaction times that are approximately a factor of two longer than those of the nonsticking model depending on l.

5.2 Mass Exchange: Dependence on Energy Damping

The correlation between the angular distribution and mass exchange in damped heavy-ion reactions discussed in the preceding section is a rather indirect one and becomes almost meaningless in the case of strong angular focusing. On the other hand, the intimate correlation between the energy loss and the angular momentum–dependent interaction time has been demonstrated. Since the energy loss is certainly the fundamental observable specifying the reaction, it is important to investigate the experimental correlation between mass exchange and energy loss.

In order to illustrate the various features of this relation, several different correlations are shown for the ^{209}Bi + ^{136}Xe reaction at a laboratory energy of 1130 MeV. A rather extensive set of data is available for this reaction (21, 28, 256, 277, 278). In all of the correlations to be discussed, the distributions are for the lighter Xe-like fragments. Corrections for the emission of neutrons emitted during the de-excitation of the fragments have also been made. The differential cross section $d^2\sigma/dZ\, d(TKE)$ is plotted as a function of total kinetic energy (TKE) for different element bins in Figure 34. For Z values near the projectile, the spectra contain sizable high-energy components, and as Z decreases or increases, the spectra become softer. Hence, the

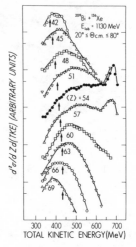

Figure 34   Plot of $d^2\sigma/dZ\,d(TKE)$ as a function of the total kinetic energy for different elements produced in the $^{209}\text{Bi} + ^{136}\text{Xe}$ reaction. Each bin contains three elements. The ordinate is a logarithmic scale. The arrows indicate Coulomb energies for touching spheres [from (256)].

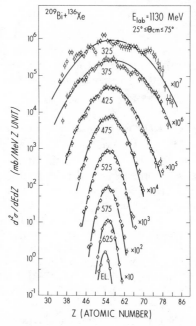

Figure 35   Plot of the differential cross section $d^2\sigma/dE\,dZ$ as a function of atomic number for various final kinetic-energy bins of 50-MeV width measured for the $^{209}\text{Bi} + ^{136}\text{Xe}$ reaction. The solid lines are Gaussian fits to the data [from (256)].

degree of energy damping is a function of the mass of the product. For Z values near the projectile, the spectra contain events of a very wide range of kinetic energies, whereas large net-Z transfers are correlated with high degrees of damping.

One of the most informative ways to examine the relationship between the charge (or mass) distribution and the final kinetic energy is a plot of the differential cross section $d^2\sigma/dZ\,d(TKE)$ as a function of Z for different final kinetic-energy bins. Such a plot is shown in Figure 35, where the energy bins are 50-MeV wide. The curves represent Gaussian fits to the experimental data. It is readily seen that the width in the charge distribution increases markedly as the kinetic energy decreases. The experimental charge distributions are suggestive of a diffusion process (109, 110, 117) and have been analyzed with the Fokker-Planck equation (Equation 29) in terms of the variance σ_Z^2. Such an analysis of the mass distribution as a function of energy is based on the concept that the intermediate system loses kinetic energy as nucleons are interchanged. Therefore, the energy loss and the number of exchanged nucleons are a measure of the reaction time. The progressive increase in the variance of the charge distribution as a function of increasing kinetic-energy loss is a feature common to all very-heavy-ion reactions studied so far (21, 28, 30, 119). A quantitative correlation between the measured variances of the charge distribution and the amount of kinetic-energy loss is shown in Figure 36 for four different heavy-ion reactions (28). Although the mass distributions are in general angle-dependent (see Figure 19), in the case of Xe-induced reactions the variance is independent of reaction angle for events of the same kinetic-energy loss (30, 255, 256). The variance of the charge distribution increases smoothly with increasing total kinetic-energy loss. However, the slope of the energy dissipation as a function of variance is largest for small variances and decreases as the variance increases.

In order to calculate the kinetic-energy loss associated with the exchange of a single nucleon, it is necessary to deduce the total number of nucleons exchanged from the experimental charge distribution. If nucleon exchange in heavy-ion reactions is a random-walk process, the total number of nucleons exchanged is related

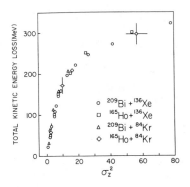

Figure 36   Total kinetic-energy loss as a function of the variance of the charge distribution σ_Z^2 of the projectile-like fragments for four heavy ions [from (28)].

to the variance σ_A^2 of the final mass distribution. It is important to differentiate between the total number of nucleons interchanged between target and projectile and the net transfer which gives the final mass distribution.

The experimental evidence presented in Section 4.3 shows that the neutron-to-proton degree of freedom equilibrates relatively fast compared to energy dissipation for heavy-ion reactions (34). Therefore, the rates of proton and neutron exchange are similar. It is then permitted to calculate the total number N of nucleons exchanged from the number $N_Z = \sigma_Z^2$ of protons exchanged according to $N = (A/Z)\sigma_Z^2$. The experimental total energy lost during the time needed for one nucleon exchange δE is obtained by differentiating the experimental energy-loss function with respect to σ_Z^2, $\delta E = (Z/A) \, dE_{loss}/d\sigma_Z^2$.

Experimental results for the energy loss per nucleon exchanged for the reactions ^{165}Ho + ^{84}Kr, ^{165}Ho + ^{136}Xe and ^{209}Bi + ^{136}Xe (28) at a bombarding energy of 8.5 MeV/amu and for the reaction ^{197}Au + ^{86}Kr at 7.2 MeV/amu (118) are plotted in Figure 37 as a function of the available energy per nucleon defined in Section 2.2. This parameter is calculated under the assumption that E_B be given by the Coulomb potential of two spherical nuclei at a separation $r = R_{SA}$. Hence, for high energy losses, this evaluation of E_{av} is questionable due to the formation of a neck and the deformation of the fragments.

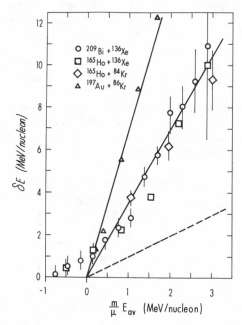

Figure 37 Plot of the energy loss per nucleon exchanged, δE, in MeV/nucleon as a function of the available energy per nucleon for the ^{136}Xe- and ^{84,86}Kr-induced reactions. The solid lines are drawn through the data points. The dashed line represents the prediction of a one-body dissipation model for the energy loss due to nucleon exchange [from (93)].

The most important observation to be made from the data is that the energy δE dissipated per exchanged nucleon decreases with decreasing available kinetic energy E_{av}, i.e. increasing energy loss or total excitation energy of the system. The data points follow a straight line intercepting the abscissa at $(m/\mu)E_{av} = 0$, as predicted by Equation 22, over an unexpectedly wide range of available energies. For small available energies the data points deviate from these lines. The part of the energy dissipation mediated by nucleon exchange, as given by Equation 19, is represented by the dashed line in Figure 37. Obviously, mass exchange induced by the average one-body field can account for $\sim 30\%$ of the energy dissipation at a bombarding energy of 8.5 MeV/amu and for $\sim 15\%$ at 7.2 MeV/amu, whereas the remainder can be explained by one-body dissipation processes such as particle-hole excitation. Data for the $^{144}Sm + {}^{84}Kr$ reaction (257) taken at three bombarding energies follow this systematics. The mechanism responsible for nucleon exchange accounting for a smaller fraction of the total energy loss as the bombarding energy is reduced is presently not understood. Plots of the kinetic-energy loss as a function of the variance (see Figure 36) for the same reaction at different bombarding energies produce correlations that are very similar for the first 75 MeV of kinetic-energy loss. Such a result is inconsistent with a simple one-body dissipation mechanism unless k_{nex}/k_{ex} depends on the bombarding energy.

The diffusion model of Nörenberg et al (110, 111) can account for large values of δE for low energy losses and, in addition, predicts a decrease of δE with decreasing available energy which is also experimentally observed (93). However, the experimentally observed dependence of δE on E_{av} is not reproduced by the above diffusion model that does not take the reaction dynamics fully into account.

Examining the predictions of the one-body proximity-dissipation mechanism (101, 177, 178), one notices that the value of the proximity-flux function $\psi(\zeta_0)$ varies from 0.74 to 2.1 for -0.75 fm $\leq \zeta_0 \leq 0.75$ fm (101). For Kr- and Xe-induced reactions on heavy targets, Equation 25 can be approximated by

$$\frac{1}{E}\frac{dE}{dt} \approx 1 \times 10^{21}\psi(\zeta_0)\ \text{sec}^{-1} \approx (0.7\text{–}2.1) \times 10^{21}\ \text{sec}^{-1}. \qquad 63.$$

The above reactions have been analyzed recently (28) with the friction model given by the energy-loss rate $-dE/dt = 2(k/\mu)E$, yielding a friction constant $k = (0.9 \pm 0.3) \cdot 10^{-43}D_Z$ MeV sec^2 fm^{-2}. Employing diffusion coefficients D_Z based on nonsticking and sticking models (277), which are discussed below, one obtains values of $(dE/dt)/E \approx (1\text{–}2) \cdot 10^{21}$ sec^{-1} that agree qualitatively with the predictions (cf Equation 63) of the one-body proximity-dissipation mechanism. The explanation of the surprising result of a friction constant in contrast to an expected friction form factor equivalent to an implicitly time-dependent coefficient $k(t)$ can be attributed to the fact that in this analysis reference was made to a diffusion model in order to estimate the reaction time (cf Section 2.3). Hence, the analysis of (28) and the one described above are similar and based on the one-body nucleon-exchange mechanism.

The above results are consistent with the view that to a large extent the kinetic-energy loss and the nucleon exchange occur on a similar time scale, contrary to what is often assumed (118, 263). The observed correlations between the kinetic-energy loss and nucleon diffusion suggest that the nucleon-diffusion mechanism operates continuously on the fast interaction-time scale of 10^{-22}–10^{-21} sec, simultaneously with other dissipation mechanisms. However, some contribution to the energy dissipation by a very fast mechanism, as has been suggested by Broglia et al (103), may be consistent with the data.

In order to determine the basic parameter of the diffusion model, the diffusion coefficient D_Z, one must employ l-dependent interaction times $\tau(l)$ discussed in Section 5.1. According to Section 2.3, the variance of the fragment-charge distribution is then also l-dependent and given by the relation $\sigma_Z^2(l) = 2D_Z(l)\tau(l)$. Experimental values of $\sigma_Z^2(l)$ as a function of $\tau(l)$ are plotted in Figure 38 for the $^{209}\text{Bi} + ^{136}\text{Xe}$ reaction (277) for both the nonsticking and sticking models. Values of the average diffusion coefficients for three heavy-ion reactions, determined by the slopes of the lines drawn through the points, are given in Table 4.

Diffusion coefficients reported by Nörenberg (110) are larger than those in Table 4 due to an underestimation of the interaction times. Moretto et al (118), assuming a sticking model and a Gaussian distribution of interaction times where the centroid varies linearly with l, have calculated diffusion coefficients by fitting such a theoretical function to experimental data. For the $^{197}\text{Au} + ^{86}\text{Kr}$ reaction at $E_{\text{Lab}} = 620$ MeV, they obtain $D_Z = 0.33 \times 10^{22}$ sec^{-1}, in reasonable agreement with the values in Table 4, which are, however, evaluated using the interaction times discussed in Section 5.1. There is an indication from the experimental data that

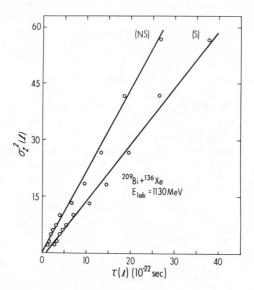

Figure 38   The variance σ_Z^2 of the experimental fragment-charge distribution plotted as a function of the interaction time τ for the $^{209}\text{Bi} + ^{136}\text{Xe}$ reaction. The two sets of results correspond to the assumption of nonsticking (NS) or sticking (S) of the two ions during the interaction [from (277)].

Table 4 Proton number (D_Z) and mass number (D_A) diffusion coefficients in units of 10^{22} sec^{-1} for Kr- and Xe-induced reactions[a]

Reaction	Sticking model		Nonsticking model	
	D_Z	D_A	D_Z	D_A
^{209}Bi + ^{136}Xe	0.75	4.8	1.1	7.0
^{209}Bi + ^{84}Kr	0.62	3.7	0.87	5.3
^{165}Ho + ^{84}Kr	0.55	3.2	0.74	4.3

[a] The proton number diffusion coefficient does *not* refer to proton diffusion alone, but to mass diffusion measured by the number of transferred protons; hence $D_Z = (Z/A)^2 D_A$ for a constant Z/A ratio. The diffusion coefficients listed in this table are calculated from the *slopes* of lines drawn through plots of $\sigma_Z^2(l)$ vs $\tau(l)$ over a range of l values. In the case of the sticking model, for example, the points for the highest l waves do not lie on a line which passes through the origin (see Figure 38). *Individual values* of $D_Z(l)$ for the sticking model are l dependent and increase initially as l decreases. The Kr- and Xe-projectile energies (lab) are 714 and 1130 MeV, respectively. The errors in the diffusion coefficients are of the order of 30%. However, the values scale with the contact radius (see Equations 41 and 58) which for the reported values is assumed to be the strong absorption radius R_{SA} discussed in Section 3.6 [from (277)].

D_A increases with $A = A_T + A_P$, as predicted by theory (111, 112). The mass-diffusion coefficient is given approximately by $D_A \approx A \cdot 10^{20}$ sec^{-1}.

The relationship between the drift coefficient v_Z (or v_A) and the diffusion coefficient D_Z (or D_A) is approximately given by the Einstein relation (see Equation 33). The ratio of v_Z to D_Z is

$$\frac{v_Z}{D_Z} = \frac{v_Z t}{D_Z t} = -\frac{1}{T}\frac{\partial}{\partial Z} U_l(Z), \qquad 64.$$

where $\partial U_l(Z)/\partial Z$ is the derivative of the ground-state potential energy of the combined system with relative angular momentum l and final fragment charge Z, and T is the appropriate nuclear temperature. The quantity $D_Z t$ is measurable and equal to $\sigma_Z^2/2$, while $v_Z t$ represents the drift in the centroid of the Z distribution. Measured values of the ratio v_Z/D_Z for the ^{197}Au + ^{86}Kr ($E_{Lab} = 620$ MeV) and Ag + ^{40}Ar ($E_{Lab} = 288$ MeV) reactions are 0.9 and 0.4, respectively (118), in reasonable agreement with expectations based on the right-hand side of Equation 64. Large energy losses for the ^{209}Bi + ^{136}Xe ($E_{Lab} = 1130$ MeV) reaction lead to values of v_Z/D_Z of about 0.1 (277), somewhat smaller than predicted for a potential with liquid-drop masses.

Ngô et al (33) have studied the width and centroid of the mass distribution as a function of angle for the ^{197}Au + ^{63}Cu reactions at 365 and 443 MeV. Their results are consistent with a diffusion process. A plot of the drift in mass $v_A t$ as a function of the width in mass defined by the variance σ_A^2 (equal to $2D_A t$) gives a straight line for each energy. These slopes for the two bombarding energies are expected to be in the ratio

$$\frac{(v_A/D_A)_{365}}{(v_A/D_A)_{443}} = \frac{T_{443}}{T_{365}}, \qquad 65.$$

if the values of $\partial U_l(A)/\partial A$ at the two energies are similar enough to cancel. Evaluation of T_{443}/T_{365} from a Fermi gas model gives a theoretical ratio of the temperatures of approximately 2, a value in close agreement with the ratio determined from the experimental slopes.

5.3 Very Large Mass and Energy Transfer: Compound-Nucleus Formation

The very strongly damped reaction processes, where large amounts of kinetic energy are lost and many nucleons are exchanged, are very difficult to distinguish from a compound-nucleus reaction process. This is especially true for the very-heavy-ion reactions, where it is only possible to set limits on the compound-nucleus cross section σ_{CN} in terms of the ratio to the total reaction cross section, σ_{CN}/σ_R. These limits may be considerably larger than the actual values for some Kr- and Xe-induced reactions (14, 204, 280). Although radiochemical mass measurements (7, 20, 27, 253) have unit-mass resolution, this method does not distinguish reaction mechanisms and in general has involved thick targets so that the results represent an integral yield over energy. For lighter systems, there is a recent compilation of fusion cross sections (262). Additional recent measurements are reported by Scobel et al (179). Many of the more reliable measurements of the fusion cross sections are based on the technique of measuring evaporation residues (279), which have to be complemented, however, by compound-nucleus fission measurements for the heavier systems. Although the compound-nucleus cross section makes up a large fraction of the reaction cross section for light heavy-ion reactions (e.g. $^{208}Pb + {}^{16}O$ at 5–10 MeV/nucleon), this is no longer true for very heavy projectiles. The limiting cross section for fusion of heavy systems is often discussed in terms of either entrance- or exit-channel restrictions. Some aspects of this subject are covered in a recent review article (262).

As a matter of practical concern, a knowledge of the fusion probability for heavy systems is important to experimental attempts to synthesize superheavy elements. However, even in the absence of compound-nucleus formation, it may be possible to produce small yields of superheavy nuclei by heavy-ion collisions through the mass-diffusion process (109, 112) discussed in Section 5.2. The probability for massive nucleon transfer to the heavier fragment to produce superheavy nuclei in the tail of the heavy-fragment mass distribution can be calculated with a knowledge of the mass-diffusion coefficient and the Fokker-Planck equation (Equation 29).

It has been emphasized above that the result of a damped reaction depends critically on the dynamical equilibrium of forces acting on various degrees of freedom. For deeply penetrating collisions that may eventually lead to compound-nucleus formation, a parameterization of the mass-asymmetry potential, as given in Equation 54, is an oversimplification of the problem. Greiner and collaborators (194, 281–290) developed the fragmentation theory of fission and heavy-ion collisions that is based on the asymmetric two-center shell model and treats mass and charge asymmetry as dynamical collective coordinates. These fragmentation variables are coupled to the coordinates of relative motion and other variables specifying the shape degrees of freedom of the intermediate system and treated in a time-dependent adiabatic calculation using various approximations. Such studies predict fragment mass and Z distributions. They also suggest (288–290) projectile-target combinations that are favorable candidates for the production of very heavy and superheavy compound nuclei, leading to a minimum excitation energy of the intermediate system. The theory has so far been in agreement with experiments carried out at Berkeley (291–293) and Dubna (294–297), in which attempts were made to synthesize very heavy elements via compound-nucleus reactions employing relatively light projectiles ranging from ^{12}C to ^{48}Ca.

6   EMISSION OF LIGHT PARTICLES AND γ RAYS IN DAMPED REACTIONS

Evidence was presented in Section 4.1 for the mainly binary reaction mode of damped collisions. However, this definition of the reaction mode allows for emission of some light particles such as neutrons, protons, and α particles prior to or after the breakup of the intermediate system. Particles emitted in the course of the interaction are expected to reflect the degree of equilibration reached in the process, both in their spectrum and their angular correlation. On the other hand, de-excitation particles and γ rays from the final fragments should yield valuable information on the partition of the total excitation energy between the two final fragments and the amount of orbital angular momentum transferred into intrinsic fragment spin.

6.1   Light-Particle Emission

Light particles may in principle be emitted from the intermediate system or, after its breakup, from the target-like or projectile-like final fragments. This results in a rather complicated appearance of these phenomena, since different emission mechanisms may be superimposed. Experimental evidence presented in Section 5 indicates that damped reaction times comprise generally only a fraction of typical rotation times. Therefore, evaporation-like features are not expected for particle emission occurring on a time scale comparable to the reaction time. Such processes have to be direct or preequilibrium ones (298). This emission mechanism predicts particle energy spectra extending to energies several times higher than the incident energy per nucleon (299–301) and distinct angular correlations for the higher energies (300, 301).

The evaporation-type features of particle decay from equilibrated compound nuclei produced in the interaction of light projectiles with many targets are very well known. However, particle emission from reactions induced by heavy ions has rarely been studied. The high spin of the nucleus formed in such reactions affects the relative yields and the angular distributions of subsequently emitted particles. Particle angular distributions may be highly anisotropic (302) and the relative particle yields are expected to be strongly dependent on the initial angular-momentum distribution and the excitation energy (303, 304).

Up to now, particle emission in damped reactions has been investigated in only a few cases. Among these are the systems Ni + ^{13}C (220), ^{93}Nb + ^{14}N (222), ^{27}Al + ^{16}O (272, 305), ^{58}Ni + ^{16}O (227, 306), ^{58}Ni + ^{40}Ar (212, 213), and ^{197}Au + ^{86}Kr (307). Emission probabilities for α particles in coincidence with fragments from the reaction

^{58}Ni+^{16}O (306) have been measured to range between 0.1 and 0.2, an estimated 50% of which are due to precompound emission. A study of the ^{197}Au+^{86}Kr reaction (307) at 720 MeV yielded for the emission probabilities of p, d, t, and α particles the values 0.29, 0.06, 0.04, and 0.34, respectively. The spectra of the coincident α particles from the reaction ^{58}Ni+^{16}O are reported (306) to have a Maxwellian shape with well-defined peak energies. However, at forward angles, an additional high-energy tail is seen. These α particles are strongly correlated with damped fragments, quasi-elastic coincidence events being strongly suppressed.

Kinematical considerations are valuable in analyzing three-body events. For

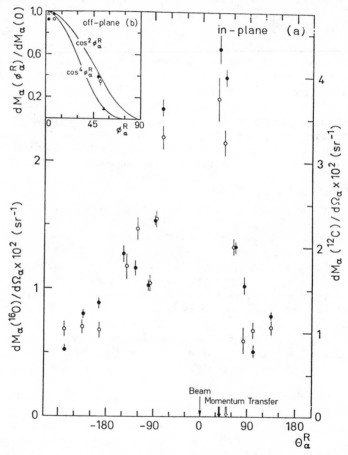

Figure 39   In-plane (*a*) and off-plane (*b*) angular correlation of the differential α multiplicity $dM_\alpha/d\Omega_\alpha$ for oxygen (*open symbols*) and carbon (*full symbols*) fragments from the reaction ^{58}Ni+^{16}O. The emission angle θ_α^R corresponds to the fragment rest frame. $\theta_\alpha^R = 0$ has been chosen for the beam direction. Negative θ_α^R corresponds to angles on the side of the fragment detector [from (306)].

example, for a sequential α decay, it is necessary that the average excitation energy of the intermediate nucleus inferred from the kinetic energies of α particle and final fragment be independent of the detection angles. This has been affirmed for the above ^{16}O-induced reactions (305, 306). The measurements are consistent with α emission from the intermediate heavy fragment, rendering a breakup of the light fragment unlikely.

Of particular importance are the angular correlations between α particle and final fragment. Figure 39 shows in-plane and off-plane correlation measurements for the reaction ^{58}Ni + ^{16}O (306), where oxygen and carbon fragments are detected at the fixed laboratory angle $\theta_{\text{Lab}} = -35°$. The differential α emission probability $dM_\alpha/d\Omega_\alpha$, normalized to the damped singles yield, is plotted vs α emission angle θ_α^R in the rest frames of the heavy fragments. The angular correlation is seen to peak strongly at angles near the beam direction. The high off-plane anisotropies suggest that α particles are emitted either from highly aligned fragments or from spatially confined regions localized at the fragment's equator. For the systems ^{58}Ni + ^{16}O and ^{93}Nb + ^{14}N, the most probable angles in the α-fragment angular correlations do not agree with the recoil angle of the heavy fragment determined by the momentum transfer. However, for the reaction ^{27}Al + ^{16}O (305), this seems to be the case. For the above three systems, a striking forward-backward asymmetry is observed, as illustrated by Figure 39. This suggests that α emission occurs on a very short time scale. Rough estimates of this time for the ^{16}O-induced reactions yielded $\tau \approx (6-20) \cdot 10^{-22}$ sec, at least one order of magnitude smaller than compound-nucleus lifetimes. These findings lend strong support to the occurrence of precompound particle emission on a time scale comparable to that of the damped reaction. It is then obvious that for a detailed analysis of the charged-particle angular correlation, the time evolution of the combined Coulomb field of the two heavy fragments attains importance (308). Of course, neutron emission from the fragments, so far not studied in heavy-ion reactions, is not sensitive to these latter effects.

Although many more experimental studies are needed to establish precompound particle emission in damped reactions, the available evidence indicates that such processes have to be considered, in addition to compound-nucleus evaporation for the interpretation of secondary radiation. Moreover, statistical-model analyses neglecting effects of the fragment spin and, for light systems, the individual level structure of the residual nucleus, cannot be expected to give an appropriate description.

6.2 Emission of γ Rays

The emission of γ rays is mainly confined to the final stages of the fragment de-excitation. Hence, information obtained from in-beam γ-ray studies is necessarily more indirect and has to be compared with the results of realistic calculations in order to draw meaningful conclusions on the reaction mechanism. Nevertheless, such studies are valuable since they can supply information on the reaction mechanism not readily accessible by other methods. Within this context, it should be mentioned that investigations of conversion electrons after damped reactions give similar information as γ-ray experiments, but for certain applications are more advantageous than the latter, due to a high sensitivity of the electron spectrum to recoil effects and the possibility of inferring the transition multipolarity from K/L or subshell intensity ratios.

Experimental results reported so far pertain to the identification of characteristic γ radiation from the final fragments and the determination of their multiplicity. It has been realized in studies of compound-nuclear reactions (303, 309–311) that the multiplicity distribution of de-excitation γ rays reflects the spin distribution of the initial states populated in the reaction. In a simple approach (310), the mean γ multiplicity $\langle M_\gamma \rangle$ is connected to the mean initial angular momentum $\langle l \rangle$ by the relation $\langle M_\gamma \rangle \approx \langle l \rangle / f$, where f is the mean angular momentum carried away by one γ ray. Higher moments of the multiplicity distribution are connected to those of the l distribution of states prior to γ emission. Knowing the reaction cross section, the mean l value characterizing the initial distribution can be calculated using Equation 6. This value has to be corrected for particle emission in order to obtain $\langle l \rangle$. Values for f obtained from compound nuclear studies in the region of rare-earth elements range from $f = (1-2)\hbar$ per γ transition (310, 311).

To estimate a proper value of f is one of the main problems encountered in γ-multiplicity studies related to damped reactions. Commonly, values for f are adopted from compound reactions involving nuclei quite different from the ones under consideration. Comparably little attention is given to α-particle emission, by which the initial spin distribution might be drastically changed. The importance of this latter effect, especially for the lighter systems, is possibly illustrated by the results of Van Bibber et al (224), who observed that the fraction of heavy fragments

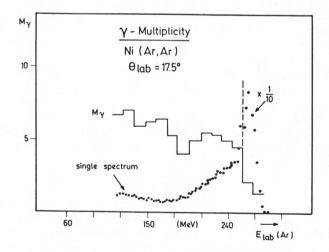

Figure 40 The multiplicity M_γ of γ rays in coincidence with Ar fragments from the reaction $^{58}\mathrm{Ni} + ^{40}\mathrm{Ar}$ at 280 MeV is plotted as a function of fragment energy. For comparison, a singles-fragment energy spectrum is indicated by the points [from (213)].

from the reaction $^{27}Al + {}^{16}O$ detected in low-lying states increases as the energy loss increases.

In-beam γ-multiplicity measurements aimed at inferring details of the damped reaction mechanism have been performed for a number of systems, ranging from $^{12}C + {}^{16}O$ to $^{197}Au + {}^{63}Cu$ (13, 213, 222, 224, 248, 312–314). Multiplicities as high as $M_\gamma = 25$ have been observed (248) for strongly damped fragments from the reaction $^{197}Au + {}^{63}Cu$. An observation common to most studies is that the γ multiplicity increases with increasing energy loss. This effect is shown in Figure 40 for Ar fragments produced in the $Ni + {}^{40}Ar$ reaction (315). From the energy-loss mass-exchange correlation discussed in Section 5.2, one expects then also a correlation between γ multiplicity and mass exchange. However, the importance of the energy-loss parameter is illustrated by Figure 40 and by the results of Glässel et al (314), who reported for low-Z fragments from the reaction $Ag + {}^{20}Ne$ an increase of the γ multiplicity with increasing reaction angle, which is also the direction of increasing energy loss. Such a behavior is consistent with what is expected from a diffusion process, where energy and angular momentum are lost as nucleons are exchanged. However, more experimental data are needed in order to establish this connection.

Considering the ambiguities in deducing the spin distribution of the final fragments from the γ-multiplicity distribution, it appears inappropriate at present to draw firm conclusions with respect to details of the damped-reaction mechanism. However, the identification of the deduced internal fragment spins with the angular-momentum loss, (cf Section 2.4) gives some hint to the evolution of the intermediate system. The large γ anisotropies observed for small energy losses (314) indicate highly aligned final fragments. Measurements for higher energy losses are not sensitive to the initial fragment-spin alignment because the evaporation of a large number of particles destroys such an alignment. The evidence presented so far indicates that the intermediate system does not stick for small energy losses. As the energy loss increases, the intermediate system approaches a rolling condition that may finally result in a sticking situation for a very high degree of damping.

7 SUMMARY

Although considerable progress has been made in understanding heavy-ion collisions in terms of a damped-reaction mechanism, there are a number of unanswered questions remaining to be solved. In this review, reactions are considered in which the bombarding energies are up to a few MeV per nucleon above the Coulomb barrier. Such energies are small compared to the kinetic energies of the Fermi motion. Therefore, the time scale in which internal degrees of freedom change is expected to be fast compared to typical collision times. Excitation of the intermediate system and particle exchange between its constituents is then also more likely to be determined by a one-body interaction due to the combined single-particle field, than by the two-body interaction of individual nucleons. The latter mechanism is realized in reactions induced by particles of incident energies of several tens of MeV per nucleon. Many more experimental and theoretical studies are needed to determine whether the heavy-ion damped-reaction and fission processes are dominated by one- or two-body dissipation.

The outstanding property of the damped-reaction mechanism is the strong correlation of experimental features with the loss of relative kinetic energy, which signifies the stage of evolution of the reaction. During the interaction, an equilibration process occurs that can be described in terms of a nucleon-diffusion mechanism. This new reaction mechanism connects in a continuous way the domains of quasi-elastic transfer and compound-nucleus formation. In reactions induced by lighter ions, the transition between these two domains occurs very abruptly with only a few l waves contributing to the damped-reaction mechanism. The latter was therefore discovered only after heavy-ion beams became available. Further experimental and theoretical studies are necessary to establish the reaction features in more detail and to investigate their transition for higher bombarding energies.

ACKNOWLEDGMENTS

The authors are grateful to a number of colleagues for helpful discussions on heavy-ion reactions and for supplying us with reprints and preprints of their publications. Special thanks are due to J. R. Birkelund and W. Wilcke for assisting us with various aspects of the manuscript.

Literature Cited

1. Kaufmann, K., Wolfgang, W. 1961. *Phys. Rev.* 121:192, 206
2. Gridnev, G. F., Volkov, V. V., Wilczyński, J. 1970. *Nucl. Phys. A* 142:385
3. Galin, J., Guerreau, D., Lefort, M., Péter, J., Tarrago, X. 1970. *Nucl. Phys. A* 159:461
4. Artyukh, A. G., Gridnev, G. F., Mikheev, V. L., Volkov, V. V., Wilczyński, J. 1973. *Nucl. Phys. A* 211:299, 215:91
5. Hanappe, F., Lefort, M., Ngô, C., Péter, J., Tamain, B. 1974. *Phys. Rev. Lett.* 32:738
6. Wolf, K. L., Unik, J. P., Huizenga, J. R., Birkelund, J. R., Freiesleben, H., Viola, V. E. 1974. *Phys. Rev. Lett.* 33:1105
7. Kratz, J. V., Norris, A. E., Seaborg, G. T. 1974. *Phys. Rev. Lett.* 33:502
8. Thompson, S. G., Moretto, L. G., Jared, R. C., Babinet, R. P., Galin, J. G., Fowler, M. M., Gatti, R. C., Hunter, J. B. 1974. *Phys. Scr. A* 10:36
9. Péter, J., Ngô, C., Tamain, B. 1975. *J. Phys. Lett.* 36:L23
10. Jacmart, J. C., Colombani, P., Doubre, H., Frascaria, N., Poffé, N., Riou, M., Roynette, J. C., Stéphan, C., Weidinger, A. 1975. *Nucl. Phys. A* 242:175
11. Colombani, P., Frascaria, N., Jacmart, J. C., Riou, M., Stéphan, C., Doubre, H., Poffé, N., Roynette, J. C. 1975. *Phys.*

Lett. B 55:45
12. Wilczyński, J., Siwek-Wilczyńska, K., Larsen, J. S., Acquadro, J. C., Christensen, P. R. 1975. *Nucl. Phys. A* 244:147
13. Albrecht, R., Dünnweber, W., Graw, G., Ho, H., Steadman, S. G., Wurm, J. P. 1975. *Phys. Rev. Lett.* 34:1400
14. Péter, J., Ngô, C., Tamain, B. 1975. *Nucl. Phys. A* 250:351
15. Gatty, B., Guerreau, D., Lefort, M., Tarrago, X., Galin, J., Cauvin, B., Girard, J., Nifenecker, H. 1975. *Nucl. Phys. A* 253:511
16. Galin, J., Moretto, L. G., Babinet, R., Schmitt, R., Jared, R., Thompson, S. G. 1975. *Nucl. Phys. A* 255:472
17. Moretto, L. G., Kataria, S. S., Jared, R. C., Schmitt, R., Thompson, S. G. 1975. *Nucl. Phys. A* 255:491
18. Moretto, L. G., Babinet, R. P., Galin, J., Thompson, S. G. 1975. *Phys. Lett. B* 58:31
19. Tamain, B., Plasil, F., Ngô, C., Péter, J., Berlanger, M., Hanappe, F. 1976. *Phys. Rev. Lett.* 36:18
20. Otto, R. J., Fowler, M. M., Lee, D., Seaborg, G. T. 1976. *Phys. Rev. Lett.* 36:135
21. Schröder, W. U., Birkelund, J. R., Huizenga, J. R., Wolf, K. L., Unik, J. P., Viola, V. E. 1976. *Phys. Rev. Lett.* 36:514

22. Cormier, T. M., Lazzarini, A. J., Neuhausen, M. A., Sperduto, A., Van Bibber, K., Videbaek, F., Young, G., Blum, E. B., Herreid, L., Thoms, W. 1976. *Phys. Rev.* C 13:682
23. Vandenbosch, R., Webb, M. P., Thomas, T. D. 1976. *Phys. Rev.* C 14:143; *Phys. Rev. Lett.* 36:459
24. Babinet, R., Moretto, L. G., Galin, J., Jared, R., Moulton, J., Thompson, S. G. 1976. *Nucl. Phys.* A 258:172
25. Moretto, L. G., Galin, J., Babinet, R., Fraenkel, Z., Schmitt, R., Jared, R., Thompson, S. G. 1976. *Nucl. Phys.* A 259:173
26. Moretto, L. G., Cauvin, B., Glässel, P.. Jared, R., Russo, P., Sventek, J., Wozniak, G. 1976. *Phys. Rev. Lett.* 36:1069
27. Kratz, J. V., Liljenzin, J. O., Norris, A. E., Seaborg, G. T. 1976. *Phys. Rev. C* 13:2347
28. Huizenga, J. R., Birkelund, J. R., Schröder, W. U., Wolf, K. L., Viola, V. E. 1976. *Phys. Rev. Lett.* 37:885
29. Eggers, R., Namboodiri, M. N., Gonthier, P., Geoffroy, K., Natowitz, J. B. 1976. *Phys. Rev. Lett.* 37:324
30. Vandenbosch, R., Webb, M. P., Thomas, T. D., Zisman, M. S. 1976. *Nucl. Phys.* A 263:1
31. Webb, M. P., Vandenbosch, R., Thomas, T. D. 1976. *Phys. Lett. B* 62:407
32. Volkov, V. V. 1976. *Nukleonika* 21:53
33. Ngô, C., Péter, J., Tamain, B., Berlanger, M., Hanappe, F. 1976. *Nucl. Phys.* A 267:181
34. Galin, J., Gatty, B., Guerreau, D., Lefort, M., Tarrago, X., Babinet, R., Cauvin, B., Girard, J., Nifenecker, H. 1976. *Z. Phys.* A 278:347
35. Péter, J., Ngô, C., Plasil, F., Tamain, B., Berlanger, M., Hanappe, F. 1977. *Nucl. Phys. A* 279:110
36. 1976. *Proc. Symp. Macroscopic Features Heavy Ion Collisions*, Argonne Natl. Lab. Rep. ANL/PHY-76-2 1976. Vol. I: Invited Papers, Vol. II: Contributed Papers
37a. 1976. *Proc. Eur. Conf. Nucl. Phys. Heavy Ions, Caen 1976. J. Phys.*, Colloque No. 5, C-5
37b. *Eur. Conf. Nucl. Phys. Heavy Ions, Caen 1976*, communications
37c. Fleury, A., Alexander, J. M. 1974. *Ann. Rev. Nucl. Sci.* 24:279
38. Wilczyński, J. 1973. *Phys. Lett. B* 47:484
39a. Swiatecki, W. J. 1975. Lawerence Berkeley Lab. LBL-4296
39b. Swiatecki, W. J. 1972. *Jour. Phys. C* 5:45
39c. Swiatecki, W. J. 1975. *Proc. Int. Sch.*

Semin. React. Heavy Ions with Nuclei Synth. New Elem., Sept–Oct p. 89
40. Huizenga, J. R. 1976. *Accounts Chem. Res.* 9:325
41. Satchler, G. R. 1976. See Ref. 36, p. 33
42. Glendenning, N. K. 1975. *Rev. Mod. Phys.* 47:659
43. Baltz, A. J. 1976. See Ref. 36, p. 65
44. Von Oertzen, W. 1975. In *Nuclear Spectroscopy*, ed. J. Cerny. New York: Academic
45. Buttle, P. J. A., Goldfarb, L. J. B. 1971. *Nucl. Phys.* A 176:299
46. Messiah, A. 1965. *Quantum Mechanics.* Amsterdam: North-Holland
47. Broglia, R. A., Winther, A. 1972. *Phys. Rep. C* 4:153
48. Broglia, R. A., Landowne, S., Malfliet, R. A., Rostokin, U., Winther, A. 1974. *Phys. Rev. C* 11:1
49. Gross, D. H. E. 1976. *Nucl. Phys. A* 260:333
50. Siemens, P. J., Becchetti, F. D. 1972. *Phys. Lett. B* 42:389
51. Frahn, W. E., Gross, D. H. E. 1976. *Ann. Phys.* 101:520
52. Nörenberg, W., Weidenmüller, H. A. 1976. *Introduction to the Theory of Heavy Ion Collisions*, Lect. Notes Phys. 51. Berlin-Heidelberg-New York: Springer
53. Frahn, W. E., Venter, R. H. 1963. *Ann. Phys.* 24:243
54. Venter, R. H. 1963. *Ann. Phys.* 25:405
55. Frahn, W. E. 1971. *Phys. Rev. Lett.* 26:568
56. Frahn, W. E. 1972. *Ann. Phys.* 72:524
57. Blair, J. S. 1954. *Phys. Rev.* 95:1218
58. Blair, J. S. 1954. *Phys. Rev.* 108:827
59. Kerlee, D. C., Blair, J. S., Farwell, G. W. 1957. *Phys. Rev.* 107:1343
60. Kanai, E. 1948. *Progr. Theor. Phys.* 3:440
61. Kostin, M. D. 1972. *J. Chem. Phys.* 57:3589
62. Kostin, M. D. 1975. *J. Stat. Phys.* 12:145
63. Albrecht, K. 1975. *Phys. Lett. B* 56:127
64. Immele, J. K., Kan, K. K., Griffin, J. J. 1975. *Nucl. Phys.* A 241:47
65. Hasse, R. W. 1975. *J. Math. Phys.* 16:2005
66. Yasue, K. 1976. *Phys. Lett. B* 64:239
67. Eck, J. S., Thompson, W. J. 1976. Prepr.
68. Hasse, R. W. 1976. See Ref. 37b, p. 143
69. Gross, D. H. E. 1975. *Nucl. Phys. A* 240:472
70. Ledergerber, T., Paltiel, Z., Pauli, H. C., Schütte, G., Yariv, Y., Fraenkel, Z. 1975. *Phys. Lett. B* 36:417
71. Schütte, G., Wilets, L. 1975. *Nucl. Phys. A* 252:21
72. Glas, D., Mosel, U. 1976. *Nucl. Phys. A* 264:268

73. Koonin, S. E., Nix, J. R. 1976. *Phys. Rev. C* 13:209
74. Pauli, H. C., Wilets, L. 1976. *Z. Phys. A* 277:83
75. Hofmann, H. 1975. 13th Int. Winter Meet. Nucl. Phys., Bormio, Italy
76. Hofmann, H., Siemens, P. J. 1976. *Nucl. Phys. A* 257:165
77. Mshelia, E. D., Scheid, W., Greiner, W. 1975. *Nuovo Cimento Ser* II, A 30:598
78. Gross, D. H. E. 1976. Prepr.; Gross, D. H. E., Kalinowski, H. 1977. 4. Sess. 'Étud. Bienn. Phys. Nucl., La Toussuire, France
79. Ko, C. M., Pirner, H. J., Weidenmüller, H. A. 1976. *Phys. Lett. B* 62:248
80. Kerman, A. K., Koonin, S. E. 1974. *Phys. Scr. A* 10:118
81. Agassi, D., Ko, C. M., Weidenmüller, H. A. 1976. Rep. MPIH-1976-V25, Heidelberg
81a. Ko, C. M., Agassi, D., Weidenmüller, H. A. 1977. Rep. MPIH-1977-V5, Heidelberg
82. Hofmann, H. 1976. *Phys. Lett. B* 61:423
83. Hofmann, H., Siemens, P. J. 1976. *Nucl. Phys. A* 275:464
84. Beck, R., Gross, D. H. E. 1973. *Phys. Lett. B* 47:143
85. Tsang, F. C. 1974. *Phys. Scr. A* 10:90
86. Albrecht, K., Stocker, W. 1977. *Nucl. Phys. A* 278:95
87. Glas, D., Mosel, U. 1974. *Phys. Lett. B* 49:301
88. Giraud, B., LeTourneux, J., Osnes, E. 1975. *Phys. Rev. C* 11:82
89. Wilczyński, J. 1976. See Ref. 36, p. 211
90. Siwek-Wilczyńska, K., Wilczyński, J. 1976. *Nucl. Phys. A* 264:115
91. Siemens, P. J., Bondorf, J. P., Gross, D. H. E., Dickmann, F. 1971. *Phys. Lett. B* 36:24
92. Brink, D. M. 1972. *Phys. Lett. B* 40:37
93. Schröder, W. U., Huizenga, J. R., Birkelund, J. R., Wolf, K. L., Viola, V. E. To be published
94. Bass, R. 1973. *Phys. Lett. B* 47:139
95. Bass, R. 1974. *Nucl. Phys. A* 231:45
96. Bondorf, J. P., Huizenga, J. R., Sobel, M. I., Sperber, D. 1975. *Phys. Rev. C* 11:1265
97. Bondorf, J. P., Liou, M. K., Sobel, M. I., Sperber, D. 1976. Prepr.
98. Gross, D. H. E., Kalinowski, H., De, J. N. 1975. *Symp. Classical Quantum Mech. Aspects Heavy Ion Collisions*, Heidelberg, 1974. Heidelberg: Springer (Heidelberg). *Lect. Notes Phys.* 33:194
99. De, J. N., Gross, D. H. E., Kalinowski, H. 1976. See Ref. 36, p. 523
100. De, J. N., Gross, D. H. E., Kalinowski,

H. 1976. *Z. Phys. A* 277:385
101. Randrup, R. 1977. Prepr.
102. Bass, R. 1976. See Ref. 37b, p. 147
103. Broglia, R. A., Dasso, C. H., Winther, A. 1976. *Phys. Lett. B* 61:113
104. Mustafa, M. G. 1975. *Phys. Lett. B* 60:15
105. Mustafa, M. G. 1976. *Phys. Rev. C* 14:2168
106. Deubler, H. H., Dietrich, K. 1975. *Phys. Lett. B* 56:241
107. Deubler, H. H., Dietrich, K. 1977. *Nucl. Phys. A* 277:493
108. Haken, H. 1975. *Rev. Mod. Phys.* 47:67
109. Nörenberg, W. 1974. *Phys. Lett. B* 52:289
110. Nörenberg, W. 1975. *Z. Phys. A* 274:241; 1976. *Z. Phys. A* 276:84. See Ref. 37a, p. C5-141
111. Ayik, S., Schürmann, B., Nörenberg, W. 1976. *Z. Phys. A* 277:299
112. Ayik, S., Schürmann, B., Nörenberg, W. 1976. *Z. Phys. A* 279:145
113. Ayik, S. 1976. *Phys. Lett. B* 63:22
114. Agassi, D., Weidenmüller, H. A. 1975. *Phys. Lett. B* 56:305
115. Bunakov, V. E., Nesterov, M. M. 1976. *Phys. Lett. B* 60:417
116. Mädler, P., Reif, R., Ropke, G., Zschau, H. E. 1976. *Phys. Lett. B* 61:427
117. Moretto, L. G., Sventek, J. S. 1975. *Phys. Lett. B* 58:26
118. Moretto, L. G., Schmitt, R. 1976. *J. Phys.* 11, C 5:109
119. Moretto, L. G., Sventek, J. S. 1976. See Ref. 36, p. 235
120. Sperber, D., private communication
121. Sperber, D. 1974. *Phys. Scr. A* 10:115
122. Bondorf, J. P. 1974. *Proc. Int. Conf. React. Complex Nuclei, Nashville*
123. Bondorf, J. P., Sobel, M. I., Sperber, D. 1974. *Phys. Rep. C* 15:83
124. Bondorf, J. P. 1975. Varenna Summer School
125. Davis, R. H. 1974. *Phys. Rev. C* 9:2411
126. Davis, R. H. 1975. Int. Symp. Cluster Struct. Nuclei Transfer React. Induced by Heavy Ions, Tokyo
127. Davis, R. H. 1976. See Ref. 36, p. 517
128. Davis, R. H. 1976. Prepr.
129. Sperber, D. 1975. *Nukleonika* 20:755
130. Seglie, E., Sperber, D., Sherman, A. 1975. *Phys. Rev. C* 11:1227
131. Lin, L., Sherman, A., Sperber, D. 1976. *Phys. Rev. Lett.* 37:327
132. Beck, F. 1976. *Phys. Lett. B* 62:385
133. Deubler, H. H., Dietrich, K., Hofmann, H. 1977. *Heavy Ion Collisions*, ed. R. Bock. Amsterdam: North-Holland. To be published
134. Hofmann, H., Ngô, C. 1976. *Phys. Lett.*

B 65:97
135. Ngô, C., Hofmann, H. 1977. *Z. Phys. A* 282:83
136. Bonche, P., Koonin, S., Negele, J. W. 1976. *Phys. Rev. C* 13:1226
137. Koonin, S. E. 1976. *Phys. Lett. B* 61:227
138. Cusson, R. Y., Smith, R. K., Maruhn, J. A. 1976. *Phys. Rev. Lett.* 36:1166
139. Cusson, R. Y., Maruhn, J. 1976. *Phys. Lett. B* 62:134
139a. Maruhn, J., Cusson, R. Y. 1976. *Nucl. Phys. A* 270:471
140. Koonin, S. E., Davies, K. T. R., Maruhn-Rezwani, V., Feldmeier, H., Krieger, S. J., Negele, J. W. 1977. *Phys. Rev. C* 15:1359
141. Bonche, P., Grammaticos, B., Jaffrin, A. See Ref. 37b, pp. 151–52
142. Maruhn-Rezwani, V., Davies, K. T. R., Koonin, S. E. 1977: *Phys. Lett. B* 67:134
143. Griffin, J. J., Kan, K.-K. 1975. *Proc. Int. Workshop Gross Prop. Nuclei Nucl. Excitations, 3rd, Hirschegg, Austria*, Rep. AED-Conf-75-009-000
144. Griffin, J. J., Kan, K.-K. 1976. *Rev. Mod. Phys.* 48:467
145. Wong, C. Y. 1976. *J. Math. Phys.* 17:1008
146. Wong, C. Y., Maruhn, J. A., Welton, T. A. 1975. *Nucl. Phys. A* 253:469
146a. Wong, C. Y., Welton, T. A., Maruhn, J. A. 1977. *Phys. Rev. C* 15:1558
147. Wong, C. Y., Welton, T. A., Maruhn, J. A. 1976. ORNL prepr.
148. Wong, C. Y., McDonald, J. A. 1976. ORNL prepr.
149. Alonso, C. T. 1974. *Int. Colloq. Drops Bubbles, Pasadena.* Rep. LBL-2993
150. Davies, K. T. R., Koonin, S. E., Nix, J. R., Sierk, A. J. 1975. See Ref. 143
151. Davies, K. T. R., Sierk, A. J., Nix, J. R. 1976. *Phys. Rev. C* 13:2385
152. Sierk, A. J., Nix, J. R. 1976. Rep. LAP-151, see Ref. 36, p. 407
153. Zint, P. G., Mosel, U. 1975. *Phys. Lett. B* 56:424
154. Satchler, G. R. 1974. *Proc. Int. Conf., React. Complex Nuclei,* p. 171. Amsterdam: North-Holland
155. Riesenfeldt, P. W., Thomas, T. D. 1970. *Phys. Rev. C* 2:711
156. Wong, C. Y. 1972. *Phys. Lett. B* 42:186; *Phys. Rev. Lett.* 31:766
157. Fuller, R. C. 1975. *Phys. Rev. C* 12:1561
158. Donnelly, T. W., Dubach, J., Walecka, J. D. 1974. *Nucl. Phys. A* 232:355
159. Birkelund, J. R., Huizenga, J. R., Freiesleben, H., Wolf, K. L., Unik, J. P.,

Viola, V. E. 1976. *Phys. Rev. C* 13:133
160. Christensen, P. R., Winther, A. 1976. *Phys. Lett. B* 65:19
161. Ball, J. B., Fulmer. C. B., Gross, E. E., Halbert, M. L., Hensley, D. C., Ludemann, C. A., Saltmarsh, M. J., Satchler, G. R. 1975. *Nucl. Phys. A* 252:208
162. Vandenbosch, R., Webb, M. P., Thomas, T. D., Yates, S. W., Friedman, A. M. 1976. *Phys. Rev. C* 13:1893
163. Huizenga, J. R., Birkelund, J. R., Johnson, M. W. See Ref. 36, p. 1
164. Huizenga, J. R. 1975. *Nukleonika* 20:291
165. Blair, J. S. 1970. *Proc. Conf. Nucl. React. Induced Heavy Ions,* ed. W. Hering, R. Bock. Amsterdam: North-Holland
166. Cramer, J. G., DeVries, R. M., Goldberg, D. A., Zisman, M. S., Maguire, C. F. 1976. *Phys. Rev. C* 14:2158
167. Wilczyński, J. 1973. *Nucl. Phys. A* 216:386
168. Michaud, G., Vogt, E. W. 1972. *Phys. Rev. C* 5:350
169. Myers, W. D. 1973. *Nucl. Phys. A* 204:465
170. Brink, D. M., Rowley, N. 1974. *Nucl. Phys. A* 219:79
171. Vary, J. B., Dover, C. B. 1973. *Phys. Rev. Lett.* 31:1510
172. Dover, C. B., Vary, J. P. 1975. *Classical and Quantum-Mechanical Aspects of Heavy Ion Collisions,* ed. H. L. Harney et al., p. 1. Berlin-Heidelberg-New York: Springer
173. Sinha, B. 1975. *Phys. Rep. C* 20:1
174. Satchler, G. R. 1975. *Phys. Lett. B* 59:121
175. Satchler, G. R. 1977. *Nucl. Phys. A* 279:493
176. Mosel, U. 1976. See Ref. 36, p. 341
177. Blocki, J., Randrup, J., Swiatecki, W. J., Tsang, C. F. 1977. *Ann. Phys.* 105:427
178. Swiatecki, W. J., private communication
179. Scobel, W., Gutbrod, H. H., Blann, M., Mignerey, A. 1976. *Phys. Rev. C* 14:1808
180. Birkelund, J. R., Huizenga, J. R. 1977. *Phys. Rev. C*
181. Bass, R. 1977. *Phys. Rev. Lett.* 39:265
182. Brink, D. M., Stancu, Fl. 1975. *Nucl. Phys. A* 243:175
183. Zint, P. G., Mosel, U. 1976. *Phys. Rev. C* 14:1488
184. Brueckner, K. A., Buchler, J. R., Kelly, M. M. 1968. *Phys. Rev.* 173:944
185. Ngô, C., Tamain, B., Beiner, M., Lombard, R. J., Mas, D., Deubler, H. H. 1975. *Nucl. Phys. A* 252:237
186. Greiner, W., Scheid, W. 1971. *J. Phys.*

32:C6–91

187. Krappe, H. J., Nix, J. R. 1974. *Proc. IAEA Symp. Phys. Chem. Fission, 3rd, IAEA, Vienna* I:159
188. Nix, J. R., Sierk, A. J. 1974. *Phys. Scr. A* 10:94
189. Möller, P., Nix, J. R. 1976. *Nucl. Phys. A* 272:502
190. Davies, K. T. R., Nix, J. R. 1976. *Phys. Rev. C* 14:1977
191. Krappe, H. J. 1976. *Nucl. Phys. A* 269:493
192. Möller, P., Nix, J. R. 1977. *Nucl. Phys. A* 272:502
193. Morović, T., Greiner, W. 1976. *Z. Naturforsch.* 31a:327
194. Zohni, O., Blann, M. 1977. Preprint
195. Arnould, M., Howard, W. M. 1976. *Nucl. Phys. A* 274:295
196. Videbaek, F., Goldstein, R. B., Grodzins, L., Steadman, S. G., Belote, T. A., Garrett, J. D. 1977. *Phys. Rev. C* 15:954
197. Colombani, P. 1974. Rep. IPNO-PhN 74-21
198. Lefort, M., Ngô, C., Peter, J., Tamain, B. 1973. *Nucl. Phys. A* 216:166
199. Bimbot, R., Gardes, D., Hahn, R. L., DeMoras, Y., Rivet, M. F. 1974. *Nucl. Phys. A* 228:85; *Nucl. Phys. A* 248:377
200. Rivet, M. F., Bimbot, R., Fleury, A., Gardes, D., Llabador, Y. 1977. *Nucl. Phys. A* 276:157
201. Agarwal, S., Galin, J., Gatty, B., Guerreau, D., Lefort, M., Tarrago, X., Babinet, R., Girard, J., Nifenecker, H. To be published
202. Nicholson, W. J., Halpern, I. 1959. *Phys. Rev.* 116:175
203. Minor, M. N., Salwin, A. E., Viola, V. E. 1972. *Nucl. Instrum. Methods* 99:63
204. Wolf, K. L., Unik, J. P., Viola, V. E., Birkelund, J. R., Schröder, W. U., Huizenga, J. R., Freiesleben, H. To be published
205. Wolf, K. L., Roche, C. T. 1976. See Ref. 36, p. 295
206. Natowitz, J. B., Namboodiri, M. N., Chulick, E. T. 1976. *Phys. Rev. C* 13:171
207. Bond, P. D., Chasman, C., Schwarzschild, A. Z. 1976. See Ref. 36, p. 475
208. Natowitz, J. B., Namboodiri, M. N., Eggers, R., Gonthier, P., Geoffroy, K., Hanus, R., Towsley, C., Das, K. 1977. *Nucl. Phys. A* 277:477
209. Gelbke, C. K., Braun-Munzinger, P., Barrette, J., Zeidman, B., LeVine, M. J., Gamp, A., Harney, H. L., Walcher, Th. 1976. *Nucl. Phys. A* 269:460
210. Braun-Munzinger, P., Gelbke, C. K.,

Barrette, J. Zeidman, B., LeVine, M. J., Gamp, A., Harney, H. L., Walcher, Th. 1976. *Phys. Rev. Lett.* 36:849
211. Cormier, T. M., Cosman, E. R., Lazzarini, A. J., Garrett, J. D., Wegner, H. E. 1976. See Ref. 36, p. 509
212. Babinet, R., Cauvin, B., Girard, J., Nifenecker, H., Agarwal, S., Galin, J., Gatty, B., Guerreau, D., Lefort, M., Tarrago, X. 1976. See Ref. 37b, p. 168
213. Albrecht, A., Back, B. B., Bock, R., Fisher, B., Gobbi, A., Hildenbrand, K., Kohl, W., Lynen, U., Rode, I., Stelzer, H., Anger, G., Galin, J., Lagrange, J. M. 1976. See Ref. 37b, p. 167
214. Vater, P., Becker, H. J., Brandt, R., Freiesleben, H. 1976. See Ref. 37b, p. 110
215. Russo, P., Schmitt, R. P., Wozniak, G. J., Cauvin, B., Glässel, P., Jared, R. C., Moretto, L. G. 1977. *Phys. Lett. B* 67:155
216. Vigdor, S. E. 1976. See Ref. 36, p. 95
217. Strutinsky, V. M. 1973. *Phys. Lett. B* 44:245; *Sov. Phys. JETP* 19:1401
218. Lefort, M., Basile, R., Galin, J., Guerreau, D., Péter, J., Tarrago, X. 1970. *Proc. Int. Conf. Nucl. React. Induced Heavy Ions, Heidelberg.* Amsterdam: North-Holland
219. Eyal, Y., Beg, K., Logan, D., Miller, J., Zebelman, A. 1973. *Phys. Rev. C* 8:1109
220. Ost, R., Sanderson, N. E., Mordechai, S., England, J. B. A., Fulton, B. R., Nelson, J. M., Morrison, G. C. 1976. *Nucl. Phys. A* 265:142
221. Wilczyński, J., Volkov, V. V., Decowski, P. 1967. *Yad. Fis.* 5:942
222. Mikumo, T., Kohno, I., Katori, K., Motobayashi, T., Nakajima, S., Yoshia, M., Kamitsubo, H. 1976. *Phys. Rev. C* 14:1458; Kamitsubo, H. 1976. IPCR-Cyclotron Rep. 38
223. Pühlhofer, F., Diamond, R. M. 1972. *Nucl. Phys. A* 191:561
224. Van Bibber, K., Ledoux, R., Steadman, S. G., Videbaek, F., Young, G., Flaum, C. 1977. *Phys. Rev. Lett.* 38:334
225. Kovar, D. G., Eisen, Y., Henning, W., Ophel, T. R., Zeidman, B., Erskine, J. R., Fortune, H. T., Sperr, P., Vigdor, S. E. 1976. See Ref. 36, p. 645
226. Albrecht, R., Demond, F. J., Dünnweber, W., Graw, G., Ho, H., Slemmer, J., Wurm, J. P. 1976. See Ref. 37b, p. 160
227. Albrecht, R., Bercks, C., Dünnweber, W., Graw, G., Ho, H., Wurm, J. P., Disdier, D., Rauch, V., Scheibling, F. 1976. See Ref. 37b, p. 159
228. McGrath, R. L., Cormier, T. M.,

Geesaman, D. V., Harris, J. W., Lee, L. L., Wurm, J. P. 1976. See Ref. 36, p. 681
229. Gelbke, C. K., Buenerd, M., Hendrie, D. L., Mahoney, J., Mermaz, M. C., Olmer, C., Scott, D. K. 1976. *Phys. Lett. B* 65:227
230. Artyukh, A. G., Avdeichikov, V. V., Erö, J., Gridnev, G. F., Mikheev, V. L., Volkov, V. V., Wilczyński, J. 1971. *Nucl. Phys. A* 160:511
231. Buenerd, M., Gelbke, C. K., Harvey, B. G., Hendrie, D. L., Mahoney, J., Menchaca-Rocha, A., Olmer, C., Scott, D. K. 1976. *Phys. Rev. Lett.* 37:1191
232. Volkov, V. V. 1976. *Proc. Symp. Classical Quantum Mech. Aspects Heavy-Ion Collisions*, ed. H. L. Harney, P. Braun-Munzinger, C. K. Gelbke, p. 253. Berlin-Heidelberg-New York: Springer
233. Young, G., Blum, E. B., Lazzarini, A. J., Neuhausen, M. A., Steadman, S. G., Tsoupas, N. 1976. See Ref. 36, p. 849
234. Halbert, M. L., Stockstad, R. G., Obenshain, F. E., Plasil, F., Hensley, D. C., Snell, A. H., Ferguson, R. L., Pleasonton, F. 1976. See Ref. 36, p. 601
235. Mikheev, V. L., Artyukh, A. G., Volkov, V. V., Gridnev, G. F. 1977. *Yad. Fiz* (*USSR*) 25:255
236. Gatty, B., Guerreau, D., Lefort, M., Pouthas, J., Tarrago, X., Galin, J., Cauvin, B., Girard, J., Nifenecker, H. 1974. *J. Phys. Lett.* 35:L117
237. Gatty, B., Guerreau, D., Lefort, M., Pouthas, J., Tarrago, X., Galin, J., Cauvin, B., Girard, J., Nifenecker, H. 1975. *Z. Phys. A* 273:65
238. Agarwal, S., Galin, J., Gatty, B., Guerreau, D., Lefort, M., Tarrago, X., Babinet, R., Cauvin, B., Girard, J., Nifenecker, H. 1976. See Ref. 37b, p. 169
239. Britt, H. C., Erkkila, B. H., Stokes, R. H., Gutbrod, H. H., Plasil, F., Ferguson, R. L., Blann, M. 1976. *Phys. Rev. C* 13:1483
240. Jacmart, J. C., Frascaria, N., Poffé, N., Colombani, P., Doubre, H., Riou, M., Roynette, J. C., Stéphan, C. 1975. *J. Phys. C* 5:107
241. Roynette, J. C., Doubre, H., Jacmart, J. C., Poffé, N. 1976. See Ref. 37b, p. 165
241a. Roynette, J. C., Doubre, H., Frascaria, N., Jacmart, J. C., Poffé, N., Riou, M. 1977. *Phys. Lett. B* 67:395
242. Colombani, P. 1974. *J. Phys.* 11:C5–75
243. Rivet, M. F., Bimbot, R., Fleury, A., Gardés, D., Llabador, Y. 1977. *Nucl.*

Phys. A 276:157
244. Reus, N., Habbestad-Wätzig, A. M., Esterlund, R. A., Patzelt, P. 1976. See Ref. 37b, p. 171
244a. Ouichaoui, S., Ngô, C., Péter, J., Plasil, F., Tamain, B., Berlanger, M., Hanappe, F. 1976. See Ref. 37b, p. 112
245. Tassangot, L., Frascaria, N., Garron, J. P., Jacmart, J. C., Poffé, N., Stéphan, C. 1976. See Ref. 37b, p. 174
246. Berlanger, M., Hanappe, F., Ngô, C., Péter, J., Plasil, F., Tamain, B. 1977. *Nucl. Phys. A* 276:347
247. Cabot, Cl., Gauvin, H., LeBeyec, Y., Lefort, M. 1976. See Ref. 37b, p. 111
248. Berlanger, M., Deleplanque, M. A., Gerschel, C., Hanappe, F., Leblanc, M., Mayault, J. F., Ngô, C., Paya, D., Perrin, N., Péter, J., Tamain, B., Valentin, L. 1977. *J. Phys. Lett.* 37:L323
249. Gauvin, H., Hahn, R. L., LeBeyec, Y., Lefort, M. 1974. *Phys. Rev. C* 10:722
250. Schmitt, R. P., Russo, P., Babinet, R., Jared, R., Moretto, L. G. 1977. *Nucl. Phys. A* 279:141
251. Schröder, W. U., Birkelund, J. R., Huizenga, J. R., Wolf, K. L., Unik, J. P., Viola, V. E. 1976. See Ref. 37b, p. 178
252. Russo, P., Schmitt, R. P., Wozniak, G. J., Jared, R. C., Glässel, P., Cauvin, B., Sventek, J. S., Moretto, L. G. 1977. *Nucl. Phys. A* 281:509
253. Kratz, J. V., Brüchle, W., Dreyer, I., Franz, G., Wirth, G., Schädel, M., Weis, M. 1976. See Ref. 37b, p. 175
254. Freiesleben, H., Volant, C., Hildenbrand, K. D., Schneider, W. F. W., Chestnut, R. P., Pühlhofer, F., Kohlmeyer, B., Pfeffer, W. 1976. See Ref. 37b, p. 140
255. Birkelund, J. R., Schröder, W. U., Huizenga, J. R., Wolf, K. L., Unik, J. P., Viola, V. E. 1976. See Ref. 36, p. 451
256. Schröder, W. U., Birkelund, J. R., Huizenga, J. R., Wolf, K. L., Unik, J. P., Viola, V. E. To be published
257. Boudrie, R. L., Wolf, K. L., Roche, C. T., Birkelund, J. R., Schröder, W. U., Huizenga, J. R., Viola, V. E. 1976. *Bull. Am. Phys. Soc.* 21:969; unpublished data
258. Russo, P., Wozniak, G. J., Schmitt, R., Jared, R., Glässel, P., Moretto, L. G. 1975. Rep. LBL-5075
259. Kaufmann, S. B., Steinberg, E. P., Wilkins, B. D., Unik, J., Gorski, A. J., Fluss, M. J. 1974. *Nucl. Instrum. Methods* 115:47
260. Bondorf, J. P., Nörenberg, W. 1973.

Phys. Lett. B 44:487
261. Wolf, K. L., Huizenga, J. R., Birkelund, J. R., Freiesleben, H., Viola, V. E. 1976. *Bull. Am. Phys. Soc.* 21:31
262. Lefort, M. 1976. *Rep. Progr. Phys.* 39:129
263. Galin, J. 1976. See Ref. 37a, p. C5-83
264. Birkelund, J. R., Freiesleben, H., Schröder, W. U., Huizenga, J. R., Wolf, K. L., Unik, J., Viola, V. E. To be published
265. Myers, W. D., Swiatecki, W. J. 1969. *Ann. Phys. NY* 55:395
266. Volkov, V. V. 1975. Workshop Effects Dynamics Close Collisions Complex Nuclei, Gif/Yvette
267. Artyukh, A. G., Volkov, V. V., Gridnev, G. F., Mikheev, V. L. 1975. *Izv. Akad. Nauk USSR Ser. Phys.* 34:2
268. Volkov, V. V. 1975. *Part. Nucl. USSR* 6:1040; 1976. *Sov. J. Part. Nucl.* 6:420
269. Bondorf, J. P., Dickmann, F., Gross, D. H. E., Siemens, P. J. 1971. *J. Phys.* 32:C6-145
270. Artukh, A. G., Gridnev, G. F., Mikheev, V. L., Volkov, V. V. 1976. See Ref. 37b, p. 162–63
271. Wilczyński, J. 1973. *Phys. Lett. B* 47:124
272. Cormier, T. M. 1976. See Ref. 36, p. 153
273. Artukh, A. G., Wilczyński, J., Volkov, V. V., Gridnev, G. F., Mikheev, V. L. 1973. *Yad. Fis. USSR* 17:1126
274. Cormier, T. M., Braun-Munzinger, P., Cormier, P. M., Harris, J. W., Lee, L. L. 1977. *Phys. Rev. C* 16:215
275. Lefort, M. 1976. *Nukleonika* 21:111
276. De, J. N. 1977. *Phys. Lett. B* 66:315
277. Schröder, W. U., Birkelund, J. R., Huizenga, J. R., Wolf, K. L., Viola, V. E. 1977. *Phys. Rev. C* 16:623
278. Birkelund, J. R., Huizenga, J. R. 1977. Private communication
279. Gutbrod, H. H., Winn, W. G., Blann, M. 1973. *Nucl. Phys. A* 213:267
280. Colombani, P., Gatty, B., Jacmart, J. C., Lefort, M., Péter, J., Riou, M., Stéphan, C., Tarrago, X. 1972. *Phys. Lett. B* 42:208
281. Lichtner, P., Drechsel, D., Maruhn, J., Greiner, W. 1973. *Phys. Lett. B* 45:175
282. Fink, H. J., Maruhn, J., Scheid, W., Greiner, W. 1974. *Z. Phys.* 268:321
283. Zohni, O., Maruhn, J., Scheid, W., Greiner, W. 1975. *Z. Phys. A* 275:235
284. Yamaji, S., Scheid, W., Fink, H. J., Greiner, W. 1976. *Z. Phys. A* 278:69
285. Gupta, R. K., Greiner, W. 1976. Int. Winter Meet. Nucl. Phys., 14th, Bormio, Italy
286. Zohni, O. 1976. See Ref. 36, p. 867

287. Gupta, R. K. To be published in *Part. Nucl. USSR*
288. Săndulescu, A., Gupta, R. K., Scheid, W., Greiner, W. 1976. *Phys. Lett. B* 60:225
289. Gupta, R. K., Săndulescu, A., Greiner, W. 1977. *Phys. Lett. B* 67:257
290. Gupta, R. K., Pârvulescu, C., Săndulescu, A., Greiner, W. 1977. Preprint
291. Ghiorso, A., Nurmia, M., Harris, J., Eskola, K., Eskola, P. 1969. *Phys. Rev. Lett.* 22:1317
292. Ghiorso, A., Nurmia, M., Eskola, K., Eskola, P. 1970. *Phys. Lett. B* 32:95
293. Ghiorso, A., Nitschke, J. M., Alonso, J. R., Alonso, C. T., Nurmia, M., Seaborg, G. T., Hulet, E. K., Longheed, R. W. 1974. *Phys. Rev. Lett.* 33:1490
294. Flerov, G. N., Oganesyan, Yu. Ts., Lobanov, Yu. V., Kuznetsov, V. I., Druin, V. A., Perelygin, V. P., Gavrilov, K. A., Tretiakova, S. P., Plotko, V. M. 1964. *Phys. Lett.* 13:73
295. Oganesian, Yu. Ts., Iljinov, A. S., Demin, A. G., Tretyakova, S. P. 1975. *Nucl. Phys. A* 239:353
296. Ter-Akopyan, G. M., Iljinov, A. S., Oganesian, Yu. Ts., Orlova, O. A., Popeko, G. S., Tretyakova, S. P., Chepigin, V. I., Shilov, B. V., Flerov, G. N. 1975. *Nucl. Phys. A* 255:509
297. Flerov, G. N., Oganesian, Yu. Ts., Pleve, A. A., Pronin, N. V., Tretyakov, Yu. P. 1976. *Nucl. Phys. A* 267:359
298. Blann, M. 1975. *Ann. Rev. Nucl. Sci.* 25:123
299. Blann, M. 1974. *Nucl. Phys. A* 235:211
300. Bertini, H. W., Gavriel, T. A., Santoro, R. T. 1974. *Phys. Rev. C* 9:522
301. Bertini, H. W., Santoro, R. T., Herrmann, O. W. 1976. *Phys. Rev. C* 14:590
302. Galin, J., Gatty, B., Guerreau, D., Rousset, C., Schlotthauer-Voos, U. C., Tarrago, X. 1974. *Phys. Rev. C* 9:1113; 1974. *Phys. Rev. C* 9:1126
303. Grover, J. R., Gilat, J. 1967. *Phys. Rev.* 157:814; 1967. *Phys. Rev.* 157:823
304. Beckerman, M., Blann, M. To be published
305. Harris, J. W., Cormier, T. M., Geesaman, D. V., Lee, L. L., McGrath, R. L., Wurm, J. P. 1977. *Phys. Rev. Lett.* 38:1460
306. Ho, H., Albrecht, R., Dünnweber, W., Graw, G., Steadman, S. G., Wurm, J. P., Disdier, D., Rauch, V., Scheibling, F. 1977. Prepr.
307. Miller, J. M., Alexander, J. M., Kaplan,

M. 1977. Private communication
308. Vandenbosch, R., Huizenga, J. R. 1973. *Nuclear Fission*. New York and London: Academic
309. Tjøm, P. O., Stephens, F. S., Diamond, R. M., de Boer, J., Meyerhof, W. E. 1974. *Phys. Rev. Lett.* 33: 593
310. Hagemann, G. B., Broda, R., Herskind, B., Ishihara, M., Ogaza, S. 1975. *Nucl. Phys. A* 245: 166
311. Sarantites, D. G., Barker, J. H., Halbert, M. L., Hensley, D. C., Dayras, R. A., Eichler, E., Johnson, N. R., Gronemeyer, S. A. 1976. *Phys. Rev. C* 14: 2138

312. Ishihara, M., Numao, T., Fukuda, T., Tanaka, K., Inamura, T. 1976. IPCR-Cyclotron Rep. 35, Wako-Shi, Saitama, Jpn.
313. Slemmer, J., Albrecht, R., Demond, F. J., Ho, H., Wurm, J. P. 1976. See Ref. 37b, p. 156
314. Glässel, P., Simon, R. S., Diamond, R. M., Jared, R. C., Lee, I. Y., Moretto, L. G., Newton, J. O., Schmitt, R., Stephens, F. S. 1977. *Phys. Rev. Lett.* 38: 331
315. Hanappe, F., Deleplanque, M. A. 1977. Sess. d'Étud. Biènn. Phys. Nucl., 4th, LaToussuire, France

AUTHOR INDEX

CUMULATIVE INDEXES

CONTRIBUTING AUTHORS VOLUMES 18 - 27

CHAPTER TITLES VOLUMES 18-27

568 CHAPTER TITLES